ERCOFTAC Series

Volume 27

ERCOFTAC (European Research Community on Flow, Turbulence and Combustion) was founded as an international association with scientific objectives in 1988. ERCOFTAC strongly promotes joint efforts of European research institutes and industries that are active in the field of flow, turbulence and combustion, in order to enhance the exchange of technical and scientific information on fundamental and applied research and design. Each year, ERCOFTAC organizes several meetings in the form of workshops, conferences and summer schools, where ERCOFTAC members and other researchers meet and exchange information. The *ERCOFTAC* series publishes the proceedings of ERCOFTAC meetings, which cover all aspects of fluid mechanics. The series comprises proceedings of conferences and workshops, and of textbooks presenting the material taught at summer schools. The series covers the entire domain of fluid mechanics, which includes physical modelling, computational fluid dynamics including grid generation and turbulence modelling, measuring techniques, flow visualization as applied to industrial flows, aerodynamics, combustion, geophysical and environmental flows, hydraulics, multi-phase flows, non-Newtonian flows, astrophysical flows, laminar, turbulent and transitional flows.

Indexed by SCOPUS, Google Scholar and SpringerLink.

More information about this series at http://www.springer.com/series/5934

Manuel García-Villalba · Hans Kuerten ·
Maria Vittoria Salvetti
Editors

Direct and Large Eddy Simulation XII

Springer

Editors
Manuel García-Villalba
Department of Bioengineering
and Aerospace Engineering
Universidad Carlos III de Madrid
Leganés, Spain

Hans Kuerten
Department of Mechanical Engineering
Eindhoven University of Technology
Eindhoven, Noord-Brabant, The Netherlands

Maria Vittoria Salvetti
Dipartimento di Ingegneria
Civile e industriale
University of Pisa
Pisa, Italy

ISSN 1382-4309 ISSN 2215-1826 (electronic)
ERCOFTAC Series
ISBN 978-3-030-42821-1 ISBN 978-3-030-42822-8 (eBook)
https://doi.org/10.1007/978-3-030-42822-8

This Springer imprint is published by the registered company Springer Nature Switzerland AG
The registered company address is: Gewerbestrasse 11, 6330 Cham, Switzerland

Committees

Organizing Committee

M. García-Villalba, Universidad Carlos III de Madrid, Spain
H. Kuerten, TU Eindhoven, The Netherlands
M. V. Salvetti, University of Pisa, Italy

Scientific Committee

V. Armenio, University of Trieste, Italy
B. J. Boersma, TU Delft, The Netherlands
D. Borello, University of Rome, Italy
M. Breuer, Helmut-Schmidt University Hamburg, Germany
W. P. Breugem, TU Delft, The Netherlands
L. Davidson, Chalmers University Technology, Sweden
M. de Marchis, Kore University, Italy
O. Flores, Universidad Carlos III de Madrid, Spain
R. Friedrich, TU Munich, Germany
J. Fröhlich, TU Dresden, Germany
K. Fukagata, Keio University, Japan
B. J. Geurts, University of Twente, The Netherlands
T. Kempe, ILK Dresden, Germany
E. Lamballais, PPRIME Poitiers, France
C. Marchioli, University of Udine, Italy
M. Meldi, PPRIME Poitiers, France
J. Meyers, University Leuven, Belgium
U. Piomelli, Queens University, Canada
J. Pozorski, Polish Academy Sciences, Poland
S. Sarkar, University of California San Diego, USA

Preface

The DLES Workshop series, which started in 1994, focuses on modern techniques to simulate turbulent flows based on the partial or full resolution of the instantaneous turbulent flow structures, as Direct Numerical Simulation (DNS), Large-Eddy Simulation (LES), or hybrid models based on a combination of LES and RANS approaches.

The present edition of the Workshop is the 12th and in these 25 years, a significant evolution of the topics can be noticed. In the first editions, most of the contributions dealt with the study of canonical problems and the development of the DNS and LES methodologies mostly in the context of single-phase flows. Nowadays, the complexity of the phenomena that are computationally tractable has increased considerably. It is now possible to handle increasingly complex geometries but the various physical phenomena require ad-hoc models: the effect of bubbles, solid particles, gas-liquid interaction, heat transfer, combustion processes, etc. These models must be sufficiently precise but at the same time computationally affordable for industrial use. Often, it is difficult to have detailed experimental data for validation purposes. As a consequence, the validation of closure models for these complex phenomena has to be done in many cases via DNS. As in other fields of scientific knowledge, the amount of data being used is increasing enormously, and there is a tendency to incorporate data assimilation and machine learning into the development of turbulence models, both in the context of LES and hybrid approaches.

The goal of the workshop series is to establish a state-of-the-art of DNS, LES, and related techniques for the computation and modeling of turbulent and transitional flows and to provide an opportunity for discussions about recent advances and applications.

The 12th edition of the bi-annual Workshop series on Direct and Large-Eddy Simulation (DLES12) was held in Madrid, Spain on June 5–7, 2019. 120 participants from 16 different countries attended this 3-day workshop. The majority of participants were from academia and research institutes, but several companies were also represented. Eight keynote lectures were given by experts in different scientific fields: atmospheric boundary layers (Juan Pedro Mellado, Max Planck

Institute for Meteorology, Germany), Rayleigh-Bénard convection (Roberto Verzicco, University of Rome "Tor Vergata", Italy), instabilities in the wake of oscillating cylinders (Ivette Rodríguez, Universitat Politècnica de Catalunya, Spain), direct numerical simulations under constant power input (Bettina Frohnapfel, Karlsruhe Institute of Technology, Germany), shallow mixing interfaces at river confluences (George Constantinescu, University of Iowa, USA), DLES of drug deposition in the human respiratory system (Stavros Kassinos, University of Cyprus, Cyprus), ocean/atmosphere transfer processes (Stephane Popinet, Sorbonne Université, France), and DNS and LES for nuclear reactor safety (Ed Komen, NRG, The Netherlands).

Next to the invited lectures, 91 oral presentations were selected by a Scientific Committee of 30 experts. This volume contains most of the contributed papers, which were submitted and further reviewed for publication. They cover advances in computational techniques and modeling approaches, and applications in several fields, namely, general turbulence, multiphase and reactive flows, convection and heat transfer, compressible flows, internal and external aerodynamics, jets, wakes and separated flows, fluid-structure interaction, and other complex flows.

The organization of DLES12 and the preparation of these proceedings would not have been possible without the help of many. The support from ERCOFTAC (SIG1) and Universidad Carlos III de Madrid is gratefully acknowledged. Finally, thanks go to the members of the Scientific Committee for their help in reviewing the submitted abstracts and the contributions to the proceedings.

Madrid, Spain
December 2019

Manuel García-Villalba
Hans Kuerten
Maria Vittoria Salvetti

Contents

Chapter 1
Using Numerical Simulations to Study the Atmospheric Boundary Layer

J. P. Mellado

Introduction

The atmospheric boundary layer (ABL) is the lower part of the atmosphere, the part that is in contact with the surface and responds to changes in surface properties in a few hours. The ABL is typically in a turbulent state in which turbulence is mechanically driven, or convectively driven, or both. Another important feature of the ABL is that it is shallow, because the stratification in the free atmosphere above it hinders the vertical displacement of air. Typical ABL depths vary between a few hundred meters and 1 or 2 km, depending on the environmental conditions. The ABL is hence a small element of the Earth system, but of paramount importance.

The ABL is an arbiter of weather and climate and therefore key for our short- and long-term planning. We live in the ABL and key sectors of our society, such as energy production, transportation, air quality, and agriculture and ecology, rely on having a good understanding and characterization of the ABL. From a larger scale perspective, the ABL modulates the fluxes between the atmosphere, land, and ocean, and thus the ABL also plays a crucial role in Earth's energy balance and in assessing the consequences of global warming.

However, understanding and, above all, quantifying some key aspects of the ABL remains a challenge because of its multi-scale, multi-physics, and polymorphous character. Using 1 km and 1 m s^{-1} as characteristic length and velocity scales of the large energy-containing motions, one finds a Reynolds number on the order of 10^8 and a length-scale separation between the large energy-containing motions and the small dissipative ones of about 6 orders of magnitude. In cases in which small-scale motions strongly influence the input of mechanical energy into the ABL,

J. P. Mellado (✉)
Division of Aerospace Engineering, Department of Physics,
Universitat Politècnica de Catalunya, C. Jordi Girona 1-3, Barcelona, Spain
e-mail: juan.pedro.mellado@upc.edu

Max Planck Institute for Meteorology, Bundesstr. 53, Hamburg, Germany

M. García-Villalba et al. (eds.), *Direct and Large Eddy Simulation XII*,
ERCOFTAC Series 27,
https://doi.org/10.1007/978-3-030-42822-8_1

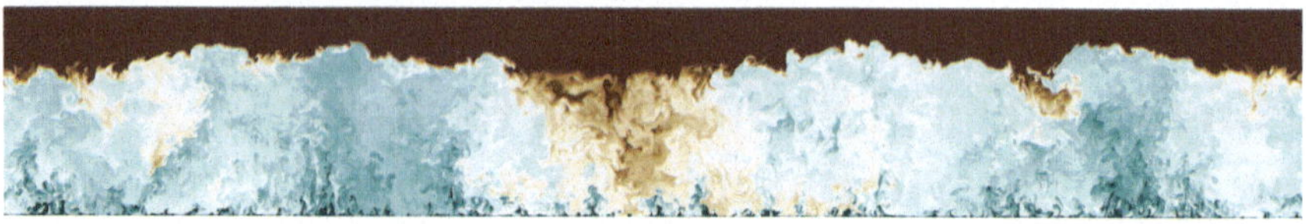

Fig. 1.1 Vertical cross-section of the moisture field in an unstable ABL. Green and brown colors indicate moist and dry air, respectively. Small plumes near the surface converge into large-scale updrafts, which impinge into the capping inversion at the ABL top and cause entrainment of dry air from the free atmosphere. From DNS data at a Reynolds number $\approx 10^4$ in Mellado et al. [31]

numerical simulations need to represent at least partly these small-scale motions. Besides, the ABL is statistically inhomogeneous in the vertical direction and there is no clear separation between the dynamics near the ABL bottom and the dynamics near the ABL top, so that numerical simulations need to represent the whole vertical extent of the ABL. Current computational capabilities allow us to perform numerical simulations with a Reynolds number on the order of 10^4 (Fig. 1.1). To what extent can we then extrapolate what we learn from these simulations to atmospheric conditions?

Moreover, numerical simulations necessarily consider configurations that are idealized with respect to physical phenomena, boundary conditions, and initial conditions. Besides turbulence, the ABL compounds radiative transfer, cloud physics, atmospheric chemistry, and the effect of surface roughness and surface heterogeneity. The interaction among these complex phenomena leads to a very rich non-linear behavior, and the ABL can appear in very different forms or regimes. The initial conditions are also important because the ABL is often in a transient state due to the unsteady forcing caused by the diurnal cycle or by inertial oscillations. What can we then learn from simulations of idealized configurations?

Research during the last 50 years has shown that we can actually learn a lot about the ABL from numerical simulations of idealized configurations at moderate Reynolds numbers. Simulations complement field measurements because they allow for control and systematic studies and provide more complete data sets. The basic configuration in simulations is an Ekman layer with a vertical stratification of buoyancy, either a stable stratification, or a neutral stratification, or an unstable stratification (see section "Simplified Formulation"). J. W. Deardorff did the first studies of the neutral and unstable ABL using large-eddy simulation (LES) [8, 9]. Previous numerical studies of the ABL had used integral models such as mixed-layer models, where results are very sensitive to closure assumptions and these assumptions had to rely exclusively on field measurements [26]. Deardorff's work was a proof of concept, informing about closure assumptions and making the flow organization visible. J. D. Deardorff was not only a pioneer in illustrating how LES can be applied to engineering and environmental problems, but he also performed seminal work on laboratory studies of the unstable ABL [11, 12, 39], and the convective scales that he introduced for the characterization of convection have been used ever since [7].

The first LES of a cloud-topped ABL was in 1980 [10], and the first LES of the weakly stable ABL was in 1990 [28]. G. Coleman and colleagues did the first direct numerical simulations (DNSs) of the weakly stable ABL and unstable ABL during the 90s [5, 6], but we had to wait 20 more years for DNS to become more routinely used, for instance, to study entrainment in unstable ABLs [17, 24], turbulence collapse in strongly stable ABLs [2, 13, 36], and entrainment in cloud-topped ABLs [30]. These are problems where meter and submeter scales become important, and one needs to represent both large and small scales with sufficient accuracy, which is the strength of DNS. Nowadays, the degree of Reynolds number similarity in DNS is good enough to extrapolate results to atmospheric conditions, making DNS a powerful tool in ABL research. After briefly introducing a minimal formulation, I will use three examples to further illustrate this point.

Simplified Formulation

In its simplest form, the ABL is often modeled as an Ekman layer in which the effect of buoyancy forces is retained. Using the Boussinesq approximation to the Navier–Stokes equations, and linearizing the equation of state that relates the density variations with the temperature and the specific-humidity variations, one finds

$$\partial_t \mathbf{v} + \nabla\cdot(\mathbf{v}\otimes\mathbf{v}) = -\nabla p + \mathrm{Re}^{-1}\nabla^2\mathbf{v} + \mathrm{Ri}\, b\mathbf{k} - \mathrm{Ro}^{-1}\mathbf{k}\times\mathbf{v} \tag{1.1}$$

$$\nabla\cdot\mathbf{v} = 0 \tag{1.2}$$

$$\partial_t b + \nabla\cdot(\mathbf{v}b) = \mathrm{Re}^{-1}\nabla^2\mathbf{v} \tag{1.3}$$

having assumed that the Prandtl number is equal to one, which is a good approximation for typical atmospheric conditions. Cloud physics and radiative transfer such as needed for the simulations discussed in section "Toward Reynolds Number Similarity in Cloud-Topped ABLs" introduce complexity and details can be found in Mellado et al. [30], but this simplified set is sufficient for most of the discussion hereafter.

The reference Reynolds number is defined as $\mathrm{Re} = U_0 L_0/\nu$, where ν is the kinematic viscosity, and U_0 and L_0 are generic velocity and length scales, to be specified depending on the configuration. For typical atmospheric conditions, one finds $\nu = 1.5\times10^{-5}\,\mathrm{m}^2\,\mathrm{s}^{-1}$. The reference Rossby number is defined as $\mathrm{Ro} = U_0/(L_0 f)$, where $f = 2\Omega_0 \sin\phi$ is the Coriolis frequency, Ω_0 is Earth's angular velocity, and ϕ is the latitude. Typical midlatitude values are $f = 10^{-4}$ rad s^{-1}. The reference Richardson number is defined as $\mathrm{Ri} = L_0 B_0/U_0^2$, where B_0 is a reference buoyancy scale. For temperature variations of a few degrees, one finds $B_0 = 0.1\,\mathrm{m\,s}^{-2}$. Instead of a Richardson number, a Froude number $\mathrm{Fr} = \mathrm{Ri}^{-1/2}$ is also used. The number of non-dimensional parameters can be reduced to two when the reference scales U_0 and L_0 are chosen to coincide with the scales imposed by the buoyancy forcing, in which case Ri = 1, or when they are chosen to coincide with the scales imposed by the geostrophic balance between the large-scale pressure gradient and the Coriolis force, in which case Ro = 1. The number of non-dimensional parameters in the evolution equations is therefore small, which facilitates a control and systematic study of different environmental conditions. However, as explained in the introduction, there are many ways in which the environmental conditions can be imposed as boundary and initial conditions, and the class of problems described by Eqs. 1.1–1.3 is relatively large. In the following, we will consider only three examples of this class of problems to illustrate the role of DNS in ABL research.

Spatial Intermittency During Turbulence Collapse in Stable ABLs

As a first example of the importance of meter scales in the ABL, we consider conditions of stable stratification, when the mean buoyancy increases with height. Such conditions are favored at night and at high latitudes, when the radiative cooling of the surface is not compensated by the solar radiative warming. Stable stratification reduces the turbulence kinetic energy (TKE) for two reasons. One reason is that part of the TKE is converted into potential energy in an irreversible manner due to mixing [21]. Another reason is that the streamwise and vertical velocity fluctuations become less correlated, which implies a reduction of the shear production of TKE [22]. Since the maximum shear production occurs only a few meters above the surface, numerical simulations need to represent these meter-scale motions near the surface in addition to the large energy-containing motions that transport the buoyancy across the whole ABL. This has proven to be a long-standing challenge under conditions of strong stratification when turbulence collapses and the flow relaminarizes [27].

Turbulence collapse implies a drastic reduction of the vertical fluxes, which has substantial implications for ABL properties and surface properties [20, 35]. For instance, under stable stratification, the turbulent flux of internal energy is downward and compensates part of the radiative cooling of the surface. When turbulence collapses, this downward flux is strongly reduced and the surface cools faster, which favors the formation of fog and frost, two phenomena that are relevant for transportation and agriculture.

One important aspect of turbulence collapse in the stable ABL is that, according to field measurements, it often occurs intermittently in space. Spatial intermittency greatly complicates the analysis of the ABL and the representation of relevant properties such as vertical fluxes in atmospheric models. However, numerical simulations of the stably stratified Ekman layer showed temporal intermittency instead of spatial intermittency, in the sense that the turbulence collapsed in the whole computational domain and became again turbulent everywhere after some time. One could conclude from this result that the spatial intermittency in nature is caused by heterogeneity in the surface or in the gravity wave field, and, since these features were not retained in the numerical simulations, spatial intermittency did not occur. However, numerical studies using other configurations such as channel flows indicated spatial intermit-

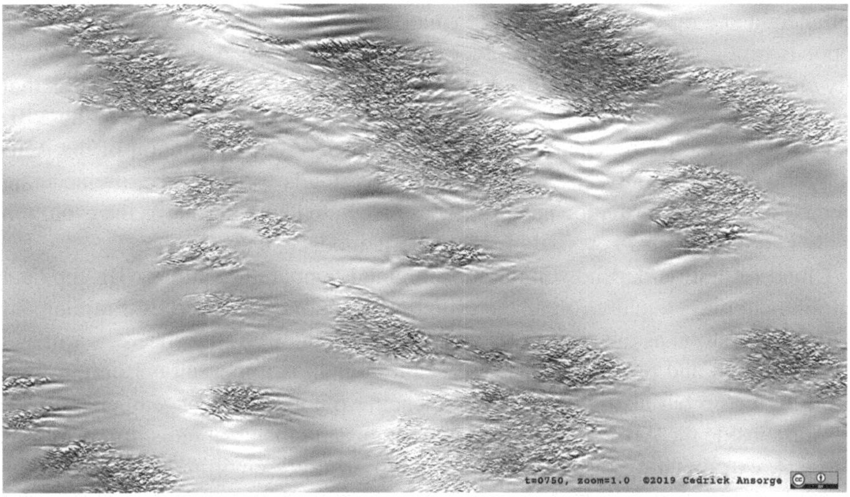

Fig. 1.2 Horizontal cross-section of the enstrophy field near the surface in a stable ABL. During turbulence collapse, turbulent and laminar regions coexist, a feature that is referred to as spatial intermittency. Courtesy of C. Ansorge, DNS data corresponding to case `ri233` in Ansorge [1]

tency during turbulence collapse [3, 4, 15, 18]. Why was spatial intermittency not present in simulations of the Ekman layer?

The reason is that the computational domain was too small. Recent work has shown that, when the computational domain is large enough, spatial intermittency also occurs in simulations of the stably stratified Ekman layer [2, 13]. The computational domain needs to be large enough for large wave-like structures to develop, and the wavelength of these structures is 2–3 times the boundary-layer depth (Fig. 1.2). At the same time, one needs to resolve the small scales of the turbulence inside the turbulence pockets. This requirement to simulate large and small scales at the same time has been a challenge until recently, but, as observed in Fig. 1.2, we start to have the computational resources for controlled and systematic studies of turbulence collapse in the stable ABL.

Entrainment Effects on Cloud Formation in Unstable ABLs

The second example considers cloud formation in conditions of unstable stratification, when the mean buoyancy decreases with height in approximately the lower half of the ABL. Such conditions are favored over land and in the absence of clouds, when the incoming solar radiation strongly warms the surface. Unstable stratification leads to convective instability, which increases TKE and maintains a well-mixed ABL where the mean profiles of conserved properties such as specific humidity and entropy are approximately constant with height. The corresponding temperature decreases with height at a rate of $\approx$10 K km^{-1}, so that, as the ABL depth grows in time, water vapor is brought to increasingly colder temperatures at the ABL top, and eventually clouds start to form. Clouds reduce the amount of incoming solar radiation that reaches the surface, which reduces the convective forcing of turbulence in the ABL and introduces a heterogeneity in this forcing, changing the evolution of the ABL as a whole and not only at the ABL top where clouds form. Hence, the accurate prediction of cloud formation and properties associated with it, such as the evolution of cloud fraction, are a key goal of atmospheric models.

Cloud formation depends on the moisture and temperature at the ABL top, and entrainment plays a major role in the evolution of these two properties. Entrainment is here understood as the process by which air from the free atmosphere, warm and dry, is brought and mixed into the ABL. Entrainment is important not only locally at the ABL top but also globally by influencing the evolution of temperature and moisture inside the ABL. Despite the large set measurements of moisture properties [38, 41], extrapolating those measurements to arbitrary environmental conditions remains a challenge. One reason is that the entrainment flux of moisture is a priori unknown. Equally important is to know what are the relevant characteristic scales in the entrainment zone, because this informs us about the accuracy that is necessary in measurements and simulations to represent entrainment properties.

Direct numerical simulations have proven very valuable in shedding light into these questions during the last years [17, 19, 29]. This work has shown that the reference Ozmidov scale

$$L_0 = (F_{B,0}/N^3)^{1/2} \tag{1.4}$$

provides an estimate of the vertical depth of the entrainment zone. In this definition, $F_{B,0}$ is an estimate of the surface buoyancy flux and N is an estimate of the buoyancy or Brunt-Väisälä frequency in the free atmosphere. In stably stratified turbulence, the Ozmidov scale is the cross-over scale that separates larger scales, which are dominated by gravity waves and are very inefficient in mixing, from smaller scales, which are dominated by turbulence and are very efficient in mixing [14, 33]. Figure 1.1 illustrates the large undulations at the ABL top that represent wave-like motions, and one can easily observe the small turbulent motions on top of them. For typical mid-day values, L_0 varies between 20 and 200 m. Hence, under environmental conditions of strong stratification in the free atmosphere or weak surface flux of buoyancy, one needs grid spacings of a few meters to accurately resolve the turbulent motions that are important for entrainment properties.

The reference Ozmidov scale also helps to answer the question of how to estimate the entrainment flux of moisture and its dependence on environmental conditions. If γ_q is an estimate of the moisture gradient in the free atmosphere, then $\gamma_q L_0$ provides an estimate of the moisture variation across the entrainment zone and $(\gamma_q L_0)(NL_0)$ provides an estimate of the entrainment flux of moisture, where NL_0 represents a velocity scale [31]. Such an expression allows us to estimate the evolution of moisture inside the ABL as a function of surface fluxes and lapse rates in the free atmosphere.

This result confirms the relevance of the reference Ozmidov scale defined by Eq. 1.4, and illustrates the importance of meter scales in the characterization of key ABL processes like cloud formation.

Toward Reynolds Number Similarity in Cloud-Topped ABLs

Resolving entrainment at the ABL top is even more demanding in stratocumulus-topped boundary layers, where the Ozmidov scale in the entrainment zone can be on the order of 1 m or less [23, 25, 32]. Stratocumulus clouds can extend over thousands of kilometers over the eastern sides of the subtropical oceans, favored by the strong contrast between the cold ocean and the warm and dry air in the descending branches of the Hadley cell. Besides, stratocumulus has a net cooling effect, because they have a larger albedo than the underlying surface of the ocean but they practically emit long-wave radiation at the same rate because they are close to the surface and therefore have a similar temperature. This combination of large area coverage and net cooling effect makes stratocumulus a key element of Earth's radiative balance, to the extent that an increase of a few percent in the area coverage of stratocumulus could offset the anthropogenic contribution to global warming [34]. Despite this relevance, quantifying stratocumulus effects on climate remains a challenge, mainly because of the difficulty to appropriately represent cloud-top entrainment in numerical simulations [29, 37, 40].

There are two key aspects about the dynamics of stratocumulus-topped boundary layers. First, cloud-top radiative cooling is the main turbulence forcing, and cloud-top radiative cooling strongly depends on the cloud amount at the ABL top. Second, the cloud amount at the ABL top depends on the balance between the drying caused by the entrainment of dry and warm air from the free atmosphere and the moistening caused by the upward flux of water vapor from the ocean surface by the large turbulent motions. This balance between the surface moistening and the entrainment drying depends on various phenomena in delicate ways [40], but turbulent mixing and entrainment is critical: if simulations fail to represent the correct amount of entrainment fluxes at the ABL top, then simulations need to be tuned to reproduce measurements.

For instance, when using grid spacings on the order of 5 m, LES with isotropic grids overestimate entrainment and the cloud disappears, whereas measurements show a continuous cloud deck [37]. The reason is that, as anticipated before, the Ozmidov scale is typically on the order of 1 m. The Ozmidov scale in the stratocumulus-topped ABLs is smaller than in cloud-free conditions because radiative cooling at the cloud top maintains an inversion which is significantly thinner, which implies a stronger stratification and hence a smaller value of L_0 according to Eq. 1.4. The solution adopted in LES is to use a grid spacing on the order of 5–10 times larger in the horizontal directions, which effectively reduces entrainment and renders a cloud fraction that is consistent with measurements. It remains unclear, however, to what extent this tuned LES can be applied to other environmental conditions.

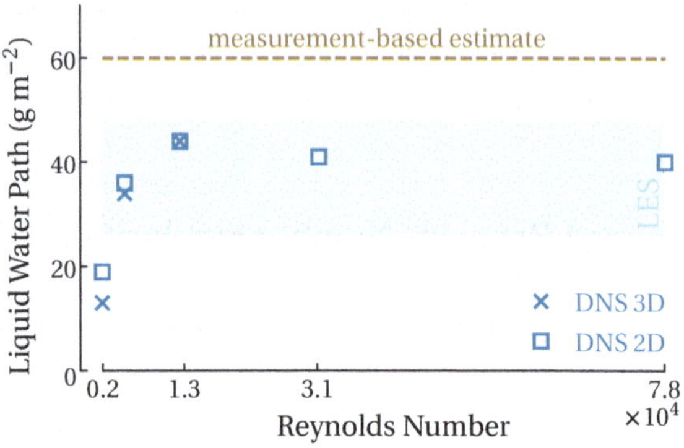

Fig. 1.3 Convergence of liquid water path (vertically integrated amount of liquid water per unit area) toward Reynolds number similarity. The light-blue region denotes the central half of the distribution of the LES ensemble in Stevens et al. [37]. Adapted from Mellado et al. [30]

DNS can help in this problem by providing reference data to validate LES in different environmental conditions, offering an alternative to field campaigns [30]. As observed in Fig. 1.3, the Reynolds numbers in current DNS are large enough to show a degree of Reynolds number similarity that allows a useful extrapolation to atmospheric conditions. Besides, since DNS is free from the uncertainty introduced by the tuning in LES, the difference between results and measurements in Fig. 1.3 can be attributed to processes that were not included in the DNS, such as sedimentation or horizontal advection. In this way, DNS can help to differentiate between uncertainties associated with mixing and uncertainties associated with other phenomena, such as cloud microphysics.

Summary and Conclusions

When three-dimensional numerical simulations of the ABL started about 50 years ago, the Reynolds numbers that were affordable at that time were too small, close to the laminar-to-turbulent transition, and LES was the natural choice [16]. In this paper, we have argued that this paradigm has changed, and, 50 years after Deardorff's pioneering work, the degree of Reynolds number similarity in DNS allows a useful extrapolation of results to atmospheric conditions. Hence, DNS has become a powerful tool for quantitative analysis and hypothesis testing of small-scale processes in the ABL, which opens new possibilities to solve key long-standing problems.

There are various ways in which DNS can be used to advance our understanding and characterization of the ABL. We have illustrated how DNS helps when meter and submeter scales become important for the evolution of large-scale ABL properties, such as during turbulence collapse in strongly stable ABLs, and during cloud formation and cloud evolution in unstable ABLs. DNS can also serve as a substitute of field measurements to tune (or calibrate) LES and assess the behavior of LES out of the configuration for which it was tuned. Last, with DNS, we can better use analytical relationships derived from the original governing equations, which is challenging in

LES because subgrid-scale contributions are modeled and because numerical contributions are difficult to estimate. In this way, DNS can complement LES and field measurements to accelerate current lines of ABL research.

References

1. Ansorge, C.: Scale dependence of atmosphere-surface coupling through similarity theory. Bound.-Layer Meteorol. **170**, 1–27 (2019)
2. Ansorge, C., Mellado, J.P.: Global intermittency and collapsing turbulence in the stratified planetary boundary layer. Bound.-Layer Meteorol. **153**, 89–116 (2014)
3. Armenio, V., Sarkar, S.: An investigation of stably stratified channel flow using large-eddy simulation. J. Fluid Mech. **459**, 1–42 (2002)
4. Brethouwer, G., Duguet, Y., Schlatter, P.: Turbulent-laminar coexistance in wall flows with Coriolis, buoyancy or Lorentz. J. Fluid Mech. **704**, 137–172 (2012)
5. Coleman, G.N., Ferziger, J.H., Spalart, P.R.: Direct simulation of the stably stratified turbulent Ekman layer. J. Fluid Mech. **244**, 677–712 (1992)
6. Coleman, G.N., Ferziger, J.H., Spalart, P.R.: A numerical study of the convective boundary layer. Bound.-Layer Meteorol. **70**, 247–272 (1994)
7. Deardorff, J.W.: Convective velocity and temperature scales for the unstable planetary boundary layer and for Rayleigh convection. J. Atmos. Sci. **27**, 1211–1213 (1970)
8. Deardorff, J.W.: A three-dimensional numerical investigation of the idealized planetary boundary layer. Geophys. Fluid Dyn. **1**, 377–410 (1970)
9. Deardorff, J.W.: Numerical investigation of neutral and unstable planetary boundary layers. J. Atmos. Sci. **29**, 91–115 (1972)
10. Deardorff, J.W.: Cloud top entrainment instability. J. Atmos. Sci. **37**, 131–147 (1980)
11. Deardorff, J.W., Willis, G.E., Lilly, D.K.: Laboratory investigation of non-steady penetrative convection. J. Fluid Mech. **35**, 7–31 (1969)
12. Deardorff, J.W., Willis, G.E., Stockton, B.H.: Laboratory studies of the entrainment zone of a convectively mixed layer. J. Fluid Mech. **100**, 41–64 (1980)
13. Deusebio, E., Schlatter, G.B.P., Lindborg, E.: A numerical study of the unstratified and stratified Ekman layer. J. Fluid Mech. **755**, 672–704 (2014)
14. Dougherty, J.P.: The anisotropy of turbulence at the meteor level. J. Atmos. Terr. Phys. **21**, 210–213 (1961)
15. Flores, O., Riley, J.: Analysis of turbulence collapse in the stably stratified surface layer using direct numerical simulation. Bound.-Layer Meteorol. **139**, 241–259 (2011)
16. Fox, D.G., Lilly, D.K.: Numerical simulation of turbulent flows. Rev. Geophys. Space Phys. **10**, 51–72 (1972)
17. Garcia, J.R., Mellado, J.P.: The two-layer structure of the entrainment zone in the convective boundary layer. J. Atmos. Sci. **71**, 1935–1955 (2014)
18. García-Villalba, M., del Álamo, J.: Turbulence modification by stable stratification in channel flow. Phys. Fluids **23**(045104), 1–22 (2011)
19. Haghshenas, A., Mellado, J.P.: Characterization of wind-shear effects on entrainment in a convective boundary layer. J. Fluid Mech. pp. 145–183 (2019)
20. Holtslag, A., Svensson, G., Baas, P., Basu, S., Beare, B., Beljaars, A., Bosveld, F., Cuxart, J., Lindvall, J., Steeneveld, G., Tjernström, M., van de Wiel, B.: Stable atmopsheric boundary layers and diurnal cycles. Bull. Am. Meteor. Soc. **94**, 1691–1706 (2013)
21. Ivey, G.N., Winters, K.B., Koseff, J.R.: Density stratification, turbulence, but how much mixing? Annu. Rev. Fluid Mech. **40**, 169–184 (2008)
22. Jacobitz, F., Sarkar, S., Atta, C.W.V.: Direct numerical simulations of the turbulence evolution in a uniformly sheared and stably stratified flow. J. Fluid Mech. **342**, 231–261 (1997)

23. Jen-La Plante, I., Ma, Y.F., Nurowska, K., Gerber, H., Khelif, D., Karpinska, K., Kopec, M.K., Kumala, W., Malinowski, S.P.: Physics of Stratocumulus Top (POST): turbulence characteristics. Atmos. Chem. Phys. **16**, 9711–9725 (2016)
24. Jonker, H.J.J., van Reeuwijk, M., Sullivan, P.P., Patton, E.G.: Interfacial layers in clear and cloudy atmospheric boundary layers. In: Proceedings of the 7th International Symposium on Turbulence, Heat and Mass Transfer, pp. 3–14 (2012)
25. Katzwinkel, J., Siebert, H., Shaw, R.: Observation of self-limiting, shear-induced turbulent inversion layer above marine stratocumulus. Bound.-Layer Meteorol. **145**, 131–143 (2012)
26. Lilly, D.K.: Models of cloud-topped mixed layers under strong inversion. Q. J. Roy. Meteorol. Soc. **94**, 292–309 (1968)
27. Mahrt, L.: Stably stratified atmospheric boundary layers. Annu. Rev. Fluid Mech. **46**, 23–45 (2014)
28. Mason, P.J., Derbyshire, S.H.: Large-eddy simulation of the stably-stratified atmospheric boundary layer. Bound.-Layer Meteorol. **53**, 117–162 (1990)
29. Mellado, J.P.: Cloud-top entrainment in stratocumulus clouds. Annu. Rev. Fluid Mech. **41**, 145–169 (2017)
30. Mellado, J.P., Bretherton, C.S., Stevens, B., Wyant, M.C.: DNS and LES of stratocumulus: better together. J. Adv. Model. Earth Syst. **10**, 1421–1438 (2018)
31. Mellado, J.P., Puche, M., van Heerwaarden, C.C.: Moisture statistics in free convective boundary layers growing into linearly stratified atmospheres. Q. J. R. Meteorol. Soc. **143**, 2403–2419 (2017)
32. Mellado, J.P., Stevens, B., Schmidt, H.: Wind shear and buoyancy reversal at the top of stratocumulus. J. Atmos. Sci. **71**, 1040–1057 (2014)
33. Ozmidov, R.V.: On the turbulent exchange in a stably stratified ocean. Izv., Atmos. Oceanic Phys. Ser. **1**, 853–860 (1965)
34. Randall, D.A., Coakley, J.A., Fairall, C.W., Kropfli, R.A., Lenschow, D.H.: Outlook for research on subtropical marine stratiform clouds. Bull. Am. Meteorol. Soc. **65**, 1290–1301 (1984)
35. Sandu, I., Beljaars, A., Bechtold, P., Mauritsen, T., Balsamo, G.: Why is it so difficult to represent stably stratified conditions in numerical weather prediction (NWP) models? J. Adv. Model. Earth Syst. **5**(2), 117–133 (2013)
36. Shah, S.K., Bou-Zeid, E.: Direct numerical simulation of Ekman boundary layers with increasing static stability: modifications to the bulk structure and second-order statistics. J. Fluid Mech. **760**, 494–539 (2014)
37. Stevens, B., Moeng, C.H., Ackerman, A.S., Bretherton, C.S., Chlond, A., de Roode, S., Edwards, J., Golaz, J.C., Jiang, H., Khairoutdinov, M., Kirkpatrick, M.P., Lewellen, D.C., Lock, A., Müller, F., Stevens, D.E., Whelan, E., Zhu, P.: Evaluation of large-eddy simulations via observations of nocturnal marine stratocumulus. Mon. Wea. Rev. **133**, 1443–1462 (2005)
38. Turner, D., Wulfmeyer, V., Berg, L.K., Schween, J.H.: Water vapor turbulence profiles in stationary continental convective boundary layers. J. Geophys. Res. Atmos **119**, 11151–11165 (2014)
39. Willis, G.E., Deardorff, J.W.: A laboratory model of the unstable planetary boundary layer. J. Atmos. Sci. **31**, 1297–1307 (1974)
40. Wood, R.: Stratocumulus clouds. Mon. Wea. Rev. **140**, 2373–2423 (2012)
41. Wulfmeyer, V., Muppa, S.K., Behrendt, A., Hammann, E., Späth, F., Sorbjan, Z., Turner, D., Hardesty, R.M.: Determination of convective boundary layer entrainment fluxes, dissipation rates, and the molecular destruction of variances: theoretical description and a strategy for its confirmation with a novel Lidar system synergy. J. Atmos. Sci. **73**, 667–692 (2016)

Chapter 2
Power-Spectral Density in Turbulent Boundary Layers on Wings

Á. Tanarro, R. Vinuesa and P. Schlatter

Introduction

Turbulent boundary layers (TBLs) subjected to various amplitudes of adverse pressure gradients (APGs) are of paramount importance in the field of fluid dynamics due to their implications in countless industrial applications in which the flow develops on a curved surface such as wings or nozzles, among others. The relevance of this type of flows is further demonstrated by the numerous works on the analysis of APG TBLs, both numerical and experimental, that have been performed over the years. Among these studies, we find the early direct numerical simulation (DNS) and experiment on pressure-gradient TBLs carried out by Spalart and Watmuff [10], who observed a reduction in inner-scaled mean velocity within the buffer layer when subjected to strong APGs. On the other hand, the experimental study by Skåre and Krogstad [9] was focused on the effect of APGs on the turbulent kinetic energy budgets, revealing that TBLs under strong APGs exhibit an outer peak in the turbulence production. More recently, due to the increase in computational power, the number of numerical studies on APG TBLs has significantly increased, adding additional insight. Some of the most relevant are the DNS of a NACA4412 wing section at $\mathrm{Re}_c = 400{,}000$ (i.e. Reynolds number based on the chord length c and free-stream velocity U_∞) by Hosseini et al. [7], and the analysis of the effect of APG at high Re performed by Vinuesa et al. [12]. There has also been some recent numerical work on turbulence control in wings, see Refs. [1, 2].

The main aim of this work is to study in more depth the effect of adverse pressure gradients and flow history on the development of the turbulent boundary layer. In order to perform such analysis, well-resolved large-eddy simulations (LESs) are

Á. Tanarro · R. Vinuesa (✉) · P. Schlatter
Linné FLOW Centre, KTH Mechanics, SE-100 44 Stockholm, Sweden
e-mail: rvinuesa@mech.kth.se
URL: https://www.kth.se/social/group/simex/

M. García-Villalba et al. (eds.), *Direct and Large Eddy Simulation XII*,
ERCOFTAC Series 27,
https://doi.org/10.1007/978-3-030-42822-8_2

Fig. 2.1 Instantaneous visualisation of the NACA0012 case showing coherent vortical structures together with the employed spectral-element mesh (where the grid points within elements are not shown). Structures coloured by their streamwise velocity from (*blue*) low to (*red*) high velocity

carried out on two wing sections: the cambered NACA4412 with 5° angle of attack and the symmetric NACA0012 with 0° angle of attack, both with Reynolds number $Re_c = 400{,}000$.

Numerical Method

The well-resolved LESs are carried out with the spectral-element code Nek5000 (Fischer et al. [6]). The spatial discretisation is performed by means of Lagrangian interpolants of polynomial order $N = 11$, and the spatial resolution near the wall is expressed in terms of wall units: $\Delta x_t^+ = 18.0$, $\Delta y_n^+ = (0.64, 11.0)$ and $\Delta z^+ = 9.0$. Note that x_t and y_n denote the directions tangential and normal to the wing surface, respectively, and z is the spanwise direction. The scaling in wall units is in terms of the friction velocity u_τ and the viscous length $\ell^* = \nu / u_\tau$. The domain in both cases is a C-mesh (as shown in Fig. 2.1 together with a visualisation of the vortical structures in the flow) with chordwise extent $L_x = 6c$, vertical extent $L_y = 4c$ and spanwise width $L_z = 0.1c$. The boundary conditions are obtained from a Reynolds-averaged Navier–Stokes (RANS) simulation except in the outlet where a stabilised outflow is implemented and in the spanwise direction where a periodic boundary condition is used. In order to simulate a TBL, the flow is tripped with a random volume forcing at $x/c = 0.1$ from the leading edge. Additional details regarding the numerical setup can be found in the work by Tanarro et al. [11]. A total of 380 million grid points were used to discretise the NACA0012 case, and the total computational cost was 2.5 million CPU hours. On the other hand, 466 million grid points were employed for the non-symmetric NACA4412 case, and the total cost was 4.3 million CPU hours.

Results

The effects of APG and flow history on the wing TBLs are characterised through power-spectral density distributions which provide a great insight on how scales of different sizes react to these effects. First, Fig. 2.2 (left) shows the one-dimensional spanwise pre-multiplied power-spectral density of the tangential velocity fluctuations of the cases simulated (i.e. the NACA4412 and the NACA0012 at $0.4c$ and $0.75c$ from the leading edge on the suction side, denoted with the subscript ss) together with the results from the zero-pressure gradient (ZPG) TBL from Eitel-Amor et al. [5] at $\mathrm{Re}_\tau = 305$ and $\mathrm{Re}_\tau = 2480$. In Fig. 2.2 (left), the power-spectral density from the ZPG TBL shows that for increasing Reynolds number, the largest scales are energised whereas the smaller scales show no significant effect from the increment in Re_τ. Nevertheless, when comparing the power-spectral density from the ZPG with the APG cases it can be observed that APGs exhibit significantly more small-scale energy in the outer region. In the analysis of the wing sections, we consider two different profiles in which the APG and the Re_τ are different. The magnitude of the APG is measured in terms of the Clauser pressure-gradient parameter $\beta = \delta^*/\tau_w \mathrm{d}P_e/\mathrm{d}x_t$, where δ^* is the displacement thickness, τ_w is the wall-shear stress, P_e is the pressure at the boundary-layer edge and x_t is the distance in the tangential direction (Clauser [4]). The NACA0012 exhibits $\beta = 0.20$ and $\mathrm{Re}_\tau = 211$ at $x_{ss}/c = 0.4$ whereas $\beta = 0.66$ and $\mathrm{Re}_\tau = 339$ at $x_{ss}/c = 0.75$. On the other hand, the NACA4412 presents $\beta = 0.67$ and $\mathrm{Re}_\tau = 240$ at $x_{ss}/c = 0.4$ whereas $\beta = 3.59$ and $\mathrm{Re}_\tau = 366$ at $x_{ss}/c = 0.75$. Figure 2.2 (left) shows that, despite the small increase in Re_τ in the wings, the differences between the profiles are considerable and the contours of power-spectral density show significantly different behaviours compared to when increasing the Reynolds number in the ZPG. The first observation that can be made is the emergence of an outer peak in the tangential fluctuation power-spectral density (i.e. $\lambda_z^+ \approx 300$) for the NACA4412 at $x_{ss}/c = 0.75$. Although this effect is not unique of APG TBLs as ZPG TBLs at very high Reynolds numbers also show a spectral outer peak (Eitel-Amor et al. [5]), the emergence of the spectral outer peak appears at much lower Re in the presence of strong APG. Apart from the spectral outer peak, there are two additional regions of relevance in the power-spectral densities shown in Fig. 2.2 (left). The first one is the region containing the largest scales of the flow (i.e. $\lambda_z^+ > 400$). The power-spectral density shows that at a higher APG, the largest scales of the flow are energised in a way similar to what is observed when increasing Re_τ, nevertheless, in the near-wall region, the APG TBLs show scales with higher energisation whereas the increase in Re does not affect significantly this region. The second region of most interest in this analysis corresponds to the one containing the smallest scales in the outer layer (i.e. top-left corner of the power-spectral density map), where a significant input of small-scale energy (i.e. $\lambda_z^+ \approx 100$) can be observed for APG TBLs, as opposed to the decrease of this energy in the ZPG when increasing Re. This result suggests that the APG induces a vertical motion within the boundary layer that advects some of the small scales in the near-wall region towards the outer region (Vinuesa et al. [12]).

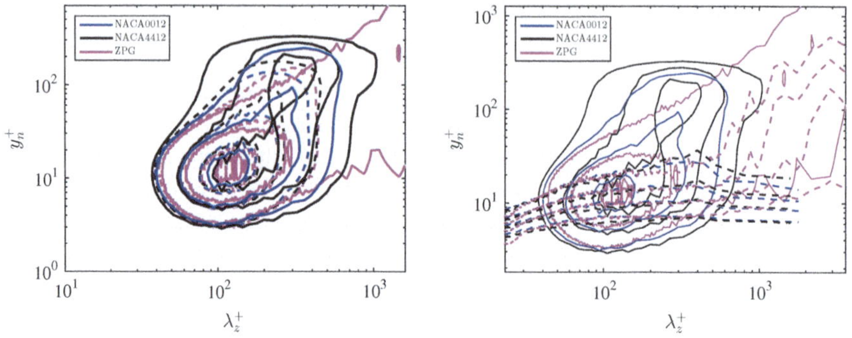

Fig. 2.2 **(Left)** Inner-scaled pre-multiplied spanwise power-spectral density of the tangential velocity fluctuations of the NACA0012 and NACA4412 wing sections at $x_{ss}/c = 0.4$ (....) and $x_{ss}/c = 0.75$ (——) together with the ZPG TBL spectra at $\mathrm{Re}_\tau = 305$ (.....) and $\mathrm{Re}_\tau = 2480$ (——) by Eitel-Amor et al. [5] and **(right)** contours of linear coherence spectra with respect to the near-wall region (......) of the tangential fluctuations on wing sections at $x_{ss}/c = 0.75$, and ZPG TBL at $\mathrm{Re}_\tau = 2480$ (Eitel-Amor et al. [5]). The contours correspond to (1.0, 2.1, 3.4) and (0.2, 0.4, 0.6, 0.8) in the plots of power-spectral density and linear coherence spectra, respectively

The analysis of the one-dimensional pre-multiplied power-spectral density has shown a significant contrast between the scales on APG TBLs and high-Re ZPG TBLs, suggesting that the energisation mechanisms of the APG and those of high Reynolds numbers are different (Vinuesa et al. [12]). In order to further analyse the effect of APG in contrast to high Reynolds numbers, the linear coherence spectrum (LCS) introduced by Baars et al. [3] is computed for the flows under study and shown in Fig. 2.2 (right) together with spanwise pre-multiplied power-spectral density for the two wings and a ZPG TBL at higher Reynolds number (Eitel-Amor et al. [5]). It can be observed that the outer spectral peak exhibited by the ZPG at high Re has a high correlation with the near-wall region, whereas the energetic outer spectral peak of the APG TBL does not show a noticeable correlation with the near-wall region. These differences indicate that in strong-APG TBLs, there is a larger independence between the outer and inner regions than in high-Re TBLs, suggesting that the large scales in the outer region become progressively less attached to the wall as the APG increases. This result shows that the energisation mechanisms of high-Re and APG TBLs are different, a conclusion that is further confirmed with the analysis of the two-dimensional power-spectral density. Figure 2.3 (left) shows the evolution of the two-dimensional power-spectral densities with increasing Re_τ, which shows consistent results with those obtained by Jiménez et al. [8] in terms of the contour bounds: for the upper part of the spectrum, there is a linear relation $\lambda_t^+ \sim \lambda_z^+$ in the shortest scales represented with the dashed line whereas there is a square-root relation $\lambda_z^+ \sim \left(\lambda_t^+\right)^{1/2}$ (dotted line) bounding the longer scales. On the other hand, Fig. 2.3 (right) confirms that the APG affects in a different manner the scales of the turbulent boundary layer. For the upper part of the spectrum, the longer scales in the ZPG at high Reynolds number follow, for $\lambda_t^+ > 100$, the relation $\lambda_z^+ \sim (\lambda_t^+)^{1/2}$. However, the APG TBLs

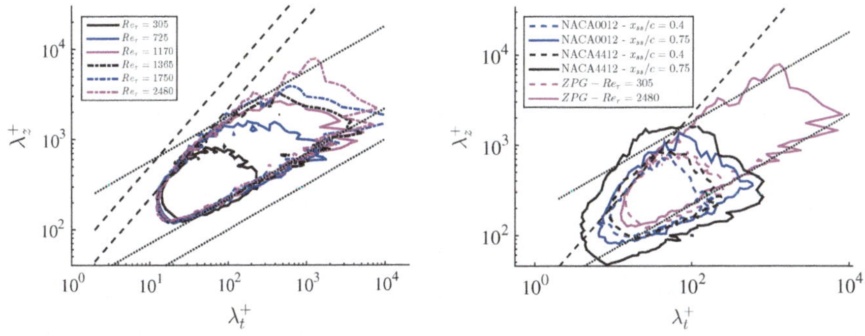

Fig. 2.3 Inner-scaled two-dimensional pre-multiplied power-spectral density of the tangential velocity fluctuations at $y_n^+ \simeq 150$ for (**left**) ZPG TBL at different Re_τ from the database by Eitel-Amor et al. [5] and (**right**) ZPG at different Re_τ and for the wing sections at $x_{ss}/c = 0.4$ and 0.75. The contours represent a value of 0.15. The dashed straight lines depict $\lambda_z^+ \sim \lambda_t^+$ and the dotted lines show $\lambda_z^+ \sim (\lambda_t^+)^{1/2}$

at the Re under investigation do not exhibit that behaviour; furthermore, the spatial and temporal energetic scales span a wider range than that of the ZPG, which is a completely different behaviour that is observed when increasing Re.

Conclusions

In this work, the power-spectral density distributions of APG TBLs and ZPG TBL are analysed in order to study the differences between the energisation mechanisms of APG and high Reynolds numbers. Starting from the one-dimensional pre-multiplied spanwise power-spectral density, several differences in the contours of the different TBLs can be observed. Among these, the most significant difference between strong-APG TBLs and high-Re ZPG TBLs is based on the small-scale energy in the outer region. In the case of the ZPG TBL, the small-scale energy in the outer region decreases slightly when increasing Re_τ whereas the APG TBLs show considerably more energetic small scales in the outer region which increases with stronger APG. This observation suggests that the APG favours the transport normal to the wall of the small scales and, in turn, a different energisation mechanism than that of high-Re TBLs. On the other hand, the computation of the linear coherence spectra confirms the previous result about APG and high-Re having different energisation mechanisms. The LCS shows that in high-Re ZPG TBLs, the spectral outer peak is highly correlated with the near-wall region, unlike the APG TBLs in which this correlation is very low and indicates that the scales in the outer region of APG TBLs may not be attached on the wall. Lastly, the comparison of the two-dimensional pre-multiplied power-spectral density for ZPG TBL with increasing Re and APG TBLs shows very different trends which further confirm that the energisation mechanisms of these two effects are different.

References

1. Albers, M., Meysonnat, P.S., Schröder, W.: Actively reduced airfoil drag by transversal surface waves. Flow Turbul. Combust. **102**, 865–886 (2019)
2. Atzori, M., Vinuesa, R., Stroh, A., Frohnapfel, B., Schlatter, P.: Assessment of skin-friction-reduction techniques on a turbulent wing section. In: Proceedings of the Engineering Turbulence Modelling and Measurements (ETMM12), September 26–28, Montpellier, France (2018)
3. Baars, W.J., Hutchins, N., Marusic, I.: Self-similarity of wall-attached turbulence in boundary layers. J. Fluid Mech. **823**, R2 (2017)
4. Clauser, F.H.: Turbulent boundary layers in adverse pressure gradients. J. Aero Sci. **21**, 91–108 (1954)
5. Eitel-Amor, G., Örlü, R., Schlatter, P.: Simulation and validation of a spatially evolving turbulent boundary layer up to $Re_\theta = 8300$. Int. J. Heat Fluid Flow **47**, 57–69 (2014)
6. Fischer, P., Lottes, J., Kerkemeier, S.: Nek5000: open source spectral element CFD solver (2008). https://nek5000.mcs.anl.gov
7. Hosseini, S.M., Vinuesa, R., Schlatter, P., Hanifi, A., Henningson, D.S.: Direct numerical simulation of the flow around a wing section at moderate Reynolds number. Int. J. Heat Fluid Flow **61**, 117–128 (2016)
8. Jiménez, J., Hoyas, S., Simens, M.P., Mizuno, Y.: Turbulent boundary layers and channels at moderate Reynolds numbers. J. Fluid Mech. **857**, 335–360 (2010)
9. Skåre, P.E., Krogstad, P.-Å.: A turbulent equilibrium boundary layer near separation. J. Fluid Mech. **272**, 319–348 (1994)
10. Spalart, P.R., Watmuff, J.H.: Experimental and numerical study of a turbulent boundary layer with pressure gradients. J. Fluid Mech. **249**, 337–371 (1993)
11. Tanarro, Á., Vinuesa, R., Schlatter, P.: Effect of adverse pressure gradients on turbulent wing boundary layers. J. Fluid Mech. **883**(A8) (2020)
12. Vinuesa, R., Negi, P.S., Atzori, M., Hanifi, A., Henningson, D.S., Schlatter, P.: Turbulent boundary layers around wing sections up to $Re_c = 1{,}000{,}000$. Int. J. Heat Fluid Flow **72**, 86–99 (2018)

Chapter 3
Effect of Free-Slip and No-Slip Boundaries on Isotropic Turbulence

J. G. Wissink and H. Herlina

Introduction

The effect of free-slip and no-slip boundaries on isotropic turbulence is studied by direct numerical simulation (DNS). In each simulation, isotropic turbulence of fixed intensity was introduced at the bottom of the computational domain. The isotropic turbulence was generated in a separate large-eddy simulation (LES) that was running concurrently with the DNS. The study was carried out for various levels of turbulence. In all simulations, high pressure at the surface was found to be correlated with splats (updrafts) and positive surface divergence while low-pressure regions correlated with anti-splats and negative surface divergence. As previously shown and explained in Flores et al. [4], also in this study the horizontal velocities were found to exhibit a $-5/3$ spectrum very close to the surface.

Turbulence is known to significantly promote heat and mass transfer across the air–water interface. One way of introducing turbulence into the water is by employing oscillating grids underneath the water surface. The movement of such grids generates (approximately) isotropic turbulence that subsequently diffuses towards the surface. Following Brumley and Jirka [2], for isotropic turbulence, the turbulent Reynolds number is defined by

$$R_T = \frac{2\Lambda u_{\mathrm{rms}}}{\nu}, \tag{3.1}$$

J. G. Wissink (✉)
MAE Department, Brunel University London, Kingston Lane, Uxbridge UB8 3PH, UK
e-mail: jan.wissink@brunel.ac.uk

H. Herlina
Institute for Hydromechanics, Karlsruhe Institute of Technology (KIT), Kaiserstr. 8, 76128 Karlsruhe, Germany
e-mail: herlina@kit.edu

M. García-Villalba et al. (eds.), *Direct and Large Eddy Simulation XII*, ERCOFTAC Series 27,
https://doi.org/10.1007/978-3-030-42822-8_3

where Λ is the integral length scale, ν is the kinematic viscosity and u_{rms} is the root mean square of the horizontal velocity fluctuations. For distances sufficiently far from the interface (i.e. larger than one-Λ, outside the so-called surface-influenced layer), the growth in Λ is matched by the decline in u_{rms} such that R_T becomes virtually independent of depth. In the context of interfacial mass transfer, Theofanous et al. [9] identified a low R_T regime and a high R_T regime. The two regimes are separated from one another by a critical turbulent Reynolds number $R_{T,\mathrm{crit}} \approx 500$. In the low R_T regime the near surface turbulence is dominated by large scales, while in the high R_T regime the smaller scales become more important. This difference in turbulent characteristics in the two regimes is reflected in, e.g. a different R_T-scaling of the gas transfer velocity

$$K_L = \frac{|j|}{C_i - C_b}, \tag{3.2}$$

where $j = D(\partial C/\partial z)_i$ is the interfacial gas flux, C is the gas concentration, while the subscripts i and b denote the interface and bulk, respectively. According to Danckwert's surface renewal model [3] $K_L \approx \sqrt{Dr}$, where r is the surface renewal rate. In the low R_T regime, r is found to depend on the large turbulent scales Λ and u_{rms}, so that $K_L \approx \sqrt{Du_{\mathrm{rms}}/\Lambda}$, while for in the high R_T regime r depends on the Kolmogorov scales, i.e. $r = \sqrt{\varepsilon/\nu}$, where ε is turbulence dissipation rate. To explore the low- and high-R_T regimes in more detail, results from a number of three-dimensional direct numerical simulations (DNS)—in which the influence of symmetry and no-slip boundary conditions on isotropic turbulence was calculated—are analysed. In the DNS, both surface boundary conditions were combined with a variety of turbulence levels and integral length scales, such that data in both the low- and high-R_T regime were obtained.

Computational Setup

Each calculation presented here consists of two concurrently running parts. A direct numerical simulation (DNS) part and a large-eddy simulation (LES) part. The purpose of the LES was to generate an instantaneous 3D isotropic turbulent flow field of which a cross section at constant z was used as a time-dependent bottom boundary condition for the DNS.

For the discretisations of the 3D incompressible Navier–Stokes equation in both DNS and LES, a fourth-order-accurate kinetic energy conserving discretisation of the convective terms [10] was combined with a fourth-order central discretisation of the dissipation. To enforce incompressibility each time step a Poisson equation for the pressure was solved using the conjugate gradient method with a simple diagonal preconditioning. Time-integration was performed using the second-order-accurate Adams–Bashforth method. The simulations were run on a massively parallel supercomputer. To optimise load balancing, the computational domain was divided into equal parts, each associated to its own processing core. Communication

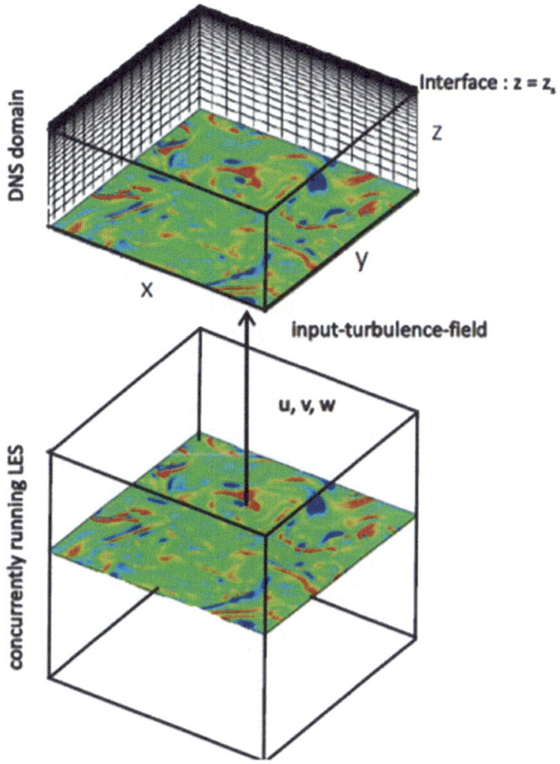

Fig. 3.1 Schematic of computational domain

between cores was performed using the standard message passing interface (MPI). A schematic of the computational domain employed in the simulations can be seen in Fig. 3.1. In the LES periodic boundary conditions were employed in all directions while in the DNS only the horizontal directions were periodic. As mentioned above, the isotropic turbulence was introduced at the bottom of the DNS while at the top either free-slip or no-slip boundary conditions were used. The standard Smagorinsky model was used to model the subgrid-scale stresses in the LES. A detailed discussion on the numerical method and the setup of the simulations can be found in Herlina and Wissink [5–7]. An overview of the simulations performed can be found in Table 3.1.

Table 3.1 Overview of simulations, FS and NS cases have a free-slip and no-slip surface boundary condition, respectively

Case	Domain size	DNS mesh	R_T
FS-500	$20L \times 20L \times 5L$	$512 \times 512 \times 300$	500
FS-1400	$20L \times 20L \times 5L$	$1024 \times 1024 \times 500$	1400
NS-50	$5L \times 5L \times 5L$	$128 \times 128 \times 300$	50
NS-800	$20L \times 20L \times 5L$	$512 \times 512 \times 300$	800

Results

Figure 3.2 shows the horizontally averaged velocity rms profiles, $u_{\text{rms}} = \sqrt{\overline{\langle u^2 \rangle}}$ and $w_{\text{rms}} = \sqrt{\overline{\langle w^2 \rangle}}$, in the computational domain. Note that $\overline{\cdot}$ and $\langle \cdot \rangle$ represent averaging in time and averaging in the horizontal (homogeneous) directions, respectively. The left pane shows the behaviour of u_{rms} and w_{rms} when approaching the free-slip interface of simulation FS-500. As expected, towards the surface w_{rms} reduces linearly to zero. As w_{rms} reduces part of its kinetic energy is transferred to u_{rms} leading to a growth in the horizontal fluctuations close to the interface (see also Perot and Moin [8]). Only very close to the surface there is a thin layer where u_{rms} becomes constant. The right pane shows the u_{rms} and w_{rms} profiles obtained in simulation NS-800. Here it can be seen that the turbulence is approximately isotropic in the lower part of the computational domain while both u_{rms} and w_{rms} decrease linearly with reducing distance to the surface (z_s). As soon as the reduction in w_{rms} increases the reduction in u_{rms} decreases and even turns into an increase. This also is a consequence of the conservation of turbulent kinetic energy. Very close to the surface u_{rms} begins to decrease again and subsequently reduces linearly to zero. Simultaneously, the reduction in w_{rms} decreases so that in a very small layer adjacent to the surface $\partial w_{\text{rms}}/\partial z_s \approx 0 \approx w_{\text{rms}}$.

Figure 3.3 shows surface plots of the pressure coloured by contours of the horizontal divergence

$$\beta = \frac{\partial u}{\partial x} + \frac{\partial v}{\partial y} \tag{3.3}$$

in the grid plane adjacent to the interface. For both free-slip and no-slip interfaces, upward moving flow creates "splats" which are identified by local maxima in the static pressure due to stagnation (see also Bodart et al. [1]). Vice versa, downward moving flow corresponds to "anti-splats" which are identified by low static pressure. Simultaneously, it can be seen that the pressure maxima also correspond to max-

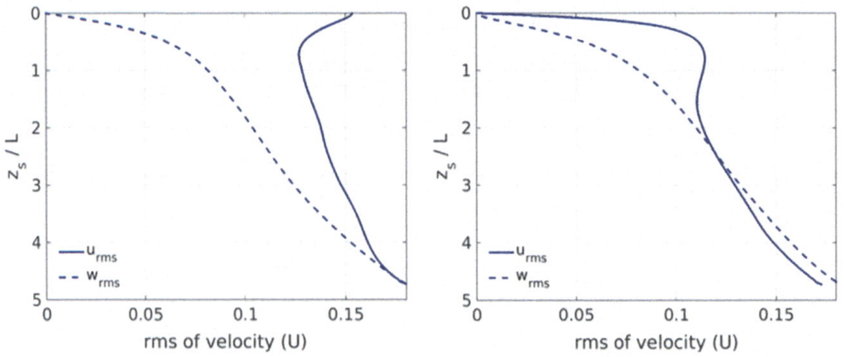

Fig. 3.2 Velocity fluctuations. Left: free-slip surface (FS-500); right: no-slip surface (NS-800)

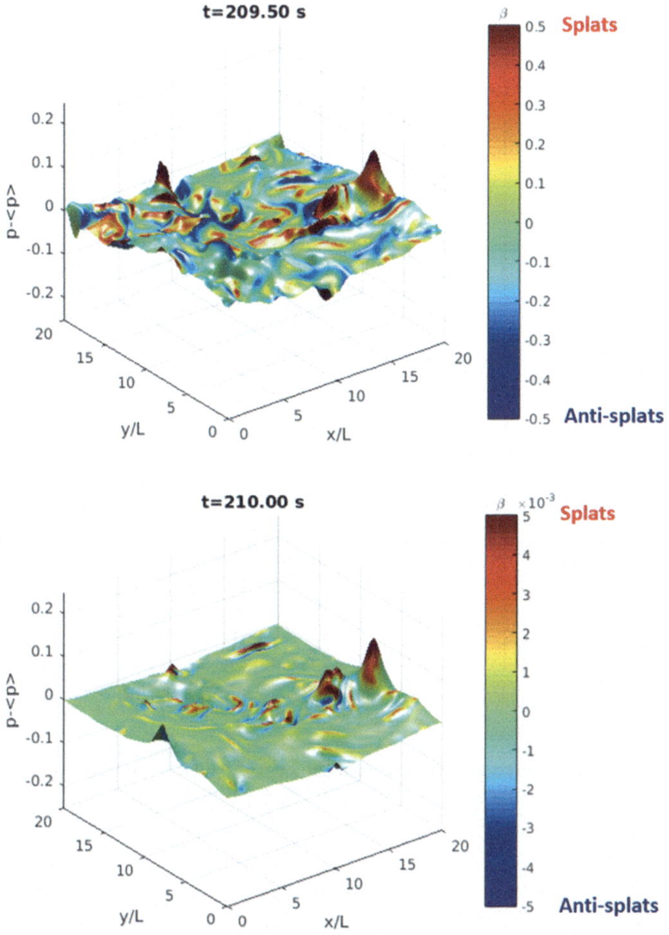

Fig. 3.3 Surface plots of pressure distribution near the interface coloured by horizontal divergence in the grid plane near the interface (upper pane: FS-500, lower pane: NS-800)

ima in the horizontal divergence. This positive divergence is a consequence of the upwards moving flow being forced to spread out horizontally when approaching the interface. The top side of a typical convection cell contains one or more of such areas with positive divergence. Where two convection cells meet, their upper flow fields converge and subsequently move downward. This leads to relatively thin regions at the surface with low pressure and negative β.

Despite the significantly lower turbulence level close to the interface in the no-slip case, similar flow structures as in the free-slip case are found for comparable turbulent Reynolds numbers. In the DNS with low R_T large vortical structures are observed, while for higher R_T a more dynamic pattern of small vortical structures

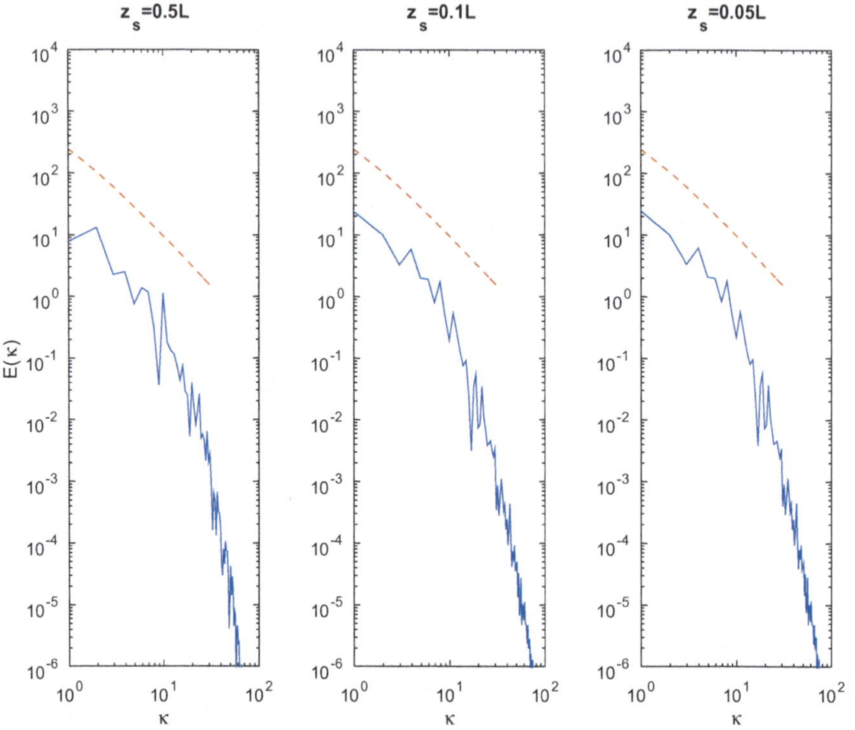

Fig. 3.4 Instantaneous energy spectra of the horizontal velocity at $x/L = 10$ of simulation FS-500 at various distances to the surface)

is found, clearly indicating the dominance of large and small eddies at low and high R_T, respectively.

As shown in Fig. 3.4, for simulation FS-500 the energy spectra of the horizontal velocity consistently exhibit a $-5/3$ behaviour when approaching the free-slip surface. This behaviour was explained by Flores et al. [4] by a strong vertical shearing of the horizontal velocity resulting in a downscale transfer of energy, similar to what was found for turbulence in strongly stratified flows. Very close to the surface, at $\zeta = 0.1L$ and $\zeta = 0.05L$, the spectra were found to be almost identical.

References

1. Bodart, J., Cazalbou, J.-B., Joly, L.: Direct numerical simulation of unsheared turbulence diffusing towards a free-slip or no-slip surface. J. Turbul. **11** (2010)
2. Brumley, B.H., Jirka, G.H.: Near-surface turbulence in a grid-stirred tank. J. Fluid Mech. **183**, 235–263 (1987)

3. Danckwerts, P.V.: Significance of liquid-film coefficients in gas absorption. Ind. Eng. Chem. **43**, 1460–1467 (1951)
4. Flores, O., Riley, J., Horner-Devine, A.: On the dynamics of turbulence near a free surface. J. Fluid Mech. **821**, 248–265 (2017)
5. Herlina, H., Wissink, J.G.: Direct numerical simulation of turbulent scalar transport across a flat surface. J. Fluid Mech. **744**, 217–249 (2014)
6. Herlina, H., Wissink, J.G.: Isotropic-turbulence-induced mass transfer across a severely contaminated water surface. J. Fluid Mech. **797**, 665–682 (2016)
7. Herlina, H., Wissink, J.G.: Simulation of air-water interfacial mass transfer driven by high-intensity isotropic turbulence. J. Fluid Mech. **860**, 419–440 (2019)
8. Perot, B., Moin, P.: Shear-free turbulent boundary layers. Part 1. Physical insights into near-wall turbulence. J. Fluid Mech. **295**, 199–227 (1995)
9. Theofanous, T.G., Houze, R.N., Brumfield, L.K.: Turbulent mass transfer at free gas liquid interfaces with applications to open channel bubble and jet flows. Int. J. Heat Mass Transf. **19**, 613–624 (1976)
10. Wissink, J.G.: On unconditional conservation of kinetic energy by finite-difference discretisations of the linear and non-linear convection equation. Comput. Fluids **33**, 315–343 (2004)

Chapter 4
Large Eddy Simulations of Rough Turbulent Channel Flows Bounded by Irregular Roughness: The Role of Geometrical Parameters

D. Saccone, E. Napoli, B. Milici and M. De Marchis

Introduction

Almost all bounded flows, in nature and industry, are bounded by rough boundaries. Important efforts have been made, in fact, to analyse the physics of the fluids over such geometries. Basically, rough walls strongly influence the total energy dissipation, with important consequences in the form of higher costs especially in industrial applications. By now, several researches were carried out to investigate the modification of turbulence structures induced by roughness on pipe, channel and/or boundary layer flows. After decades of laboratory and numerical experiments, the energy dissipation induced by a rough surface is not completely understood. What is well known is the evident effect of the roughness in decreasing the mean streamwise velocity profile in the logarithmic region. This shift is known as roughness function ΔU^+ (the symbol $^+$ represents quantities made non-dimensional using the friction velocity u_τ, or the viscous length scale ν/u_τ). The mean velocity profile in rough condition can be predicted with the law [1]:

$$U^+ = \frac{1}{\kappa}\ln(x_3^+) + B - \Delta U^+ \tag{4.1}$$

where x_3^+ indicates the wall normal direction κ is the classical von Kàrmàn constant and B is a specific constant. In order to predict the mean velocity profile, at least in the logarithmic region, the roughness function must be estimated. Several researches (see among others [2–5] and literature therein cited) analysed the roughness function

D. Saccone · B. Milici · M. De Marchis (✉)
Faculty of Engineering and Architecture, Kore University of Enna, Enna, Italy
e-mail: mauro.demarchis@unikore.it

E. Napoli
Department of Civil, Environmental, Aerospace, Materials Engineering, University of Palermo, Palermo, Italy

M. García-Villalba et al. (eds.), *Direct and Large Eddy Simulation XII*, ERCOFTAC Series 27,
https://doi.org/10.1007/978-3-030-42822-8_4

in light of statistical moments of the roughness geometry, as the averaged absolute deviation k, the rms of wall oscillations k_{rms}, the skewness s_k, the kurtosis k_u and, more recently, the so called *effective slope* (ES) [2, 6–9]. The effective slope, as demonstrated by [8, 10], has a good correlation with the effects on the turbulence and is able to predict the downward shift of the mean velocity profile in regular or irregular rough surfaces as well as in 2D or 3D rough configurations. The above researches, unfortunately, were carried out for a single friction Reynolds number and focusing the attention on single geometrical parameters. The results cannot be thus considered universally valid and further investigations are required. The effective slope does not takes into account the friction Reynolds number of the mean flow. On the other hand, it is universally accepted that flow field having higher Reynolds number is characterised by higher energy dissipation, hence higher roughness function. To find a universal correlation between the roughness geometry and the induced roughness function, the best choice is to take into account two or more geometrical parameter, including the effective slope and a parameter able to consider the friction Reynolds number. The objective of the proposed research is to investigate the correlation between ΔU^+ and a series of combination of two geometrical parameters.

Numerical Simulations

Turbulent flow is resolved using DNS or LES approaches, based on the numerical solution of the filtered mass and momentum conservation. The momentum and continuity equations are resolved using the finite-volume numerical code called PANORMUS, which is second-order accurate both in time and space (available at the site www.panormus3d.org). The Adams–Bashforth method is used for the time advancement of the solution, with the classical fractional-step technique. The pressure Poisson equation is solved through a line-SOR technique in conjunction with a multigrid V-cycle accelerator [11]. LES was performed on the domain reported in Fig. 4.1. The grids are refined at the solid boundaries in order to resolve steep velocity gradients. Basically, the grid is refined thus to have the first grid at a distance from the solid boundary, in wall units (x_3^+), less than 1. The computational domain consists of a channel delimited by two parallel and fixed slabs irregularly roughened through the superimposition of different sinusoidal functions, both in the streamwise and spanwise directions by means of the equation:

$$r(x_1, x_2) = \sum_{i=1}^{n_1} A_i \sin\left(\frac{2i\pi x_1}{L_{x_1}/2}\right) \cdot \sum_{j=1}^{n_2} B_j \sin\left(\frac{2j\pi x_2}{L_{x_2}/2}\right) \tag{4.2}$$

The domain was discretised by a curvilinear structured mesh. In order to analyse the effect of rough walls on turbulent flow, fluid dynamic simulations were performed changing both the mean roughness height and the friction Reynolds number. Specifi-

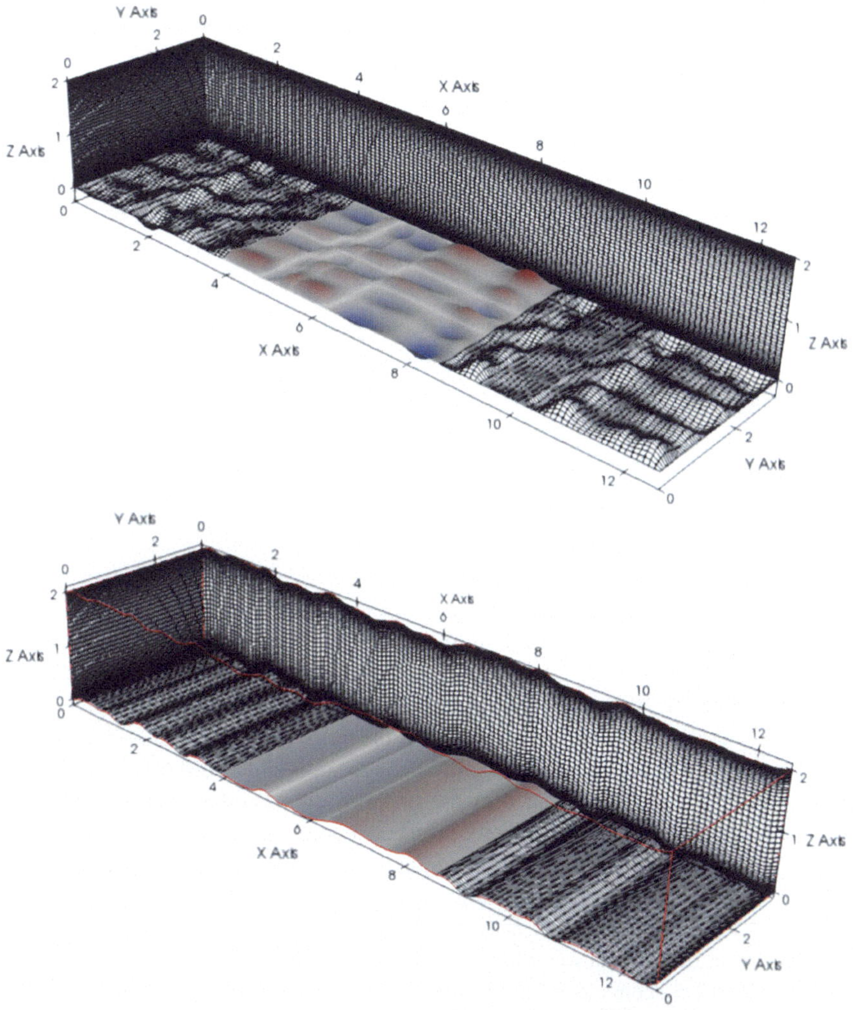

Fig. 4.1 Layout of the computational domain of rough channel. Top panel: 3D rough configuration; Bottom panel: 2D irregular roughness

cally, two different mean roughness heights, falling in transitional and fully turbulent regimes, were simulated at increasing Re_τ: 180, 395, 590 and 1000.

In Table 4.1 details on the numerical simulations are reported.

Table 4.1 Details on the roughness type and the related geometric parameters. k_s^+: equivalent sand grain roughness. k^+: mean roughness height. k_{max}^+: highest roughness peak. k_{rms}^+: root mean squares of the wall roughness

Case	Simulation	Roughness	Re_τ	k^+	k_{max}^+	k_s^+	k_{rms}^+	ES
1	DNS	2D	180	9.00	20.52	32.40	10.62	0.206
2	LES	2D	395	9.48	27.65	23.70	11.45	0.101
3	LES	2D	395	19.75	56.48	75.05	23.70	0.206
4	LES	3D	395	9.48	67.15	23.70	13.03	0.100
5	LES	3D	395	19.75	146.15	138.00	27.65	0.241
6	LES	2D	590	14.16	41.30	29.00	16.52	0.090
7	LES	2D	590	29.50	82.60	118.00	35.40	0.203
8	LES	2D	1000	24.00	70.00	28.00	28.70	0.092
9	LES	2D	1000	50.00	140.00	200.00	60.00	0.203

Results

In this study, the roughness function ΔU^+ was analysed starting from the value of some geometrical parameters. Following the recent literature findings [8], k_{max}^+, highest roughness peak; k^+, mean roughness height; k_{rms}^+, root mean squares of the wall roughness; ES, effective slope; k_s^+, equivalent sand grain roughness, were analysed. Literature results showed that k_s^+ is the more suitable parameter to predict the roughness function, at least in the fully rough regime. Nevertheless, as pointed out by [6], k_s^+ is not a physical measure of the roughness topography and it can be estimated only if the mean velocity profile is known. In Fig. 4.2 the dependence of ΔU^+ on geometrical parameters is plotted. The simulations were performed at different friction Reynolds and different mean roughness height.

Figure 4.2 shows that ΔU^+, at least for the simulations here performed, does not have a specific correlation with the proposed geometrical parameters. Also the effective slope ES seems to be influenced by the increase of the friction Reynolds number Re_τ. Looking at the correlation of ΔU^+ versus ES, in fact, it is possible to observe a good correlation when the same friction Reynolds number is considered. On the contrary, looking at the data highlighted with the dotted grey box, simulations having the same geometry but different Re_τ have different trends. Basically, all the data seem to be aligned in a vertical line. This is due to the fact that increasing the friction Reynolds number, increases ΔU^+ even though the effective slope is kept constant.

This result suggests that the roughness function must be predicted using another type of parameter. In order to this, several combinations have been considered. In Fig. 4.3 it is plotted the dependence of the Roughness function with ES $* k_{rms}^+$. The trend is clearly similar to that achieved with the k_s equivalent sand grain roughness, with the advantage that both ES and k_{rms}^+ are real geometrical parameters of the wall roughness and can be a-priori estimated.

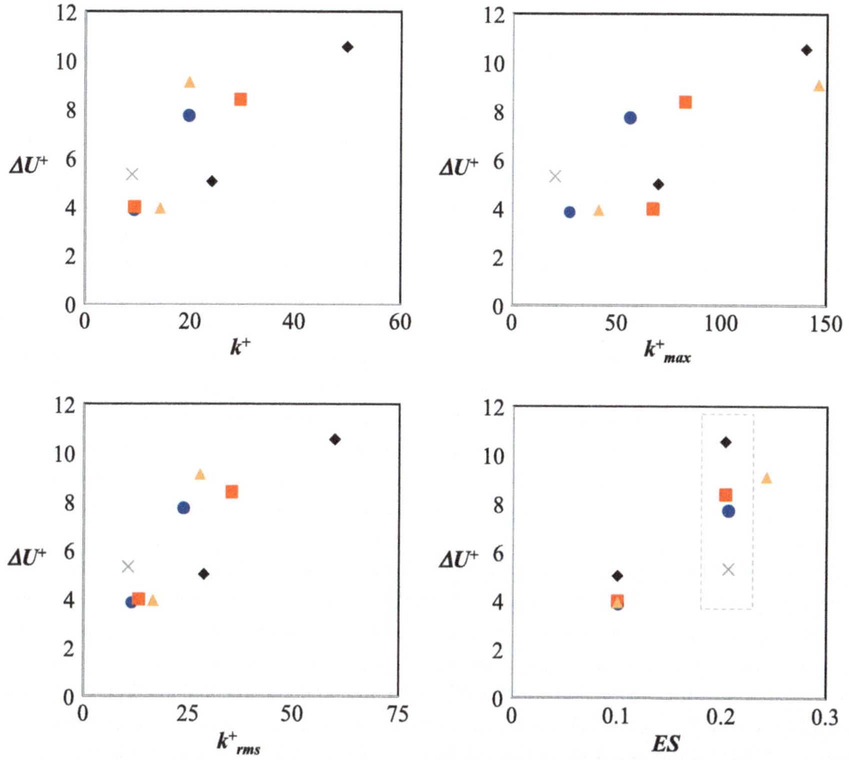

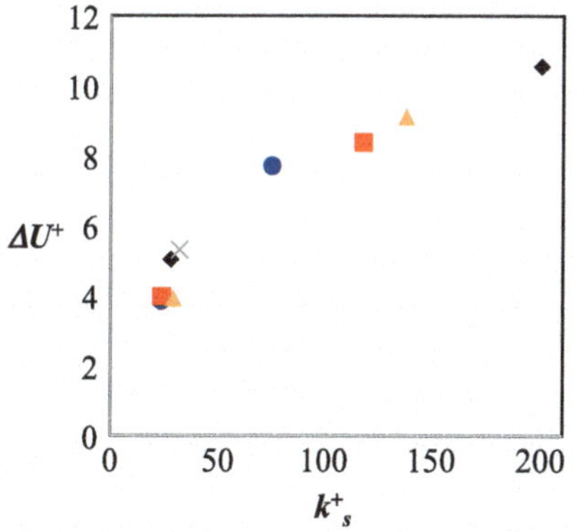

Case	Re_τ		ΔU^+	
1	180	2D	5.30	×
2	395	2D	3.83	●
3	395	2D	7.73	●
4	395	3D	4.00	▲
5	395	3D	9.12	▲
6	590	2D	3.95	■
7	590	2D	8.40	■
8	1000	2D	5.00	◆
9	1000	2D	10.56	◆

Fig. 4.2 Dependence of the roughness function ΔU^+ on geometrical parameters. k^+_{max}: highest roughness peak; k^+: mean roughness height; k^+_{rms}: root mean squares of the wall roughness; ES: effective slope; k^+_s: equivalent sand grain roughness

Fig. 4.3 Dependence of the roughness function ΔU^+ on $ES * k^+_{rms}$ ♦, $Re_\tau = 1000$—2D roughness; ■, $Re_\tau = 590$—2D roughness; •, $Re_\tau = 395$—2D roughness; ▲, $Re_\tau = 395$—3D roughness; ×, $Re_\tau = 180$—2D roughness

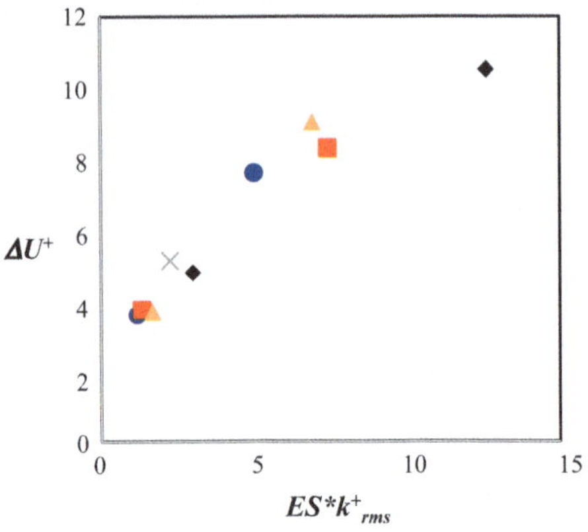

Conclusions

LES and DNS up to $Re_\tau = 1000$, have been performed to investigate on the mean velocity reduction induced by irregular wall roughness. The roughness function was correlated with some geometrical parameters: k_s^+: the equivalent sand grain roughness, k^+: the mean roughness height, k^+_{max}: the highest roughness peak, k^+_{rms}: the root mean squares of the wall roughness and ES: the effective slope. The analysis pointed out that the roughness function cannot be properly estimated with a single geometry feature of the roughness, rather a combination of the above parameters is the key to predict the roughness function. The numerical simulations here performed suggest that the combination of ES and k^+_{rms} is a geometrical feature able to predict ΔU^+. Further numerical experiments must be carried out to confirm the validity of the law also at higher Reynolds number.

References

1. Pope, S.: Turbulent Flow. Cambridge University Press, Cambridge
2. Yuan, J., Piomelli, U.: Estimation and prediction of the roughness function on realistic surfaces. J. Turbul. **15**(6), 350–365 (2014)
3. Rao, V.N., Jefferson-Loveday, R., Tucker, P.G., Lardeau, S.: Large eddy simulations in turbines: influence of roughness and free-stream turbulence. Flow Turbul. Combust. **92**(1–2), 543–561 (2014)
4. Forooghi, P., Stroh, A., Magagnato, F., Jakirlic, S., Frohnapfel, B.: Towards a universal roughness correlation. ASME J. Fluids Eng. **139**(12), 12121 (2017)
5. Forooghi, P., Stroh, A., Frohnapfel, B.: A systematic study of turbulent heat transfer over rough walls. Int. J. Heat Mass Transf. **127**(part c), 1157–1168 (2018)
6. Schultz, M.P., Flack, K.A.: Turbulent boundary layers on a systematically varied rough wall. Phys. Fluids **21**(015104), 1–9 (2009)

7. Mejia-Alvarez, R., Christensen, K.T.: Wall-parallel stereo particle-image velocimetry measurements in the roughness sublayer of turbulent flow overlying highly irregular roughness. Phys. Fluids **25**, 115109 (2013)
8. De Marchis, M.: Large eddy simulations of roughened channel flows: estimation of the energy losses using the slope of the roughness. Comput. Fluids **140**, 148–157 (2016)
9. Milici, B., De Marchis, M.: Statistics of inertial particle deviation from fluid particle trajectories in horizontal rough wall turbulent channel flow. Int. J. Heat Fluid Flow **60**, 1–11 (2016)
10. Napoli, E., Armenio, V., De Marchis, M.: The effect of the slope of irregularly distributed roughness elements on turbulent wall-bounded flows. J. Fluid Mech. **613**, 385–394 (2008)
11. De Marchis, M., Milici, B., Napoli, E.: Numerical observations of turbulence structure modification in channel flow over 2D and 3D rough walls. Int. J. Heat Fluid Flow **56**, 108–123 (2015)

Chapter 5
Roughness Effects on Scalar Transport

Z. Hantsis and U. Piomelli

Introduction

Scalar transport occurs frequently in engineering and the natural sciences, in problems ranging from the transport of particles or pollutants in a flow to heat transfer in mechanical systems. It is a classic topic that has been studied for many decades and is the subject of many textbooks [9]. Because of the similarity between the transport equations for the velocity and scalar, the Reynolds Analogy was proposed [18], which postulates, in its simplest form, that the ratio between the dimensionless scalar-transfer coefficient, the Stanton number, and the friction coefficient, is constant. The Reynolds analogy can be extended to include similarity of the velocity and scalar profiles (properly normalized) and even of the turbulent fluctuations. Most of the studies of scalar transport in wall-bounded flows with smooth surfaces found that the Reynolds analogy holds very well, when no significant pressure gradients exist, for Prandtl numbers of order unity [9].

However, all surfaces are rough when the Reynolds number Re becomes large enough. While the effects of roughness on momentum transport are well known (see, for instance, [6, 17]), its effects on scalar transport have not been investigated as widely. Still, it has been known for over 50 years that the Reynolds analogy fails in the presence of roughness: in the fully rough regime, the drag coefficient is dominated by the form drag and becomes independent of Re, while scalar transport remains fully dependent on the diffusive effects [13]; their ratio cannot, therefore, remain constant. The failure of the Reynolds analogy for rough walls has been documented by experimental [1, 2] and, more recently, numerical studies [4, 10, 14].

Z. Hantsis (✉) · U. Piomelli
Department of Mechanical and Materials Engineering, Queen's University,
Kingston, ON K7L 3N6, Canada
e-mail: zvi.hantsis@queensu.ca

U. Piomelli
e-mail: ugo@queensu.ca

M. García-Villalba et al. (eds.), *Direct and Large Eddy Simulation XII*, ERCOFTAC Series 27,
https://doi.org/10.1007/978-3-030-42822-8_5

The purpose of the present paper is to examine the mechanisms that drive the exchange of scalar in the roughness sublayer, comparing them with those that govern the momentum transfer. This will be achieved by performing direct numerical simulations (DNS) of the flow over a rough wall. The problem will be formulated in the next section. Then, some results will be described; concluding remarks will close the paper.

Problem Definition

We consider the transport of passive scalars in an incompressible, fully developed turbulent flow. The governing equations are the conservation of mass, momentum, and scalar:

$$u_{j,j} = 0 \tag{5.1}$$

$$u_{i,t} + \left(u_i u_j\right)_{,j} = -P_{,j}\delta_{ij} + \nu u_{i,jj} + \Pi\delta_{i1} + F_i \tag{5.2}$$

$$\theta_{,t} + \left(u_j\theta\right)_{,j} = \alpha\theta_{,jj} + Q + F_\theta \tag{5.3}$$

where $i = 1, 2$, and 3 (or x, y, and z) are the streamwise, wall-normal, and spanwise directions, respectively, u_i are the velocity components, θ is the normalized scalar (which will be defined momentarily), P is the density-normalized hydrodynamic pressure and ν and α are the kinematic viscosity and diffusivity. The subscripts , j and t, denote differentiation with respect to x_j and time, respectively. $\Pi\delta_{i1}$ is the driving pressure gradient, Q is a scalar source term, also defined below, and F_i, F_θ are forcing terms used to represent the roughness geometry. Periodic boundary conditions are applied on all quantities in the streamwise and spanwise direction, no-slip conditions are used for u_i at the walls and inside the roughness, and homogeneous Dirichlet conditions $\theta = 0$ are used for the scalar.

Following the approach of [7], we define $\theta = T_w - T$, where T is the scalar, and its value at the wall is T_w. Constant heat flux at both walls in a fully developed channel results in the time-averaged T_w increasing linearly with x. For Prandtl number of order unity, the scalar fluctuations at the wall are very small [8], so that a constant heat flux condition corresponds to $\theta = 0$. The source term resulting from the transformations is $Q = u_1(\mathrm{d}\overline{T}_w/\mathrm{d}x)$, where overline denotes time averaging.

The roughness is modeled using an Immersed Boundary Method (IBM), based on the Volume-of-Fluid approach. Following [19], the surface geometry is a distribution of randomly oriented ellipsoids with semiaxes k, $1.4k$, and $2k$, where k was 4% of the channel halfwidth δ.

The equations of motion are solved using a second-order accurate discretization in space and time on a staggered mesh. The numerical model has been widely validated for rough-wall flows [21, 22]. The Prandtl number is Pr = 1. The Reynolds number, based on friction velocity u_τ and δ, is $\mathrm{Re}_\tau = 1000$ for the smooth-wall case and 1700 for the rough-wall one; the equivalent sand grain height, in wall units, is 90, so that the flow is fully rough. $640 \times 530 \times 320$ grid points are used to discretize a domain of dimensions $4\delta \times 2\delta \times 2\delta$. Although this domain is somewhat smaller than those commonly used, it is significantly larger than the minimal channel used by [10] in similar calculations.

Results

Several averaging operators can be used in rough-wall channel flows. Time averaging is indicated by an overbar. In the (x, z)-plane we define the *intrinsic* spatial average $\langle\cdot\rangle$, carried out only in the fluid domain, and the *superficial* average $\langle\cdot\rangle_s$, carried out over the whole plane [12].

A flow quantity can then be decomposed into a time- and space-averaged (or *double-averaged*) component, the spatial variation of the time average, denoted by a tilde, and the fluctuation, denoted by a prime:

$$\theta(x, y, z, t) = \overline{\theta}(x, y, z) + \theta'(x, y, z, t) = \langle\overline{\theta}\rangle(y) + \widetilde{\theta}(x, y, z) + \theta'(x, y, z, t). \tag{5.4}$$

$\widetilde{\theta}$ is known as the "wake field", or "form-induced perturbation".

Figure 5.1 shows the first- and second-order statistics; d is the location of the resultant force acting on the roughness sublayer [5]. The velocity is normalized by the friction velocity u_τ, the scalar by its equivalent, θ_τ. Both velocity and scalar exhibit a logarithmic layer; their slopes are, however, slightly different: the Von-Kàrmàn constant is 0.4 for the velocity and 0.46 for the scalar, in agreement with

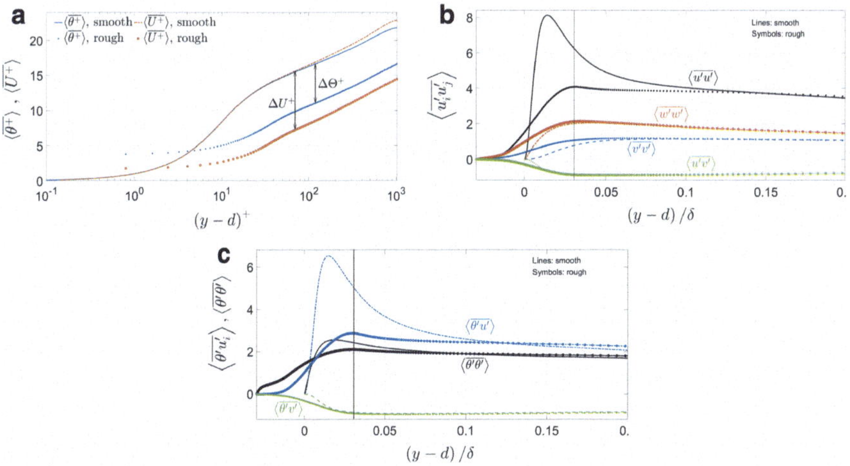

Fig. 5.1 **a** Mean velocity and scalar profiles in wall units; **b** Reynolds stresses and **c** scalar variance and scalar-velocity correlations. Lines: smooth wall; lines and symbols: rough wall. The vertical line indicates the position of the roughness crest

other DNS results [15]. The downward shift of the logarithmic region, the roughness function, is much lower for the scalar than for the velocity, also consistent with previous results [10, 14].

Figure 5.1b, c shows the second-order moments of velocity and scalar fluctuations. The peak value of $\langle\overline{u'u'}\rangle_s$, located below the roughness crest, is reduced considerably by roughness while the $\langle\overline{v'v'}\rangle_s$ and $\langle\overline{w'w'}\rangle_s$ components slightly increase, thereby reducing the anisotropy of the Reynolds stresses [20]. While $\langle\overline{\theta'u'}\rangle_s$ mirrors the damping of the streamwise fluctuations, $\langle\overline{\theta'\theta'}\rangle_s$ is reduced much less, as also observed in [14], although to a lesser degree. Townsend similarity hypothesis (i.e., the smooth- and rough-wall statistics, normalized by u_τ and θ_τ, collapse away from the crest [3]), holds for both velocity and scalar moments.

The budget of $\left\langle\overline{u_i'u_j'}\right\rangle_s$ is

$$0 = \mathscr{T}_{ij} + \mathscr{A}_{ij} + \nu\left\langle\nabla^2\left(\overline{u_i'u_j'}\right)\right\rangle_s + \mathscr{P}_{ij} + \Pi_{ij} - \varepsilon_{ij} \tag{5.5}$$

where the terms on the right-hand-side of (5.5) are, respectively, the turbulent transport, wake transport, viscous diffusion, production, pressure work (or velocity–pressure-gradient), and the dissipation [12, 16]. The production term for the normal components ($i = j = \beta$, no summation on β) can be written as [11, 12, 17] follows:

$$\mathscr{P}_{\beta\beta} = \left[-2\left\langle\overline{u_\beta'u_k'}\right\rangle_s\langle\overline{u_\beta}\rangle_{,k}\right] + \left[-2\left\langle\overline{u_\beta'u_k'}\right\rangle_s\langle\widetilde{u}_{\beta,k}\rangle\right] + \left[-2\left\langle\widetilde{\overline{u_\beta'u_k'}}\,\widetilde{u}_{\beta,k}\right\rangle_s\right] \tag{5.6}$$

where the first term in (5.6) is the shear production $\mathscr{P}^s_{\beta\beta}$, also present in smooth-wall cases, and the other two depend on the wake field and, therefore, only occur where the wall is rough. Combined, they give the "Form-induced production" $\mathscr{P}^{fi}_{\beta\beta} = \mathscr{P}^m_{\beta\beta} + \mathscr{P}^w_{\beta\beta}$, which plays a significant role in the redistribution of energy between the normal Reynolds stresses [22]. Following the same procedure, we write the budget of $\langle\overline{\theta'\theta'}\rangle_s$

$$0 = \mathscr{T}_\theta + \mathscr{A}_\theta + \alpha\left\langle\nabla^2\left(\overline{\theta'\theta'}\right)\right\rangle_s + \mathscr{P}_\theta + Q_\theta - \varepsilon_\theta, \tag{5.7}$$

where the terms on the right-hand-side of (5.7) are, respectively, the scalar turbulent transport, wake transport, conductive diffusion, production, source-term contribution, and the dissipation. The scalar production term can also be decomposed into a shear contribution and a form-induced part: $\mathscr{P}_\theta = \mathscr{P}^s_\theta + \mathscr{P}^{fi}_\theta$.

Figure 5.2 shows the budgets of $\langle\overline{u'u'}\rangle_s$ and $\langle\overline{\theta'\theta'}\rangle_s$. When the wall is smooth they are remarkably similar, indicating that the Reynolds Analogy also extends to the budget terms. However, when the wall is rough, the dissipation ε_{11} is much lower in magnitude compared with the smooth case, while the pressure work Π_{11} becomes a significant negative term; this reflects the role of the form drag. In the scalar budget, on the other hand, ε_θ remains the leading loss term, particularly outside of the roughness sublayer where the diffusion terms vanish. The form-induced production does not

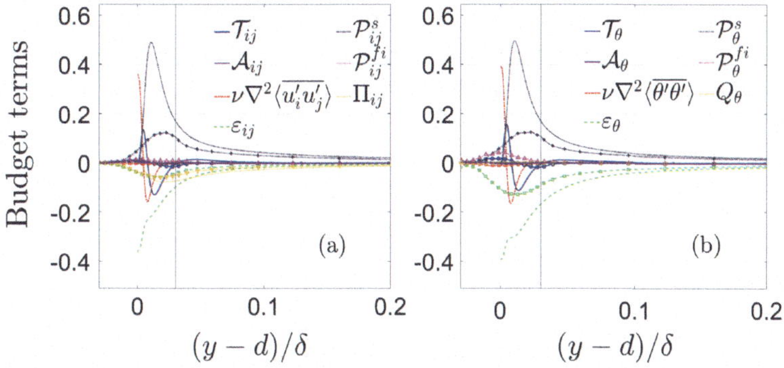

Fig. 5.2 Budgets of **a** $\langle\overline{u'u'}\rangle_s$ and **b** $\langle\overline{\theta'\theta'}\rangle_s$. Lines: smooth wall; lines and symbols: rough wall

contribute much to the budget of $\langle\overline{u'u'}\rangle_s$ (although it plays an important role in those of $\langle\overline{v'v'}\rangle_s$ and $\langle\overline{w'w'}\rangle_s$) while it is much more significant in the scalar-variance budget. This difference may be one of the causes of the less significant decrease of $\langle\overline{\theta'\theta'}\rangle_s$ observed above.

Figure 5.3 shows contours of various quantities in a plane parallel to the wall, inside the roughness sublayer. $\widetilde{u}$ is characterized by channeling between roughness elements and $\widetilde{\theta}$ mirrors this behavior. $\widetilde{v}$, on the other hand, is significant near the stagnation points and in the recirculation regions behind elements; its length scales are

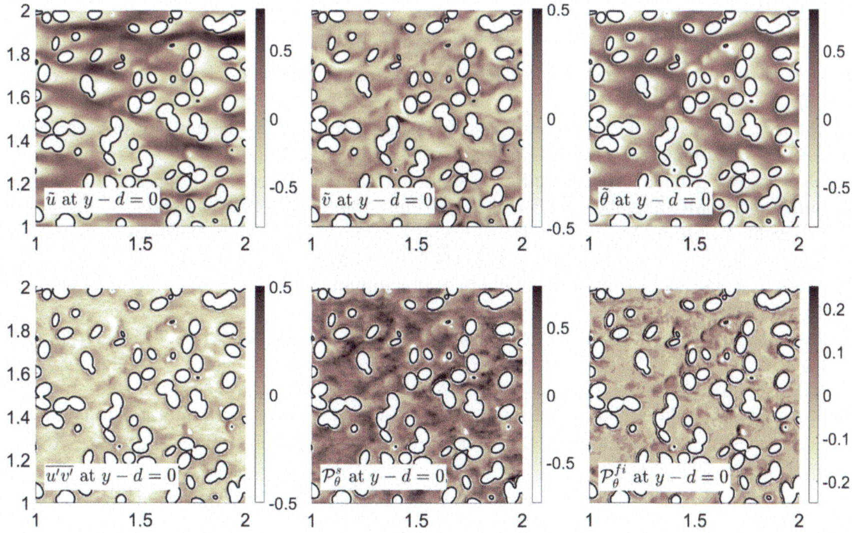

Fig. 5.3 Contours of various quantities in the $y-d=0$ horizontal plane; Absica is x/δ and ordinate is z/δ. All quantities are normalized by their maximum value

much smaller than those of $\widetilde{u}$ and $\widetilde{\theta}$. The Reynolds stress $\overline{u'v'}$ is mostly concentrated in empty regions between elements, where the solid bodies do not inhibit the velocity fluctuations. Locally, the shear production resembles $\overline{u'v'}$, while the form-induced production has much smaller scales, and seems to be largely associated with the stationary ejections and sweeps caused by the roughness elements.

Conclusions

A DNS of channel flow with constant scalar wall-flux was performed to investigate the effects of roughness on the transport of passive scalars with Pr = 1.0. The log-layer shift existing in the mean velocity profile was also seen in the mean scalar profile, however to a lesser degree. Roughness damps the peak of $\langle\overline{u'u'}\rangle_s$ and, to a lesser extent, that of $\langle\overline{\theta'\theta'}\rangle_s$. This may be due to the fact that the form-induced production is much larger in the budget of $\langle\overline{\theta'\theta'}\rangle_s$ than in that of $\langle\overline{u'u'}\rangle_s$. This indicates that the geometry details play a significant role in explaining the differences between scalar and momentum transport.

Acknowledgements Research supported by Hydro-Québec and the Natural Science and Engineering Research Council of Canada (NSERC) under the Collaborative Research and Development program (Grant No. CRDPJ 418786 - 11).

References

1. Bons, J.: A critical assessment of Reynolds analogy for turbine flows. J. Heat Transfer-T. ASME **127**, 472–485 (2005)
2. Dipprey, D.F., Sabersky, R.H.: Heat and momentum transfer in smooth and rough tubes at various Prandtl numbers. Int. J. Heat Mass Transfer **6**, 329–353 (1963)
3. Flack, K.A., Schultz, M.P., Shapiro, T.A.: Experimental support for Townsend's Reynolds number similarity hypothesis on rough walls. Phys. Fluids **17**, 035102 (2005)
4. Forooghi, P., Stripf, M., Frohnapfel, B.: A systematic study of turbulent heat transfer over rough walls. Int. J. Heat Fluid Flow **127**, 1127–1168 (2018)
5. Jackson, P.: On the displacement height in the logarithmic velocity profile. J. Fluid Mech. **111**, 15–25 (1981)
6. Jiménez, J.: Turbulent flows over rough walls. Annu. Rev. Fluid Mech. **36**, 173–196 (2004)
7. Kasagi, N., Tomita, Y., Kuroda, A.: Direct numerical simulation of passive scalar field in a turbulent channel flow. J. Heat Transfer-T. ASME **114**(3), 598–606 (1992)
8. Kasagi, N., Kuroda, A., Hirata, M.: Numerical investigation of near-wall turbulent heat transfer taking into account the unsteady heat conduction in the solid wall. J. Heat Transfer-T. ASME **111**(2), 385–392 (1989)
9. Kays, W.M., Crawford, M.E.: Convective Heat and Mass Transfer. McGraw-Hill, New York (1980)
10. MacDonald, M., Hutchins, N., Chung, D.: Roughness effects in turbulent forced convection. J. Fluid Mech. **861**, 138–162 (2019)
11. Mignot, E., Barthelemy, E., Hurter, D.: Double-averaging analysis and local flow characterization of near-bed turbulence in gravel-bed channel flows. J. Fluid Mech. **618**, 279–303 (2009)

12. Nikora, V., McEwan, I., McLean, S., Coleman, S., Pokrajac, D., Walters, R.: Double-averaging concept for rough-bed open-channel and overland flows: theoretical background. J. Hydr. Eng. **133**(8), 873–883 (2007)
13. Owen, P.R., Thomson, W.R.: Heat transfer across rough surfaces. J. Fluid Mech. **15**(3), 321–334 (1963)
14. Peeters, J.W.R., Sandham, N.D.: Turbulent heat transfer in channels with irregular roughness. Int. J. Heat Fluid Flow **138**, 454–467 (2019)
15. Pirozzoli, S., Bernardini, M., Orlandi, P.: Passive scalars in turbulent channel flow at high Reynolds number. J. Fluid Mech. **788**, 614–639 (2016)
16. Pope, S.B.: Turbulent Flows. Cambridge University Press, Cambridge (2000)
17. Raupach, M.R., Antonia, R.A., Rajagopalan, S.: Rough-wall boundary layers. App. Mech Rev. **44**(1), 1–25 (1991)
18. Reynolds, O.: On the extent and action of the heating surface for steam boilers. Manchester Lit. Phil. Soc. **14**, 7–12 (1874)
19. Scotti, A.: Direct numerical simulation of turbulent channel flows with boundary roughened with virtual sandpaper. Phys. Fluids **18**, 031701–1–4 (2006)
20. Shafi, H.S., Antonia, R.A.: Anisotropy of the reynolds stresses in a turbulent boundary layer on a rough wall. Exp. Fluids **18**, 213–215 (1995)
21. Yuan, J., Piomelli, U.: Numerical simulations of sink-flow boundary layers over rough surfaces. Phys. Fluids **26**, 015113–1–28 (2014)
22. Yuan, J., Piomelli, U.: Roughness effects on the Reynolds stress budgets in near-wall turbulence. J. Fluid Mech. **760**, R1–1–12 (2014)

Chapter 6
LES of Coupled Poiseuille–Couette (Air–Water) Flows: Sensitivity to Scaling

S. López Castaño, B. J. Geurts and V. Armenio

Abstract Scaling parameters governing the physics of two-phase flow turbulence are presented, in particular for the study of coupled Poiseuille–Couette flows where waves are well under the capillary regime. Different approaches can be followed to either simplify the analysis of the results or computational requirements to run such case. Such simplification implies the modification of the comprising fluids' properties in a way that reduces the local Reynolds number to a one-to-one relationship. Here an analysis is made to show that, regardless the real properties of the comprising fluids, a proper scaling of the comprising governing equations motion and compatibility conditions at the interface between the fluids render a solution suitable for the study of the physics of turbulence. Nevertheless, quantitative differences exist specially for quantities regarding TKE and Reynolds stresses.

Keywords LES · Coupled Poiseuille–Couette flow · Air–water interface

Introduction

The nature of turbulence in the vicinity of the interface between air and water has become of importance in science and engineering, since the interaction between the comprising fluid flows differ with respect to the case where said flows are considered separately. Previous works [2] have shown that in coupled Poiseuille–Couette flows nonlinear interactions across the interface increase intermittency, vertical turbulent mixing, inclined vortex cores, and different patterns of TKE distribution, as measured by the pressure–strain term.

For the study of canonical, non-miscible, incompressible, and coupled flows a particular way of scaling is necessary. In particular, notice that the *shear velocity*

S. López Castaño (✉) · V. Armenio
Department of Engineering & Architecture, Università degli studi di Trieste, Trieste, Italy
e-mail: santilopez01@gmail.com

B. J. Geurts
Department of Applied Mathematics, University of Twente, Enschede, Netherlands

M. García-Villalba et al. (eds.), *Direct and Large Eddy Simulation XII*,
ERCOFTAC Series 27,
https://doi.org/10.1007/978-3-030-42822-8_6

$$u_\tau = \sqrt{\frac{\tau_\Gamma}{\rho}}$$

is not unique for the simulation, since it depends on the density ρ, even if the deviatoric *interfacial shear stress* τ_Γ is unique. Additional to the scaling of the incompressible Navier–Stokes Equations, scaling for the dynamic compatibility condition

$$(\tau_{\text{air}} - \tau_{\text{water}}) \cdot \mathbf{n}_\Gamma + (P_{\text{water}} - P_{\text{air}}) = 0 \tag{6.1}$$

at the interface with surface-normal $\mathbf{n}_\Gamma$, with thermodynamic pressure P and stress τ, and scaling for the tangential velocities u the kinematic compatibility condition

$$u_{\text{air}} = u_{\text{water}} \tag{6.2}$$

has to be performed. Notice that for the former equation, scaling is trivial since the scaling parameter, τ_Γ, is unique. On the other hand, scaling of the kinematic condition leads to the following expression:

$$u^*_{\text{air}} = \sqrt{\frac{\rho_{\text{air}}}{\rho_{\text{water}}}} u^*_{\text{water}}. \tag{6.3}$$

Notice that the dynamic condition reduces to the following identity in the case of incompressible flows:

$$\tau_{\text{air}} = \tau_{\text{water}},$$

which leads to a relation between the *shear Reynolds numbers* of the comprising fluids

$$\begin{aligned}
\frac{\mu_{\text{air}}(u_\tau)_{\text{air}}}{\delta}\left(\frac{\partial U}{\partial y}\right)_{\text{air}} &= \frac{\mu_{\text{water}}(u_\tau)_{\text{water}}}{\delta}\left(\frac{\partial U}{\partial y}\right)_{\text{water}}, \\
\frac{\mu_{\text{air}}}{\mu_{\text{water}}}\sqrt{\frac{\rho_{\text{water}}}{\rho_{\text{air}}}}\left(\frac{\partial U}{\partial y}\right)_{\text{air}} &= \left(\frac{\partial U}{\partial y}\right)_{\text{water}}, \\
\text{Re}_{\text{air}} &= \frac{\mu_{\text{air}}}{\mu_{\text{water}}}\sqrt{\frac{\rho_{\text{water}}}{\rho_{\text{air}}}}\text{Re}_{\text{water}},
\end{aligned} \tag{6.4}$$

where δ is a measure of the boundary layer thickness in channel-like flows. Note that for high differences in the density, like in the case of air and water, relation (6.4) may impose a rather high Re in the air side. This may present some computational challenges when running DNS and LES, specially if the focus of investigation is on the water side. Thus, some authors [2, 3] have opted for modifying the comprising fluids' properties in a way that reduces relation (6.4) into

$$\text{Re}_{\text{air}} = \text{Re}_{\text{water}}.$$

The only physically valid way of doing so is by taking the properties of air and water at a higher ambient temperature ($T \approx 323$ K). However, it is unclear whether the aforementioned simplifying assumption may affect the physics of the flows considered. Previous studies using real fluid properties of air and water [2] have argued about the validity of said assumption, without formally making a comparison between the two scenarios.

The article will be divided as follows: section "Numerical Approach" contains a brief description of the numerical approach used, the cases being studied, and details on the simulation. In section "Statement of the Problem" a summary of the results obtained from the two scenarios described and a discussion is presented.

Numerical Approach

The non-dimensional, space-filtered, governing equations of fluid motion, for the velocity u_i and pressure p, are

$$\begin{aligned}\frac{\partial \overline{u_i}}{\partial x_i} &= 0,\\ \frac{\overline{u_i}}{\partial t} + \frac{\partial \overline{u_i}\,\overline{u_j}}{\partial x_j} + \frac{\partial \overline{p}}{\partial x_i} &= -\frac{\partial \overline{\tau}_{ij}^{sgs}}{\partial x_j} + \frac{1}{\mathrm{Re}}\frac{\partial^2 \overline{u}}{\partial x_k \partial x_k},\\ \overline{\tau}_{ij}^{sgs} &= \left(\overline{u_i u_j} - \overline{u_i}\,\overline{u_j}\right),\end{aligned} \tag{6.5}$$

in which the only physically relevant parameter for describing turbulence is the *Reynolds number*, $\mathrm{Re} = u_\tau \delta/\nu$, where ν is the kinematic viscosity of the fluid, and $\overline{\tau}_{ij}^{sgs}$ is the Sub-Grid scale (SGS) tensor. The indices $i = 1, 2, 3$ indicate the streamwise, vertical, and spanwise direction. Details on the modeling of the SGS fluxes are given in [1]. Details on the numerical integration of the aforementioned equations, considering also the coupling between the fluids' flows via relations (6.1) and (6.3), are given in [2].

Statement of the Problem

Consider a classical coupled Poiseuille–Couette flow between two non-miscible fluids in which waves are well within the capillary regime, as shown in Fig. 6.1. The upper subdomain is driven by a constant pressure gradient Π, and will correspond to the air side of the coupled system. The lower waterside will be driven by the interface stress field τ_Γ produced by the air side, resembling a Couette flow. The computational grids for each subdomain have dimensions of $2\pi\delta \times 2\delta \times \pi\delta$ for the water side, and for the air side is $2\pi\delta \times \delta \times \pi\delta$. Furthermore, each of the grids is

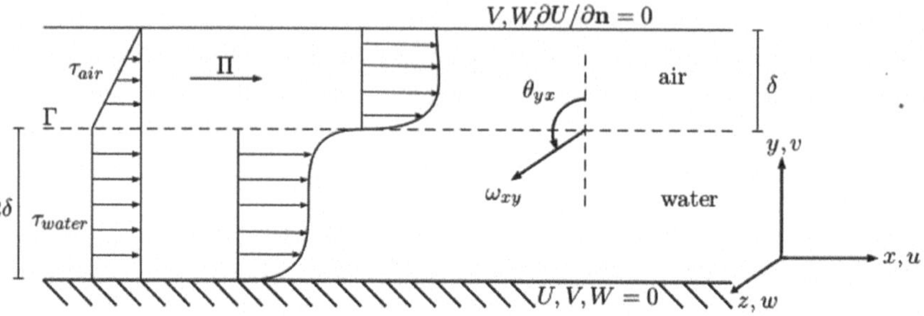

Fig. 6.1 Sketch of archetypal Poiseuille–Couette flow. Here, the velocity components are labeled (u, v, w), and the vorticity is referred as ω

split into $128 \times 96 \times 96$ hexahedra, and where hyperbolic tangent functions were used in order to refine the grids near the interface/wall and guarantee a *resolved* LES.

Two cases are considered: (1) one in which the properties of water and air are modified by increasing the temperature to 323 K, thus rendering

$$\mathrm{Re}_{\mathrm{air}} = \mathrm{Re}_{\mathrm{water}},$$

and (2) another where the properties at standard temperature (293 K) are used, therefore leading to

$$\mathrm{Re}_{\mathrm{air}} \approx 2.21\mathrm{Re}_{\mathrm{water}}.$$

Hereafter, the former scenario will be referred to as Equal Reynolds Poiseuille–Couette (ERPC), and the latter as Different Reynolds Poiseuille–Couette (DRPC). For both cases, the shear Reynolds number used for the waterside is

$$\mathrm{Re}_{\mathrm{water}} = 171.$$

Results

The present work will focus on the water side, where most of the differences with respect to archetypal non-coupled flows are present [2]. Mean velocities and shear Reynolds stress of the flow in the waterside for the ERPC and DRPC cases are presented in Fig. 6.2. Notice that some differences exist in the mean velocity profile for the boundary layer near the interface. In particular, the boundary layers on the DRPC case are closer to each other. The resolved Reynolds shear stress shows an increased turbulent vertical mixing close to the interface in the DRPC case, whereas the asymmetry across the mid-channel is weakly present in the ERPC case. Figure 6.3 shows

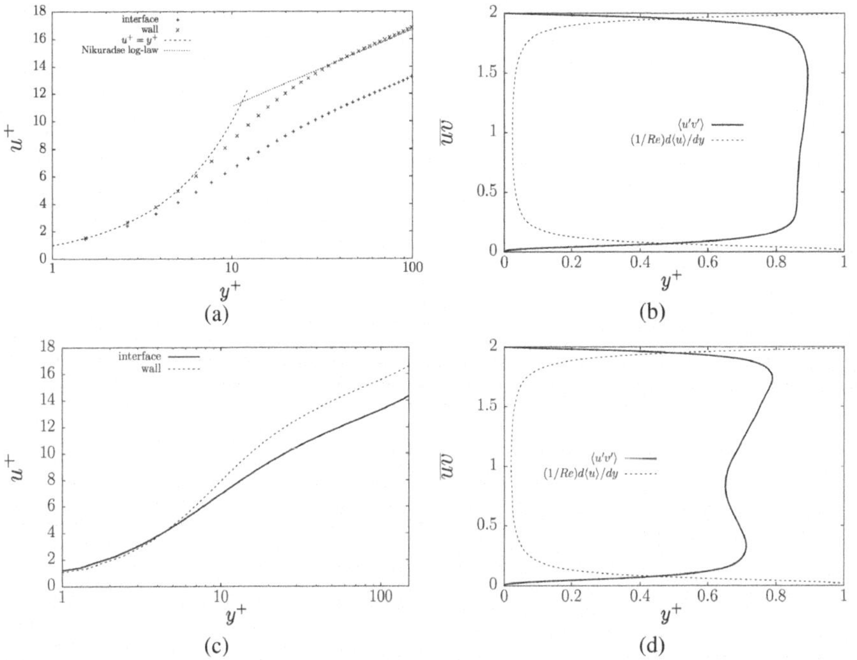

Fig. 6.2 First-order statistics on the water side. **ERPC**: **a** mean velocity profile, **b** mean Reynolds shear stress. **DRPC**: **c** mean velocity profile, **d** mean Reynolds shear stress. The word "interface" means the interface boundary layer on the water side, whereas "wall" refers to the wall boundary layer in the same domain

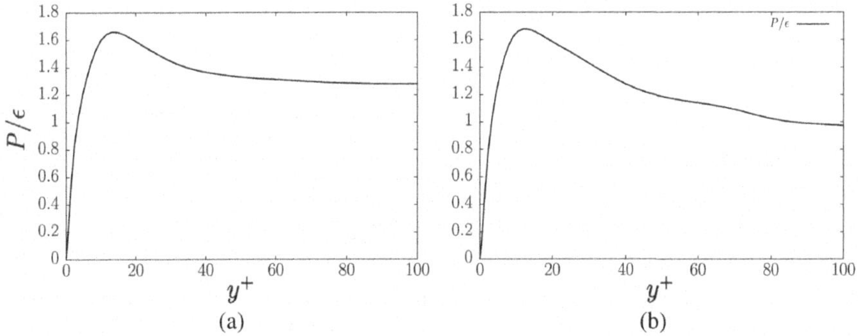

Fig. 6.3 Mean TKE production-to-dissipation ratio for **a ERPC** and **b DRPC**

the ratio of turbulent kinetic energy production-to-dissipation near the interface, a measure that shows how much excess energy is transported away from it. Notice that much of the energy produced in the ERPC case is transported toward the center of the channel, whereas in the DRPC case all energy produced in the buffer region is destroyed in the log region as expected.

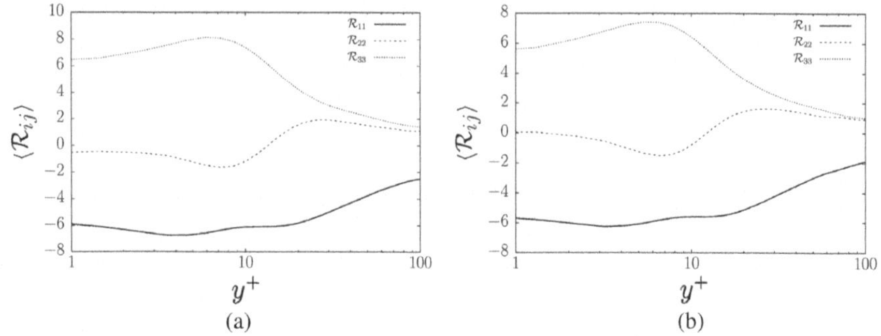

Fig. 6.4 Mean energy redistribution rate, studied via the pressure–strain term, for **a ERPC** and **b DRPC**

The pressure-strain correlation may shed light on the fate of the TKE being produced at the liquid-gas interface. By analyzing the diagonal contributions of pressure–strain tensor of the Reynolds stress budget

$$\mathscr{R}_{ij} = \left\langle p' \frac{\partial u_i}{\partial x_j} \right\rangle \tag{6.6}$$

it is possible to determine the direction in which energy is being drained. Figure 6.4 shows that near the interface energy is drained from the streamwise and spanwise directions toward the vertical in the ERPC case. However, said differences are relatively small as one can see that most of the energy drained comes from the streamwise direction. On the other hand, in DRPC energy is only being drained from the streamwise direction.

Conclusions

The present approach has shown to be a reliable tool for the study of coupled non-mixed fluid flows sharing a flat interface. But contrary to previous consensus, the phenomena of turbulence in the near interface region on water traced by wind is affected by the ambient temperature of the system. Despite the differences only being quantitative for some of the flow statistics, quantities such as turbulent kinetic energy transport, redistribution, and Reynolds stress behave differently in ERPC and DRPC. In the future, an in-depth analysis of the viability of such assumption should be made for a broader range of flows and Reynolds numbers.

References

1. Armenio, V., Piomelli, U.: A Lagrangian mixed subgrid-scale model in generalized coordinates. Flow Turbul. Combust. **65**, 51–81 (2000)
2. López Castaño, S., Armenio, V., Geurts, B.J.: An investigation of strong backflow events at the interface of air-water systems using large-eddy simulation. J. Turbul. **19**(7) (2018)
3. Lombardi, P., De Angelis, V., Banerjee, S.: Direct numerical simulation of near-interface turbulence in coupled gas-liquid flow. Phys. Fluids **8**(6) (1996)

Chapter 7
Wakes and Instabilities of Static and Freely Vibrating Cylinders

I. Rodríguez, O. Lehmkuhl, D. Pastrana, J. C. Cajas and G. Houzeaux

Introduction

The flow around static and freely vibrating cylinders are of interest in understanding different phenomena encountered in many practical applications. In the particular case of a freely vibrating cylinder, it might also be of special interest as a clean source of energy based on vortex-induced vibrations for energy harvesting in the range of frequencies where flow induced vibrations originate a strong coupling between the oscillating body and the fluctuating wake.

For the static cylinder case, the flow involves a complex physical phenomena with a strong dependency with the Reynolds number characterised by flow separations, transition to turbulence and the shedding of vortices (see for instance [1–3]). Roshko [4], based on the characteristics of the flow, classified it into different regimes: subcritical, critical, super-critical and transcritical. The subcritical regime, $\mathrm{Re} = 10^3 - 2 \times 10^5$, is mainly characterised by separation of the boundary layer around $\theta = 90°$ (measured from the front stagnation point), transition to

I. Rodríguez (✉)
Universitat Politècnica de Catalunya, Barcelona, Spain
e-mail: ivette.rodriguez@upc.edu

O. Lehmkuhl · D. Pastrana · G. Houzeaux
Barcelona Supercomputing Center, Barcelona, Spain
e-mail: oriol.lehmkuhl@bsc.es

D. Pastrana
e-mail: daniel.pastrana@bsc.es

G. Houzeaux
e-mail: guillaume.houzeaux@bsc.es

J. C. Cajas
Universidad Nacional Autónoma de México, Mérida, Mexico
e-mail: carlos.cajas@enesmerida.unam.mx

M. García-Villalba et al. (eds.), *Direct and Large Eddy Simulation XII*,
ERCOFTAC Series 27,
https://doi.org/10.1007/978-3-030-42822-8_7

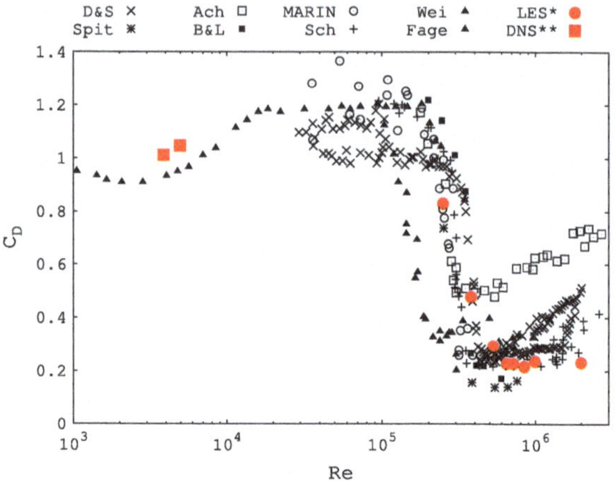

Fig. 7.1 Drag coefficient in cylinders from subcritical to super-critical regime. Comparison with the literature. (red solid circles) DNS [13, 14] and LES [7, 12]; (D&S) [15]; (Spit) [16]; (Ach) [6]; (B&L) [17]; (MARIN) [18]; (Sch) [11]; (Wei) [19]; (Fage) [20]

turbulence in the separated shear layers and an approximately constant drag coefficient $C_D \approx 1$ (see Fig. 7.1). This range is also characterised by a coherent wake and an almost constant shedding of vortices at $\mathrm{St} = f_{vs}D/U_{ref} \approx 0.21$. The critical regime, $2 \times 10^5 < \mathrm{Re} < 5 \times 10^5$, is characterised by a sudden decrease in the drag coefficient accompanied by the appearance of a laminar separation bubble, first on one side of the cylinder which produces a positive lift coefficient, and then on both sides of the cylinder [5–7]. At $\mathrm{Re} > 5 \times 10^5$, the flow enters the super-critical regime with the drag coefficient reaching its minimum value of $C_D \approx 0.2$. Traditionally, the super-critical wake has been described as incoherent [1, 3, 4, 8, 9], with turbulent transition at separation. However, different authors have measured coherent vortex shedding in this regime (see for instance the experimental works of [5, 10, 11]), and as was pointed out in Lehmkuhl et al. [7] and Rodriguez et al. [12], vortex shedding was detected at super-critical Reynolds numbers between $\mathrm{Re} = 6.5 \times 10^5 - 8.5 \times 10^5$. In these two studies, the wake has been described as a coherent vortex street more compact than that observed at subcritical Reynolds numbers (see Fig. 7.2). The non-dimensional vortex shedding frequency measured in the super-critical regime was around $\mathrm{St} = f_{vs}D/U_{ref} = 0.45$ in agreement with experimental measurements of [5, 11].

All these features get more complicated when the cylinder is free to vibrate in two-degrees of freedom (2dof), as the wake characteristics are not only affected by the Reynolds number, i.e. whether the flow is in the subcritical regime or in the super-critical regime, but also by the structural response of the system as a function of the reduced velocity $U^* = U_\infty/(f_n D)$ (i.e. the inverse of the natural frequency) non-dimensionalised with the free stream velocity U_∞ and the diameter of the cylinder D (see for instance the reviews of [21–24]). In this contribution, some characteristics of the wake and instabilities of a 2dof freely vibrating cylinder are described. To do this, results from large-eddy simulations of the flow performed at different Reynolds numbers and reduced velocities are used.

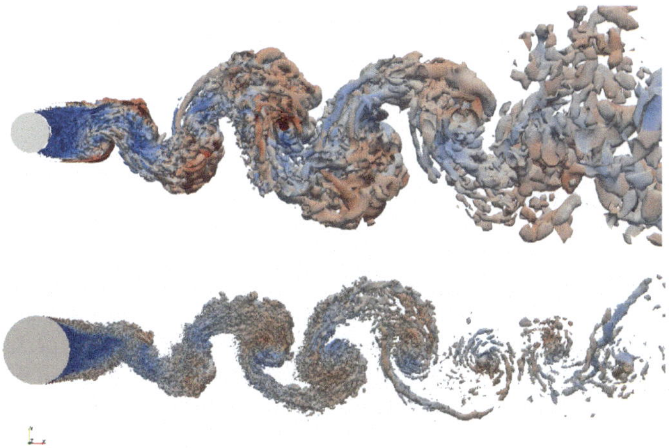

Fig. 7.2 The wake of a static circular cylinder (top) subcritical regime Re $= 3900$, (bottom) supercritical regime Re $= 7.2 \times 10^5$

Relevant Results

The system under consideration consists of a rigid cylinder elastically mounted with springs that allow the cylinder to move in the in-line and cross-flow directions. The problem is governed by the Reynolds number together with three other parameters which characterise the structure: the fluid-structure mass ratio, the reduced velocity and the damping ratio,

$$m^* = 4m/(\rho_f \pi D^2); \qquad U^* = U_\infty/(f_n D); \qquad \zeta = c/(2\sqrt{km}) \qquad (7.1)$$

where m is the mass of the cylinder, ρ_f the density of the fluid and f_n the natural frequency of the body $f_n = (2\pi)^{-1}\sqrt{k/m}$ with k the structural stiffness and c is the structural damping. The body vibrations are driven by the transient fluid forces. The in-line and cross-flow force coefficients are defined as $\mathbf{C} = 2\mathbf{F}/(\rho_f D U_\infty^2)$, $\mathbf{C} \equiv (C_x, C_y)$ and $\mathbf{F} \equiv (F_x, F_y)$ being the in-line and cross-flow sectional forces. The system is solved using the code Alya [25]. The low-dissipation methodology for solving the flow is described in detail in Lehmkuhl et al. [26] and the methodology used for the resolution of the fluid-structure interaction system can be found in Cajas et al. [27] and Pastrana et al. [28].

The most relevant features regarding the wake configuration and instabilities that appear in 2dof cylinders are hereafter commented. For the analysis, large-eddy simulations at Reynolds numbers in the range of Re $= 3900 - 7.5 \times 10^5$ are considered. Notice that this range covers both the subcritical and super-critical regimes. In order to analyse the response of the system, reduced velocity ranges are $U^* = 3.0–12$ for the subcritical regime and $U^* = 0.5–5.0$ for the super-critical regime. In all simulations, a low-mass ratio $m^* = 2.6$ and zero damping $\zeta = 0$ system is considered. Moreover, in this work, the structural stiffness and damping are equal in both directions ($k_x = k_y = k$ and $c_x = c_y = c$).

The Subcritical Wake

In Fig. 7.3, the maximum cross-flow and in-line non-dimensional amplitudes of vibration for different Reynolds numbers up to Re $= 1 \times 10^5$, corresponding to the subcritical regime, are plotted. In the figure, the experimental results of Jauvtis and Williamson (JW04) [29] for a low-mass ratio cylinder $m^* = 2.6$ and in the range of Re $= 1000-15{,}000$ are given. In addition, the numerical results by Gsell et al. (G10) [30] at a Reynolds number of Re $= 3900$ are also included. At these Reynolds numbers, the low-mass ratio 2dof cylinder response is characterised by three branches: the initial (*I*), the super-upper (SU) and the lower (*L*) branches. Each of these branches have different characteristics and amplitude of response.

The *I* branch, where the synchronisation or lock-in begins, has very small in-line vibrations (see Fig. 7.3) and the wake is characterised by the typical von Kármán street vortex shedding mode or 2S vortex formation mode (following the nomenclature of Williamson and Roshko [31]), i.e. 2 vortices shed each cycle. This wake configuration can be seen in Fig. 7.4 (top), where ensemble average vorticity contours at $U^* = 3.0$ are depicted. Actually, in the 2S mode, and in spite of the cross-flow fluctuation of the cylinder, the vortices shed are in-line with the wake centreline in a similar fashion to the static cylinder as can be seen in the figure.

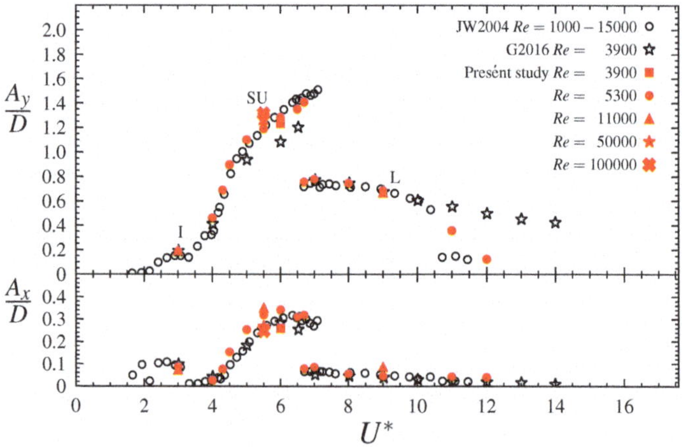

Fig. 7.3 Non-dimensional cross-flow and in-line maximum amplitudes of oscillation as a function of the reduced velocity. Present LES results at different Reynolds numbers compared with the experimental data of JW04 and the computational data of G10

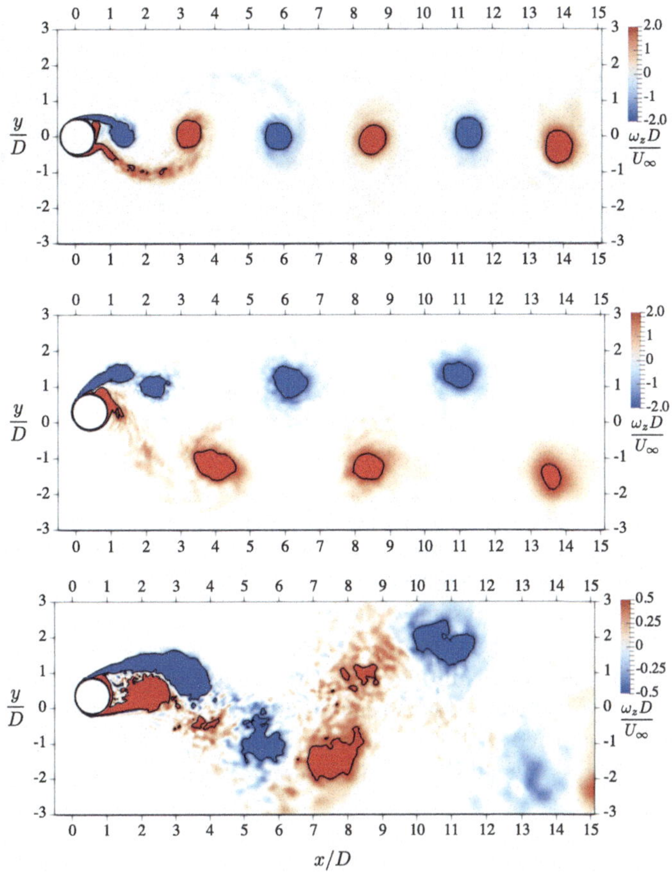

Fig. 7.4 The 2dof freely vibrating cylinder wake. Ensemble average vorticity contours in the initial branch $U^* = 3.0$ (top), the super-upper branch $U^* = 5.5$ (middle) and, the lower branch $U^* = 9.0$ (bottom)

Of particular interest is the *SU* branch, where in-line and cross-flow amplitudes of fluctuation peak. This branch is not present in 1dof systems and is characteristic only of 2dof with low-mass ratio $m^* < 6$ [29]. In this branch, the wake is broader (see Fig. 7.4 (middle)) and vortex shedding mode changes from a 2S pattern to a 2T one, i.e. 2 triplets of vortices shed each shedding cycle. This mode was first described by JW04. Actually in Fig. 7.4, the footprint of two out of three vortices shed once the cylinder reaches its maximum elongation in the cross-flow direction is observed. However, as the flow moves downstream, the triplets of vortices shed each cycle merge into a single vortex located off the wake centreline. Finally, the *L* branch (see Fig. 7.4 (bottom)), with a low level of fluctuation in the in-line direction and with decreasing cross-flow amplitudes, as the reduced velocity increases up until the de-synchronisation of the system around $U^* = 12$ occurs. In this branch, the vortex

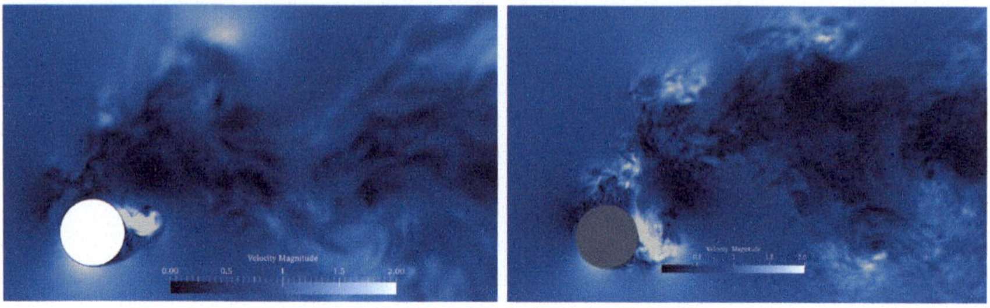

Fig. 7.5 Instantaneous flow in the subcritical regime in the super-upper branch at $U^* = 5.5$. Re = 5300 (left) and Re = 10^5 (right)

shedding mode is different than in the other two branches and is characterised by two pair of vortices shed each cycle (also known as 2P mode), as it can be seen in Fig. 7.4 (bottom).

Last but not the least, attention should be also paid to the fact that along the whole subcritical regime, cylinder amplitudes are only function of the reduced velocity and independent on the Reynolds number (see Fig. 7.3). In fact, the cylinder in-line and cross-flow positions at each instant of the the cycle (not shown here) are almost the same regardless of the Reynolds number, which points out to the similarities with the static cylinder in which along the whole subcritical regime, the wake characteristics are nearly the same. Actually, if the instantaneous flow at Re = 5300 and Re = 10^5 for $U^* = 5.5$ is compared at the maximum elongation position, one can see the same large scale structures of the flow and wake configuration for both Reynolds numbers (see Fig. 7.5).

The Super-Critical Wake

In Fig. 7.6, the instantaneous flow at Re $= 7.5 \times 10^5$ at $U^* = 2.5$ is depicted together with the non-dimensional cross-flow and in-line amplitudes of fluctuation as a function of the reduced velocity. Similar to the static cylinder, the super-critical wake of the 2dof vibrating cylinder is narrower than in the subcritical regime. In a similar manner, the boundary layer separates past the cylinder apex, on both sides of the cylinder. As a result, the separated shear layers are closer to each other and thus the fluctuations produced due to their interaction are also small. As a consequence of this super-critical flow configuration, synchronisation and de-synchronisation are attained at lower reduced velocities than in the subcritical regime, as can also be seen in Fig. 7.6 (right). Although the range of reduced velocities covered by the present study is quite discrete, it seems that in the super-critical regime, only initial and lower branches are present, with a behaviour (in terms of the amplitude of fluctuations which are almost negligible in the in-line direction) similar to that observed in high-mass ratio 1dof cylinders (see for instance [32]).

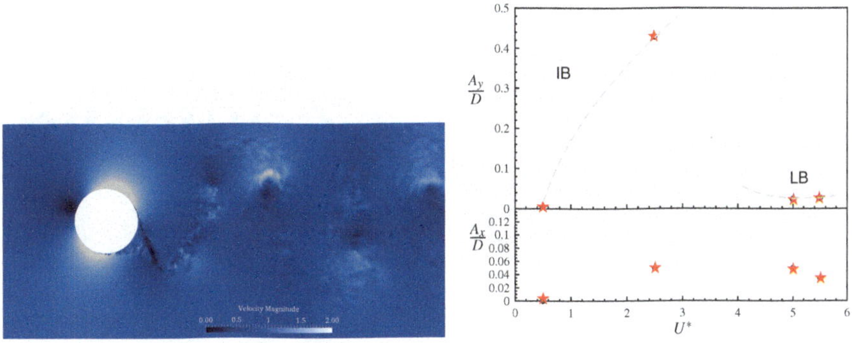

Fig. 7.6 The super-critical wake of a 2dof cylinder at Re $= 7.5 \times 10^5$ (right) and non-dimensional cross-flow and in-line amplitudes as a function of the reduced velocity in the super-critical regime

Instabilities in the Cylinder Surface

Similar to the differences observed in the wake of the cylinder, three-dimensional structures triggered and developed are also different depending on the branch of response (see Fig. 7.7). However, both initial and lower branches present some similarities. In these two branches, three-dimensional instabilities are triggered in the separated shear layers; the boundary layer being laminar during the whole fluctuation cycle. Nonetheless, a noticeable difference between these two branches is related with the length of the vortex formation zone. While in the former, shear layers roll-up close to the cylinder surface, in the latter both shear layers are almost in phase with little interaction between each other. As a consequence, the inception of instabilities that the vortex-induced vibration (VIV) produces occurs farther from the cylinder and thus, vortex formation zone is larger than in the initial branch. A fundamental change is produced in the super-upper branch, where the oscillation of the cylinder and the increase in the local angular velocity trigger a centrifugal instability in the boundary layer, forming Görtler-like structures. These structures form and dissipate periodically twice every cycle depending on the position of the cylinder (see also Fig. 7.8) and can be seen as counter-rotating vortical structures covering the whole span. Similar structures have been observed before in cylinders with forced oscilla-

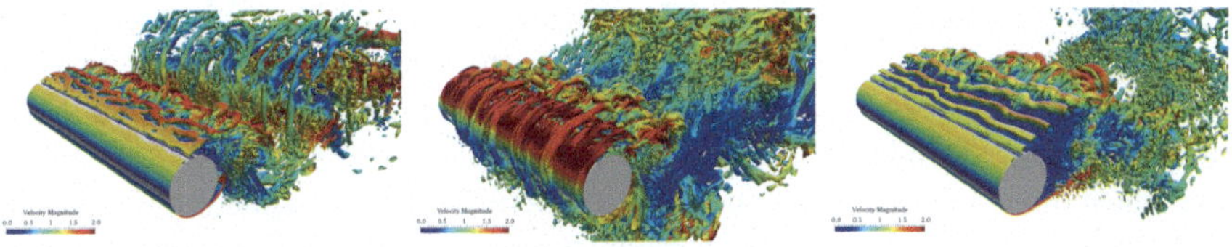

Fig. 7.7 Instantaneous vortical structures close to the cylinder surface. Initial branch (left), super-upper branch (middle), lower branch (right)

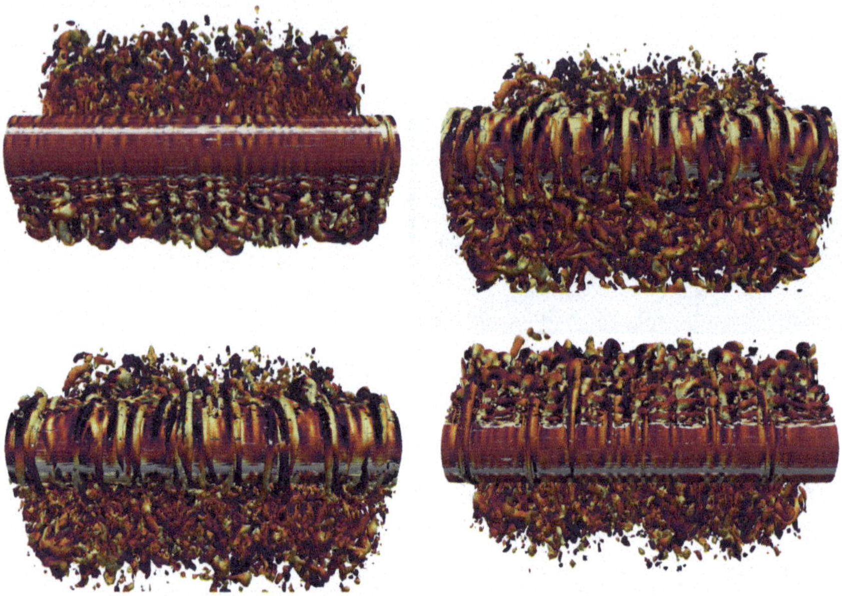

Fig. 7.8 Görtler-like structures observed in the super-upper branch. The image represents different positions of the cylinder during one cycle. The Görtler-like structures are observed (top right and bottom left) in the zone where the cylinder attains its maximum elongation

tions in a fluid at rest [33, 34], in the flow over a rotating cylinder [35] and in the flow over a cylinder performing rotating oscillations [36], but have never been reported on the VIV context.

The spanwise length of the Görtler-like structures can be determined by applying the Hilbert transform to the signal of probes located in lines parallel to the cylinder axis in the boundary layer. The local wavelength of the signal $w(z)$ is obtained, $kD = \frac{\mathrm{d}\phi}{\mathrm{d}z}$, ϕ being its local amplitude. Thus, the spanwise length is defined as $\lambda_z/D = 2\pi/(kD)$. The mean value $\overline{\lambda_z}/D$ is determined as the temporal average of the mean of the 5% most frequent wavelengths. These structures are observed in the super-upper branch, in this work, for reduced velocities $U^* = 4.3,\ 5.5,\ 6.5$; the average spanwise length is $\overline{\lambda_z}/D = 0.280,\ 0.322$ and 0.311, respectively. For the sinusoidal oscillatory flow around a circular cylinder, Sarpkaya [34] using data from experimental measurements obtained the correlation $\lambda_z/D = 22\beta^{-3/5}$, $\beta = fD^2/\nu$ being the Stokes number. Using the correlation, the size of the structures obtained is $\lambda_z/D = 0.255,\ 0.284,\ 0.276$, which is comparable to the values measured in the freely oscillating cylinder. Thus, it can be stated that in 2dof freely vibrating cylinders, when the cylinder reaches its maximum amplitudes of fluctuation (i.e. in the *SU* branch), centrifugal instabilities might trigger three-dimensional structures in the boundary layer. These structures have similar characteristics to those observed in systems such as forced sinusoidal oscillating cylinders.

Summary

The main characteristics of the wake of 2dof freely vibrating cylinders are described using results from large-eddy simulations of the flow in the range of Reynolds numbers of $Re = 3900 - 7.5 \times 10^5$, which covers both the subcritical and the supercritical regimes. The behaviour of 2dof vibrating cylinders remains almost the same along the whole subcritical regime ($Re = 3900 - 10^5$) and the three branches of response of this system is well captured by the LES. In the super-critical regime, the wake is narrower than that observed in the subcritical regime and only two branches of response (I and L) are observed, with in-line fluctuations almost negligible and a behaviour similar to that observed in high-mass ratio 1dof systems. In the SU branch, a centrifugal instability appears and as a consequence, Görtler-like structures form in the boundary layer twice every cycle; the size of these structures being comparable to the values measured in forced oscillating cylinders in a fluid at rest.

Acknowledgements This work has been partially financially supported by the Ministerio de Economía y Competitividad, Secretaría de Estado de Investigación, Desarrollo e Innovación, Spain (Ref. TRA2017-88508-R). We also acknowledge Red Española de Surpercomputación (RES) for awarding us access to the MareNostrum IV machine based in Barcelona, Spain (Ref. FI-2017-2-0016) and PRACE for awarding us access to SuperMUC (Project ViValdi Ref. 2017174222). D. Pastrana acknowledges support of the CONACyT-SENER graduate fellowship program to study abroad 278102/439162.

References

1. Roshko, A.: Perspectives on bluff body aerodynamics. J. Wind. Eng. Ind. Aerodyn. **49**(1), 79–100 (1993)
2. Williamson, C.H.K.: Vortex dynamics in the cylinder wake. Annu. Rev. Fluid Mech. **28**(1), 477–539 (1996)
3. Zdravkovich, M.: Flow around circular cylinders. Volume I: Fundamental. J. Fluid Mech. **350**(1), 375–378 (1997)
4. Roshko, A.: Experiments on the flow past a circular cylinder at very high Reynolds number. J. Fluid Mech. **10**(3), 345–356 (1961)
5. Bearman, P.W.: On vortex shedding from a circular cylinder in the critical Reynolds number regime. J. Fluid Mech. **37**, 577–585 (1969)
6. Achenbach, E., Heinecke, E.: On vortex shedding from smooth and rough cylinders in the range of Reynolds numbers 6e3 to 5e6. J. Fluid Mech. **109**, 239–251 (1981)
7. Lehmkuhl, O., Rodríguez, I., Borrell, R., Chiva, J., Oliva, A.: Unsteady forces on a circular cylinder at critical Reynolds numbers. Phys. Fluids **26**(12), 125110 (2014)
8. Lienhard, J.: Synopsis of lift, drag and cortex shedding frequency data for rigid circular cylinders. Tech. Rep. Bulletin 300, College of Engineering. Research Division (1966)
9. Shih, W., Wang, C., Coles, D., Roshko, A.: Experiments on flow past rough circular cylinders at large Reynolds numbers. J. Wind. Eng. Ind. Aerodyn. **49**, 351–368 (1993)
10. Achenbach, E.: Distribution of local pressure and skin friction around a circular cylinder in cross-flow up to Re = 5e6. J. Fluid Mech. **34**, 625–639 (1968)
11. Schewe, G.: On the force fluctuations acting on a circular cylinder in crossflow from subcritical up to transcritical Reynolds numbers. J. Fluid Mech. **133**, 265–285 (1983)

12. Rodríguez, I., Lehmkuhl, O., Chiva, J., Borrell, R., Oliva, A.: On the flow past a circular cylinder from critical to super-critical Reynolds numbers: wake topology and vortex shedding. Int. J. Heat Fluid Flow **55**, 91–103 (2015)
13. Lehmkuhl, O., Rodríguez, I., Borrell, R., Oliva, A.: Low-frequency unsteadiness in the vortex formation region of a circular cylinder. Phys. Fluids **25**, 085109 (2013)
14. Aljure, D., Lehmkhul, O., Rodríguez, I., Oliva, A.: Three dimensionality in the wake of the flow around a circular cylinder at Reynolds number 5000. Comput. Fluids **147**, 102–118 (2017)
15. Delany, N., Sorensen, N.: Low-speed drag of cylinders of various shapes. Tech. Rep., NACA (1953)
16. Spitzer, R.: Measurements of unsteady pressures and wake fluctuations for flow over a cylinder at supercritical Reynolds number. Ph.D. thesis, California Institute of Technology (1965)
17. Bursnall, W., Loftin, L.J.: Experimental investigation of the pressure distribution about a yawed circular cylinder in the critical Reynolds number range. Tech. rep., NACA (1951)
18. Vaz, G., Mabilat, C., van der Wal, R., Gallagher, P.: Viscous flow computations on smooth cylinders: a detailed numerical study with validation. In: 26th International Conference on Offshore Mechanics and Artic Engineering. OMAE2007, San Diego, California (2007)
19. Wieselsberger, C.: New data on the laws of fluid resistance. Tech. Rep. TN 84, NACA (1921)
20. Fage, A.: Drag of circular cylinders and spheres. Tech. Rep., Aeronautical Research Council (1930)
21. Sarpkaya, T.: Vortex-induced oscillations: a selective review. J. Appl. Mech. **46**, 241–258 (1979)
22. Sarpkaya, T.: A critical review of the intrinsic nature of vortex-induced vibrations. J. Fluids Struct. **19**(4), 389–447 (2004)
23. Williamson, C.H.K., Govardhan, R.: A brief review of recent results in vortex-induced vibrations. J. Wind. Eng. Ind. Aerodyn. **96**, 713–735 (2008)
24. Bearman, P.W.: Circular cylinder wakes and vortex-induced vibrations. J. Fluids Struct. **27**(5–6), 648–658 (2011)
25. Vázquez, M., Houzeaux, G., Koric, S., Artigues, A., Aguado-Sierra, J., Arís, R., Mira, D., Calmet, H., Cucchietti, F., Owen, H., Taha, A., Burness, E.D., Cela, J.M., Valero, M.: Alya: multiphysics engineering simulation towards exascale. J. Comput. Sci. **14**, 15–27 (2016)
26. Lehmkuhl, O., Houzeaux, G., Owen, H., Chrysokentis, G., Rodriguez, I.: A low-dissipation finite element scheme for scale resolving simulations of turbulent flows. J. Comput. Phys. **390**, 51–65 (2019)
27. Cajas, J., Houzeaux, G., Vázquez, M., García, M., Casoni, E., Calmet, H., Artigues, A., Borrell, R., Lehmkuhl, O., Pastrana, D., Yáñez, D., Pons, R., Martorell, J.: Fluid-structure interaction based on HPC multicode coupling. SIAM J. Sci. Comput. **40**(6), C677–C703 (2018)
28. Pastrana, D., Cajas, J.C., Lehmkuhl, O., Rodríguez, I., Houzeaux, G.: Large-eddy simulations of the vortex-induced vibration of a low mass ratio two-degree-of-freedom circular cylinder at subcritical Reynolds numbers. Comput. Fluids **173**, 118–132 (2018)
29. Jauvtis, N., Williamson, C.H.K.: The effect of two degrees of freedom on vortex-induced vibration at low mass and damping. J. Fluid Mech. **509**, 23–62 (2004)
30. Gsell, S., Bourguet, R., Braza, M.: Two-degree-of-freedom vortex-induced vibrations of a circular cylinder at Re=3900. J. Fluids Struct. **67**, 156–172 (2016)
31. Williamson, C.H.K., Roshko, A.: Vortex formation in the wake of an oscillating cylinder. J. Fluids Struct. **2**(4), 355–381 (1988)
32. Williamson, C., Govardhan, R.: Vortex-induced vibrations. Annu. Rev. Fluid Mech. **36**, 413–455 (2004)
33. Honji, H.: Streaked flow around an oscillating circular cylinder. J. Fluid Mech. **107**, 509–520 (1981)
34. Sarpkaya, T.: Experiments on the stability of sinusoidal flow over a circular cylinder. J. Fluid Mech. **457**, 157–180 (2002)

35. Aljure, D.E., Rodríguez, I., Lehmkuhl, O., Pérez-Segarra, C.D., Oliva, A.: Influence of rotation on the flow over a cylinder at Re = 5000. Int. J. Heat Fluid Flow **55**, 76–90 (2015)
36. D'Adamo, J., Godoy-Diana, R., Wesfreid, J.E.: Centrifugal instability of Stokes layers in crossflow: the case of a forced cylinder wake. Proc. R. Soc. Lond. A Math. Phys. Sci. **471**(2178), 20150011 (2015)

Chapter 8
Direct Numerical Simulation of an Oblique Jet in a Particle-Laden Crossflow

G. Agati, D. Borello, G. Camerlengo, F. Rispoli and J. Sesterhenn

Introduction

Jet in crossflow is a classic fluid dynamics problem widely studied in the last decades because of the big quantity of natural and industrial processes in which it is encountered [7]. The present study focuses on the interaction between solid suspended particles and gas turbines film cooling that is a commonly used coolant technique aiming at generating a protective film of cold fluid around the blade profile. Effective cooling systems are crucial to increase turbine inlet gas temperature and to protect turbine blade surfaces from the huge thermal stress generated. For this kind of application, to reduce the jet penetration into the crossflow, low values of jet to crossflow velocities and blowing ratio are generally adopted, with values generally ranging from 0.5 to 2 [1]. The blowing ratio (BR) is defined as the jet to the main flow mass flux ratio. In the jet in crossflow interaction, coherent vortex systems are generated in the zone of injection and, depending on the characteristic of the problem, dominate the phe-

G. Agati (✉) · D. Borello · F. Rispoli
Dipartimento di Ingegneria Meccanica e Aerospaziale, Università degli Studi di Roma La Sapienza, Via Eudossiana 18, 00184 Rome, Italy
e-mail: giuliano.agati@uniroma1.it

D. Borello
e-mail: domenico.borello@uniroma1.it

F. Rispoli
e-mail: franco.rispoli@uniroma1.it

G. Camerlengo · J. Sesterhenn
Institut für Strömungmechanik und Technische Akustik, Technische Universität Berlin, Muller-Breslau-Str. 15, 10623 Berlin, Germany
e-mail: gabriele.camerlengo@tu-berlin.de

J. Sesterhenn
e-mail: joern.sesterhenn@tu-berlin.de

M. García-Villalba et al. (eds.), *Direct and Large Eddy Simulation XII*,
ERCOFTAC Series 27,
https://doi.org/10.1007/978-3-030-42822-8_8

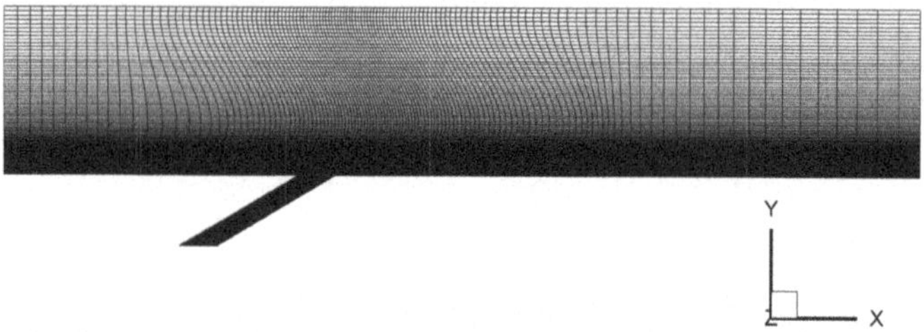

Fig. 8.1 Grid consists in about 542 mln grid points. Here it is shown in the symmetry plane and, for the sake of clarity, only one on eight points is plotted in all the dimensions. A grid clustering is performed in the y- and x-directions

nomenon. When low values of BR are used then the interaction of jets in crossflow is dominated by hairpin vortices [11, 16]. Solid suspended particles are often encountered in the flow evolving in gas turbines due to sand ingestion when operating in hard environments or to the not complete hydrocarbon combustion (soot) [6]. Dispersed particles can deposit on gas turbines blades and vanes interacting with the cooling system and eventually provoking a reduction of its effectiveness [5]. As discussed in [14], particle deposition and entrainment in wall-bounded flows are strongly related to near-wall turbulent coherent structures. Because of the inherent complexity of the problem, a Direct Numerical Simulation of a 30° inclined cylindrical laminar jet issuing in particle-laden crossflow was performed.

Numerical and Simulation Details

The carrier fluid flow is described by means of skew-symmetric Navier–Stokes equations [10] discretized on a multi-block curvilinear grid illustrated in Fig. 8.1. Time integration is performed through a low-storage fourth-order Runge–Kutta scheme while, as concerns the space discretization, a sixth-order Summation By Parts finite difference scheme is adopted. The parallelization (MPI) is achieved by a hybrid approach using a multi-block code with a ghost layer synchronization as well as block internal decomposition into CPUs as explained in [15]. The dispersed phase is simulated by adopting a Lagrangian two-way coupling point-particle approach. In order to make computationally more efficient the particles tracing scheme, particle equation is solved in a uniform and orthonormal computational space different from the one where fluid flow equations are discretized. In this space, the algorithm tracing particles do not need iterative methods since an explicit correspondence exists between the coordinates of a point and the grid cell containing it. When the coupling between fluid and particle properties has to be accomplished, a given particle position in the c-space is transformed into its physical coordinate following the trilinear interpolation as suggested in [12]. Wall-particle interaction was also considered by adopting a generalized version [4] of the elastoplastic impact/adhesion model firstly

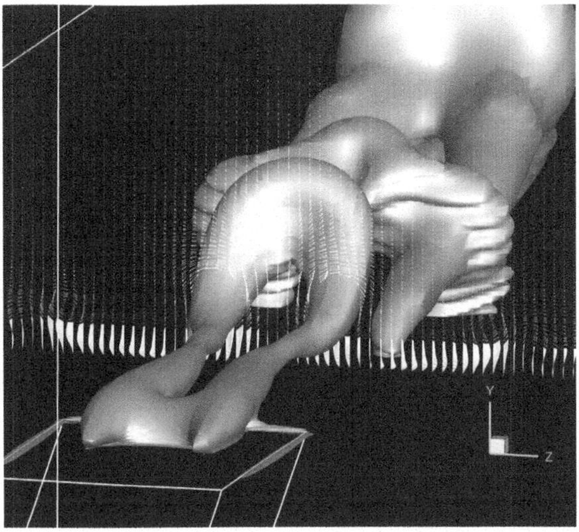

Fig. 8.2 Pressure isosurfaces are used to capture hairpin-shaped vortices. Planar velocity vectors are represented on a transversal section

proposed by Thornton and Ning [2, 3]. To obtain the desired circular section of the jet, porous media, acting as solid wall boundaries, were also included by means of a volume penalization method in the formulation presented in [13].

Jet in crossflow blowing ratio based on the bulk jet velocity and density, respectively, and on the same quantities defined for the undisturbed crossflow field, is set equal to 0.5. Mach number of the crossflow is equal to 0.8, while the Reynolds number, defined using the jet orifice diameter and on the jet bulk velocity, is 1420. The crossflow is initialized with a Blasius laminar boundary layer with a $\delta_{99\%} = 1.32d$ at the jet exit centre (being d the pipe jet diameter), while for the jet a fully developed laminar profile based on Hagen–Poiseuille flow is set along the pipe length as initial condition. Particles are randomly seeded at the crossflow inlet section at every simulation timestep. A log-normal distribution is assumed to characterize particle diameters distribution with a count median diameter CMD $= 5\,\mu$m and a geometric standard deviation GSD $= 1.6$ resulting in a mean value $d_m = 5.58\,\mu$m. Stokes particles number (St) ranges from 1 to 175 and the different particle sizes are accounted for within the same flow field.

Results and Discussion

Instantaneous observation of the three-dimensional field demonstrated that, as expected, the fluid flow is dominated by hairpin-shaped vortices shed periodically into the crossflow. As illustrated in Fig. 8.2 hairpin legs are associated to counter rotating vortex pairs (CVP) that persist in the far field of the jet. These coherent structures influence temperature distribution (Fig. 8.3): hot air is entrained into the

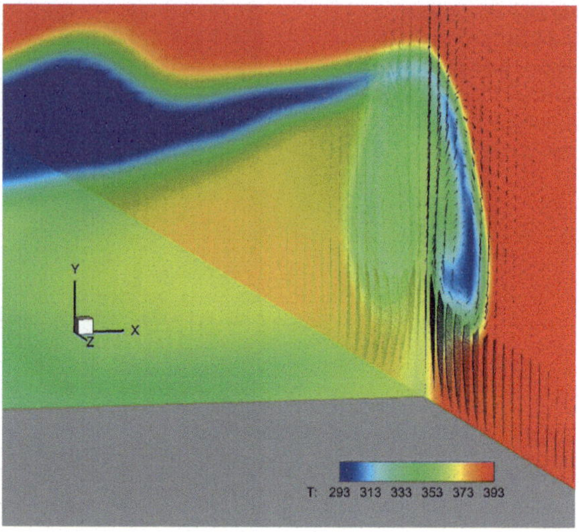

Fig. 8.3 Contour plot of temperature in the symmetry and in a cross-sectional plane where also velocity vectors indicating the presence of the CVP are drawn

core of the jet resulting in some 'hot spots' that can be observed along the wall surface.

In Fig. 8.4 (left) jet trajectory, defined as in [8], i.e. as the streamline originating from the centre of the jet exit on the symmetry plane, is depicted. Jet trajectory presents a growing trend for the whole length of the crossflow region even if jet penetration is always lower than $1.5d$. Jet temperature along its trajectory is also plotted in the same figure. Figure 8.4 (right) shows the variation of film cooling effectiveness (Φ) on a line crossing the centre of the bottom wall of the flat plate. Film cooling effectiveness is a commonly used dimensionless parameter to measure the quality of the cooling and it is defined as the gas-to-wall temperature difference over the gas-to-coolant temperature difference $\Phi = \frac{T_\infty - T_W}{T_\infty - T_c}$. After $x/d = 5$, Φ is less than 0.4, while for $x/d = 10$, $\Phi = 0.2$ and the jet influence on the crossflow temperature field can be considered negligible.

Particle spatial distribution is also strongly influenced by coherent fluid structures. Figure 8.5 shows how a high concentration of particles is created on two symmetric regions extending along the lateral sides of hairpin vortices.

Following an analysis similar to [9], in Fig. 8.6 and in Fig. 8.7 particles are coloured by the wall-normal velocity. Particles that are found in the proximity of the hairpin counter rotating legs are transported towards the wall while the particles entrapped by the head of the hairpins are pushed outwards. All the particles seem to be subject to the entrapping effect of the counter-rotating vortex but, as further analysed beyond, inertia modulates this phenomenon.

A similar behaviour can be also observed in Fig. 8.8 where a map of particle impacts on the bottom wall is shown. Statistics on particle impacts were carried out by means of probability density functions (PDFs) (not presented here). Only the 2% of the total number of impacts occurs in the zone upstream from the jet exit

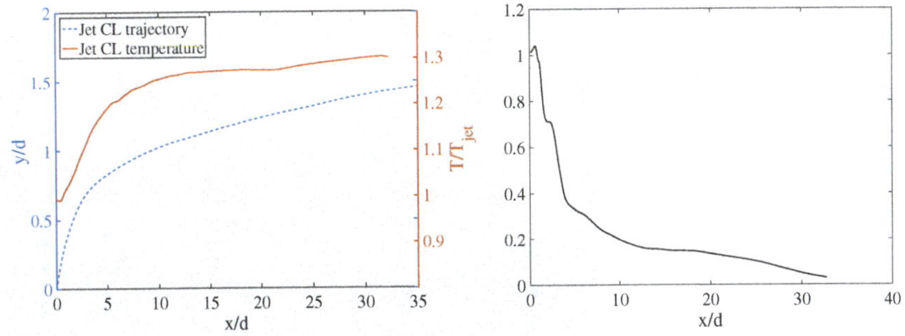

Fig. 8.4 On the left figure: trajectory and temperature variation along jet centreline. On the right figure: film cooling effectiveness on the centreline of the flat plate. Horizontal axis origin is here located on the centre of jet pipe exit

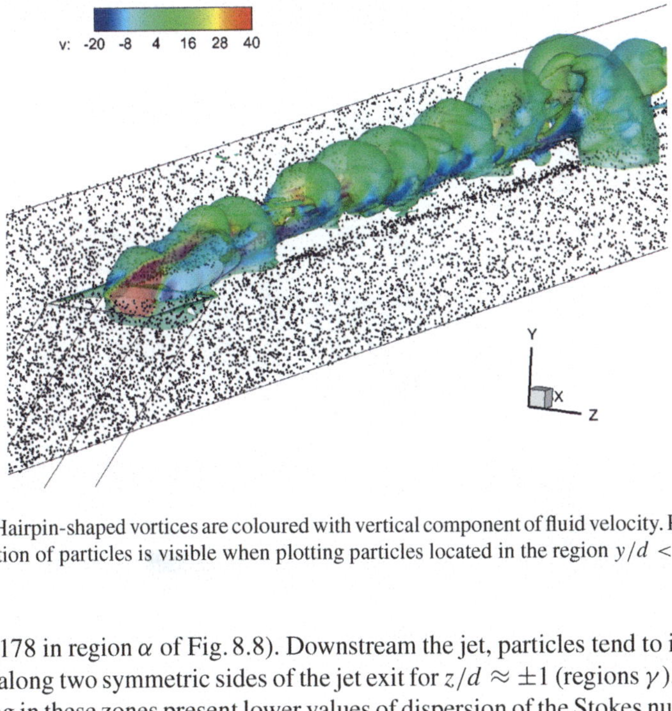

Fig. 8.5 Hairpin-shaped vortices are coloured with vertical component of fluid velocity. Preferential concentration of particles is visible when plotting particles located in the region $y/d < 0.5$

($n_{imp} = 178$ in region α of Fig. 8.8). Downstream the jet, particles tend to impact on the wall along two symmetric sides of the jet exit for $z/d \approx \pm 1$ (regions γ). Particles impacting in these zones present lower values of dispersion of the Stokes number (i.e. lower standard deviation σ_{St}) and a mean value $St_m \simeq 19.68$. These are the particles that are trapped into fluid coherent structures and are transported towards the wall. In the external zones (regions β of Fig. 8.8), particles with a wider class of diameters impact the wall generally presenting bigger values of the St ($St_m = 37.64$). These particles are subject to the effect of the CVP but do not fully follow vortices due to stronger inertia effects. Statistics carried out are summarized in Table 8.1.

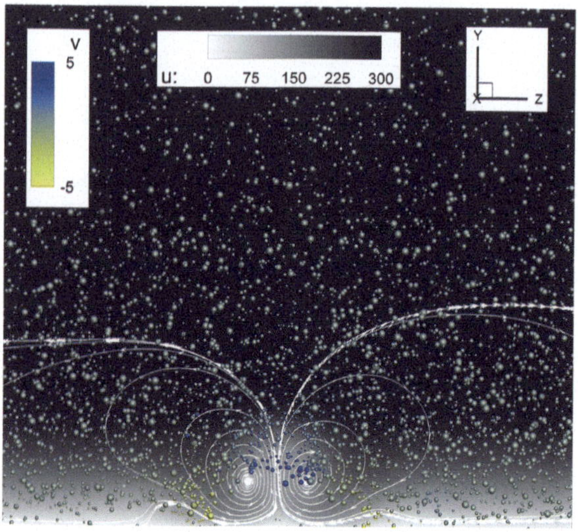

Fig. 8.6 Transversal section at $x = 8d$ from the jet exit. The contour plot and the planar streamlines are taken from the mean field. Particles size varies with the St number

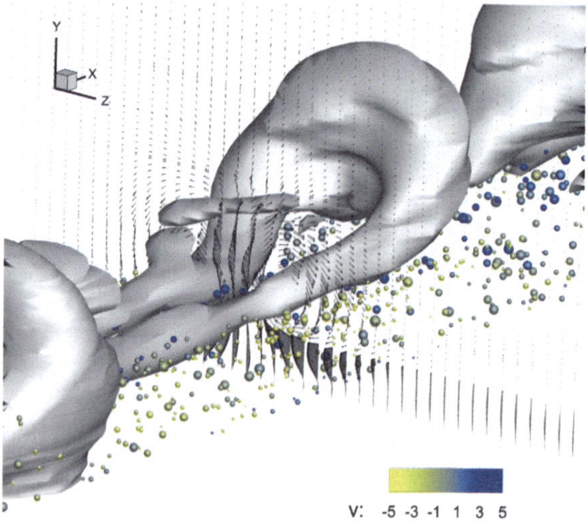

Fig. 8.7 Influence of CVP associated with hairpin legs on particle spatial distribution. Particles size varies with the St number

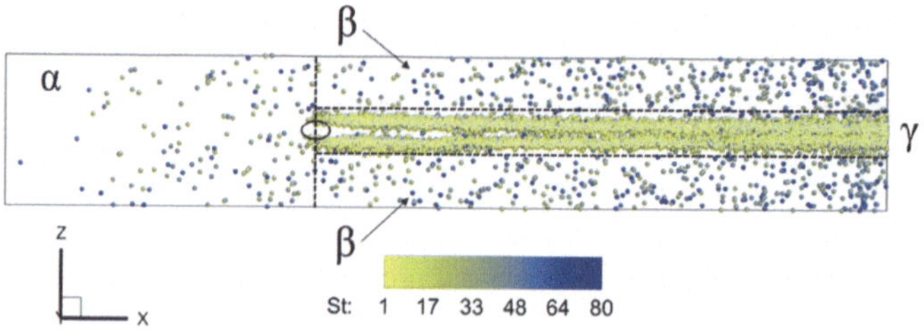

Fig. 8.8 Impact points on the wall coloured with particle Stokes number

Table 8.1 Statistics on particle impacts on the wall for the different regions illustrated in Fig. 8.8

Regions	n_{imp}	St_m	σ_{St}
Overall	9794	22.34	16.04
Region α	178	36.96	24.25
Region β	1329	37.64	21.90
Region γ	7547	19.68	12.89

Conclusions

A DNS of a jet in a particle-laden crossflow was performed. As expected, the fluid flow is dominated by hairpin vortices shed periodically from the pipe into the crossflow. The hairpin legs are counter rotating and this streamwise vorticity has a strong influence on temperature distribution and induces a preferential concentration on the dispersed phase field. Spatial distributions characterizing the impact regions were computed. Downstream from the jet exit particles characterized by low values of the St number tend to impact the wall in two symmetric sides of the jet exit for $z/d = \pm 1$ while in the external regions particles with higher values of St more probably impact the wall.

Acknowledgements We acknowledge the CINECA award under the ISCRA initiative, for the availability of high performance computing resources and support.

References

1. Acharya, S., Tyagi, M., Hoda, A.: Flow and heat transfer predictions for film cooling. Ann. N. Y. Acad. Sci. **934**, 110–125 (2001)
2. Agati, G., Borello D., Rispoli, F., Venturini, P.: An innovative approach to model temperature influence on particle deposition in gas turbines. In: ASME Turbo Expo: Power for Land, Sea, and Air, Volume 5C: Heat Transfer (2016)
3. Borello, D., Rispoli, F., Venturini, P.: An integrated particle-tracking impact/adhesion model for the prediction of fouling in a subsonic compressor. J. Eng. Gas Turbines Power **134**, 9 (2012)
4. Camerlengo, G., Borello, D., Salvagni, A., Sesterhenn, J.: DNS study of dust particle resuspension in a fusion reactor induced by a transonic jet into vacuum. Flow Turbul. Combust. **101**(1), 101–247 (2018)
5. Lawson, S.A., Thole, K.A.: Effects of simulated particle deposition on film cooling. J. Turbomach. **133**(2) (2010)
6. Lefebvre, A.H., Ballal, D.R.: Gas Turbine Combustion: Alternative Fuels and Emissions, 3rd edn. CRC Press (2010)
7. Mahesh, K.: The interaction of jets with crossflow. Annu. Rev. Fluid Mech. **45**(1), 379–407 (2013)
8. Muppidi, S., Mahesh, K.: Study of trajectories of jets in crossflow using direct numerical simulations. J. Fluid Mech. **530**, 81–100 (2005)

9. Prevel, M., Vinkovic, I., Doppler, D., Pera, C., Buffat, M.: Direct numerical simulation of particle transport by hairpin vortices in a laminar boundary layer. Int. J. Heat Fluid Flow **43**, 2–14 (2013)
10. Reiss, J., Sesterhenn, J.: A conservative, skew-symmetric finite difference scheme for the compressible Navier-Stokes equations. Comput. Fluids **101**, 208–219 (2014)
11. Sau, R., Mahesh, K.: Dynamics and mixing of vortex rings in crossflow. J. Fluid Mech. **604**, 389–409 (2008)
12. Schafer, F., Breuer, M.: Comparison of c-space and p-space particle tracing schemes on high-performance computers: accuracy and performance. Int. J. Numer. Methods Fluids **39**(4), 277–299 (2002)
13. Schulze, J., Sesterhenn, J.: Optimal distribution of porous media to reduce trailing edge noise. Comput. Fluids **78**, 41–53 (2013)
14. Soldati, A., Marchioli, C.: Physics and modelling of turbulent particle deposition and entrainment: review of a systematic study. Int. J. Multiph. Flows **35**, 827–839 (2009)
15. Stein, L., Sesterhenn, J.: An acoustic model of a Helmholtz resonator under a grazing turbulent boundary layer. Acta Mech. **230**(6), 2013–2029 (2019)
16. Tyagi, M., Acharya, S.: Large eddy simulation of film cooling flow from an inclined cylindrical jet. J. Turbomach. **125**(4), 734–742 (2003)

Chapter 9
Large Eddy Simulation of Circular Impinging Jet for Heat Transfer Applications

K. Rönnberg and C. Duwig

Introduction

Impinging jets are being used in a wide range of applications, e.g., food processing, turbine blade cooling, and automobile windscreen defrosting. While early studies are about half a century old, impinging jets are still being explored and new features are revealed constantly through experimental and numerical investigations [1–3].

Thanks to recent advances in CFD and to the increase of available computational power, further studies will be carried out with the aim to better understand the heat transfer features and also to optimize performances.

This paper will present high-fidelity simulations of a circular impinging jet inspired by the experimental study by Violato et al. [2].

Configuration, Setup, and Grid Dependence Analysis

The simulated geometry consists of a circular inlet facing a flat plate. The distance between the inlet and the plate is $H/D = 4$. A schematic representation of the simulated geometry is shown in Fig. 9.1 together with a snapshot of the results. The snapshot shows normalized velocity distribution ($|U|/U_b$), where U_b is the bulk inlet velocity, in a plane and Nusselt number distribution.

K. Rönnberg (✉) · C. Duwig
KTH Royal Institute of Technology, Stockholm, Sweden
e-mail: kriron@kth.se

C. Duwig
e-mail: duwig@kth.se

K. Rönnberg
ABB AB, Corporate Research, Västerås, Sweden

M. García-Villalba et al. (eds.), *Direct and Large Eddy Simulation XII*,
ERCOFTAC Series 27,
https://doi.org/10.1007/978-3-030-42822-8_9

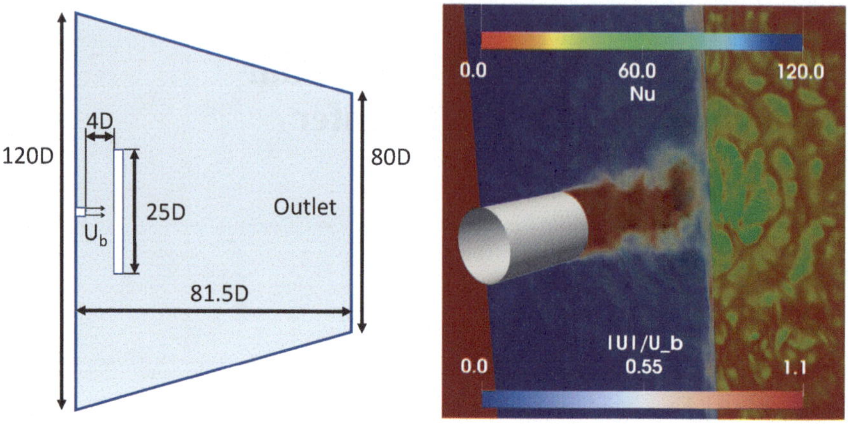

Fig. 9.1 Schematic representation of geometry configuration and result snapshot. The snapshot shows normalized velocity distribution ($|U|/U_b$) at the jet axis and Nusselt number distribution on the plate surface

The simulated flow is characterized by Re = 5000, calculated as $\mathrm{Re} = \frac{\rho U_b D}{\mu}$ where ρ is the density, μ is the dynamic viscosity, and D is the nozzle diameter. The thermal properties of the fluid simulated are characterized by a Prandtl number of 0.7.

The bulk inlet flow velocity, U_b, is such that $M = (U_b/c) \approx 1.5 \cdot 10^{-2} \ll 0.3$, which ensures that the flow can be regarded as incompressible. In this context, c refers to the speed of sound in the fluid. The relative difference between the inlet fluid temperature, T_F, and the hot wall temperature, T_W, were specified so that $\Delta T/T_F \ll 1$. The temperatures considered were such that $c_p(T_F)/c_p(T_W) \approx 1$ and $\rho(T_F)/\rho(T_W) \approx 1$, where c_p is the specific heat at constant pressure. Thus, the flow can be considered to have constant properties and the temperature can be modeled as a passive scalar.

The metric of local heat transfer at the plate surface was obtained through the temperature gradient normal to the impingement surface. Using the retrieved temperature gradient a formulation for the local Nusselt number can be made as

$$\mathrm{Nu} = -\left(\frac{D}{T_W - T_F}\right)\frac{\partial T}{\partial n} \tag{9.1}$$

where the temperature gradient $\frac{\partial T}{\partial n}$ is retrieved from the simulations. D, T_W, and T_F are as defined above.

The problem was solved numerically using OpenFOAM [4] version 3.0.1, a FVM (Finite Volume Method) based code. The algorithm used was PISO with adaptive time-stepping. The time step was controlled by limiting the Courant number so that values around 0.3–0.4 are obtained at the plate surface and around the stagnation region of the jet. Temporal discretization was performed using a second-order implicit

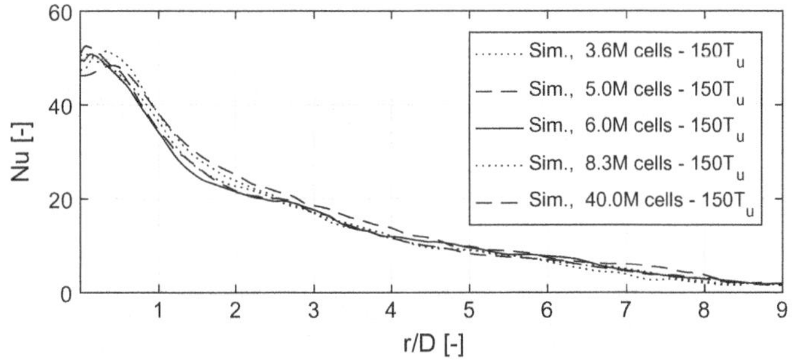

Fig. 9.2 Nusselt number distributions obtained from grid dependence analysis

scheme. Second-order schemes were used for discretization of the divergence of the velocity and the temperature. A solver for compressible flows was used.

The mesh employed was a hexahedral dominant mesh with a boundary layer at the plate. The majority of the cells were located at the plate and around the jet core.

The subgrid-scale model used was the Wall-Adapting Local Eddy-viscosity (WALE) model [5] using the default coefficients provided in the implementation in OpenFOAM 3.0.1.

The simulations described in this paper all employed a *flat* velocity profile with magnitude U_b unless otherwise stated. No velocity fluctuations were prescribed at the inlet.

A grid dependence investigation was performed. Grids with cell counts from 3.6 to 40 million were used. The Nusselt number response was chosen as the metric for the grid sensitivity study. The temperature gradient at the plate surface was sampled along a line and translated according to (9.1).

The Nusselt number distributions obtained from the different meshes are presented in Fig. 9.2. The data presented here is averaged over 150 T_u, where $T_u = D/U_b$. Over these 150 T_u around 700 sampling line readings were collected. Figure 9.2 indicate that grids perform fairly equal in terms of rendering heat transfer results.

For a balance between run-time and resolution the 6M cell case was chosen. To assess the suitability of this mesh the y^+, at the plate, and resolved kinetic energy criteria ("Pope criteria"), introduced in [6], were used. For the 6M cell grid a maximum y^+ of 1.4 and an average y^+ of 0.22 was recorded over 30 time samples, evenly distributed over the 150 T_u, used for averaging. The 6M mesh was also dense enough to resolve, on average, over 80% of the kinetic energy near the plate surface where jet-induce heat transfer occurs further indicating the suitability of the 6M mesh.

Validations

The obtained simulation results concerning heat transfer were validated against two data sources; the experimental results presented in [2] and correlations for a single round nozzle jet presented in [7]. The correlation used reads

$$\begin{aligned} Nu_{SRN} &= \frac{1-1.1/r^*}{r^*+0.1(h^*-6)} \cdot F(\mathrm{Re}) \cdot Pr^{0.42} \\ F(\mathrm{Re}) &= 2\left[\mathrm{Re}\left(1+0.005 \cdot \mathrm{Re}^{0.55}\right)\right]^{0.5} \end{aligned} \tag{9.2}$$

where $r^* = r/D$ (normalized radial position) and $h^* = H/D$ (normalized nozzle-to-plate distance), and was established experimentally for a round jet impinging on a flat plate. What is obtained from Eq. 9.2 is *the mean Nusselt number over a circle with radius* r^*. The correlation is reported to be valid in the ranges $2.5 \leq r^* \leq 7.5$, $2.0 \leq h^* \leq 12$ and $2000 \leq \mathrm{Re} \leq 400{,}000$. The flow conditions considered in this paper thus falls within the validity range ($\mathrm{Re} = 5000$ and $h^* = H/D = 4$).

As the experimental results presented in [2] are azimuthally averaged and correlation data from (9.2) is for average over a circle, the simulation data had to be treated in a similar fashion. Thus, simulation data was azimuthally averaged in order to render data to be compared with [2]. For comparison with the correlation data, simulation and experimental data were processed to render average values over circles with various radii.

A comparison between simulation data and experimental data from [2] is presented in Fig. 9.3. A noticeable difference between the simulation data and the experimental data can be seen in Fig. 9.3. Since the round nozzle configuration in [2] served as inspiration for the setup of the simulation case, the discrepancy was (and still is) somewhat puzzling. The observed discrepancy led to reasoning around potential differences between simulation and reality which could explain the difference. Among the more plausible, the inlet profile velocity shape and the approximation of the plate as a constant temperature boundary were identified. Of these two, the inlet veloc-

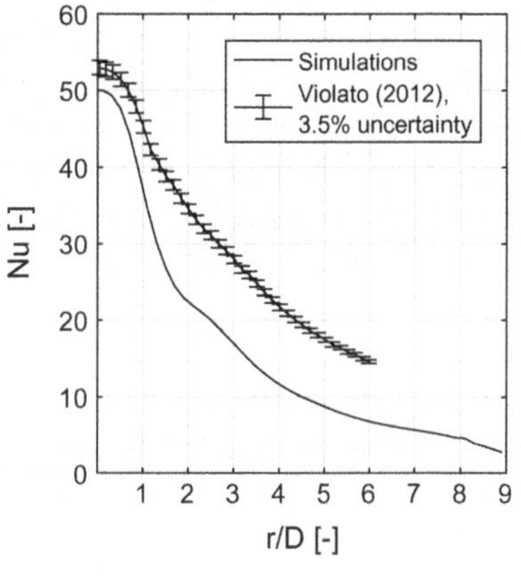

Fig. 9.3 Azimuthally averaged Nu. Experimental data from [2]

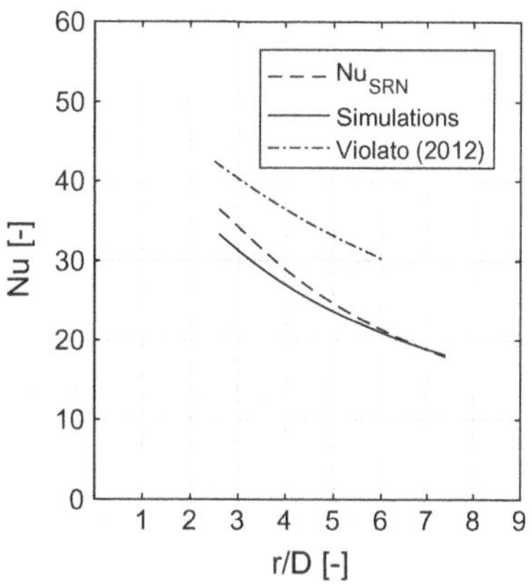

Fig. 9.4 Average Nu over circle with radius r/D. Experimental data from [2] and correlation from [7]

ity profile shape was investigated which is elaborated on in section "Effect on Heat Transfer of Inlet Velocity Profile".

As discrepancies were observed, additional sources of data for comparisons were sought. This led to finding the correlation for a single round nozzle jet in [7], restated here as Eq. (9.2). Comparison with simulation, correlation and experimental data is presented in Fig. 9.4. From this comparison, it can be observed that the correlation and simulation render more similar results than the correlation and data from [2]. The original source to the correlation, [8], has at the time of writing not been investigated for a potential explanation to why it appear to correspond better with the simulation results than more recent experimental data [2].

Effect on Heat Transfer of Inlet Velocity Profile

Due to the discrepancies between simulations and measurements observed, described in section "Validations", an investigation on the impact on the heat transfer due to inlet velocity profile was performed. Two profiles, in addition to the flat profile, were simulated, using the 6M cell mesh, and compared. The profiles chosen are one describing a laminar flow and a turbulent flow, described by $U_{\text{lam}} = 2U_{\text{avg}}\left(1 - r^2/R^2\right)$ and $U_{\text{turb}} = U_{\max}\left(1 - r/R\right)^{1/n}$. For U_{turb}, $n = 7$ was chosen. U_{avg} and $U_{\max}$ were chosen such that $\text{Re} = 5000$ was achieved. The profiles are illustrated in Fig. 9.5.

In Fig. 9.6 it can be seen that the Nusselt number in the region $r/D < 1$ increases the more parabolic the inlet profile becomes. Hence, there could be a possibility that an inlet profile which is intermediate between the flat and turbulent profiles, shown in Fig. 9.5, would lead to an even closer match to the experimental results in $r/D < 1$. However, for the region $r/D > 1.5$ all employed inlet profiles render lower Nusselt number values than the experimental results. Thus, it cannot be concluded that the

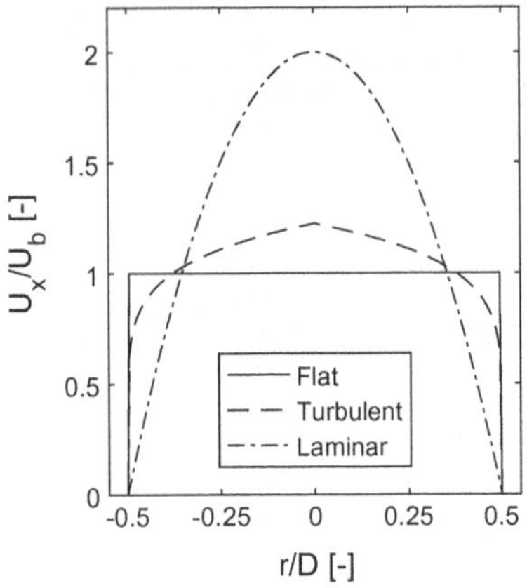

Fig. 9.5 Inlet velocity profiles employed

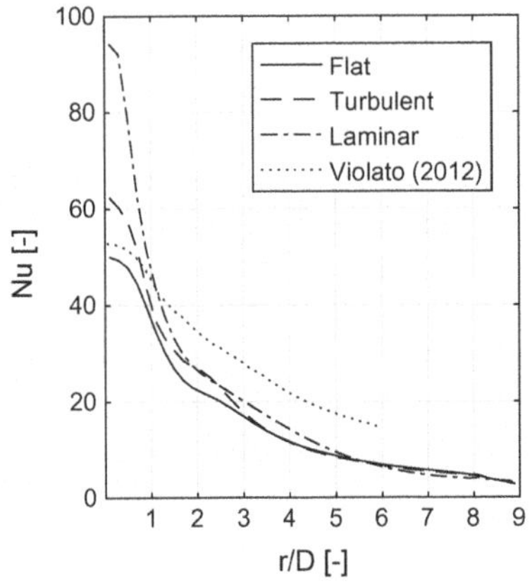

Fig. 9.6 Azimuthally averaged Nu obtained with different inlet velocity profiles

inlet velocity profile is the only source for the discrepancies between simulations and experiments observed here.

A comprehensive discussion on observed vortex structures and their local vorticity, as well as their relation to heat transfer was presented in [3]. There, the flow conditions studied were Re $= 10{,}000$ and the nozzle-to-plate distance $H/D = 2$.

The data obtained from the present simulations, employing the flat profile and the 6M cell grid, was analyzed with a similar approach regarding vortex structures. A Q-criterion of $Q = 4\,(U_b/D)^2$ was used to capture the structures. While being qualitative, Fig. 9.7 is intended to show one observation made; that vortex structures moving radially outward appear to have trailing bands of low Nusselt numbers. According to the reasoning put forward in [3] high heat transfer should correlate with the azimuthal vorticity associated with the vortex structures.

Additional analysis of the simulation data is necessary in order to establish quantitative relationships between flow structures and Nusselt number distribution.

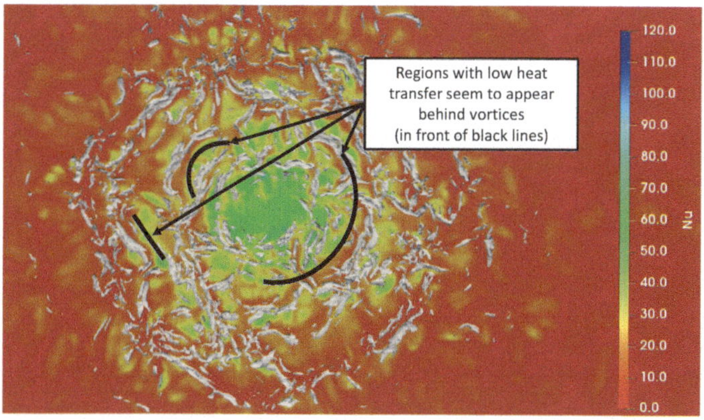

Fig. 9.7 Near-plate vortex structures (clipped at 0.1D above the plate) and Nusselt number distribution

Acknowledgements Simulations were performed on computational resources provided by ABB AB, Corporate Research, in Västerås, Sweden. The work is partially funded by the Swedish Foundation for Strategic Research.

References

1. Du, X., Yang, Z., Jin, Z., Xia, C., Bao, D.: A comparative study of passive control on flow structure evolution and convective heat transfer enhancement for impinging jet. Int. J. Heat Mass Transf. **126**, 256–280 (2018)
2. Violato, D., Ianiro, A., Cardone, G., Scarano, F.: Three-dimensional vortex dynamics and convective heat transfer in circular and chevron impinging jets. Int. J. Heat Mass Transf. **37**, 22–36 (2012)
3. Dairay, T., Fortun, V., Lamballais, E., Brizzi, L.E.: Direct numerical simulation of a turbulent jet impinging on a heated wall. J. Fluid Mech. **764**, 362–394 (2015)
4. Weller, H.G., Tabor, G., Jasak, H., Fureby, C.: A tensorial approach to computational continuum mechanics using object-oriented techniques. Comput. Phys. **12**(6), 620–631 (1998). https://doi.org/10.1063/1.168744. https://aip.scitation.org/doi/abs/10.1063/1.168744
5. Nicoud, F., Ducros, F.: Subgrid-scale stress modelling based on the square of the velocity gradient tensor. Flow Turbul. Combust. **62**(3), 183–200 (1999)
6. Pope, S.B.: Ten questions concerning the large-eddy simulation of turbulent flows. New J. Phys. **6**(1), 35–35 (2004)
7. VDI Heat Atlas, 2nd ed. VDI-Buch (2010)
8. Schlünder, E., Gnielinski, V.: Wärme- und stoffübertragung zwischen gut und aufprallendem düsenstrahl. Chem. Ing. Tech. **39**, 578584 (1967)

Chapter 10
Large Eddy Simulation of Precession of a Non-swirling Turbulent Jet in a Counterflow

M. Rovira, K. Engvall and C. Duwig

Introduction

For many industrial applications, jets with high mixing performance are essential. While effective, the complexity of active control for enhanced mixing makes it less desirable than passive techniques. To that extent, the jet in counterflow is a relatively unexplored alternative, which has been shown to improve mixing compared to other jet configurations [1]. Additionally, the jet in counterflow is a fundamental free-shear flow problem, meaningful from a scientific perspective.

A jet in counterflow describes a flow configuration in which a jet is issuing against a larger opposing uniform stream. Depending on the jet-to-counterflow velocity ratio ($\alpha = U_j/U_0$), a stable and an unstable flow condition is observed. Above $\alpha = 1.4$, symmetric vortex shedding develops into chaotic low-frequency oscillations in the axial and radial directions [2].

Bernero and Fiedler [3] studied the characteristic axial and radial low-frequency jet motions by isolating these patterns through Proper Orthogonal Decomposition (POD). These highly energetic fundamental flow patterns were previously detected from flow visualization and were thought to be the cause of improved mixing [1]. Nevertheless, the two-dimensional nature of the tools employed could potentially be hiding other three-dimensional flow patterns.

M. Rovira (✉) · C. Duwig
Department of Engineering Mechanics, KTH Royal Institute of Technology, Stockholm, Sweden
e-mail: marrs@kth.se

C. Duwig
e-mail: duwig@kth.se

K. Engvall
Department of Chemical Engineering, KTH Royal Institute of Technology, Stockholm, Sweden
e-mail: kengvall@kth.se

M. García-Villalba et al. (eds.), *Direct and Large Eddy Simulation XII*,
ERCOFTAC Series 27,
https://doi.org/10.1007/978-3-030-42822-8_10

Currently, numerical studies on the subject are scarce [4, 5]. Owing to the lack of comprehensive computational studies and leveraging the importance of detailed three-dimensional results, the present study aims at applying Large Eddy Simulations (LES) and POD for the study of a jet in counterflow.

Model Description

For unsteady, incompressible flow, the governing equations read

$$\frac{\partial \bar{u}_i}{\partial x_i} = 0\,, \qquad \frac{\partial \bar{u}_i}{\partial t} + \bar{u}_j \frac{\partial \bar{u}_i}{\partial x_j} = -\frac{1}{\rho}\frac{\partial \bar{p}}{\partial x_i} + \frac{\partial}{\partial x_j}\left(\nu \frac{\partial \bar{u}_i}{\partial x_j} - \tau_{ij}^{\mathrm{SGS}}\right) \tag{10.1}$$

$$\frac{\partial \bar{\theta}}{\partial t} + \bar{u}_j \frac{\partial \bar{\theta}}{\partial x_j} = \frac{\partial}{\partial x_j}\left(\frac{\nu}{Sc}\frac{\partial \bar{\theta}}{\partial x_j} - q_j^{\mathrm{SGS}}\right) \tag{10.2}$$

where $\bar{u}_i$, $\bar{p}$, and $\bar{\theta}$ are the filtered velocity, pressure and passive scalar, respectively. Sc is the Schmidt number, ν is a constant kinematic viscosity, ρ is the density, and τ_{ij}^{SGS} and q_j^{SGS} are the subgrid-scale stress tensor and scalar flux evaluated with the Smagorinsky closure model and a constant turbulent Schmidt number.

A cross section of the control volume employed is illustrated in Fig. 10.1. The jet is injected from a nozzle which has an exit diameter $d = 5\,\mathrm{mm}$ and replicates the geometry by Tsunoda and Saruta [6]. The nozzle protrudes $15\,d$ into the domain and jet exit is at $x = 0$. The jet inlet is defined at $-6\,d$ where a fluctuating velocity is obtained by the Synthetic Eddy Method [7] with an average value that yielded $\alpha = 5.1$ near the jet exit. The bulk flow velocity is $U_0 = -0.21$ m/s and the width and height of the channel is $D = 40\,d$. The Reynolds number, based on the jet exit diameter, is $\mathrm{Re}_d = 5300$ and the passive scalar is set to 1 in the jet inlet and 0 in the counterflow. No-slip boundary conditions were defined at the walls. In the outlet, the velocity and passive scalar were set to have zero gradient.

The problem was solved computationally using the Finite Volume Method (FVM) as implemented in OpenFOAM-2.4.0. The PISO algorithm with automatic time stepping was employed for pressure-velocity coupling. Centered second-order interpolation and NVD-derived interpolation were employed for the discretization of the divergence of velocity and the passive scalar, respectively.

The total simulation time was $10\,T_{U_0} = 2040\,T_{U_j}$, where $T_{U_0} = 40\,d/U_0$ and $T_{U_j} = d/U_j$. In order to avoid the effects of the transient jet development, flow statistics were obtained by averaging between $2.5\,T_{U_0}$ and $10\,T_{U_0}$. A maximum local Courant number of $Co_{\max} = 0.2$ for the jet was fixed throughout the simulation yielding a time step of approximately $\Delta t \approx 1 \cdot 10^{-5}$ s.

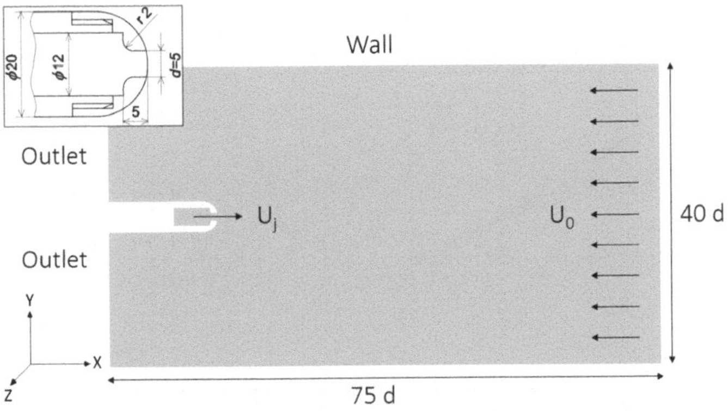

Fig. 10.1 Representation of the computational domain. *Insert* Nozzle dimensions extracted from Tsunoda and Saruta [6]

Results and Discussion

Simulations were performed using three hexa-dominant unstructured meshes with 1.5, 3, and 6 million cells. All meshes have the same distribution of mesh elements with most being near the jet exit and along the shear layer. The normalized minimum cell element length for the coarse, medium, and fine mesh is $d/l_{\min} = 57, d/l_{\min} = 75$, and $d/l_{\min} = 96$. To evaluate the suitability of the meshes, an posteriori measure of turbulent resolution [8] was computed to verify if the meshes were able to resolve at least 80% of the turbulent kinetic energy in the flow. All meshes met this criterion except for a thin region smaller than $0.5 \cdot d$ along the shear layer in the coarse mesh.

Results were compared with experimental data obtained by Tsunoda and Saruta [6] and Chan [9] for a round jet with $\alpha = 5.1$ and $\alpha = 5.1$. At a distance of $x/d = 0.44$ from the nozzle exit, all meshes reproduce the experimental top hat profile and the velocity ratio accurately. All cases are also in reasonable agreement with the turbulent intensity in the core jet, reported to be 5.5% by Tsunoda and Saruta [6] and 4.5% for the present LES simulations. However, the coarse mesh slightly underpredicts the turbulent intensity in this region.

Figure 10.2 shows downstream variations of the axial mean velocity and rms velocity fluctuation. All simulations reasonably predict the shape of the velocity decay, being in excellent agreement with the measurements of Chan [9] and showing a better penetration length estimation than that of the other LES study [5]. The penetration length l_p is defined as the distance between the stagnation point and the nozzle and is $l_p/d = 13.33,\ 13.45,\ 13.35$ and $l_p/d = 13.8,\ 14.5$ for the coarse, medium, and fine mesh and the experimental data from Tsunoda and Saruta [6] and Chan [9], respectively. For the rms velocity fluctuation, the LES cases are able to capture the double peak in the downstream direction as well as the regions of increasing and decaying turbulence. As is the case for the axial mean velocity, LES results show better agreement with the data provided by Chan [9]. Figure 10.3 presents several

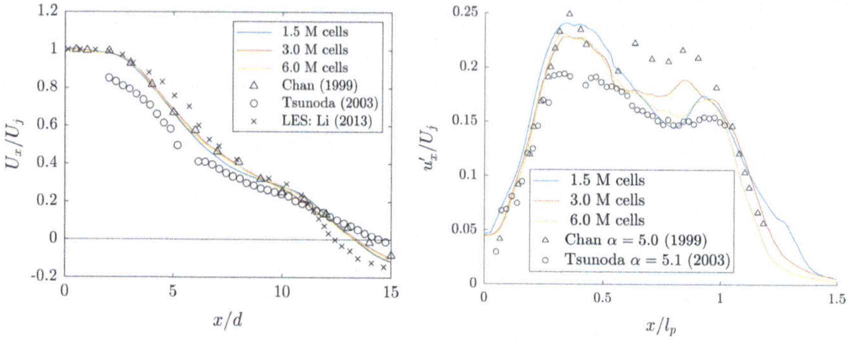

Fig. 10.2 Downstream variation of the normalized axial mean velocity $U_x/|U_0|$ (*left*) and the axial rms velocity fluctuation $u'_x/|U_0|$ (*right*) along the jet centerline

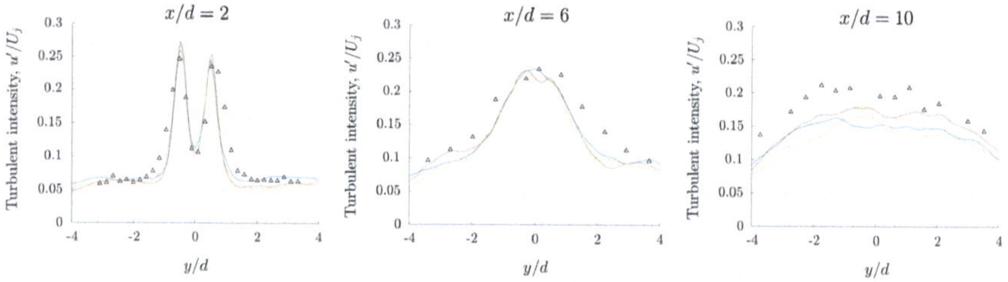

Fig. 10.3 Profiles of turbulent intensity at different locations. Legend as in Fig. 10.2

profiles of turbulent intensity and are compared with results from Chan [9]. Closer to the jet exit ($2 \leq x/d \leq 6$) considerable agreement with the experimental results is obtained, with LES results predicting both the location and magnitude of the peaks of turbulent intensity. At $x/d = 10$, LES simulations underpredict the turbulence levels shown in the experiment.

A comparison between the mean scalar field is presented in Fig. 10.4. After a relatively undisturbed core (i.e. the region with normalized scalar concentrations $0.9 \leq \theta \leq 1$), the jet breaks up. In the results by Tsunoda and Saruta [6], Chan [9] and Li [5] this occurs earlier ($x/l_p \approx 0.15$) than in the LES simulation ($x/l_p \approx 0.25$). Gradually, the jet loses axial momentum and is subsequently deflected. At this point, the passive scalar is transported at the bulk flow speed while continuously being diffused. Radial penetration of the scalar field is underpredicted by the LES simulations when compared to Tsunoda and Saruta [6]. Nevertheless, the simulation captures the relevant physics that is responsible for mixing.

Figure 10.5 shows an instantaneous visualization of the vortices produced by the jet by using the Q-criterion ($\sim 10^4$) colored by the non-dimensional axial velocity. The coherent turbulent structures generated by the Synthetic Eddy Method are inside the nozzle, growing in size as the jet develops downstream. The longest structures are convected downstream by the counterflow and therefore occur past the stagnation

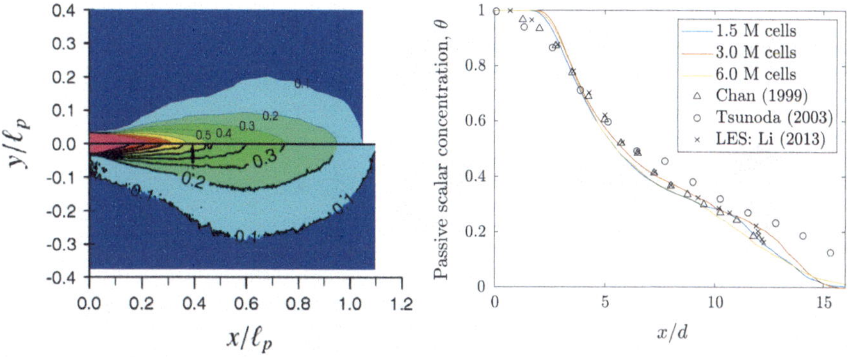

Fig. 10.4 *Left*: comparison between two normalized mean scalar fields in a vertical plane along the jet axis. *Top*: 3M mesh. *Bottom*: Tsunoda and Saruta [6]. *Right*: downstream variation of the passive scalar concentration along the jet centerline

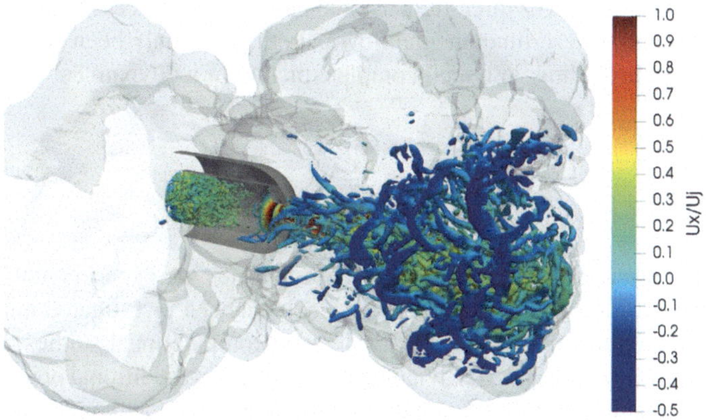

Fig. 10.5 Jet vortices visualized by Q-criterion and colored by non-dimensional axial velocity. The surrounding cloud corresponds to the isocontour $\theta = 1000$ ppm

point. Additionally, the isocontour of $\theta = 1000$ ppm is shown. The ample radial extent of this gray cloud shows the capacity of the jet in counterflow to mix efficiently.

Finally, the POD method was used to analyze dynamical flow structures, as described elsewhere [10]. The snapshot POD method was applied to $N \approx 2800$ flow instances of the passive scalar to obtain spatial eigenfunctions and eigenvalues. These represent fundamental flow patterns or modes and their relative importance, as can be seen in Fig. 10.6. Mode 0 is the mean field and has been removed from the calculation of the cumulative sum of eigenvalues. The chaotic nature of the jet in counterflow is evidenced as the sum of almost 500 modes is required to recover 80% of the energy content of the flow. Nevertheless, the first 5 modes account for 25% of the energy and could thus be considered as dominant. The isosurface representation of these most relevant modes is also shown in Fig. 10.6. There are two pairs within

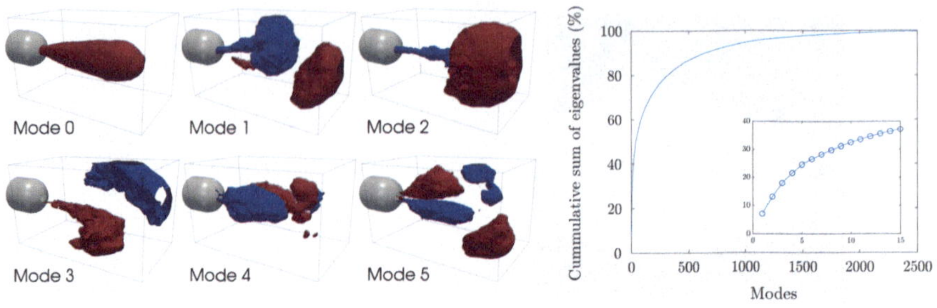

Fig. 10.6 *Left*: visual representation of the first 6 modes. *Right*: cumulative sum of eigenvalues as percentage of the total sum; *inset*: close-up of the first 15 modes

this set of modes, namely modes 1 and 2 and modes 4 and 5. These pairs exhibit a similar spatial distribution that is rotated by $\pi/2$. Hence, the jet alternates between both states which evidences three-dimensional flapping and precession and enhances mixing. Previously, these fundamental flow patterns have only been studied in two dimensions [3], providing with an incomplete picture of jet dynamics.

Conclusions

A jet issuing into a uniform counterflow that replicates the experimental set-up of Tsunoda and Saruta [6] at velocity ratio $\alpha = 5.1$ has been investigated numerically with LES simulations. Due to the lack of detailed computational studies and three-dimensional evaluations of the jet dynamics in counterflow, additional progress could be expected. Results for the LES simulations are in excellent agreement with experimental data, notably with that of Chan [9]. This shows that present simulations are able to capture in detail the physics of the jet in counterflow. Q-criterion visualization and the POD method were used to study three-dimensional dynamical flow structures. The chaotic nature and complexity of the jet in counterflow is illustrated. Additionally, POD allowed evidencing large-scale flapping and precession modes which bring about mixing.

Acknowledgements This work has been funded by Formas (Swedish Research Council for Sustainable Development). Simulations were performed on resources provided by the Swedish National Infrastructure for Computing (SNIC).

References

1. Yoda, M., Fiedler, H.E.: The round jet in a uniform counterflow: flow visualization and mean concentration measurements. Exp. Fluids **21**, 427–436 (1996)

2. Knig, O., Fiedler, H.E.: The structure of round turbulent jets in counterflow: a flow visualization study. In: Advances in Turbulence, vol. 3, pp. 61–66. Springer, Berlin, Heidelberg (1991)
3. Bernero, S., Fiedler, H.E.: Application of particle image velocimetry and proper orthogonal decomposition to the study of a jet in a counterflow. Exp. Fluids **29**, S274–S281 (2000)
4. Amamou, A., Habli, S., Sad, N.M., Bournot, P., Palec, G.L.: Computational study of mixing behaviour of a turbulent jet issuing in a uniform counterflow at low velocity ratios. J. Turbul. **17**, 237–251 (2016)
5. Li, Z., Huai, W., Qian, Z.: Large eddy simulation of a round jet into a counterflow. Sci. China Technol. Sci. **56**, 484–491 (2013)
6. Tsunoda, H., Saruta, M.: Planar laser-induced fluorescence study on the diffusion field of a round jet in a uniform counter-flow. J. Turbul. **4**, N13 (2003)
7. Kornev, N., Kroger, H., Hassel, E.: Synthesis of homogeneous anisotropic turbulent fields with prescribed second-order statistics by the random spots method. Commun. Numer. Methods Eng. **24**, 875–877 (2007)
8. Pope, S.B.: Ten questions concerning the large-eddy simulation of turbulent flows. New J. Phys. **6**(1), 35 (2004)
9. Chan, H.C.: Investigation of a round jet into a counterflow. Ph.D. thesis, Dept. of Civil Engineering, University of Hong Kong, Hong Kong (1993)
10. Berkooz, G., Holmes, P., Lumley, J.L.: The proper orthogonal decomposition in the analysis of turbulent flows. Annu. Rev. Fluid Mech. **25**(1), 539–575 (1993)

Chapter 11
Flow Around a 5:1 Rectangular Cylinder: Effects of the Rounding of the Upstream Corners

A. Mariotti, B. Rocchio, E. Pasqualetto, C. Mannini and M. V. Salvetti

Introduction

The high Reynolds number flow around a rectangular cylinder, having chord-to-depth ratio equal to 5, is the object of the benchmark BARC (Benchmark on the Aerodynamics of a Rectangular 5:1 Cylinder). This configuration is of practical interest, e.g. in civil engineering, and, in spite of the simple geometry, the related flow dynamics and topology is complex. The experimental and numerical results obtained by the contributors during the first four years of activity were reviewed in [1]. Good agreement between different results in terms of near-wake flow, base pressure, and drag coefficient was found. However, it was observed that some quantities of interest, as the standard deviation of the lift coefficient or the distribution of mean and fluctuating pressure on the cylinder sides, are affected by a significant dispersion, both in experiments and in simulations.

Sensitivity analyses carried out by the BARC contributors were not conclusive to explain the observed dispersion; rather, in some cases, they led to controversial

A. Mariotti · B. Rocchio · E. Pasqualetto · M. V. Salvetti (✉)
Department of Civil and Industrial Engineering, University of Pisa, Pisa, Italy
e-mail: mv.salvetti@ing.unipi.it

A. Mariotti
e-mail: alessandro.mariotti@for.unipi.it

B. Rocchio
e-mail: benedetto.rocchio@ing.unipi.it

E. Pasqualetto
e-mail: elena.pasqualetto@phd.unipi.it

C. Mannini
Department of Civil and Environmental Engineering, CRIACIV,
University of Florence, Florence, Italy
e-mail: claudio.mannini@unifi.it

M. García-Villalba et al. (eds.), *Direct and Large Eddy Simulation XII*,
ERCOFTAC Series 27,
https://doi.org/10.1007/978-3-030-42822-8_11

results (see e.g. [2–6]). In particular, a crucial quantity is the length of the mean recirculation regions forming on the lateral sides of the cylinder, which in turns are related with the pressure distribution on the cylinder side and the oscillating loads. Wind tunnel tests [5] clearly showed that for a low intensity of the freestream turbulence, the time-averaged location of the point of flow reattachment is close to the downstream edges. On the other hand, the study carried out in [2] indicated that increasing grid resolution in the spanwise direction led to a significant reduction of the recirculation region mean length and thus to a deterioration of the agreement with the experiments. This was recently confirmed in [4] by a stochastic analysis of the sensitivity of LES results to grid resolution in the spanwise direction and to the amount of subgrid scale (SGS) dissipation. Indeed, it was observed that numerical simulations tend to underpredict significantly the distance from the upstream corners at which the mean flow reattachment occurs, either for a fine discretization in the spanwise direction or for a low SGS dissipation.

The present work aims at shedding some light on this counter-intuitive behavior of the numerical simulations. In particular, it was shown in [4] that the length of the mean recirculation bubble on the cylinder side is strictly connected with the location at which the shear layers separating from the upstream corners lose coherence and roll-up in vortical structures. Simulations with low SGS dissipation and highly refined grids were characterized by early roll-up and by small vortical structures. Following this result, additional simulations are carried out to investigate the effect of a small rounding in the upstream corners of the rectangular cylinder. We want to investigate whether the presence of perfectly sharp corners can generate some disturbances which are not damped in numerical simulations characterized by low numerical and SGS dissipation as the present ones, which may in turn lead to a premature instability of the shear layers.

Numerical Methodology and Simulation Setup

LES simulations are carried out for the incompressible flow around a fixed rectangular cylinder with a chord-to-depth ratio, B/D, equal to 5. The angle of attack is zero. The computational domain is sketched in Fig. 11.1. A uniform velocity profile is imposed at the inflow (no turbulence), while no-slip conditions are applied at the solid walls. Periodic conditions are imposed in the spanwise direction, while traction-free boundary conditions are used at the outflow and on the remaining lateral sides of the computational domain. Finally, the Reynolds number based on the freestream velocity and on the cylinder depth, Re, is equal to 40,000. The simulations are carried out through an open-source code, Nek5000, based on a high-order accurate spectral-element method. The order of the Legendre polynomials used as basis functions inside each element is kept herein constant $N = 6$. The grid resolution in the streamwise and lateral directions is $\Delta x = \Delta y = 0.125D$. The grid resolution in the spanwise direction, defined in terms of the average element size, is $\Delta z = 0.558D$, and the weight of the explicit filter, $w = 0.05$. These values are the same as in one of

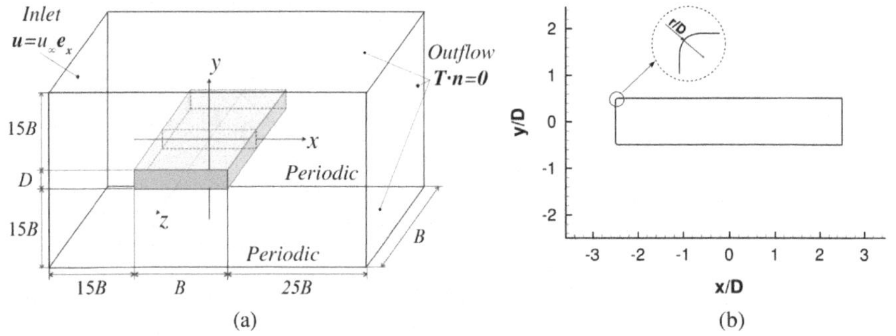

Fig. 11.1 Sketch of the computational domain (**a**) and of the upstream corner rounding (**b**)

the 13 simulations in [4] having a short mean recirculation region. The parameters chosen for the sensitivity analysis is the rounding of the upstream corners, r/D. The values $r/D = 0.003$ and $r/D = 0.05$ are considered.

As for the LES formulation, a simple approach based on the application of a low-pass explicit filter in the modal space, which is characterized by a cut-off k_c, here equal to $N - 3$, and by a weight w, is adopted (see [4] for more details). This modal filter provides a dissipation in the resolved modes that are higher than the cut-off value, and can be interpreted as a SGS dissipation.

Results and Discussion

The numerical results are first compared to the experimental ones in terms of mean pressure coefficient distribution and its standard deviation over the lateral side of the cylinder (Fig. 11.2). The pressure coefficient is defined as follows:

$$C_p = \frac{p - p_\infty}{1/2 \rho U_\infty^2}$$

where p is the local pressure, ρ is the density, and p_∞ and U_∞ are the freestream pressure and velocity, respectively. The average of C_p in time, in the spanwise direction and between the upper and lower half perimeters of the cylinder is denoted as $t - \text{avg}(C_p)$; the standard deviation, $t - \text{std}(C_p)$ is also computed by the same averaging procedure. The local abscissa, s, is the distance from the cylinder stagnation point, which is located at $x = -B/2$, $y = 0$, and $z = 0$ in Fig. 11.1, measured along the cylinder side. The results of the simulations carried out with $r/D = 0.0037$ and $r/D = 0.05$ are characterized by a mean pressure distribution occurring significantly more downstream compared to the simulation with the same numerical setup and sharp corners. In particular, the simulation having the smaller value of the rounding is in good agreement with the ensemble of the BARC experiments, whereas the mean

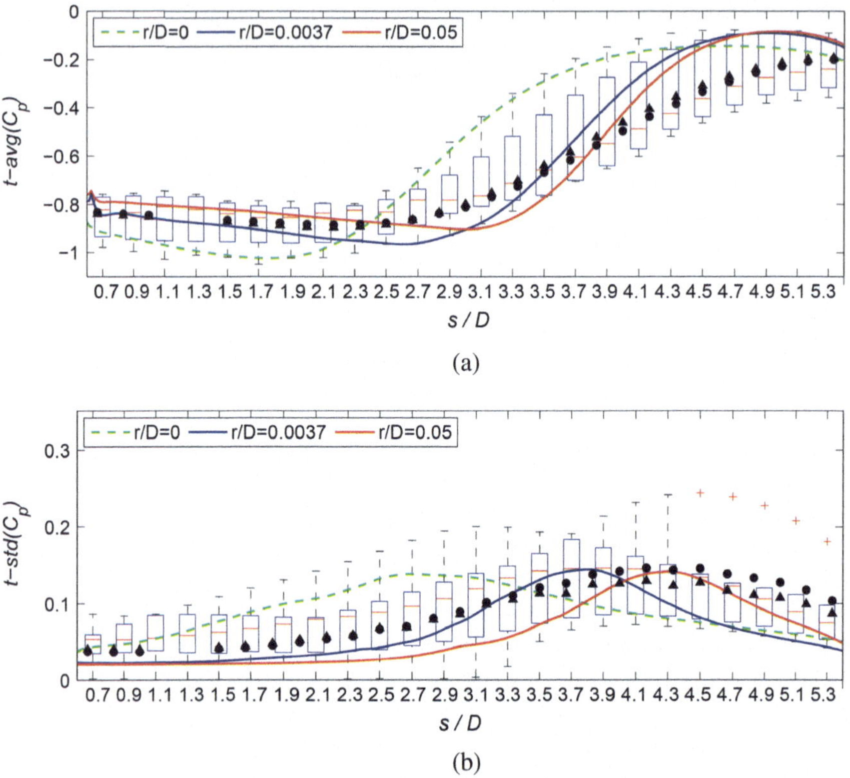

Fig. 11.2 Mean pressure coefficient (**a**) and standard deviation (**b**) on the lateral sides of the cross section for the different values of the corner roundings (both averaged in spanwise direction). A comparison is provided with the case with sharp corners [4], the ensemble statistics of the BARC experiments [1], and the experimental data in [5] (circles and triangles refer, respectively, to Re = 56,700 and 112,200)

pressure recovery is moved too downstream in the simulation with large roundings. The position of the maximum of $t - \mathrm{std}(C_p)$ along the lateral side of the cylinder, which is directly related with the length of the mean recirculation region, is also moved downstream for increasing rounding values.

The increase of the rounding curvature radius produces an increased length of the main recirculation region on the cylinder sides, as can be seen in Fig. 11.3 by means of the time-averaged vortex indicator λ_2. It should also be noted that a very small corner rounding, i.e. $r/D = 0.0037$, is enough to significantly change the shear layer dynamics compared with sharp edges, $r/D = 0$, as can be seen in Figs. 11.4, 11.5, and 11.6. First, in the sharp-edge case, the turbulent kinetic energy in the detaching shear layers rapidly grows immediately downstream the corners (see Fig. 11.4a); conversely, in the simulations with roundings, the turbulent fluctuations start to be significant only downstream (see Fig. 11.4b, c). This behavior is related to the

(a) (b) (c)

Fig. 11.3 Isosurfaces of the time-averaged vortex indicator λ_2: $r/D = 0$ (**a**) from [4], $r/D = 0.0037$ (**b**) and $r/D = 0.05$ (**c**)

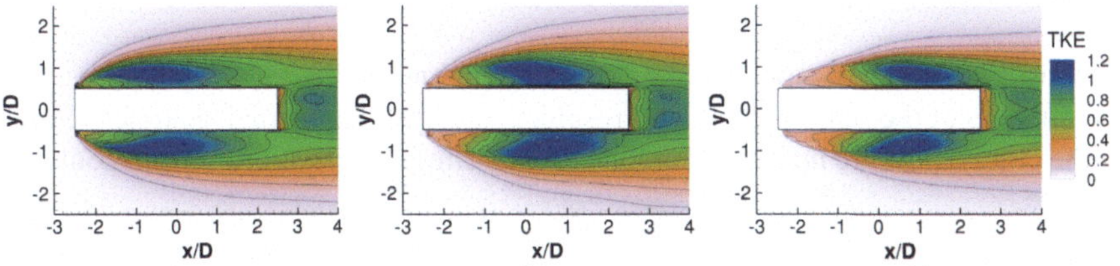

(a) (b) (c)

Fig. 11.4 Time- and spanwise-averaged turbulent kinetic energy: $r/D = 0$ (**a**) from [4], $r/D = 0.0037$ (**b**) and $r/D = 0.05$ (**c**)

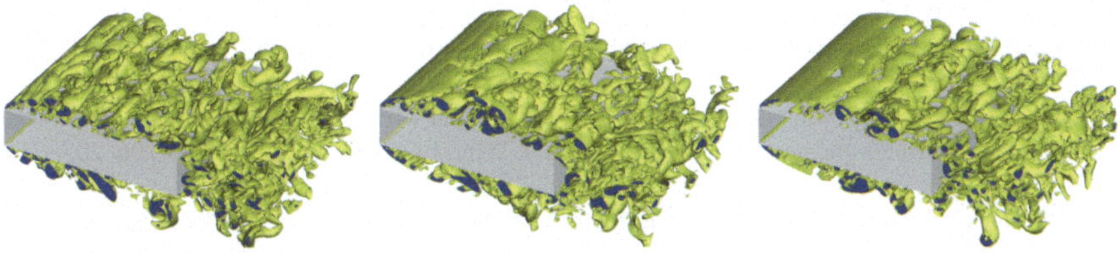

(a) (b) (c)

Fig. 11.5 Isosurfaces of the instantaneous vortex indicator λ_2: $r/D = 0$ (**a**) from [4], $r/D = 0.0037$ (**b**) and $r/D = 0.05$ (**c**)

dynamics of the vorticity contained in the shear layers. Indeed, from the analysis of the isosurfaces of the instantaneous vortex indicator λ_2 (see Fig. 11.5) and of its distribution in the plane section $z = 0$ (see Fig. 11.6), it is evident that increasing the corner roundings the detaching shear layers remain coherent up to a significantly more downstream position than the one obtained with sharp edges,in which they roll up in small vortical structures already near the upstream corners.

A very small value of the curvature radius of the corner roundings—which would be hard to be detected in experiments unless ad-hoc diagnostic techniques are employed—is enough to change the scenario and to significantly improve the agreement with the experiments. Indeed, a premature instability of the shear layers, leading to a too short mean recirculation zone, may be originated by small perturbations introduced by non-realistic perfectly sharp corners, which in highly resolved

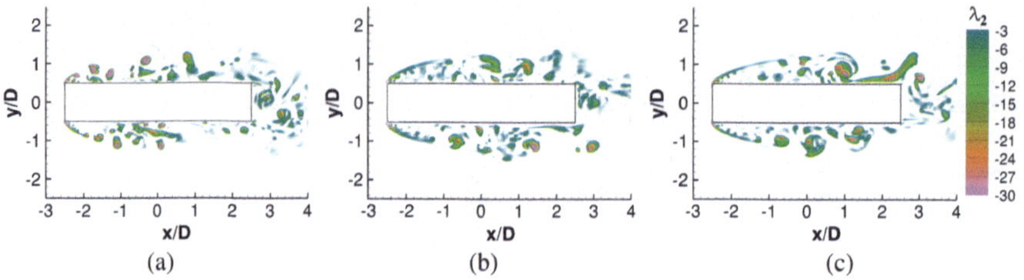

Fig. 11.6 Isosurfaces of the instantaneous vortex indicator λ_2 on the plane $z = 0$: $r/D = 0$ (**a**) from [4], $r/D = 0.0037$ (**b**) and $r/D = 0.05$ (**c**)

LES are not damped by numerical or SGS dissipation. In this framework, a systematic analysis of the sensitivity of LES results to the upstream corner sharpness could be the object of future investigations and, to this aim, a stochastic approach could be suitable.

References

1. Bruno, L., Salvetti, M.V., Ricciardelli, F.: Benchmark on the aerodynamics of a rectangular 5:1 cylinder: and overview after the first four years of activity. J. Wind Eng. Ind. Aerodyn. **126**, 87–106 (2014)
2. Bruno, L., Coste, N., Fransos, D.: Simulated flow around a rectangular 5:1 cylinder: spanwise discretisation effects and emerging flow features. J. Wind Eng. Ind. Aerodyn. **104–106**, 203–215 (2012)
3. Mariotti, A., Salvetti, M.V., Shoebi-Omrani, P., Witteveen, J.A.S.: Stochastic analysis of the impact of freestream conditions on the aerodynamics of a rectangular 5:1 cylinder. Comput. Fluids **136**, 170–192 (2016)
4. Mariotti, A., Siconolfi, L., Salvetti, M.V.: Stochastic sensitivity analysis of large-eddy simulation predictions of the flow around a 5:1 rectangular cylinder. Eur. J. Mech. B Fluid **62**, 149–165 (2017)
5. Mannini, C., Marra, A.M., Pigolotti, L., Bartoli, G.: The effects of free-stream turbulence and angle of attack on the aerodynamics of a cylinder with rectangular 5:1 cross section. J. Wind Eng. Ind. Aerodyn. **161**, 42–58 (2017)
6. Mannini, C., Mariotti, A., Siconolfi, L., Salvetti, M.V.: Benchmark on the aerodynamics of a 5:1 rectangular cylinder: further experimental and LES results. In: Salvetti, M., Armenio, V., Fröhlich, J., Geurts, B., Kuerten H. (eds.) Direct and Large-Eddy Simulation XI. ERCOFTAC Series, vol. 25, pp. 427–432. Springer, Cham (2019)

Chapter 12
Direct Numerical Simulations of a Turbulent Flow over Wall-Mounted Obstacles—A Comparison of Different Numerical Approaches

K. Schäfer, P. Forooghi, S. Straub, B. Frohnapfel and A. Stroh

Introduction and Methodology

An accurate representation of complex geometries in numerical simulations is of great interest. For the numerical implementation, there are two possibilities to represent the structure of the immersed geometry. Firstly, the simulation domain can be represented by body-conforming grids following the shape of the body. On the other hand, the simulation domain can be described by regular grids in which the body is introduced by means of the immersed boundary method (IBM). While the former method has the advantage to accurately impose the no-slip condition at the body's surface, the imposition of the boundary condition of the IBM strongly depends on the forcing method. In contrast, the computational cost and grid generation is less demanding for the immersed boundary methods, while this is a major drawback of body-conforming grid methods [9].

In the current study, we employ the compact finite-difference code INCOMPACT3D [8] and the pseudo-spectral code SIMSON [1] and compare their implemented immersed boundary methods for the turbulent flow over a transverse bar. As reference simulation, the same flow configuration is considered with a body-conforming grid in the spectral-element code NEK5000 [4]. The transverse bar is placed at the bottom wall of a turbulent open channel flow. The schematic of the numerical domain is depicted in Fig. 12.1. Periodic boundary conditions are applied in streamwise (x) and spanwise (z) directions, while the wall-normal (y) extension of the domain is bounded by no-slip boundary conditions at the lower domain wall ($y = 0$) and symmetry boundary conditions ($v = 0$, $\partial u/\partial y = \partial w/\partial y = 0$) at the upper boundary ($y = \delta$), where δ is the channel height. The domain size is $(L_x \times L_y \times L_z) = (8\delta \times \delta \times 4\delta)$ and the bar is located at position $x_b = 4\delta$. The bar

K. Schäfer (✉) · P. Forooghi · S. Straub · B. Frohnapfel · A. Stroh
Institute of Fluid Mechanics, Karlsruhe Institute of Technology (KIT),
Karlsruhe, Germany
e-mail: kay.schaefer@kit.edu

M. García-Villalba et al. (eds.), *Direct and Large Eddy Simulation XII*,
ERCOFTAC Series 27,
https://doi.org/10.1007/978-3-030-42822-8_12

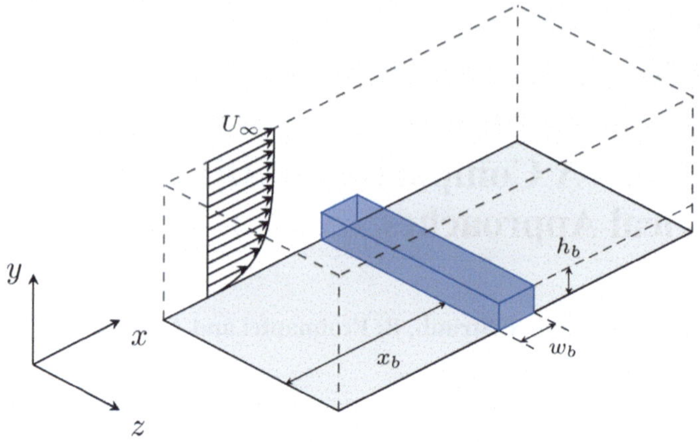

Fig. 12.1 Schematic representation of the numerical test scenario with transverse bar (not to scale)

Table 12.1 Simulation parameters for INCOMPACT3D and SIMSON

Case	$N_x \times N_y \times N_z$	N_{grid}	Δx^+	Δy^+_{min}	Δy^+_{max}	Δz^+
inc1	$256 \times 65 \times 128$	$2.13 \cdot 10^6$	5.63	1.57	5.03	5.63
inc2	$256 \times 129 \times 128$	$4.23 \cdot 10^6$	5.63	0.79	2.51	5.63
inc3	$512 \times 129 \times 128$	$8.54 \cdot 10^6$	2.81	0.79	2.51	5.63
sim1	$256 \times 129 \times 128$	$4.23 \cdot 10^6$	5.63	0.027	2.21	5.63
sim2	$512 \times 129 \times 128$	$8.54 \cdot 10^6$	2.81	0.027	2.21	5.63
sim3	$1024 \times 384 \times 128$	$50.33 \cdot 10^6$	1.41	0.003	0.78	5.63

width and height are $w_b = h_b = 0.1\delta$. The simulations are carried out under constant pressure gradient condition with $\mathrm{d}p/\mathrm{d}x$ corresponding to $\mathrm{Re}_\tau = u_\tau\delta/\nu = 180$ of a smooth turbulent channel flow. All subsequent positions are normalized by δ.

Two different types of IBM in two different numerical codes are considered. In the incompressible pseudo-spectral solver SIMSON [1], the representation of the spanwise bar is realized by an external volume force field [5] according to the continuous forcing approach proposed by Goldstein et al. [7]. In the open-source code INCOMPACT3D, which solves the Navier–Stokes equation with a sixth-order compact finite-difference scheme [8], the transverse bar is represented by an IBM approach based on the direct forcing method proposed by Fadlun et al. [2]. The simulation parameters for the different SIMSON and INCOMPACT3D configurations are given in Table 12.1. The grid spacing is given in wall units indicated by the superscript $^+$. Due to the results of the first simulations, higher grid resolutions were tested for the SIMSON simulation. As a result of the different stretching techniques of both IBM codes, the distance of the first grid point away from the wall is smaller for SIMSON than INCOMPACT3D.

Table 12.2 Simulation parameters for NEK5000

Case	N_{el}	N_{grid}	Δx^+	Δy^+	Δz^+
nek	13,056	$22.56 \cdot 10^6$	0.23–4.90	0.16–0.81	0.82–4.09

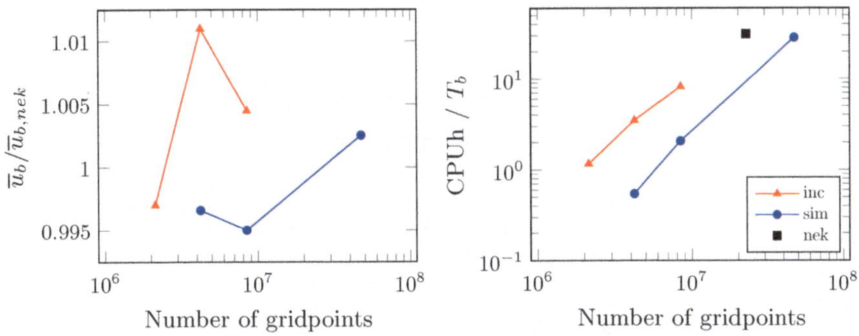

Fig. 12.2 Left image: mean bulk velocity $\overline{u}_b$ normalized by mean bulk velocity of NEK5000 simulation over number of grid points. Right image: computational cost over number of grid points

The reference simulation is conducted with the open-source spectral-element code NEK5000 [4] which combines the flexibility of finite-element methods with the accuracy of spectral methods by means of high-order polynomial basis functions. The velocity and pressure field is represented by a N-th order polynomial, which is the so-called $\mathscr{P}_N$–$\mathscr{P}_N$ formulation. For the reference simulation, the polynomial order of the simulation is $N = 11$ corresponding to 12 number of points per element and direction. The simulation parameters for the NEK5000 simulation are given in Table 12.2. The grid for the NEK5000 simulation is refined in the vicinity of the bar in x-direction as well as in the y-direction at the bottom wall. The maximum number of elements in each direction is $N_{\mathrm{el},x,\max} = 50$, $N_{\mathrm{el},y,\max} = 11$ and $N_{\mathrm{el},z} = 24$ resulting in a total number of elements $N_{\mathrm{el}} = 13056$.

Results

The mean bulk velocity $\overline{u}_b$ over the number of grid points normalized by the corresponding reference simulation value $\overline{u}_{b,\mathrm{nek}}$ is shown in the left image of Fig. 12.2. As can be seen, all simulation configurations of SIMSON and INCOMPACT3D match the mean bulk velocity within a relative error of around 1%. In the right image of Fig. 12.2, the computational costs of the simulations, measured in CPU hours per time bulk unit T_b, defined as $T_b = \delta/\overline{u}_b$, are presented over the number of grid points. The simulations of SIMSON with the implemented IBM are faster than the IBM simulations of INCOMPACT3D with respect to the used number of grid points. For example, the configuration sim2 and inc3, having the same number of grid points, show that

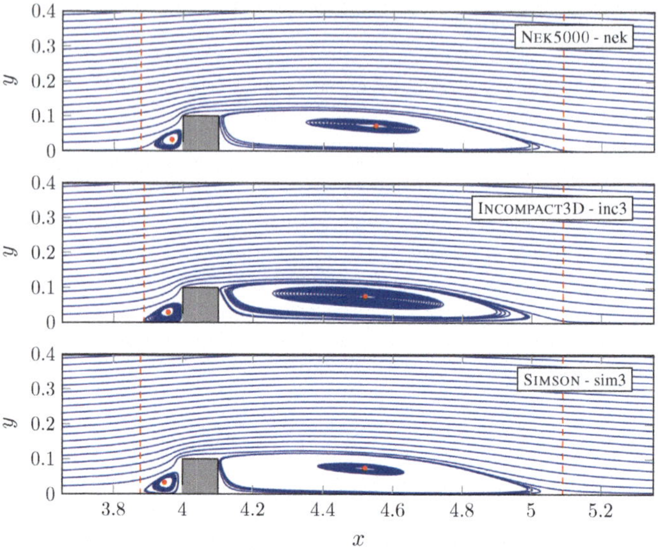

Fig. 12.3 Streamlines based on the time- and spanwise-averaged velocities in the x-y-plane. Red dashed lines mark the streamwise location of flow detachment and reattachment. Red circles show the location of vortex centers

SIMSON is roughly four times faster than INCOMPACT3D. For comparison, the computational costs of the reference simulation with NEK5000 is shown and interestingly the costs per number of degree of freedom is in a similar range as INCOMPACT3D for the present Reynolds number.

In Fig. 12.3, the recirculation zones around the bar are presented by means of the streamlines of the time- and spanwise-averaged velocities in the x-y-plane. Table 12.3 presents the position of the detachment and reattachment point of the recirculation zones as well as the center position of each recirculation zone for all simulations. The finest simulations of SIMSON, sim3, and of INCOMPACT3D, inc3, predict the correct reattachment and detachment point and give a good approximation of the recirculation's center position. The simulation sim2 also predicts the correct recirculation positions, except that the reattachment point lies further downstream.

The comparison of the mean streamwise velocity profiles between the IBM simulations is presented at the streamwise position $x = 4.5$, where the largest streamwise velocity differences occur, in Fig. 12.4. There is a good agreement between the finest IBM simulations of SIMSON, sim3, and INCOMPACT3D, inc3, with the NEK5000 simulation, whereas this is no longer valid for the coarser IBM simulations in the near-wall region.

The profiles of the mean wall-normal velocity directly behind the bar at position $x = 4.15$ are displayed in Fig. 12.5. For the INCOMPACT3D simulations only the finest configuration inc3 is capable of representing the reference simulation. The two finest simulations of SIMSON match the profile in the outer region, while there

Table 12.3 Location of characteristic points for considered simulations

Case	$x_{det.}$	$x_{reatt.}$	Upstream vortex center	Downstream vortex center
nek	3.88	5.09	(3.97, 0.036)	(4.55, 0.075)
inc1	3.88	5.09	(3.93, 0.022)	(4.53, 0.071)
inc2	3.87	5.09	(3.97, 0.028)	(4.53, 0.070)
inc3	3.89	5.09	(3.96, 0.032)	(4.52, 0.077)
sim1	3.84	5.16	(3.93, 0.030)	(4.59, 0.074)
sim2	3.88	5.16	(3.95, 0.031)	(4.56, 0.078)
sim3	3.86	5.09	(3.95, 0.036)	(4.52, 0.076)

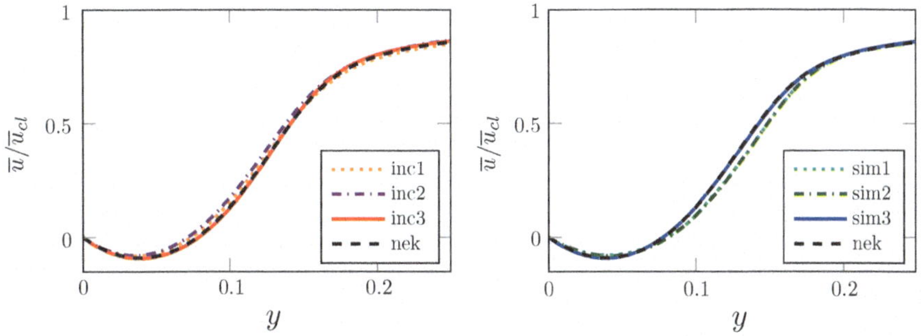

Fig. 12.4 Wall-normal profile of mean streamwise velocity $\overline{u}$ normalized by mean streamwise centerline velocity $\overline{u}_{cl}$ at position $x = 4.5$. Left image: simulations with INCOMPACT3D. Right image: simulations with SIMSON. Black dashed line indicates the NEK5000 reference simulation

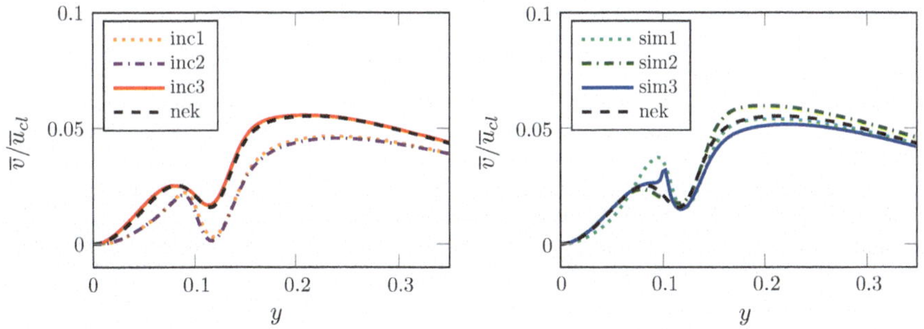

Fig. 12.5 Wall-normal profile of mean wall-normal velocity $\overline{v}$ normalized by mean streamwise centerline velocity $\overline{u}_{cl}$ at position $x = 4.15$. Left and right image as in Fig. 12.4

are discrepancies close to the wall, in particular at the bar height $y = 0.1$ where a discontinuity in the profile is observable for simulation sim3. This is the known Gibbs phenomenon [3], which is introduced by the discontinuous body forcing in spectral methods. The previous results indicate, however, that this has no influence on the global flow statistics.

Conclusion and Outlook

The current comparison for the simulation of a turbulent open channel flow with a transverse bar shows that the results of the finest SIMSON and INCOMPACT3D simulations are in good agreement with the NEK5000 reference simulation. The global flow properties, such as the recirculation region and the mean streamwise velocity profiles, are well reproduced by both IBM simulations. The computational performance of the IBM simulation of SIMSON outperforms the IBM simulation of INCOMPACT3D for the same number of grid points. However, the SIMSON simulation requires a larger amount of grid points compared to the INCOMPACT3D simulation to achieve similar flow results. Summarizing, both IBM simulation codes operate within a similar computational performance range to represent the transverse bar in a turbulent open channel flow. The computational performance of INCOMPACT3D could be further improved by the use of the recently developed and more accurate IBM for INCOMPACT3D [6] based on polynomial reconstruction of the obstacle.

Acknowledgements This work was performed on the computational resources ForHLR Phase II and bwUniCluster funded by the Ministry of Science, Research and the Arts Baden-Württemberg, and the German Research Foundation (DFG) within the framework program bwHPC. Support through the DFG in the framework of the Collaborative Research Center/Transregio 150 (TP-B02) is gratefully acknowledged.

References

1. Chevalier, M., Schlatter, P., Lundbladh, A., Henningson, D.S.: SIMSON: a pseudo-spectral solver for incompressible boundary layer flows. Tech. Rep. TRITA-MEK 2007-07, KTH Stockholm, Sweden (2007)
2. Fadlun, E.A., Verzico, R., Orlandi, P., Mohd-Yusof, J.: Combined immersed-boundary finite-difference methods for three-dimensional complex flow simulations. J. Comput. Phys. **161**(1) (2000)
3. Fang, J., Diebold, M., Higgins, C., Parlange, M.B.: Towards oscillation-free implementation of the IBM with spectral-like methods. J. Comput. Phys. **230**(22) (2011)
4. Fischer, P.F., Lottes, J.W., Kerkemeier, S.G.: Nek5000: open source spectral element CFD solver (2008). http://nek5000.mcs.anl.gov
5. Forooghi, P., Stroh, A., Schlatter, P., Frohnapfel, B.: Direct numerical simulation of flow over dissimilar, randomly distributed roughness elements. Phys. Rev. Fluids **3**(4) (2018)
6. Gautier, R., Laizet, S., Lamballais, E.: A DNS study of jet control with microjets using an immersed boundary method. Int. J. Comput. Fluid Dyn. **28**(6–10), 393–410 (2014)
7. Goldstein, D., Handler, R., Sirovich, L.: Modeling a no-slip flow boundary with an external force field. J. Comput. Phys. **105**(2) (1993)
8. Laizet, S., Lamballais, E.: High-order compact schemes for incompressible flows: a simple and efficient method with quasi-spectral accuracy. J. Comput. Phys. **228**(16) (2009)
9. Mittal, R., Iaccarino, G.: Immersed boundary methods. Annu. Rev Fluid Mech. **37**(1) (2005)

Chapter 13
Effects of Spanwise-Discontinuous Contoured Transverse Grooves on Flow Separation and Vortex Shedding

A. Mariotti, E. Pasqualetto, G. Buresti and M. V. Salvetti

Introduction

The delay of boundary layer separation over curved solid surfaces is of great importance in many engineering applications. The proposed idea is to introduce small and suitably-shaped grooves transverse to the flow (i.e., *contoured transverse grooves*) to passively generate local steady flow recirculations. They produce on the outer flow an effect analogous to a local relaxation of the no-slip boundary condition (see, e.g., [1–5]). Effective grooves should have a depth smaller than the thickness of the upstream boundary layer and have no rear corners in their shape. These characteristics are fundamental to obtain a steady passive recirculation and to avoid the presence of self-sustained cavity oscillations (see, e.g., [6]).

The effectiveness of the proposed technique in delaying separation has been verified both in internal and external flows. First, grooves were applied to the improvement of the pressure recovery in plane diffusers. The numerical results of those investigations show that significant increases in the diffuser efficiency are obtained both in laminar and in turbulent flow conditions (see [1–3]). Then, in [4], the same technique was applied to different axisymmetric boat-tailed bluff bodies, with flow separations occurring over the boat-tail lateral surface. The boat tail is a strategy

A. Mariotti (✉) · E. Pasqualetto · G. Buresti · M. V. Salvetti
Dipartimento di Ingegneria Civile e Industriale, Università di Pisa,
Via G. Caruso 8, 56122 Pisa, Italy
e-mail: alessandro.mariotti@for.unipi.it

E. Pasqualetto
e-mail: elena.pasqualetto@phd.unipi.it

G. Buresti
e-mail: g.buresti@ing.unipi.it

M. V. Salvetti
e-mail: mv.salvetti@ing.unipi.it

M. García-Villalba et al. (eds.), *Direct and Large Eddy Simulation XII*,
ERCOFTAC Series 27,
https://doi.org/10.1007/978-3-030-42822-8_13

to reduce the drag of bluff bodies consisting in a gradual reduction in the cross-sectional upstream of the base, d. For a given main body diameter, D, the pressure recovery increases along the boat-tail by decreasing d/D, but boat-tail is effective as long as the flow remains almost attached up to the base [7]. Through a synergic use of experiments and LES simulations, it was shown that the introduction of transverse contoured grooves in the boat-tail lateral surface allowed flow separation to be delayed and thus aerodynamic drag to be reduced. Finally, in [5] grooves were applied to a two-dimensional boat-tailed bluff body with a cross-sectional profile similar to one of the axisymmetric bodies. Compared to the axisymmetric cases, the main difference is that the two-dimensional body is characterized by a strong pressure gradient along the boat tail lateral surface and by high pressure and velocity fluctuations, also along the boat-tail lateral surface, caused by the alternate vortex shedding in the wake. Also for this body, the introduction of the groove produced a reduction of the wake width and of the pressure drag. However, an undesirable effect at the groove is observed, i.e., an enhanced increase of the spanwise correlation of the vortex shedding (see [5]), which is in turn connected with a straighter flow separation line in the spanwise direction induced by the presence of the groove. In the present paper, a groove geometry discontinuous in the spanwise direction is analyzed through Variational MultiScale (VMS) LES numerical simulations. The aim is to assess whether such an arrangement is able to reduce the vortex shedding correlation in spanwise direction and at the same time to keep good performance in delaying separation and in reducing drag.

Geometry Definition, Simulation Set-up, and Numerical Method

The two-dimensional boat-tailed body to which spanwise-discontinuous grooves are applied is the same as in [5]. It has a 3:1 elliptical forebody and a rectangular main part followed by a circular-arc boat-tail (see Fig. 13.1a). The ratio between the main body cross-flow dimension, D, and its overall streamwise length, L, is $D/L = 0.175$. The boat-tail length is $D/2$ and the ratio between its final dimension, d, and that of the main body is $d/D = 0.791$. The spanwise dimension of the body, l, is chosen to be equal to L. The freestream Mach number is equal to 0.1, whereas the Reynolds number is $\mathrm{Re} = D \cdot u_\infty/\nu = 9.6 \times 10^4$. At the considered Reynolds number and for laminar freestream conditions, when no groove is present, the boat-tail geometry is characterized by a significant flow separation along its lateral surface.

One suitably contoured transverse groove is introduced in the boat-tail lateral wall. The groove is extruded in spanwise direction for two equally long portions of the total spanwise length of the boat-tail, each one of $B/4$ (see Fig. 13.1a). The groove geometry is sketched in Fig. 13.1b. It starts with a sharp edge, has an upstream part with a semi-elliptical shape, and ends with a spline tangential to the boat-tail lateral surface. The position and depth of the grooves are the optimal ones found in [5], i.e., $s/D = 0.065$, $h/D = 0.024$, and $t/D = 0.129$.

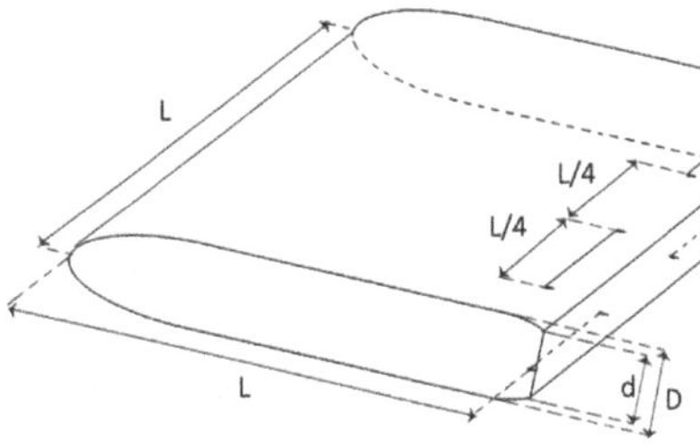

(a) Position of the spanwise-interrupted groove

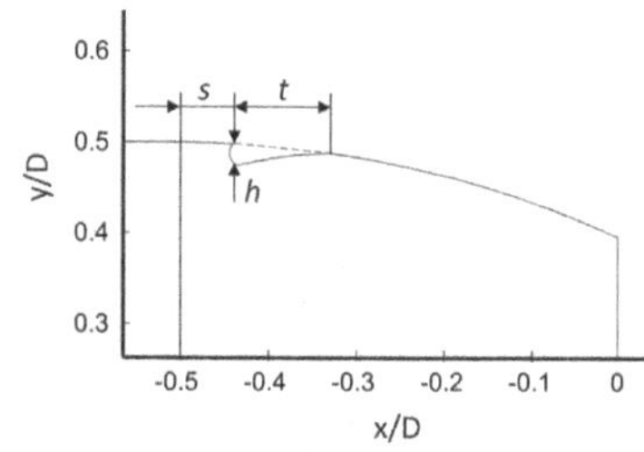

(b) Detail of the groove geometry

Fig. 13.1 Sketch of the boat-tailed body with spanwise-interrupted grooves

VMS-LES have been carried out through a proprietary compressible flow solver (AERO) based on a mixed finite-volume/finite-element method applicable to unstructured grids for space discretization and on a linearized implicit time advancing. The code has been previously used for the simulation of bluff body flows (see, e.g., [4, 8, 9]) and it is the one also employed in [5]. The accuracy of the numerical method is second order both in space and time. The Smagorinsky model is used as subgrid scale model in order to close the VMS-LES equations. The computational domain is rectangular, with a cross section of $5.71D \times 150D$ (in the z and y directions, respectively) and a length of $50D$. Periodic boundary conditions are imposed in the spanwise direction; characteristic-based boundary conditions (see, e.g., [8]) are used at the inflow, outflow, and upper and lower surfaces of the computational domain, while no-slip is imposed at the body surface. The computational domain is discretized with an unstructured grid with approximately 2×10^6 nodes. The grid resolution around the body and in the near wake is the same as in [5].

Results and Discussion

The introduction of the spanwise-interrupted groove leads to a reduction of the mean drag of about 18.8%, substantially due to a decrease of the pressure suctions acting on the boat-tail base (see Table 13.1). The drag reduction obtained for the spanwise-

Table 13.1 Boat-tail total drag coefficient, $C_{D,\text{tot}}^{\text{bt}}$, pressure and viscous contributions to the total drag coefficient, $C_{D,p}^{\text{bt}}$ and $C_{D,v}^{\text{bt}}$. Pressure contributions to the drag coefficient: boat-tail lateral surface, $C_{D,p}^{\text{ls}}$, and base, $C_{D,p}^{\text{base}}$. Average base pressure coefficient, $\overline{C_p}_{\text{base}}$

	$C_{D,\text{tot}}^{\text{bt}}$	$C_{D,p}^{\text{bt}}$	$C_{D,v}^{\text{bt}}$	$C_{D,p}^{\text{ls}}$	$C_{D,p}^{\text{base}}$	$\overline{C_p}_{\text{base}}$
Boat-tail without groove	0.4191	0.4183	0.0008	0.0853	0.3330	−0.421
Boat-tail with extruded groove [5]	0.3783	0.3778	0.0005	0.0835	0.2942	−0.372
Boat-tail with spanwise-discontinuous groove	0.3404	0.3399	0.0005	0.0942	0.2457	−0.311

discontinuous groove is higher of the one of the boat-tail with the extruded groove used in [5]. Compared to the latter case, the pressure drag is further reduced on the base while it slightly increased on the boat-tail lateral surface, as shown by the single contributions in Table 13.1.

Since the groove geometry is discontinuous in the spanwise direction, three streamwise sections are considered for the analysis of the velocity and pressure fields, i.e., $z/D = 0$, $z/D = 0.714$ and $z/D = 1.428$, corresponding to the center of the boat-tail portion without the groove, to the edge of the groove, and to its central section, respectively. The three sections are sketched in color in Fig. 13.2a, whereas the mean streamwise velocity component for the three sections is shown in Fig. 13.2b, together with the mean flow streamlines. The separation point, which in [5] was delayed from $x_{sep}/D = -0.325$ in the boat-tailed body without the groove to $x_{sep}/D = -0.269$ in the boat-tail with the extruded groove, is now slightly further moved downstream in the range $x_{sep}/D = -0.20$ to -0.15. It should be noted that the separation point is delayed not only in the portions downstream of the groove, but globally in the whole boat-tail. The delay of flow separation has a significant effect on the shape and curvature of the mean streamlines near the body and this, in turn, affects the pressure distribution. This can be observed in Fig. 13.3, where

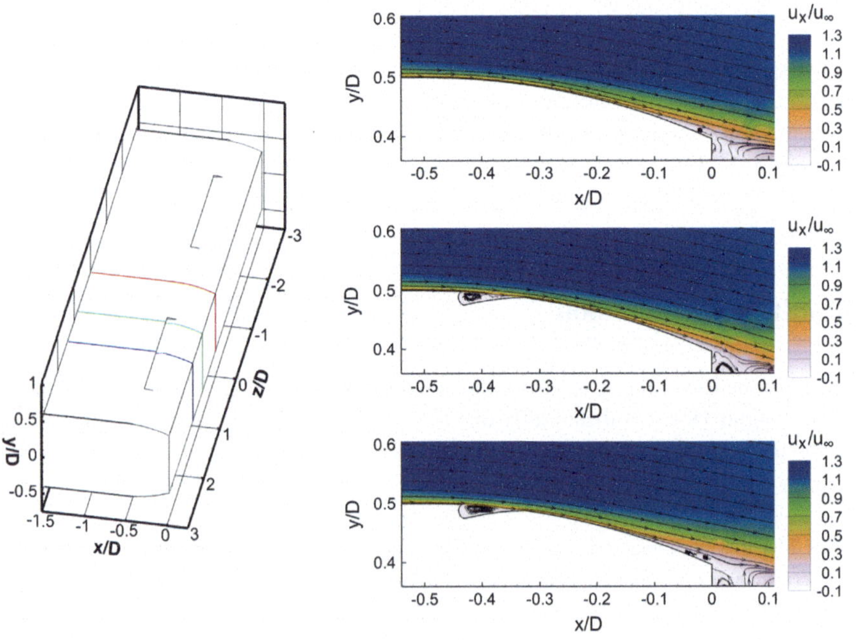

(a) Considered sections (b) Mean $x-$velocity field and flow streamlines

Fig. 13.2 Sketch of the considered streamwise sections and relative mean velocity field and mean flow streamlines: $z/D = 0$, $z/D = 0.714$ and $z/D = 1.428$ (from top to bottom)

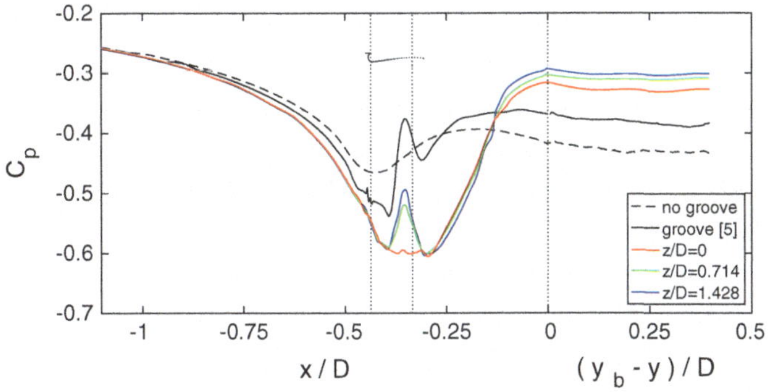

Fig. 13.3 Mean pressure coefficient over the lateral surface and the base of the boat-tail at the streamwise sections $z/D = 0$, $z/D = 0.714$ and $z/D = 1.428$

the distributions for the three considered streamwise sections of the mean pressure coefficient, defined as $C_p = (p - p_\infty)/(1/2\rho U_\infty^2)$, are plotted. In particular, the left part of Fig. 13.3 shows the C_p distributions over the lateral surface (the boat-tail starts at $x/D = -0.5$), while the right part is relative to the base surface starting from its border. A sketch indicating the position of the groove is also present. Higher suctions are present along the boat-tail lateral surface for the discontinuous groove, due to a larger curvature of the streamlines, and this leads to the previously observed increased contribution of the boat-tail lateral surface to the pressure drag. On the other hand, due to the further delay of flow separation, for all the considered spanwise sections, the discontinuous groove is characterized by a stronger pressure recovery in the last part of the boat-tail and this leads to larger pressures on the boat-tail base than the extruded one.

Finally, the use of a spanwise-discontinuous groove has a significant effect also on the vortex shedding. This can be seen, for instance, by comparing the instantaneous three-dimensional iso-surfaces of the vortex indicator λ_2 for the extruded groove from [5] (Fig. 13.4a, c) with those for the discontinuous one (Fig. 13.4b, d). In both cases, the wake is characterized by the periodic alternate shedding of vortices, but for the discontinuous groove the vortex shedding is less correlated in the spanwise direction producing more distorted vortical structures also in the very near wake. The reduction of the vortex shedding correlation in spanwise direction is generally desirable in practical applications.

Summarizing, the present study indicates the introduction of a spanwise-discontinuous groove has similar efficiency in delaying separation and in reducing drag to that of a continuous groove, while largely reducing the spanwise correlation of the vortex shedding. A more systematic analysis of the effects of different geometrical arrangements for the spanwise-discontinuous groove could be the object of future work.

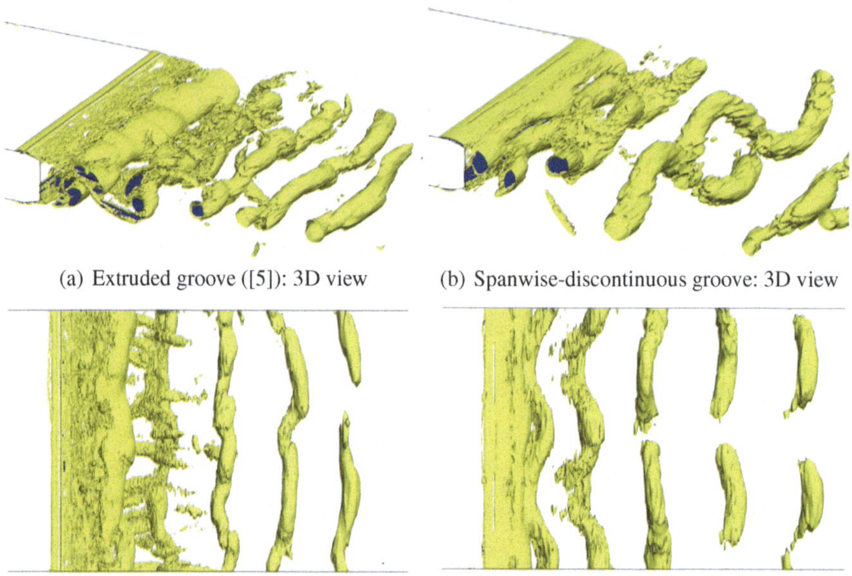

(a) Extruded groove ([5]): 3D view

(b) Spanwise-discontinuous groove: 3D view

(c) Extruded groove ([5]): top view

(d) Spanwise-discontinuous groove: top view

Fig. 13.4 Iso-surfaces of the vortex indicator λ_2

References

1. Mariotti, A., Grozescu, A.N., Buresti, G., Salvetti, M.V.: Separation control and efficiency improvement in a 2D diffuser by means of contoured cavities. Eur. J. Mech. B Fluid **41**, 138–149 (2013)
2. Mariotti, A., Buresti, G., Salvetti, M.V.: Control of the turbulent flow in a plane diffuser through optimized contoured cavities. Eur. J. Mech. B Fluid **48**, 254–265 (2014)
3. Mariotti, A., Buresti, G., Salvetti, M.V.: Use of multiple local recirculations to increase the efficiency in diffusers. Eur. J. Mech. B Fluid **50**, 27–37 (2015)
4. Mariotti, A., Buresti, G., Gaggini, G., Salvetti, M.V.: Separation control and drag reduction for boat-tailed axisymmetric bodies through contoured transverse grooves. J. Fluid Mech. **832**, 514–549 (2017)
5. Mariotti, A., Buresti, G., Salvetti, M.V.: Separation delay through contoured transverse grooves on a 2D boat-tailed bluff body: effects on drag reduction and wake flow features. Eur. J. Mech. B Fluid **74**, 351–362 (2019)
6. Rockwell, D., Naudascher, E.: Review—self-sustaining oscillations of flow past cavities. J. Fluid Eng. Trans. ASME **100**, 152–165 (1978)
7. Mair, W.A.: Reduction of base drag by boat-tailed afterbodies in low-speed flow. Aeronaut. Quaterly **XX**, 307–320 (1969)
8. Camarri, S., Salvetti, M.V., Koobus, B., Dervieux, A.: A low-diffusion MUSCL scheme for LES on unstructured grids. Comput. Fluids **33**, 1101–1129 (2004)
9. Mariotti, A., Buresti, G., Salvetti, M.V.: Connection between base drag, separating boundary layer characteristics and wake mean recirculation length of an axisymmetric blunt-based body. J. Fluid Struct. **55**, 191–203 (2015)

Chapter 14
DNS of Unequal Size Droplets Collision Using a Moving-Mesh/Level-Set Method

A. Amani, N. Balcázar, E. Gutiérrez and A. Oliva

Introduction

The dynamics of binary droplets collision is of huge importance in different fields, from multiphase reactors [11], raindrop formation, ink-jet printing, spray combustion, emulsion stability, turbine blade cooling, spray coating [5], and to drug delivery. Due to the complexity of the nature of droplets collision, this topic is one of the most challenging areas in the field of fluid dynamics. The outcome of the droplets collision can profoundly affect the overall performance of many systems. Experimental studies are providing us with different correlations to understand the droplets collision. The main parameters are the surface tension coefficient σ, droplet viscosity μ_d, droplet density ρ_d, droplets relative velocity U_{rel}, and the impact parameter I. The following non-dimensional parameters are thus defined by most researchers to characterize the droplets collision:

$$\text{We} = \frac{\rho_d {U_{\text{rel}}}^2 D_s}{\sigma}, \quad \text{Re} = \frac{\rho_d U_{\text{rel}} D_s}{\mu_d}, \quad \text{Oh} = \frac{\mu_d}{\sqrt{\rho_d \sigma D_s}}, \quad \Delta = \frac{D_s}{D_b}, \quad I = \frac{b}{D_s} \tag{14.1}$$

where We is the Weber number presenting the ratio between inertial forces and surface tension, Re is the Reynolds number representing the ratio of the inertial and viscous forces, and Oh is the Ohnesorge number representing the ratio of viscous

A. Amani · E. Gutiérrez · A. Oliva (✉)
Heat and Mass Transfer Technological Center (CTTC),
Polytechnic University of Catalonia (UPC),Terrassa, Barcelona, Spain
e-mail: oliva@cttc.upc.edu

A. Amani
e-mail: ahmad@cttc.upc.edu

N. Balcázar
TermoFluids S.L.,Avda Jacquard 97 1-E,08222 Terrassa, Barcelona, Spain
e-mail: nestor@cttc.upc.edu

M. García-Villalba et al. (eds.), *Direct and Large Eddy Simulation XII*,
ERCOFTAC Series 27,
https://doi.org/10.1007/978-3-030-42822-8_14

forces and the combined effect of inertial forces and surface tension. In this formulation D_s and D_b are the initial diameter of smaller and bigger droplets, respectively, where subscript **d** is the abbreviation of the word *droplet*. The impact parameter I characterizes the eccentricity of the collision with b as the inter-center distance of the droplets in direction normal to the collision. Equal-size droplets collision has been the topic of numerous investigations, including experimental, analytical, and numerical studies [3]. Studies in the field of collision of unequal size droplets started with the work of [8], where they studied the interactions of water droplets. Ashgriz and Poo [5] performed an extensive experimental investigation of the binary collision dynamics of unequal size water droplets. They found the boundaries between both of the separating collisions and coalescence collision for head-on collision of unequal size droplets with different size ratios. Anilkumar et al. [4] studied the momentumless coalescence of drops of the same liquid and provided images of internal mixing of unequal size droplets. Doumas et al. [10] studied the collision of miscible unequal size droplets with different materials. Despite all these studies, scientific community lacks a clear understanding on the head-on and off-center collision of unequal size droplets.

In addition to the challenges of numerical simulation of the collision of equal-sized droplets [3], the considerable displacement of the whole system of droplets in the collision of unequal-sized droplets adds to the computational cost of the simulations. In the present DNS study, collision of unequal-sized droplets is studied using a moving-mesh/level-set method. The problem of displacement of the outcome droplet is solved by using a moving-mesh technique.

Mathematical Formulation and Numerical Methods

Navier–Stokes equations are used to describe the conservation of mass and momentum of two incompressible immiscible Newtonian fluids on a spacial domain Ω with boundary $\partial\Omega$ as following [3, 12]:

$$\frac{\partial}{\partial t}(\rho \mathbf{v}) + \nabla \cdot (\rho \mathbf{v}(\mathbf{v} - \mathbf{v_d})) = -\nabla p + \nabla \cdot \mu \left(\nabla \mathbf{v} + (\nabla \mathbf{v})^T\right) + \rho \mathbf{g} + \sigma \kappa \mathbf{n} \delta_\Gamma \tag{14.2}$$

$$\nabla \cdot \mathbf{v} = 0 \tag{14.3}$$

where ρ and μ are density and dynamic viscosity of the fluids, $\mathbf{v}$ is the velocity field, $\mathbf{v_d}$ the velocity of the domain, p pressure field, $\mathbf{g}$ gravitational acceleration, and δ_Γ is the Dirac delta function concentrated at the interface (Γ). In this formulation, $\mathbf{n}$ is the unit normal vector outward to interface, κ is the interface curvature and σ is the interface tension coefficient. Taking into account that mass, density, and viscosity are constant within each fluid, they can be defined as scalar-fields inside the whole domain as follows:

$$\rho = \rho_1 H + \rho_2(1 - H) \quad \mu = \mu_1 H + \mu_2(1 - H) \tag{14.4}$$

where H is the Heaviside step function taking the value one in dispersed phase and zero elsewhere. In this research, conservative level-set (CLS) method as introduced by Balcázar et al. [6] in the context of a finite-volume method and unstructured meshes is used. This method employs a regularized indicator function ϕ in which is being advected by velocity vector field, $\mathbf{v}$, obtained from solution of Navier–Stokes equations transformed to the conservative form [2, 6] as

$$\frac{\partial \phi}{\partial t} + \nabla \cdot \phi(\mathbf{v} - \mathbf{v_d}) = 0 \tag{14.5}$$

An additional re-initialization equation is used to keep the profile and thickness of the interface constant:

$$\frac{\partial \phi}{\partial \tau} + \nabla \cdot \phi\,(1 - \phi)\,\mathbf{n}_{\tau=0} = \nabla \cdot \varepsilon \nabla \phi \tag{14.6}$$

This equation which is advanced in pseudo-time τ consists of a compressive flux $(\phi(1-\phi)\mathbf{n}_{\tau=0})$ which keeps the level-set function compressed onto the interface along the normal vector $\mathbf{n}$, and a diffusion term $(\nabla \cdot \varepsilon \nabla \phi)$ which keeps the profile in prescribed characteristic thickness of ε. Normal vector $\mathbf{n}$ and curvature κ of the interface are obtained using [6] $\mathbf{n} = \frac{\nabla \phi}{\|\nabla \phi\|}$ and $\kappa(\phi) = -\nabla \cdot \mathbf{n}$, respectively. Finite-volume (FV) approach is used to discretise the Navier–Stokes and level-set equations on a collocated grid [6]. At discretized level, physical properties are regularized in the context of the CLS method. Therefore, a linear average is used for density as $\rho = \rho_1 \phi + \rho_2(1-\phi)$, and viscosity as $\mu = \mu_1 \phi + \mu_2(1-\phi)$. The solution procedure is as follows:

1. Physical properties, interface geometric properties, and velocity field are initialized.
2. Velocity of the mesh is calculated.
3. Maximum allowable time step is calculated as used by [3].
4. The advection Eq. (14.5) and then re-initialization Eq. (14.6) are integrated in time with a third order accurate TVD Runge–Kutta scheme.
5. Physical properties in the domain (density and viscosity) and geometrical properties at the interface (curvature and interface normal) are updated from the level-set field.
6. The velocity and pressure fields are calculated using a classical fractional-step method first introduced by [9].

$$\frac{\rho \mathbf{v}^* - \rho^n \mathbf{v}^n}{\Delta t} = \frac{3}{2}(R_h^v)^n - \frac{1}{2}(R_h^v)^{n-1}, \quad \frac{\rho \mathbf{v}^{n+1} - \rho \mathbf{v}^*}{\Delta t} = -\nabla_h \mathbf{P}^{n+1} \tag{14.7}$$

where $R_h^v = -\mathbf{C}_h(\rho\mathbf{v}) + \mathbf{D}_h(\mathbf{v}) + \sigma\kappa\nabla_h(\phi)$ with $\mathbf{C}_h(\rho\mathbf{v}) = \nabla_h \cdot (\rho\mathbf{v}\mathbf{v})$ as the convective operator, $\mathbf{D}_h(\mathbf{v}) = \nabla_h \cdot \left(\mu(\nabla_h \mathbf{v} + \nabla_h^T \mathbf{v})\right)$ as the diffusive operator and ∇_h as the gradient operator. By applying the incompressibility constraint $(\nabla \cdot \mathbf{v} = 0)$, the corrector step in Eq. 14.7 changes to a Poisson equation as follows:

$$\nabla_h \cdot \left(\frac{1}{\rho}\nabla_h(\mathbf{P}^{n+1})\right) = \frac{1}{\Delta t}\nabla_h \cdot (\mathbf{v}^*) \tag{14.8}$$

The obtained linear system is solved using a preconditioned conjugated gradient method. At the end, the velocity $\mathbf{v}^{n+1}$ is corrected using

$$\mathbf{v}^{n+1} = \mathbf{v}^* - \frac{\Delta t}{\rho}\nabla_h(\mathbf{P}^{n+1}) \tag{14.9}$$

7. Move the mesh.
8. repeat steps 3–7 to reach the desired time.

Validations and verifications of the numerical methods in the context of conservative level-set method used in this work have been reported in [1–3, 6, 7, 12].

Numerical Results and Discussion

Extensive validations and verifications of the numerical methods in capturing the physics of collision of equal-sized droplets are presented in our previous publications [1, 3]. This work is developed to increase the general understanding of the application of the presented numerical tools in capturing the physics of the head-on and off-center collisions of unequal-sized droplets. Two initially departed droplets in a lighter environment are being collided with a relative velocity of U_{rel} in a domain with $(L_x, L_y, L_z) = (5D_a, 3.5D_a, 3.5D_a)$. The droplets are initially set in a way that the mass center of the droplets is in the center of the domain. Throughout the simulations, the confluence of these two points is preserved. Unless otherwise is mentioned, The density and viscosity ratios are $\rho_d/\rho_m = 666$ and $\mu_d/\mu_m = 120$, corresponding to Tetradecane as droplet and Air as the matrix. A grid size of $h = D_0/35$ is being used to discretize the domain. Figure 14.1 illustrates the topological changes of two unequal-sized water droplets colliding in air matrix compared with the experimental results of [5] where good agreement is seen.

The presented numerical tools can be used to study the internal mixing in the collision of droplets of the same or different materials. Instead of introducing extra equations for massless particles and tracking them in the domain, to study the mixing of the droplets or to follow droplets with different materials, one can define an extra scalar-field related to a specific droplet. This scalar-field then needs to be advected with the Eq. 14.5, where the internal mixing of droplets (with the same or different materials) could be solved. Figure 14.2 illustrates the simulation of internal mixing of unequal-sized droplets with a size ratio of $\Delta = 2.08$ compared with the experimental results of [4] where good agreement is seen.

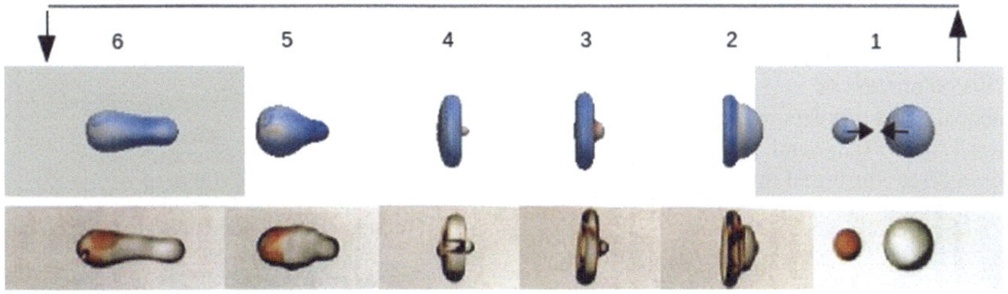

Fig. 14.1 Topological changes of head-on unequal-sized collision of water droplets in air matrix with the initial diameter ratio of $\Delta = 2.0$, Weber number of 56, and Reynolds number of 1740. Top: numerical simulation of current study, using the moving-mesh/level-set method with velocity magnitude contours. Bottom: the experimental results of [5]. Schematic presentation of the domain movement in the sequences of 1 and 6 are illustrated as well

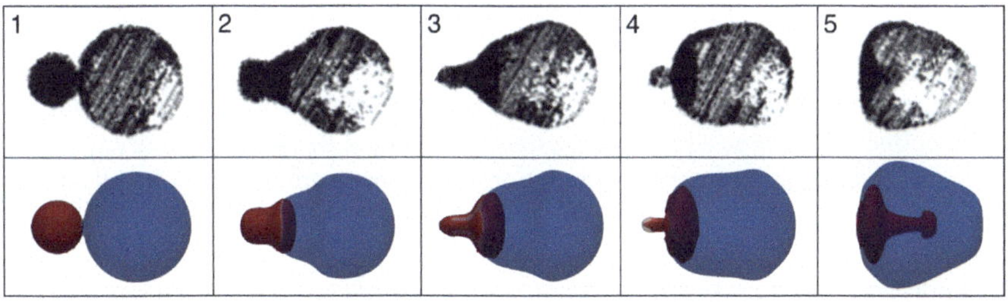

Fig. 14.2 Topological changes of head-on unequal-sized collision of droplets in an immiscible matrix with the initial size ratio of $\Delta = 2.08$, and Oh $= 0.012$. Top: the experimental results of [4], bottom: numerical simulation of current study, using the moving-mesh/level-set method with an extra scalar-field to track the internal mixing of the droplets

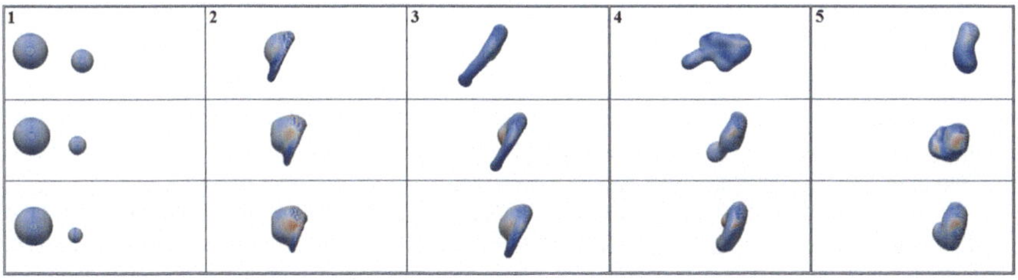

Fig. 14.3 Collision sequence results of two unequal off-center droplets with velocity magnitude contours for first row: $\Delta = 1.5$, $I = 0.33$ and We $= 70$, second row: $\Delta = 2.0$, $I = 0.33$ and We $= 70$, and third row: $\Delta = 2.5$, $I = 0.33$ and We $= 70$

The presented numerical tools can be of huge help in saving up the computational costs of numerical simulations. Figure 14.3 represent the collision sequence for two unequal-sized droplets with the size ratios of $\Delta = 1.5$, 2.0, and 2.5. The simulations of this kind can be used to efficiently extract maps on collision outcomes of different size ratios.

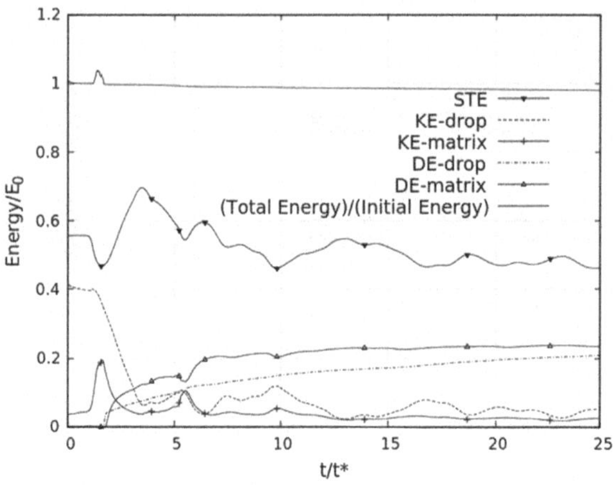

Fig. 14.4 Normalized energy budget of the collision process of unequal-sized droplets with $\Delta = 1.5$, $I = 0.33$, and We $= 70$ as illustrated in Fig. 14.3

Besides, the presented tools can be used to perform the analysis on energy budgets in the collision process, which can be used to evaluate the kinetic energy recovery factor, with application in spray simulations, etc. For example, Fig. 14.4 illustrates the energy budget of the collision process of a case with $\Delta = 1.5$, $I = 0.33$.

Acknowledgements We acknowledge *Ministerio de Economía y Competitividad, Secretaría de Estado de Investigación, Desarrollo e Innovación*, Spain (ENE2017-88697-R). A. Amani acknowledges the financial support of *(AGAUR) of Generalitat de Catalunya* research scholarship (2016 FI_B 01059). N. Balcázar acknowledges the Programa Torres Quevedo MINECO (PTQ-14-07186).

References

1. Amani, A., Balcázar, N., Naseri, A., Oliva, A.: A study on binary collision of GNF droplets using a conservative level-set method. In: 7th European Conference on Computational Fluid Dynamics (ECFD 7), June 2018, Glasgow (2018)
2. Amani, A., Balcázar, N., Castro, J., Oliva, A.: Numerical study of droplet deformation in shear flow using a conservative level-set. Chem. Eng. Sci. **207**, 153–177 (2019). https://doi.org/10.1016/j.ces.2019.06.014
3. Amani, A., Balcázar, N., Gutiérrez, E., Oliva, A.: Numerical study of binary droplets collision in the main collision regimes. Chem. Eng. J. **370**, 477–498 (2019). https://doi.org/10.1016/j.cej.2019.03.188
4. Anilkumar, A.V., Lee, C.P., Wang, T.G.: Surface-tension-induced mixing following coalescence of initially stationary drops. Phys. Fluids A **3**, 2587–2591 (1991)
5. Ashgriz, N., Poo, J.Y.: Coalescence and separation in binary collisions of liquid drops. J. Fluid Mech. **221**, 183–204 (1990)
6. Balcázar, N., Jofre, L., Lehmkuhl, O., Castro, J., Rigola, J.: A finite-volume/level-set method for simulating two-phase flows on unstructured grids. Int. J. Multiph. Flow **64**, 55–72 (2014)
7. Balcázar, N., Lehmkuhl, O., Rigola, J., Oliva, A.: A multiple marker level-set method for simulation of deformable fluid particles. Int. J. Multiph. Flow **74**, 125–142 (2015)
8. Brazier-Smith, P.R., Jennings, S.G., Latham, J.: The interaction of falling water drops: coalescence. Proc. R. Soc. A: Math. Phys. Eng. Sci. **1566**, 393–408 (1972)

9. Chorin, A.: Numerical solution of the Navier-Stokes equations. Math. Comput. **22**, 745 (1968)
10. Doumas, A., Van Den Broek, P., Affolter, M., Monod, M.: Characterization of the prolyldipeptidyl peptidase gene (dppIV) from the Koji mold As-pergillusoryzae. Appl. Environ. Microbiol. **64**, 4809–4815 (1998)
11. Dudukovic, M.P., Larachi, F., Mills, P.L.: Multi-phase reactors-revisited. Chem. Eng. Sci. **54**, 1975–1995 (1999)
12. Gutiérrez, E., Favre, F., Balcázar, N., Amani, A., Rigola, J.: Numerical approach to study bubbles and drops evolving through complex geometries by using a level set—moving mesh—immersed boundary method. Chem. Eng. J. **349**, 662–682 (2018)

Chapter 15
Direct Numerical Simulation of Turbulent Channel Flow with Condensation Using a Cluster-Based Droplet Deposition Model

P. Bahavar and C. Wagner

Introduction

Heat and mass transfer in flows with a condensable component are important characteristics in applications ranging from large-scale atmospheric flows to microchip cooling systems [4]. In the context of cabin ventilation, the particular combination of volume flux, oversaturation, and density differences between liquid condensate and vapor give rise to separated timescales that need to be considered for an efficient simulation of the interplay between phase transition and the turbulent flow.

To investigate the fundamental changes to the flow caused by condensation, a simplified approach to the modeling of the phase transition is combined with single-phase *direct numerical simulations* (DNS) of turbulent flow in a vertical channel. By adding specific aspects of the phase transition to the condensation model, their influence on the underlying turbulent flow can be studied free from the influence of additional turbulence modeling. In this work, buoyancy effects from differences in both temperature and vapor concentration are included, as well as the release of latent heat at the phase transition and the subsequent secondary buoyancy from the associated increase in fluid temperature. Additionally, the modification of the wall surface due to the deposition of condensate droplets influences the flow. Here, the fully resolved data on the spatial and temporal distribution of condensation rates is used to motivate the placement and growth of these droplets on longer timescales by applying a clustering algorithm to identify contiguous regions of high condensation mass flux.

P. Bahavar (✉)
German Aerospace Center (DLR), Cologne, Germany
e-mail: philipp.bahavar@dlr.de

C. Wagner
German Aerospace Center (DLR) & Technische Universität Ilmenau,
Cologne, Germany

M. García-Villalba et al. (eds.), *Direct and Large Eddy Simulation XII*,
ERCOFTAC Series 27,
https://doi.org/10.1007/978-3-030-42822-8_15

Governing Equations

The fluid is described as a homogeneous mixture of the vapor phase of a condensable trace species in a non-condensable carrier fluid [1]. The Boussinesq approximation is applied to the incompressible Navier–Stokes equations, resulting in

$$\nabla \cdot \mathbf{u} = 0 \tag{15.1}$$

$$\frac{\partial \mathbf{u}}{\partial t} + (\mathbf{u} \cdot \nabla)\mathbf{u} = -\frac{1}{\rho}\nabla p + \nu \nabla^2 \mathbf{u} - \mathbf{B}, \tag{15.2}$$

where $\mathbf{u}$ is the velocity field, ρ and ν are the density and kinematic viscosity of the fluid mixture, assumed as constant, and ∇p is the pressure gradient driving the flow. The force term $\mathbf{B} = \beta_T(T - T_{ref})\mathbf{g} + \beta_c(c - c_{ref})\mathbf{g}$ models the buoyant force caused by variations of the active scalar fields temperature T and vapor fraction c, with the gravitational acceleration $\mathbf{g}$ and expansion coefficients β_T and β_c, respectively [3]. The evolutions of these scalar fields are given by

$$\frac{\partial T}{\partial t} + \mathbf{u} \cdot \nabla T = \kappa \nabla^2 T + \sigma_T, \tag{15.3}$$

$$\frac{\partial c}{\partial t} + \mathbf{u} \cdot \nabla c = D \nabla^2 c + \sigma_c, \tag{15.4}$$

where κ is the thermal diffusivity of the mixture and D the mass diffusivity of the trace species with respect to the carrier. The source terms σ_T and σ_c account for the changes in the scalar fields caused by phase transition.

Removal of vapor by condensation results in the release of latent heat, causing an increased temperature of the fluid mixture. Therefore, the source terms are related by $\sigma_T = \sigma_c\, h/c_p$, with the specific enthalpy of the vapor h and the specific heat capacity of the mixture c_p. The governing equations are fully coupled with the addition of condensation. Temperature and concentration are convected by the velocity field and in turn exert influence via buoyancy. The temperature determines the saturation level and thereby the magnitude of the condensation mass flux, which again modifies the temperature via the release of latent heat. The explicit calculation of the mass transfer rate is addressed in section "Condensate Accumulation".

Simulation Setup

The flow is simulated in a vertical channel measuring $6\pi\delta \times 2\delta \times 2\pi\delta$ along the streamwise (x), wall-normal (y), and spanwise (z) directions. Periodic boundary conditions in the spanwise direction and inflow–outflow boundary conditions in the streamwise direction are applied to allow the flow to evolve along the channel. At the walls, no-slip and impermeability conditions are imposed. To provide fully developed turbulent flow at the channel inlet, a cyclic precursor domain with length $4\pi\delta$ is coupled to the primary domain [6]. Isothermal flow is continuously simulated in this precursor. A prescribed volume flux is sustained using a self-correcting global pressure gradient to drive the flow. At each timestep, the velocity field is mapped from a plane perpendicular to the mean flow direction of the precursor domain to the inlet of the primary domain.

Along with the velocity field obtained from the coupled simulation, the temperature T_{in} and vapor fraction c_{in} are fixed at the inlet. The vapor content at the inlet can be expressed as a dewpoint temperature $T_{d,in} = T_{sat}(c_{in})$. To facilitate condensation, one wall of the channel is set at a temperature $T_c < T_{d,in} < T_{in}$, such that a thermal boundary layer develops along the channel. The temperature at the opposing channel wall is set equal to the inlet temperature, eliminating temperature gradients. Condensation therefore occurs only inside the subcooled region of the thermal boundary layer at the cooled wall. Zero-gradient boundary conditions are employed for the vapor fraction field.

The finite volume DNS is performed in OpenFOAM, using second-order central differencing and an explicit, second-order accurate Leapfrog-Euler time integration scheme to solve the governing equations [5]. A projection method is used for the coupling of the velocity and pressure fields.

Both computational domains are discretized into hexahedral cells, with a uniform resolution of $\Delta x^+ = 5.3$ and $\Delta z^+ = 2.7$ in wall units along the streamwise and spanwise direction, and according to a hyperbolic tangent distribution along the wall-normal direction, resulting in $\Delta y^+ = 0.2$ directly at the channel walls and $\Delta y^+ = 3.3$ in the bulk region.

A flow with a Reynolds number $\mathrm{Re} = u_b\delta/\nu = 2000$ based on the bulk velocity u_b is considered in this investigation. Gravity is parallel to the mean flow direction and gives rise to buoyant forces, which are characterized via the Grashof number Gr. The two contributions for thermal and solutal buoyancy,

$$\mathrm{Gr}_T = \frac{g\delta^3}{\nu^2}\beta_T \Delta T = 38000, \qquad \mathrm{Gr}_c = \frac{g\delta^3}{\nu^2}\beta_c \Delta c = 1500, \tag{15.5}$$

act in the same direction, resulting in a combined $\mathrm{Gr} = 39500$. The reference scales for the temperature and vapor fraction differences are given by the maximum range of possible values, $\Delta T = T_{in} - T_c$ and $\Delta c = c_{in} - c_{sat}(T_c)$. The comparative strength of forced and natural convection is expressed by the Richardson number, $\mathrm{Ri} = \mathrm{Gr}/\mathrm{Re}^2 = 0.01$, signifying that the flow is dominated by forced convection, with only small corrections due to the buoyant forces. The accelerating effect of the buoyancy at the cooled wall causes a slight asymmetry of the Reynolds number based on the local friction velocity u_τ, with $\mathrm{Re}_\tau = u_\tau\delta/\nu = 135$ at the cooled wall. Diffusion of temperature and vapor concentration are set to represent a working fluid of humid air, with Prandtl number $\mathrm{Pr} = \kappa/\nu = 0.73$ and Schmidt number $\mathrm{Sc} = D/\nu = 0.65$. The relation between sensible heat differences and the amount of latent heat released during the phase transition is given by the Jakob number

$\mathrm{Ja} = c_p \Delta T / h = 0.012$, with the specific heat capacity of the gas mixture c_p and the latent heat of condensation h chosen to represent the transition from water vapor to liquid water.

Condensate Accumulation

The vapor source term σ_c can be derived from the Hertz–Knudsen–Schrage equation for the condensation mass flux [7],

$$J = \frac{2s}{2-s}\sqrt{\frac{M}{2\pi RT}}(p_v - p_{sat}), \tag{15.6}$$

with the molar mass of the vapor M, universal gas constant R, condensation accommodation coefficient s, and the partial pressure of the vapor $p_v = cp$. For water vapor at the oversaturation levels set by the flow parameters listed above, the timescale $\tau_c = \Delta c/\sigma_c$ obtained from this expression for the mass transfer at the phase boundary is smaller than the simulation timestep Δt by two orders of magnitude. Δt is governed by the stability criterion for the Leapfrog-Euler integration scheme [9] for a given spatial resolution, which is in turn determined by the requirement to resolve the turbulent length scales. It is therefore justified to treat condensation as instantaneous in the flow configuration considered in this work. The vapor source term representing condensation then simplifies to

$$\sigma_c = (c - c_{sat}(T))/\Delta t \quad \text{for } c > c_{sat}(T), \quad \text{else } 0, \tag{15.7}$$

meaning that any vapor exceeding the saturation level is removed instantly at every timestep.

Using this approach, flow through the cooled vertical channel including condensation can be simulated with negligible additional computational costs compared to flow of only non-condensable fluid. The interaction between turbulence and phase transition is analyzed after five convective time units $\Delta t^+ = \delta/u_\tau$, when a statistically steady state was reached.

Figure 15.1 shows the instantaneous spatial distribution of condensation directly at the surface of the cooled channel wall. Apart from the initial contact of the fluid with the cooled wall at the inlet, where high condensation rates are found along the whole span of the channel, strong inhomogeneity is observed. A condensation pattern exhibiting spanwise periodicity emerges. This structure is a consequence of the turbulent flow responsible for the convective transport of vapor. The organization of sweeps and ejections into streaks causes a footprint of high condensation rates where sweeps impinge upon the wall, carrying vapor into the subcooled region and causing condensation.

Growth of condensate droplets could be simulated by tracking the accumulating condensate mass determined by the DNS at every timestep. To have a non-negligible

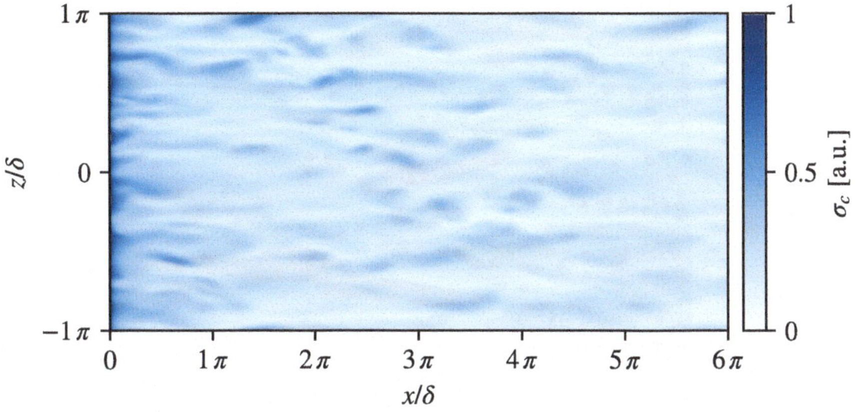

Fig. 15.1 Instantaneous distribution of the vapor source term σ_c at the cooled channel wall. Darker colors represent higher condensation rates

effect on the flow, the droplet size d needs to be at least comparable to the thickness of the viscous boundary layer [2]. For the given set of parameters, a lower bound for the timescale for the growth of the droplets is given by

$$\tau_d = \frac{d}{\Delta c u_b} \frac{\rho_l}{\rho_v}. \tag{15.8}$$

Since the ratio of condensate density ρ_l to vapor density ρ_v is much larger than unity and $\Delta c \ll 1$, τ_d is of the order of $10^3 \Delta t^+$. Directly simulating droplet growth would therefore be associated with prohibitively high computational costs.

For an alternative approach, *density-based spatial clustering for applications with noise* (DBSCAN) [8] is employed to identify contiguous regions of high mass transfer. The amount of condensate that would accumulate in each of these clusters over timescales comparable to τ_d is then extrapolated either from the instantaneous rates or from an average taken over an interval $t_{avg} \ll \tau_d$. Droplets of sufficient size are then distributed across the clusters (Fig. 15.2). By adjusting the relative size distribution as well as the number of droplets, condensate mass can be conserved per cluster, thereby leveraging the information on the spatial distribution of condensate obtained from the DNS. Information on the influence of the wall surface properties on droplet shapes and spatial frequency within clusters can be obtained from experimental measurements in similar flows. Thus, a plausible result for the modification of the wall surface due to condensate accumulation can be found while avoiding exceedingly long simulation runtimes. On this basis, DNS can then be performed to investigate the influence of the deformed wall on the turbulent flow.

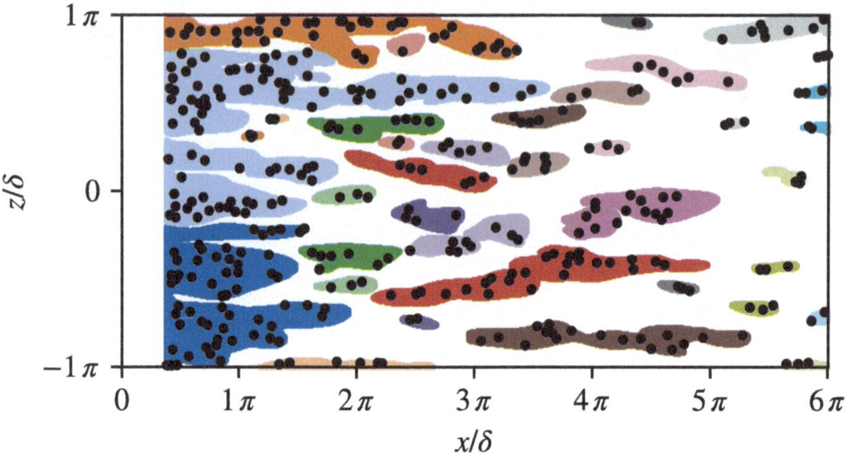

Fig. 15.2 Illustration of droplet sites distributed randomly across color-coded clusters identifying regions with high instantaneous condensation rates

Conclusions

DNS of turbulent flow through a differentially cooled vertical channel is performed for a mixture of a condensable vapor species and a non-condensable carrier fluid. Temperature and vapor fraction are considered as active scalars, coupled to the velocity fields via thermal, and solutal buoyancy. Interconnected source terms for the scalar fields model the loss of vapor during condensation as well as the temperature increase due the release of latent heat. Effects of the streak structure of the turbulent flow are observed in the condensation pattern directly at the wall, resulting from the convective transport of vapor in sweeps. Motivated by the strongly separated timescales of the problem, a clustering approach leverages this information to approximate a realistic distribution of condensate droplets.

Future work will investigate the mutual interplay between condensate accumulation and turbulent flow on the basis of these surface modifications.

References

1. Bahavar, P., Wagner, C.: Direct numerical simulation of convective turbulent channel flow of fluid mixtures. In: Salvetti, M.V. et al. (eds.) Direct and Large Eddy Simulations IX, vol. 25, pp. 285–290 (2019)
2. Chan-Braun, C., García-Villalba, M., Uhlmann, M.: Force and torque acting on particles in a transitionally rough open-channel flow. J. Fluid Mech. **684**, 441–474 (2011)
3. Gray, D.D., Giorgini, A.: The validity of the Boussinesq approximation for liquids and gases. Int. J. Heat Mass Transf. **19**, 545–551 (1976)

4. Huang, J., Zhang, J., Wang, L.: Review of vapor condensation heat and mass transfer in the presence of non-condensable gas. Appl. Therm. Eng. **89**, 469–484
5. Kath, C., Wagner, C.: Highly resolved simulations of turbulent mixed convection in a vertical plane channel. In: Dillman, A. et al. (eds.) New Results in Numerical and Experimental Fluid Mechanics X, **132**, 515–524 (2016)
6. Lund, T.S., Wu, X., Squires, K.D.: Generation of turbulent inflow data for spatially-developing boundary layer simulations. J. Comput. Phys. **140**, 233–258 (1998)
7. Marek, R., Straub, J.: Analysis of the evaporation coeffcient and the condensation coeffcient of water. Int. J. Heat Mass Transf. **44**, 39–53 (2001)
8. Pendragosa, F., et al.: Scikit-learn: machine learning in python. J. Mach. Learn. Res. **12**, 2825–2830 (2011)
9. Shishkina, O., Wagner, C.: Stability conditions for the Leapfrog-Euler scheme with central spatial discretization of any order. Appl. Numer. Anal. Comput. Math. **1**, 315–326 (2004)

Chapter 16
DNS of Drag-Force and Reactive Mass Transfer in Gravity-Driven Bubbly Flows

N. Balcázar, O. Antepara, J. Rigola and A. Oliva

Mass transfer processes in bubbly flows are relevant in both nature and industry. Multiple examples arise from the so-called unit operations of the chemical engineering, where bubble columns are used for separation processes, or as chemical and biochemical reactors. In this context, the main motivation of this research is to compute drag-force coefficients and mass transfer coefficients in gravity-driven bubbly flows, using direct numerical simulation of Navier–Stokes equations with a conservative level-set (CLS) [2, 15] methodology for interface capturing in bubble swarms [9].

Mathematical Formulation and Numerical Methods

The Navier–Stokes equations for incompressible two-phase flows are presented in the framework of the so-called one-fluid formulation which introduces a singular source term for the surface tension force at the interface (Γ) [2, 9]:

N. Balcázar (✉)
Termo Fluids S.L, Terrassa, Spain
e-mail: nestorbalcazar@yahoo.es; nestor@cttc.upc.edu

O. Antepara · J. Rigola · A. Oliva
Heat and Mass Transfer Technological Center (CTTC),
Universitat Politècnica de Catalunya - BarcelonaTech, Terrassa, Spain
e-mail: oscar@cttc.upc.edu

J. Rigola
e-mail: quim@cttc.upc.edu

A. Oliva
e-mail: oliva@cttc.upc.edu

M. García-Villalba et al. (eds.), *Direct and Large Eddy Simulation XII*,
ERCOFTAC Series 27,
https://doi.org/10.1007/978-3-030-42822-8_16

$$\frac{\partial}{\partial t}(\rho \mathbf{v}) + \nabla \cdot (\rho \mathbf{v}\mathbf{v}) = -\nabla p + \nabla \cdot \mu (\nabla \mathbf{v}) + \nabla \cdot \mu (\nabla \mathbf{v})^T + (\rho - \rho_0)\mathbf{g} + \mathbf{f}_\sigma, \tag{16.1}$$

$$\nabla \cdot \mathbf{v} = 0, \tag{16.2}$$

where $\mathbf{v}$ is the velocity, p is the pressure, $\mathbf{g}$ is the gravitational acceleration, μ is the dynamic viscosity, ρ is the fluid density, $\mathbf{f}_\sigma$ is the surface tension force per unit volume acting on Γ, subscripts d and c represents the continuous phase and dispersed phase, respectively. Physical properties are constant with a jump discontinuity at Γ: $\rho = \rho_d H_d + \rho_c H_c$, $\mu = \mu_d H_d + \mu_c H_c$. Here H_c denotes the Heaviside step function, which is one at the fluid c (in Ω_c) and zero elsewhere, whereas $H_d = 1 - H_c$. Since periodic boundary conditions are employed, the force $-\rho_0 \mathbf{g}$ with $\rho_0 = V_\Omega^{-1} \int_\Omega (\rho_d H_d + \rho_c H_c) \, dV$, avoids the acceleration of the flow field in the downward vertical direction [3, 6, 9].

A multiple marker approach [3, 14], as introduced by [9] in the context of the unstructured CLS method [2] for mass transfer in bubble swarms, is used in this research. Indeed, the ith bubble interface is the 0.5 iso-surface of a CLS function $\phi_i = (1/2)\,(tanh\,(d_i(\mathbf{x}, t)/(2\varepsilon)) + 1)$ [2], where d_i is a signed distance function, $i = \{1, \ldots, n_d\}$ and n_d is the total number of bubbles. Given the incompressible constraint Eq. (16.2), the ith interface transport equation can be written in conservative form. Moreover, a re-initialization equation is solved up to steady state, to keep a sharp and constant level-set profile [2]:

$$\frac{\partial \phi_i}{\partial t} + \nabla \cdot \phi_i \mathbf{v} = 0, \quad \frac{\partial \phi_i}{\partial \tau} + \nabla \cdot \phi_i (1 - \phi_i) \mathbf{n}_i = \nabla \cdot \varepsilon \nabla \phi_i, \quad i = \{1, \ldots, n_d\}. \tag{16.3}$$

where $\varepsilon = 0.5h^{0.9}$ and h is the grid size [2, 9, 10]. Normal vectors ($\mathbf{n}_i$) and curvatures (κ_i) are computed as follows: $\mathbf{n}_i(\phi_i) = \nabla\phi_i \|\nabla\phi_i\|^{-1}$ and $\kappa_i(\phi_i) = -\nabla \cdot \mathbf{n}_i$. The continuous surface force model [11] is employed for computation of surface tension forces, which has been extended to the multiple marker CLS method in [3, 9, 10] as follows: $\mathbf{f}_\sigma = \sum_{i=1}^{n_d} \sigma \kappa_i(\phi_i) \nabla \phi_i$, with σ defined as the surface tension coefficient. Finally, fluid properties (ρ and μ) are regularized by using a global level-set function, i.e., $H_d = \phi$, defined as [3, 9, 10]: $\phi = min\{\phi_1, \ldots, \phi_{n_d}\}$. Some advantages of the unstructured CLS method employed in this work include mass conservation of the fluid-phases, numerical stability at high Reynolds numbers and high physical properties ratios, accurate computation of surface tension forces, and efficient parallelization [2–4, 6–10].

This research focuses on bubbly flows with external mass transfer. As a consequence, a convection–diffusion–reaction equation is used for computing the distribution of chemical species concentration (C) in Ω_c:

$$\frac{\partial C}{\partial t} + \nabla \cdot (\mathbf{v}C) = \nabla \cdot (\mathscr{D} \nabla C) + \dot{r}(C), \tag{16.4}$$

where $\mathscr{D}$ is the diffusion coefficient, which is equal to $\mathscr{D}_c$ in Ω_c, $\dot{r} = -k_1 C$ denotes the overall chemical reaction rate, and k_1 is the first-order reaction rate constant.

Thermodynamic equilibrium is assumed, therefore $C_{\Gamma,c} = HC_{\Gamma,d}$, where H is the so-called Henry constant. The discontinuity introduced by Henry's law ($H \neq 1$), can be treated by rescaling the concentration and diffusion coefficient to obtain a continuous solution. Indeed, in the present research the value of C is constant inside the bubble [1, 16], whereas its value on the bubble interface ($C_{\Gamma,c}$) is imposed like a Dirichlet boundary condition. The reader is referred to [2, 9] for further technical details on the finite-volume discretization of transport equations and global algorithms implemented in the unstructured multiphase solver.

Numerical Experiments and Results

Validations and extensions of the unstructured CLS method [2] are reported in our previous works: buoyancy-driven motion of single bubbles on unconfined domains [2, 4, 5], binary droplet collision with bouncing outcome [3], drop collision against a fluid interface [3], isothermal bubbly flows in vertical channels [6], thermocapillary migration of deformable droplets [10], and mass transfer from bubbles rising on unconfined and confined domains [8, 9]. Therefore, this research is a further step for computing drag-force coefficient and mass transfer in turbulent bubbly flows at $100 < Re < 1000$.

Following, [9, 13], the hydrodynamics of bubbly flows is characterized by the Morton number $Mo = g\mu_c^4(\rho_c - \rho_d)\rho_c^{-2}\sigma^{-3}$, Eötvös number $Eo = gd^2\Delta\rho\sigma^{-1}$, density ratio $\eta_\rho = \rho_c/\rho_d$, viscosity ratio $\eta_\mu = \mu_c/\mu_d$, bubble fraction $\alpha = V_{bubbles}/V_\Omega$, and Reynolds number $Re = \rho_c U_T d/\mu_c$, where U_T is the vertical component (aligned with $\mathbf{g}$) of the drift velocity defined as the difference between the bubble velocity and continuous phase velocity. Reactive mass transfer is characterized by the Damköler (Da) number $Da = k_1 d^2/\mathscr{D}_c$, Schmidt number ($Sc = \mu_c/(\rho_c \mathscr{D}_c)$) or Peclet number ($Pe = ReSc$), and Sherwood number ($Sh = k_c d/\mathscr{D}_c$) where k_c is the mass transfer coefficient at Ω_c. The objective in this research is to compute the drag-coefficient $C_D = C_D(Eo, Re, \alpha)$ and the Sherwood number $Sh = Sh(Re, Sc, Da, \alpha)$. Further technical details on the computation of mass transfer coefficients and drag-force coefficients in bubble swarms can be found in our previous work [9].

Ω is a full-periodic cubic domain of length-side L_Ω, which is discretized by meshes of $\{100^3, 150^3, 200^3\}$ uniform hexahedral cells, distributed in $\{48, 192, 512\}$ CPU-cores, respectively. In the beginning, bubbles are distributed in Ω following a random pattern, whereas both fluid-phases are quiescent. In what follows, dimensionless numbers $(Eo, Mo) = (3.125, 5 \times 10^{-11})$ will be used unless otherwise stated. Figure 16.1 demonstrates the grid convergence of Re, Sh and $C_c = V_c^{-1}\int_{\Omega_c} CdV$, for $(\alpha, Sc, Da) = \{(9.8\%, 1, 141.8), (9.8\%, 10, 1418)\}$. It is concluded that the mesh 150^3 can be employed as a compromise between numerical accuracy and computational cost. Figure 16.2a depicts the effect of density ratio and viscosity ratio on Re, Sh, and the total interfacial area ($A^* = A_0^{-1}\sum_{i=1}^{n_b}\int_\Omega ||\nabla\phi_i||dV$), normalized by its value at $t = 0$ (A_0). These computations show that $(\eta_\rho, \eta_\mu) = (100, 100)$ lead to

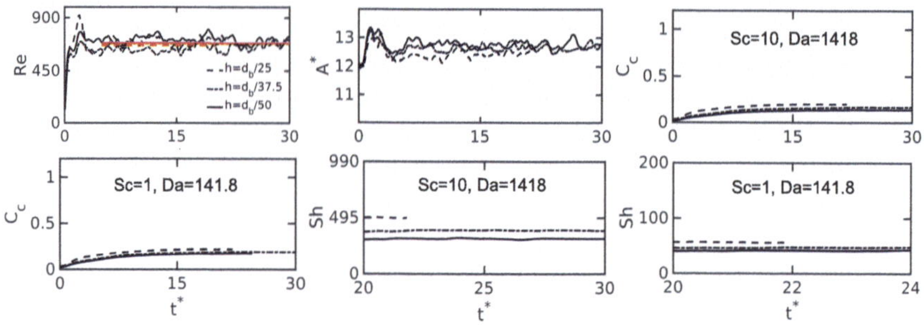

Fig. 16.1 Grid convergence. $Eo = 3.125$, $Mo = 5 \times 10^{-11}$, $\eta_\mu = \eta_\rho = 100$, $\alpha = 9.8\%$ ($N_b = 12$ bubbles), $(Sc, Da) = \{(1, 141.8),\ (10, 1418)\}$. Grid sizes $h = \{d_b/25(--),\ d_b/37.5\ (-\cdot),\ d_b/50(-)\}$ correspond to $\{100^3, 150^3, 200^3\}$ control volumes (CVs), respectively. Red lines are used for time-average Re of bubble swarms

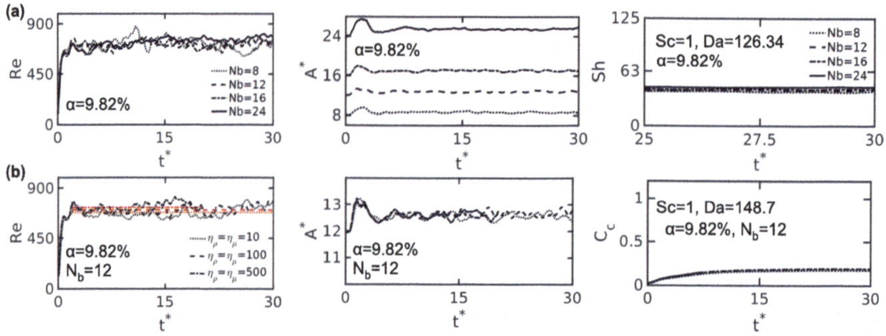

Fig. 16.2 $Eo = 3.125$, $Mo = 5 \times 10^{-11}$, $\alpha = 9.8\%$. **a** Effect of the number of bubbles (N_b), $(Sc, Da) = (1, 126.3)$, Mesh 200^3 CVs. **b** Effect of density and viscosity ratios (η_ρ, η_μ), $(Sc, Da) = (1, 148.7)$, $N_b = 12$. Mesh 150^3 CVs

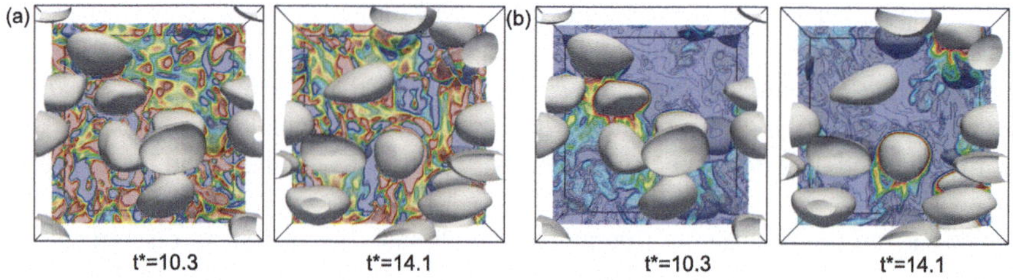

Fig. 16.3 $Eo = 3.125$, $Mo = 5 \times 10^{-11}$, $\eta_\mu = \eta_\rho = 100$, $\alpha = 9.8\%$ ($N_b = 12$ bubbles), $(Sc, Da) = (1, 141.8)$. Ω is a full-periodic cube. Grid size $h = d_b/37.5$, with 150^3 control volumes. **a** Vorticity ($\omega_z = \hat{\mathbf{e}}_z \cdot (\nabla \times \mathbf{v})$) on the plane $x - y$. **b** Concentration on the plane $x - y$

similar results than those obtained by higher physical properties ratios. Figure 16.2b illustrates the effect of the number of bubbles (N_b) when α is kept constant. It is observed that 12 bubbles is a representative sample of the swarm in the full-periodic domain (Fig. 16.3).

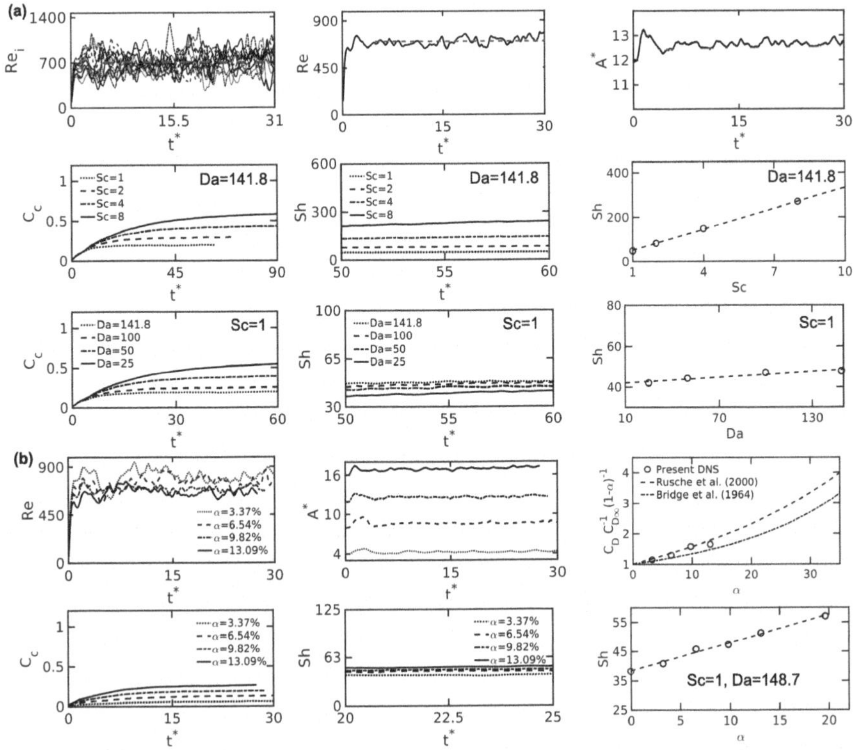

Fig. 16.4 $Eo = 3.125$, $Mo = 5 \times 10^{-11}$, $\eta_\mu = \eta_\rho = 100$. Mesh 150^3 CVs. **a** Effect of Sc and Da. **b** Effect of volume bubble fraction (α %)

Finally, $Sh = Sh(Sc)$, $Sh = Sh(Da)$, $Sh = Sh(\alpha)$ and drag-force coefficient $C_D = C_D(\alpha)$ are researched, using numerical parameters previously tuned. Figure 16.4a depicts the instantaneous distribution of vorticity $\mathbf{e}_z \cdot (\nabla \times \mathbf{v})$ generated by the interaction of bubbles, as well as the concentration field of the chemical species on the plane $x - y$. This figure also illustrates the mixing of chemical species promoted by the bubble induced turbulence. Figure 16.4a shows the effect of Sc and Da, on C_C and Sh, keeping constant $\alpha = 9.8\%$ and $N_b = 12$. It is illustrated a linear relation for $Sh = Sh(Sc)$, as well as for $Sh = Sh(Da)$. Figure 16.4b shows the effect of α, on the normalized drag-force coefficient, whereas these DNS results are compared with empirical correlations from the literature [12, 17]. Furthermore, the effect of α on C_C and Sh is researched, keeping constant the bubble radius r_b, $Sc = 1$ and $Da = 141.8$. A linear relation is observed for $Sh = Sh(\alpha)$.

Conclusions

DNS of drag-force and reactive mass transfer from a turbulent bubbly flow ($100 < Re < 1000$) has been performed using a parallel multiple marker CLS method introduced by [3, 6, 9]. The method avoids the unphysical merging of bubble interfaces, which is an artifact inherent to standard interface capturing methods, allowing for long time simulations of bubbly flows. Interactions of bubbles include a repulsion effects as well as the so-called drafting–kissing–tumbling mechanism with strong deformations of bubbles. Turbulence induced by the interaction of bubbles promotes the mixing of chemical species, whereas the chemical species concentration (averaged in Ω_c) tends to the steady state, as a consequence of the balance between the chemical reaction in Ω_c and mass transfer from bubbles to Ω_c. Numerical results show an increment of Sh as the magnitude of α or Sc increase, whereas Sh increases slightly as the magnitude of Da increases. Furthermore, C_D increases with α.

Acknowledgements This work has been financially supported by MINECO (ENE2015-70672-P), and by Termo Fluids S.L., Spain. Néstor Balcázar acknowledges financial support of the *Programa Torres Quevedo* MINECO (PTQ-14-07186), Spain. Simulations were executed using computing time awarded by PRACE 14th-Call (project 2016153612), and RES (projects $FI - 2019 - 1 - 0039$ and $IM - 2019 - 2 - 0014$), on the supercomputer MareNostrum IV based in Barcelona, Spain.

References

1. Aboulhasanzadeh, B., Thomas, S., Taeibi-Rahni, M., Tryggvason, G.: Multiscale computations of mass transfer from buoyant bubbles. Chem. Eng. Sci. **75**, 456–467 (2012)
2. Balcázar, N., Jofre, L., Lehmkhul, O., Castro, J., Rigola, J.: A finite-volume/level-set method for simulating two-phase flows on unstructured grids. Int. J. Multiph. Flow **64**, 55–72 (2014)
3. Balcázar, N., Lehmkhul, O., Rigola, J., Oliva, A.: A multiple marker level-set method for simulation of deformable fluid particles. Int. J. Multiph. Flow **74**, 125–142 (2015)
4. Balcázar, N., Lemhkuhl, O., Jofre, L., Oliva, A.: Level-set simulations of buoyancy-driven motion of single and multiple bubbles. Int. J. Heat Fluid Flow **56**, 91–107 (2015)
5. Balcázar, N., Lehmkhul, O., Jofre, L., Rigola, J., Oliva, A.: A coupled volume-of-fluid/level-set method for simulation of two-phase flows on unstructured meshes. Comput. Fluids **124**, 12–29 (2016)
6. Balcázar, N., Castro, J., Rigola, J., Oliva, A.: DNS of the wall effect on the motion of bubble swarms. Procedia Comput. Sci. **108**, 2008–2017 (2017)
7. Balcázar, N., Castro, J., Chiva, J., Oliva, A.: DNS of Falling Droplets in a Vertical Channel. Int. J. Comput. Methods Exp. Meas. **6**(2), 398–410 (2018)
8. Balcázar-Arciniega, N., Rigola, J., Oliva, A.: DNS of mass transfer from bubbles rising in a vertical channel. In: Rodrigues, J. et al. (eds.) Computational Science ICCS, ICCS 2019. Lecture Notes in Computer Science, vol. 11539. Springer, Cham (2019)
9. Balcázar-Arciniega, N., Antepara, O., Rigola, J., Oliva, A.: A level-set model for mass transfer in bubbly flows. Int. J. Heat Mass Transf. **138**, 335–356 (2019)
10. Balcázar, N., Rigola, J., Castro, J., Oliva, A.: A level-set model for thermocapillary motion of deformable fluid particles. Int. J. Heat Fluid Flow, Part B **62**, 324–343 (2016)
11. Brackbill, J.U., Kothe, D.B., Zemach, C.: A continuum method for modeling surface tension. J. Comput. Phys. **100**, 335–354 (1992)

12. Bridge, A., Lapidus, L., Elgin, J.: The mechanics of vertical gasliquid fluidized system I: countercurrent flow. AIChE J. **10**(6), 819–826 (1964)
13. Clift, R., Grace, J.R., Weber, M.E.: Bubbles, Drops and Particles. Academin Press, New York (1978)
14. Coyajee, E., Boersma, B.J.: Numerical simulation of drop impact on a liquidliquid interface with a multiple marker front-capturing method. J. Comput. Phys. **228**(12), 4444–4467 (2009)
15. Olsson, E., Kreiss, G.: A conservative level set method for two phase flow. J. Comput. Phys. **210**, 225–246 (2005)
16. Roghair, I., Van Sint Annaland, M., Kuipers, J.A.M.: An improved Front-Tracking technique for the simulation of mass transfer in dense bubbly flows. Chem. Eng. Sci. **152**, 351–369 (2016)
17. Rusche, H., Issa, R.: The effect of voidage on the drag force on particles, droplets and bubbles in dispersed two-phase flow. In: Japanese European Two-Phase Flow Meeting. Tsukuba, Japan (2000)

Chapter 17
Breakup of Agglomerates in Turbulent Flows: An Euler–Lagrange LES Study

M. Breuer and A. Khalifa

Abstract Breakup models were developed for the Euler–Lagrange framework taking breakup by turbulence, drag, and rotation into account. The modeling includes the post breakup treatment and a time lag between successive breakup processes. The prediction of the size distribution based on this four-way coupled Eulerian–Lagrangian approach delivers a reasonable agreement with experimental data.

Introduction

Particles dispersed in turbulent flows are encountered in many technical applications (e.g., coal combustors or medical inhalers) and natural processes (e.g., erosion). Thus, there is a high demand for appropriate simulation methodologies allowing to predict such challenging multiphase flows. The Euler–Lagrange approach relying on the LES methodology offers on the one hand the chance to appropriately predict the continuous phase with reasonable computational effort. On the other hand, the disperse phase can be described based on first principles tracking individual particles/agglomerates through the turbulent flow while taking various important physical effects such as collisions, agglomeration, and breakup events separately into account. The latter two play a significant role for the particle size distribution, which again strongly influences the motion of the disperse phase and finally also the continuous phase. The present study concentrates on breakup processes taking place in turbulent flows and appropriate models to describe these phenomena.

M. Breuer (✉) · A. Khalifa
Professur für Strömungsmechanik, Helmut-Schmidt-Universität Hamburg, Hamburg, Germany
e-mail: breuer@hsu-hh.de

M. García-Villalba et al. (eds.), *Direct and Large Eddy Simulation XII*, ERCOFTAC Series 27,
https://doi.org/10.1007/978-3-030-42822-8_17

Applied Methodology

The main ingredients are a finite-volume CFD solver for block-structured grids [2, 3] predicting the continuous phase based on LES. Furthermore, an efficient particle tracking scheme including a deterministic collision detection algorithm [1, 4] forms the basis to track a huge number of particles through the flow field taking the feedback effect of the disperse phase on the continuous phase and the SGS fluctuations (Langevin model) into account. Thus, a four-way coupled simulation environment results which allows to consider further highly relevant physical phenomena.

Recently, enhanced agglomeration [5, 6] and particle-wall adhesion models [7] for dry, electrostatically neutral particles were developed in the framework of the hard-sphere model. In order to describe the size distribution of the disperse phase correctly, breakup models for agglomerates are missing.

Models for Breakup in Turbulent Flows

Besides other mechanisms such as collisions between agglomerates and particles or the impaction of agglomerates at walls, fluid forces play a significant role in the fragmentation of agglomerates. The investigation concentrates on this latter phenomenon. Thus, in a recent study [8], a literature review was carried out describing the state of the art concerning the modeling of breakup processes. Based on the idea to describe these processes relying on first principles within an Euler–Lagrange point-particle LES approach, modeling ideas from the literature were revisited and extended in order to be applied in a four-way coupled simulation but with the restriction to compact dry powder agglomerates with fractal dimensions close to three. The methodology comprises three possible mechanisms for brittle instantaneous breakup: **1**. turbulent stresses, **2**. drag (inertia) stresses, and **3**. rotary stresses.

Turbulent stresses are related to fluctuations of the fluid velocities inducing pressure differences and velocity gradients that may lead to the rupture of agglomerates. The description of the velocity differences varies for different eddy size subranges of turbulence, i.e., viscous, transition, and inertial. The size ranges are defined with respect to the Kolmogorov length scale. The relations for the different regimes are all based on the local energy dissipation rate, the kinematic viscosity of the fluid, and the agglomerate size [10, 11]. The condition for breakup is satisfied when the turbulent stress is larger than the strength of the agglomerate. Following Kusters [10], in case of multiple stresses solely the dominating stress is assumed to be responsible for breakup in the present study.

Concerning breakup due to drag-induced stresses, the model is extended by examining the tensile stress evolving in planes within the agglomerate. Evaluating the inertial and fluid forces acting on portions of the accelerating agglomerate, the number of primary particles eligible for breakup is determined as a part of the model. Thus, an erosion of only one or a small group of primary particles can take place.

The third stress contribution originates from the rotation of agglomerates, which can be either induced by the fluid motion or by collisions with rigid walls or other agglomerates. For this rotary stress, a new relation was derived relying on the stress field in a sphere steadily rotating around its axis and the corresponding Mises stress. The derived relation for the stress weakly depends on the Poisson's ratio of the material but otherwise is similar to the classic formulation [12].

Furthermore, special emphasis is put on the post-breakup treatment, i.e., models for the sizes and velocities of the generated fragments. According to the different breakup phenomena the post-breakup treatment takes the theory behind each mechanism separately into account. For breakup by turbulent stresses, it is assumed that the agglomerate splits into two (nearly) identical fragments. The direction is given by the most extensional eigenvector of the strain rate tensor. For breakup by drag stresses erosion from the agglomerate is allowed and fragments are going to split along the axis of the relative velocity. Finally, for breakup by rotary stresses, the agglomerate is assumed again to break into two identical fragments separating in centrifugal direction.

The determination of the separation velocities of the fragments relies on an energy and momentum balance formulated differently for the three breakup mechanisms, see [8, 9] for further details. After breakup, the fragments are tracked through the flow field and are not allowed to further breakup within a certain time interval. This time lag is determined based on the physical argument that breakup processes are not infinitely fast and thus require a designated breakup time. That renders the problem independent of the numerical time-step size and avoids unphysical collisions and re-agglomerations. For further details, the reader is referred to [9].

Application: Dry Powder Disperser

The proposed modeling strategy is applied to a test case inspired by the experimental study of Weiler [13] for the deagglomeration in a dry powder disperser. The configuration consists of a duct with a square cross-section and a funnel protruding into the duct discretized by a block-structured grid with about 18 or 115 million CVs for the low-Re and the high-Re case, respectively. The resolution allows to resolve the near-wall regions and thus a wall-resolved LES is carried out. The inflow data are generated by an auxiliary simulation of a single duct configuration with periodic boundary conditions. According to the feeder in the experimental setup, the powder discharges at a mass flow rate of $\dot{m}_{\text{powder}} = 10\,\text{mg/s}$. It is assumed that the silica powder particles are agglomerated at the time of release. Since no information is available concerning the size distribution, mono-disperse agglomerates consisting of 100 primary particles are assumed. Three different powders (particle diameter d_p = 0.97 μm (**A**), 2.47 μm (**B**), and 5.08 μm (**C**)) and two different flow rates (25 [low-Re case: Re $\approx$ 8700] and 200 norm-liter/min [high-Re case: Re $\approx$ 69,600]) are investigated.

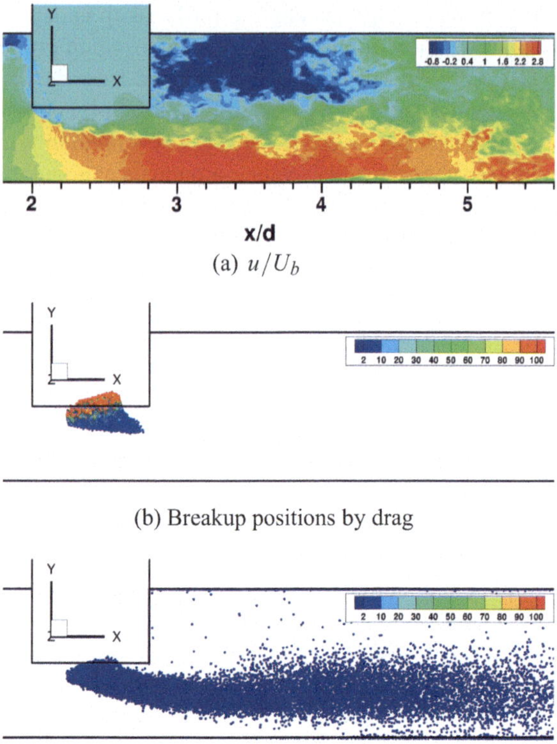

(a) u/U_b

(b) Breakup positions by drag

(c) Breakup positions by rotation

Fig. 17.1 High-Re case with powder **A**: **a** Instantaneous streamwise velocity u/U_b, Breakup positions by drag (**b**) and by rotation (**c**). For **b**, **c** the colors show agglomerate sizes at breakup in terms of the number of primary particles included

The typical flow field in the device for the high-Re case is shown in Fig. 17.1a by contours of the instantaneous streamwise velocity at an arbitrarily chosen snapshot. As visible, the funnel protrudes into the duct acting as an obstacle and thus leading to a strong contraction of the cross-section. Consequently, the flow is significantly accelerated below the funnel generating strong shear layers with high magnitudes of the strain rate tensor and the vorticity vector (not depicted here). Furthermore, a recirculation region is observed behind the obstacle.

For powder **A** at the high-Re case the breakup positions due to the drag stress are shown in Fig. 17.1b, whereas Fig. 17.1c depicts the breakup positions due to the rotary stress. Obviously, the breakup process starts with breakup events due to the drag stress. For this purpose, large relative velocities between the fluid and the agglomerates are required which are found at the funnel exit. Breakup due to rotary stress starts further downstream requiring a fast rotation of the agglomerate which is either induced by the rotation of the fluid or by inter-particle or particle-wall collisions, where overall the collisions are found to be the dominating effect.

Comparing the breakup behavior of powder **A** for the low-Re and the high-Re case, significant difference can be already observed by looking at arbitrary snapshots of the agglomerates in the disperser depicted in Fig. 17.2. In the low-Re case, the

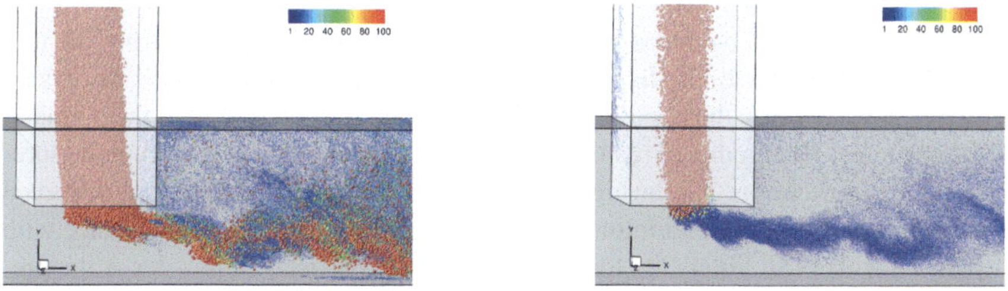

(a) Low-Re case (b) High-Re case

Fig. 17.2 Snapshot of the agglomerates (powder **A**) in the disperser for **a** the low-Re case and **b** the high-Re case (color = number of primary particles)

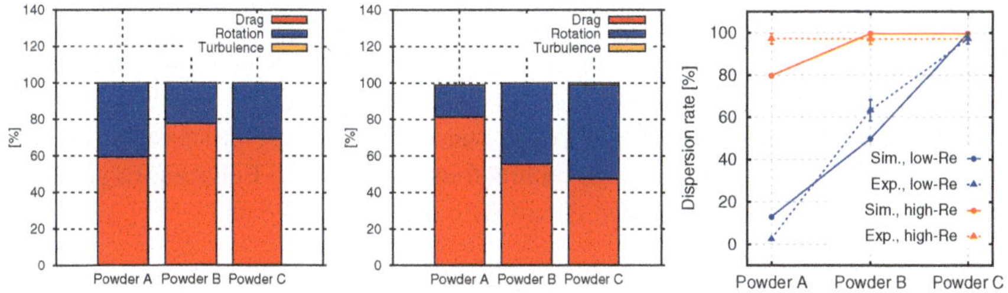

(a) Low-Re case (b) High-Re case (c) Both Re

Fig. 17.3 **a** Share of different breakup mechanisms for the low-Re and **b** high-Re case; **c** Dispersion rates for both Re. Experimental data by Weiler [13]

breakup rate is much lower than in the high-Re case leading to the situation that a high number of agglomerates pass the critical regions below the funnel exit without experiencing breakup. Thus, a major part of the released agglomerates leaves the exit of the deagglomerator with their initial size or only partially fragmented. On the contrary, in the high-Re case most of the agglomerates are broken up directly beneath the funnel exit.

Figure 17.3a and b display the contributions of the three breakup mechanisms considered for all three powders and both Re numbers. In the low-Re case, the dominating portion of all breakups is due to the drag stress followed by the rotary stress. Breakup due to turbulence is not observed at all. Comparing the different powders a clear trend concerning the share of the drag stress is not visible, since it first increase from about 59% for the smallest primary particles of powder **A** to the intermediate size of powder **B** with about 77%, but then drops again to a percentage rate of about 69% for the largest primary particles of powder **C**.

In the high-Re case, the situation is different. Here, a clear trend is obvious. The percentage rate of the breakup by drag decreases with the size of the primary particles from about 81% (powder **A**) to 56% (powder **B**) and finally to 48% (powder **C**). Breakup due to turbulent stresses is now observed but still marginal (0.002% (**A**), 0.02% (**B**) and 1.0% (**C**)). Consequently, the share of breakup by rotation increases with the primary particle size. Finally, Fig. 17.3c depicts the predicted dispersion rates in comparison to the measurements of Weiler [13]. As expected from the theory, for the low-Re case the dispersion rate significantly increases with increasing primary particles, since the bonding van-der-Waals force decreases with the diameter of the particles. The predicted results are in reasonable agreement with the measurements [13]. Furthermore, for the high-Re case a nearly total disintegration is observed at the outlet of the deagglomerator for powders **B** and **C**. That is in accordance with the experiments [13]. Solely for the smallest primary particles of powder **A** the prediction estimates a dispersion rate of about 80%, whereas in the measurements a full disintegration is found.

Conclusions

Breakup models were developed for the Euler–Lagrange framework taking breakup by turbulence, drag, and rotation into account. The modeling includes the post-breakup treatment and a time lag between successive breakup processes. The prediction of the size distribution based on this four-way coupled Eulerian–Lagrangian approach delivers a reasonable agreement with experimental data.

Acknowledgements Financial supported by Deutsche Forschungsgemeinschaft: BR 1847/13-2.

References

1. Alletto, M., Breuer, M.: One-way, two-way and four-way coupled LES predictions of a particle-laden turbulent flow at high mass loading downstream of a confined bluff body. Int. J. Multiph. Flow **45**, 70–90 (2012)
2. Breuer, M.: Large-eddy simulation of the sub-critical flow past a circular cylinder: Numerical and modeling aspects. Int. J. Numer. Meth. Fluids **28**, 1281–1302 (1998)
3. Breuer, M.: A challenging test case for large-eddy simulation: High Reynolds number circular cylinder flow. Int. J. Heat Fluid Flow **21**, 648–654 (2000)
4. Breuer, M., Alletto, M.: Efficient simulation of particle-laden turbulent flows with high mass loadings using LES. Int. J. Heat Fluid Flow **35**, 2–12 (2012)
5. Breuer, M., Almohammed, N.: Modeling and simulation of particle agglomeration in turbulent flows using a hard-sphere model with deterministic collision detection and enhanced structure models. Int. J. Multiph. Flow **73**, 171–206 (2015)
6. Almohammed, N., Breuer, M.: Modeling and simulation of agglomeration in turbulent particle-laden flows: A comparison between energy-based and momentum-based agglomeration models. Powder Technol. **294**, 373–402 (2016)
7. Almohammed, N., Breuer, M.: Modeling and simulation of particle-wall adhesion of aerosol particles in particle-laden turbulent flows. Int. J. Multiph. Flow **85**, 142–156 (2016)
8. Breuer, M., Khalifa, A.: Revisiting and improving models for the breakup of compact dry powder agglomerates in turbulent flows within Eulerian-Lagrangian simulations. Powder Technol. **348**, 105–125 (2019)
9. Breuer, M., Khalifa, A.: Refinement of breakup models for compact powder agglomerates exposed to turbulent flows considering relevant time scales, Comput. Fluids **194**, 104315 (2019)
10. Kusters, K.: The Influence of Turbulence on Aggregation of Small Particles in Agitated Vessels. Ph.D. thesis, TU Eindhoven, The Netherlands (1991)

11. Neeße, T., Mühle, K., Ivanauskas, A.: Zur physikalisch begründeten Modellierung des Flockungsprozesses Teil V: Flockenaufbruch und Flockenerosion in turbulenter Strömung. Chem.-Ing.-Techn. **39**, 292–295 (1987)
12. Rumpf, H., Raasch, J.: Desagglomeration in Strömungen. in: 1. Europ. Symp. Zerkleinern 10.-13.4.1962, Verlag Chemie, Weinheim und VDI-Verlag, Düsseldorf, 151–159 (1962)
13. Weiler, C.: Generierung leicht dispergierbarer Inhalationspulver mittels Sprühtrocknung. Ph.D. thesis, Univ. Mainz, Germany (2008)

Chapter 18
LES of a Taylor Bubble in Co-current Turbulent Pipe Flow

E. M. A. Frederix, B. Mikuž and E. M. J. Komen

Introduction

In the analysis of Nuclear Reactor Safety (NRS), modeling of multiphase flow plays an important role. Most relevant NRS scenarios can only be simulated within a feasible computational time using engineering models. These models, such as the RELAP5 and TRACE system codes, as well as Reynolds Averaged Navier–Stokes (RANS) models, must be tuned to specific situations. A practical way of doing this is by performing "high fidelity" simulation, i.e., DNS and LES, of relevant subset problems of the NRS scenario at hand. The high fidelity simulation data may then be leveraged for improving the quality of the engineering models which, in turn, may be applied to more integral settings. For this strategy to work, a high quality in the DNS or LES data must be established. In particular in the multiphase flow field, this is a challenging task, which will be addressed in this work, for complex separated turbulent two-phase flow.

In NRS, two-phase flow may exist in several regimes, ranging from bubbly flow to stratified flow or annular flow. A modeling approach for relevant NRS scenarios should be capable of capturing all those regimes, and the transition between them. In developing such a universal modeling approach, it is deemed necessary to first focus on the individual regimes separately. For separated two-phase flow, in which a large scale interface between two phases is present and modeled as such, work is already done [2]. However, this work remains limited to quasi two-dimensional settings in which the interface is relatively flat. In practice, the interface, in particular in turbulent flow, behaves much more irregularly. Therefore, we select the setting of

E. M. A. Frederix (✉) · E. M. J. Komen
Nuclear Research and Consultancy Group, Westerduinweg 3,
1755 LE Petten, The Netherlands
e-mail: frederix@nrg.eu

B. Mikuž
Jožef Stefan Institute, Jamova 39, 1000 Ljubljana, Slovenia

M. García-Villalba et al. (eds.), *Direct and Large Eddy Simulation XII*,
ERCOFTAC Series 27,
https://doi.org/10.1007/978-3-030-42822-8_18

co-current turbulent Taylor bubble flow as a validation platform for our two-phase high fidelity simulation approach. Within this setting, the goal of the current work is to perform simulation of turbulent co-current Taylor bubble flow and to present high fidelity LES data which may be leveraged for the development of two-phase engineering models.

This paper presents a validated LES strategy of co-current turbulent Taylor bubble flow. The LES strategy is applied to the experimental setting of [6]. Preliminary simulations show that the problem is inherently transient, i.e., the applied model predicts a continuous loss of void of the Taylor bubble due to strong shear stresses to which the Taylor bubble tail is exposed. This leads to a gradual decrease in Taylor bubble size, effectively changing the topology of the problem. Therefore, a consistent strategy of temporal ensemble and spatial averaging is applied. The temporal averaging is limited by the mesh-dependent rate at which the Taylor bubble loses void. We show that characteristic features of the toroidal vortex in the wake of the Taylor bubble are captured well. However, close to the Taylor bubble an underestimation of the fluctuating velocities is observed.

Simulation of Turbulent Taylor Bubble Flow

A two-phase mixture of gas and liquid is modeled using the Volume of Fluid (VOF) approach. The void fraction α is introduced, and also acts as a marker function, adhering to

$$\partial_t \alpha + \mathbf{u} \cdot \nabla \alpha = 0, \tag{18.1}$$

with partial time derivative ∂_t, velocity vector $\mathbf{u}$ and gradient operator ∇. The fluid mixture is described by the incompressible Navier–Stokes equations, i.e.,

$$\nabla \cdot \mathbf{u} = 0 \tag{18.2}$$

and

$$\partial_t(\rho \mathbf{u}) + \nabla \cdot (\rho \mathbf{u}\mathbf{u}) = -\nabla p + \rho \mathbf{g} + \nabla \cdot (2\mu_{\text{eff}} \mathbf{D}) + \sigma \kappa \delta(n) \mathbf{n}, \tag{18.3}$$

with mass density ρ, pressure p, gravitational acceleration $\mathbf{g}$, deformation tensor $\mathbf{D}$, effective mixture viscosity μ_{eff}, interface curvature κ, interface normal unit vector $\mathbf{n}$, interface Dirac delta function $\delta(n)$ and surface tension σ. The effective mixture viscosity μ_{eff} consists of a laminar and sub-grid scale component, of which the latter is modeled using the Vreman LES model [7]. We solve the set of equations (18.1)–(18.3) using a modification of OpenFOAM's standard VOF solver (interFoam). This modification includes (1) Diagonally Implicit Runge–Kutta (DIRK) time integration and (2) a removal of a numerically dissipative term specific to OpenFOAM. This approach is validated by comparing against both experimental [4] and numerical [1] data of a Taylor bubble in laminar flow.

We consider the turbulent Taylor bubble setting which was experimentally studied in [6]. In their experiments, air was injected into rising turbulent water pipe flow. Measurements of the flow field in the wake of the Taylor bubble were performed for various pipe diameters and water flow rates, with which both the single-phase and Taylor bubble Reynolds numbers can be controlled. We focus on the setting in which the single-phase Reynolds number (based on liquid superficial velocity and pipe diameter) is $\mathrm{Re} = 8250$, which is sufficiently moderate in order to perform LES within reasonable computational time.

In order to simulate the rising Taylor bubble, we develop a Moving Frame of Reference (MFR) approach which is attached to the bubble. Each numerical time step, the velocity of the MFR is adjusted such that the bubble velocity is zero. Accordingly, the wall velocity is imposed as the current liquid velocity subtracted by the superficial liquid velocity. In order to have a fully developed turbulent flow ahead of the Taylor bubble, flow is recycled at the inlet from a pipe cross section at a distance of $5D$ from the inlet. This is indicated schematically in Fig. 18.1. The meshing strategy is as follows:

- For radius $0 < r < 0.3D$, an unstructured hex mesh is used with roughly uniformly sized cells in the pipe cross section
- For radius $r > 0.3D$, a structured mesh is used with refinement towards the wall such that at the wall the mesh size is one wall unit
- The mesh is axially refined with a factor 2 in the direction toward the bubble, both ahead and below the bubble.

Three meshes are realized, i.e., a coarse, medium, and fine mesh, which have 1.87M, 5.38M, and 11.3M cells, respectively. The coarse mesh has a spanwise cell size in the bulk of 30 wall units, the medium mesh of 20 wall units, and the fine mesh of 15 wall units. Far away from the bubble, the axial cell size is twice those of the spanwise cell size, for each mesh. As said, toward the bubble the axial cell size is refined with a factor 2. The simulation strategy contains the following steps:

1. Single-phase turbulence at $\mathrm{Re} = 8250$ is simulated, until a quasi-steady state is achieved
2. At some developed time t_i, a bubble with length $3.5D$ is introduced into the flow
3. Next, temporal averaging is started after 1 Flow-Through Time (FTT) of the liquid in the pipe, which is $17D$ long
4. Finally, temporal averaging is stopped once the bubble has lost $D/2$ in length.

Step 2–4 are repeated a number of times, each repetition starting with an uncorrelated and unique initial flow field. The collection of repetitions is ensemble-averaged and finally circumferentially averaged. Step 4 assures that the flow topology within the temporal averaging window remains roughly similar and prevents that averaged quantities are smeared due to loss of void (see Fig. 18.1). Figure 18.2 shows the instantaneous loss of void (expressed in terms of the displacement of the ensemble-averaged centerline Taylor bubble tail position) as a function of time, for the three meshes. It can be seen that the rate at which void is lost is mesh-dependent, and that the finest mesh has the least loss of void. This allows for a longer temporal averaging

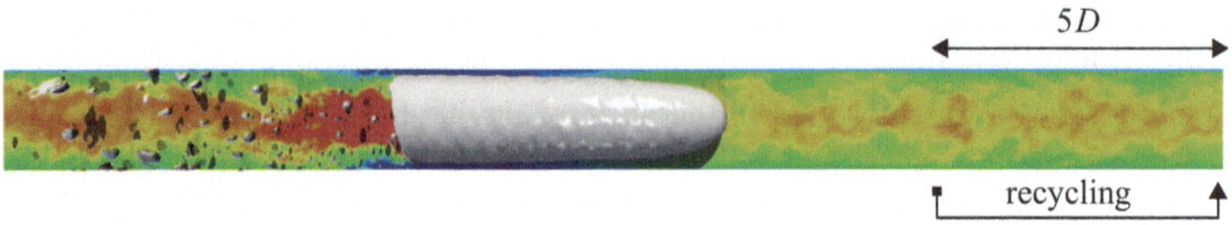

Fig. 18.1 Iso-surface of the instantaneous void fraction at 0.5. Color indicates the axial velocity component in the MFR

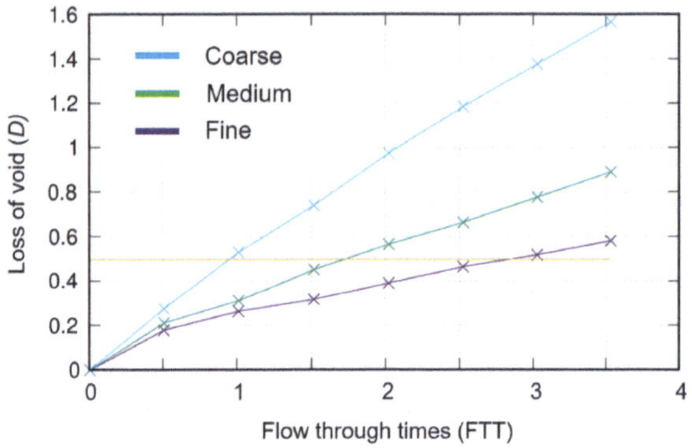

Fig. 18.2 The loss of void as a function of time, for the three meshes

window and, in turn, fewer repetitions in the ensemble. Therefore, for the coarse mesh we use 16 repetitions, for the medium mesh 8, and for the fine mesh 5. It was found that this gives, for each mesh, statistically converged results. We note that the final simulation strategy consists of a combination of temporal, ensemble, and spatial averaging in the circumferential direction.

Preliminary simulations showed that the production of small bubbles in the wake of the Taylor bubble is strongly dependent on the mesh when using a central discretization scheme for the convective term of the momentum equation. When using a more dissipative TVD scheme, such as the van Leer scheme, the production of bubbles is suppressed and less mesh-dependent. However, such a TVD scheme is much too dissipative for the single-phase turbulence, e.g., see [3]. Therefore, we use a blended scheme which applies the Van Leer scheme near an interface while reverting back to the central scheme elsewhere. We adopt the following free surface (fs) indicator:

$$\phi_{\mathrm{fs}} = \frac{1}{2}\tanh(a\Delta x[|\nabla\alpha| - |\nabla\alpha^{\star}|]) + \frac{1}{2}, \tag{18.4}$$

with $a = 100$, $|\nabla\alpha^{\star}| = 1/(5\Delta x)$ and Δx the local cell size. With these parameters, ϕ_{fs} attains a value close to unity near the free surface while moving to zero away from the free surface. This allows the blended scheme to have the schematic form

$$\text{blended scheme} = \phi_{\mathrm{fs}} \times \text{van leer scheme} + (1 - \phi_{\mathrm{fs}}) \times \text{central scheme}. \tag{18.5}$$

Results

Figure 18.1 shows an iso-surface of the instantaneous gas void fraction, using the medium mesh. It can be seen that the Taylor bubble is losing void due to the formation of bubbles which break away from the trailing edge of the bubble and are entrained into the turbulent liquid. The bubble formation is primarily observed in the "skirt" of the Taylor bubble, close to the wall where the shear velocity and surface curvature of the gas-liquid interface are largest. It was found that some bubbles coalesce into bigger bubbles, and that these bigger bubbles, due to an increased buoyant upward motion, can coalesce back into the Taylor bubble. Most bubbles, however, are permanently entrained by the liquid.

Figure 18.3 shows radial profiles of the axial liquid velocity component in the wake of the Taylor bubble, at a distance $0.6D$, $1D$, and $2D$ below the Taylor bubble. The liquid velocity component is computed as the liquid void fraction-weighted average velocity. This average is expected to relate closely to the PIV measurements as performed in the experiment. Generally, good agreement is observed with the experimental data of [6] and trends are captured well. Moreover, results are quite mesh-independent. Figure 18.4 shows similar profiles for the RMS of the axial velocity component. However, in particular for the profile closest to the Taylor bubble, the turbulent fluctuations are shown to be severely under-predicted. Moreover, the RMSs appear to be much more sensitive to the mesh. For distances further below the Taylor bubble, it is observed that the predicted fluctuating velocity profiles agree much better with the experimental ones. This suggests that the under-predicted RMS is the result of the presence of the Taylor bubble interface. Therefore, it is possible that the lower RMS is caused by the blended scheme, which introduces numerical dissipation due to the presence of the interface. We also compare axial profiles of the radial velocity in the wake with the experimental data and observe good agreement there (not shown here). This implies that the wake of the Taylor bubble, which is a toroidal vortex, is captured well in the numerical method. Nevertheless, the underestimation of the fluctuating velocities in the tail requires further investigation.

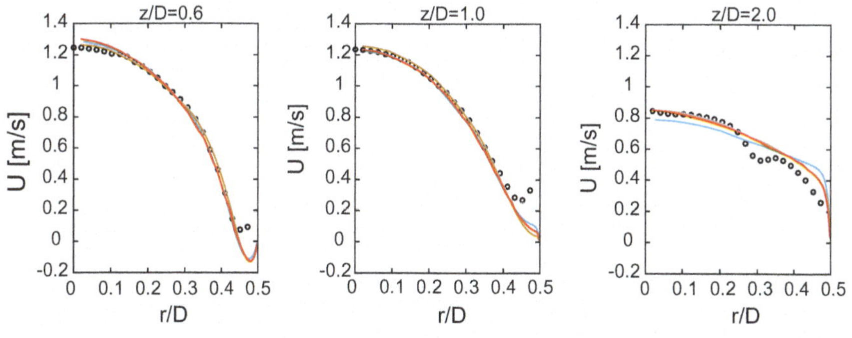

Fig. 18.3 Radial profiles of the axial velocity component at a distance $0.6D$ (left), $1D$ (middle), and $2D$ below the Taylor bubble, for the coarse mesh (blue), medium mesh (yellow), and fine mesh (red). Also the experimental data of [6] is shown (black dots)

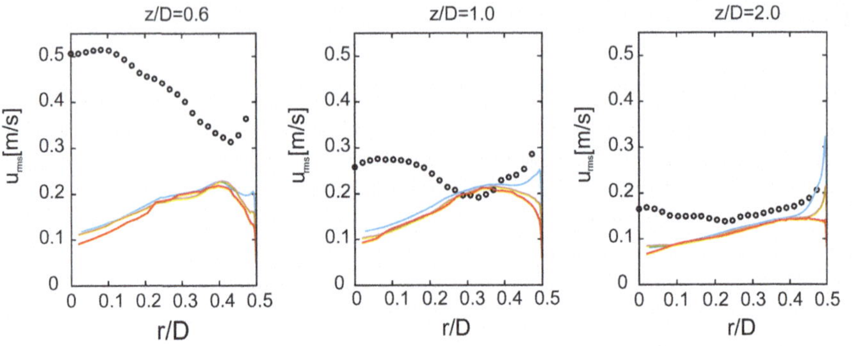

Fig. 18.4 Radial profiles of the axial velocity RMS at a distance $0.6D$, $1D$, and $2D$ below the Taylor bubble, for the coarse mesh (blue), medium mesh (yellow), and fine mesh (red). Also the experimental data of [6] is shown (black dots)

Conclusions

In this work, we presented LES of a turbulent co-current Taylor bubble flow at single-phase Reynolds number Re $= 8250$. An MFR approach was developed, in which a consistent strategy of time, ensemble and spatial averaging were used to post-process the computed instantaneous flow variables. The temporal averaging window was limited by the loss of void of the Taylor bubble. The loss of void leads to a gradual decrease in Taylor bubble size, effectively changing the topology of the problem. Good agreement was found with experimental data for the mean velocity, however, close to the Taylor bubble the fluctuating velocity components are underestimated. Nevertheless, we note that [5] observe very similar behavior.

Acknowledgements This work was carried out on the Dutch national e-infrastructure with the support of SURF Cooperative, grant Multiscale Modeling, and Simulation SH061.

References

1. Araújo, J.D.P., Miranda, J.M., Pinto, A.M.F.R., Campos, J.B.L.M.: Wide-ranging survey on the laminar flow of individual Taylor bubbles rising through stagnant Newtonian liquids. Int. J. Multiph. Flow **43**, 131–148 (2012)
2. Frederix, E.M.A., Mathur, A., Dovizio, D., Geurts, B.J., Komen, E.M.J.: Reynolds-averaged modeling of turbulence damping near a large-scale interface in two-phase flow. Nuc. Eng. Des. **333**, 122–130 (2018)
3. Komen, E.M.J., Camilo, L.H., Shams, A., Geurts, B.J., Koren, B.: A quantification method for numerical dissipation in quasi-DNS and under-resolved DNS, and effects of numerical dissipation in quasi-DNS and under-resolved DNS of turbulent channel flows. J. Comp. Phys. **345**, 565–595 (2017)

4. Nogueira, S., Riethmuller, M.L., Campos, J.B.L.M., Pinto, A.M.F.R.: Flow patterns in the wake of a Taylor bubble rising through vertical columns of stagnant and flowing Newtonian liquids: an experimental study. Chem. Eng. Sci. **61**, 7199–7212 (2006)
5. Shaban, H., Tavoularis, S.: Detached eddy simulations of rising Taylor bubbles. Int. J. Multiph. Flow **107**, 289–300 (2018)
6. Shemer, L., Gulitski, A., Barnea, D.: On the turbulent structure in the wake of Taylor bubbles rising in vertical pipes. Phys. Fluids **43**, 035108 (2007)
7. Vreman, B.: An eddy-viscosity subgrid-scale model for turbulent shear flow: algebraic theory and applications. Phys. Fluids **16**(10), 3670–3681 (2004)

Chapter 19
Droplet Homogeneous Nucleation in a Turbulent Vapour Jet in the Two-Way Coupling Regime

A. Gallegati, F. Battista, P. Gualtieri and C. M. Casciola

Introduction

Homogeneous nucleation of liquid droplets in hot vapour stream, mixing with a cooler and dry external environment, occurs in many technological applications, ranging from the generation of filter test particles [9] to the control of fugitive emissions [10] from industrial sources (refineries), up to the young discipline of Particle Engineering in the biotech industries [19].

However, a Direct Numerical Simulation (DNS) of a vapour jet is still missing, despite the multitude of experiments and its relevance for applications, which could benefit from a better understanding of such multi-physics turbulent flows.

Classical Nucleation Theory (CNT) prescribes rates and critical diameters at which droplets nucleate, depending on the local thermodynamical state. Because of the strongly nonlinear interplay between homogeneous nucleation and turbulent fluctuations, it is crucial not only to take into account all the relevant scales of turbulence, but even all the cross-coupling phenomena involved.

A. Gallegati (✉) · P. Gualtieri · C. M. Casciola
Department of Mechanical and Aerospace Engineering, Sapienza University of Rome, Via Eudossiana 18, 00184 Rome, Italy
e-mail: andrea.gallegati@uniroma1.it

P. Gualtieri
e-mail: paolo.gualtieri@uniroma1.it

C. M. Casciola
e-mail: carlomassimo.casciola@uniroma1.it

F. Battista
ENEA, Casaccia Research Centre, Via Anguillarese 301, 00123 Rome, Italy
e-mail: francesco.battista@enea.it

M. García-Villalba et al. (eds.), *Direct and Large Eddy Simulation XII*, ERCOFTAC Series 27,
https://doi.org/10.1007/978-3-030-42822-8_19

DNS allows to capture, without any modelling, the turbulence underlying the *carrier phase* dynamics. In the two-way coupling regime, the *disperse phase* back-reaction is then accounted within the point-particle approach.

The relevance of these effects on the whole process of the phase-change, i.e. droplets nucleation, condensation and evaporation, will be discussed. In particular, it will be pointed out how much the droplets back-reaction, on the thermodynamics (especially due to the phase-change), does affect the subsequent droplets nucleation rate.

Classical Nucleation Theory

Nucleation rate sensitivity to the underlying turbulent flow has been investigated in many experiments using Dibutyl phthalate (DBP), an organic compound with low toxicity and wide liquid range [10], or water [9, 11, 16] as condensing species. In particular, the strongly nonlinear interplay between homogeneous nucleation and turbulent fluctuations, within a multiphase flow, leads in a while to a non-trivial physical configuration. Thus, poor methodologies fail, because of their lack in taking into account so many cross-coupling phenomena, occurring at the same time and influencing each other.

It is therefore necessary to recall at least some basic notions of Nucleation Theory, such as the nucleation rate and the critical radius.

Becker-Döring Theory

According to [2, 8], the Classical Nucleation Theory (CNT) prescribes a rate for the droplets homogenous nucleation, rewritten here below in a more suitable form [9]

$$J(Y,\theta) = \frac{(\rho Y)^2}{\rho_p}\sqrt{\frac{2\sigma}{\pi m^3}}\,\exp\left[-\frac{16\pi}{3}\left(\frac{\sigma}{k_B\theta}\right)^3\left(\frac{m}{\rho_p \ln S}\right)^2\right], \qquad (19.1)$$

together with the different critical radius r_p^* at which, eventually, each single droplet nucleates, depending on the local thermodynamical state, when a stable molecules cluster size is reached

$$r_p^* = \frac{2\,\sigma\,m}{\rho_p\,k_B\theta\,\ln S}\,. \qquad (19.2)$$

The above expressions are nonlinear functions of the vapour mass fraction Y, temperature θ and saturation ratio S, readily found from the Clausius–Clapeyron relation. Surface tension σ, depending on the local temperature, highly affects the above rate. Liquid droplets density is indicated by ρ_p, while molecules mass by m.

Droplet Laden Vapour Flow in the Two-Way Coupling Regime

Since droplets nucleation usually happens within turbulent flows, it is quite natural to rely on DNS to capture the turbulent fluctuations. Nonetheless, in such cases, to take into account the droplets back-reaction reveals to be crucial.

A full description of the droplets nucleation, condensation and evaporation, within a turbulent vapour flow, requires to include at least the two-way coupling effects, i.e. mass, momentum and energy transfer between the carrier and the disperse phases.

This can be done solving simultaneously the system of equation for the carrier phase, with source terms representing the N_p droplets back-reaction on the fluid,

$$\frac{\partial \rho}{\partial t} + \nabla \cdot (\rho \mathbf{u}) = -\sum_{p=1}^{N_p} \dot{m}_p g_p(t)$$

$$\rho \left(\frac{\partial \mathbf{u}}{\partial t} + \mathbf{u} \cdot \nabla \mathbf{u} \right) = -\frac{1}{\gamma_0} \nabla p + \frac{1}{\mathrm{Re}} \nabla \cdot \Sigma - \sum_{p=1}^{N_p} \left(\mathbf{D}_p + \dot{m}_p \mathbf{v}_p \right) g_p(t - \varepsilon_R)$$

$$\rho c_p \left(\frac{\partial \theta}{\partial t} + \boldsymbol{u} \cdot \nabla \theta \right) = \frac{1}{\mathrm{Re\,Pr}} \nabla^2 \theta - \frac{\left(c_p^v - c_p^g \right)}{\mathrm{Re\,Sc}} \nabla \cdot (\rho \theta \, \nabla Y) - \sum_{p=1}^{N_p} \left(\dot{Q}_p + \dot{m}_p h_s^v \right) g_p(t - \varepsilon_R)$$

$$\rho \left(\frac{\partial Y}{\partial t} + \mathbf{u} \cdot \nabla Y \right) = \frac{1}{\mathrm{Re\,Sc}} \nabla \cdot (\rho \nabla Y) - \sum_{p=1}^{N_p} \dot{m}_p g_p(t - \varepsilon_R)$$

and a set of equations for each droplet dynamics: mass, position and temperature along its trajectory, with the same (although opposite) contributions found in the carrier phase equations, i.e. the coupling terms, according to [17]

$$\frac{\mathrm{d}r_p}{\mathrm{d}t} = \frac{1}{4\pi r_p^2 \rho_p} \dot{m}_p \tag{19.3}$$

$$\frac{\mathrm{d}\boldsymbol{v}_p}{\mathrm{d}t} = \frac{1}{m_p} \mathbf{D}_p \,, \quad \frac{\mathrm{d}\boldsymbol{x}_p}{\mathrm{d}t} = \boldsymbol{v}_p \tag{19.4}$$

$$\frac{\mathrm{d}\theta_p}{\mathrm{d}t} = \frac{1}{m_p c_l} \left(\dot{Q}_p + \dot{m}_p L_v \right). \tag{19.5}$$

Several DNS have been carried out here, exploiting an hybrid Eulerian–Lagrangian approach for the *carrier* mixture and the *disperse* droplets, following the point-particle approach [1]. The low Mach number formulation of the Navier–Stokes equations, accounts for the density variations, due to the thermodynamics coupling. The singular droplets back-reaction [17], has been regularised and time-delayed in a physically consistent way [7], with a Gaussian function and a *regularisation time* ε_R

$$g_p(t) = g_p(\mathbf{x} - \mathbf{x}_p(t)) = \frac{1}{(4\pi\ \nu\varepsilon_R)^{3/2}} \exp\left[-\frac{\|\mathbf{x} - \mathbf{x}_p(t)\|^2}{4\nu\varepsilon_R} \right]. \tag{19.6}$$

The expression for inter-phases mass, momentum and energy flux [12–14] reads

$$\dot{m}_p = 2\pi r_p(t) \frac{\mathrm{Sh}}{\mathrm{Re\,Sc}} \left(\tilde{Y} - Y^s_{eq}\right) \tag{19.7}$$

$$\boldsymbol{D}_p = 6\pi r_p(t)\ \frac{1}{\mathrm{Re}}\ \left(\tilde{\boldsymbol{u}} - \boldsymbol{v}_p\right), \quad \dot{Q}_p = 2\pi r_p(t) \frac{\mathrm{Nu}}{\mathrm{RePr}} \left(\tilde{\theta} - \theta_p\right) \tag{19.8}$$

Droplet Homogeneous Nucleation in a Vapour Jet

A parallel code for unsteady incompressible fluid flows, employing second-order central finite differences and a fourth-order Runge–Kutta scheme was used [3], to carry out the DNS. The inlet condition to the turbulent jet is provided by a developed turbulent pipe flow that runs simultaneously. The computational domain is cylindrical and dimensions are given in terms of pipe diameter: its height is $20 \times d_0$ while the free-jet diameter is $15 \times d_0$; the grid is uniform in the stream-wise direction and non-uniform in the radial one.

Traction-free boundary conditions on the lateral sides let the jet *being* open, ensuring the entrainment of *dry air*, while periodical and no-slip boundary conditions are prescribed, respectively, along the pipe axial and azimuthal directions. A bulk $\mathrm{Re} = 3000$ in the jet corresponds to a friction Reynolds number $\mathrm{Re}_\tau = 110$ in the pipe.

In the current study, several DNS reproduced the flow configuration addressed by the experimental setup [10], already investigated by RANS/LES in [4, 6, 18]. The droplets dynamics was there *characterised* with a completely different approach [20], the so-called Population Balance Equation (PBE) – sometimes referred to as General Dynamics Equation (GDE), in the aerosol literature – for the evolution of the particle size distribution. Various inlet conditions (DBP molar fraction $X_{jet} = 1.8 \div 3.2 \times 10^{-4}$) were investigated. A representative one has been chosen, to quantify how much the droplets back-reaction affects subsequent nucleation (Fig. 19.1), in addition to the effect of turbulent fluctuations, discussed in [5, 15].

Despite the particles back-reaction slightly affects (decreasing) the saturation ratio, because of the aforementioned nonlinear dependence, an unquestionable effect is found on the nucleation rate (see Fig. 19.1b). In fact, without considering the particle back-reaction on the fluid (Fig. 19.1e), the nucleation rate is overestimated by a 30%. It is worth noting that an overestimation of the same amount is found looking at the so-called *RANS-like rate* (Fig. 19.1c), that is the usual rate (Eq. 19.1), but providing as input the *actual* mean fields, averaging our DNS instantaneous fields. Thus it seems that the main effect of turbulent fluctuations is not only to

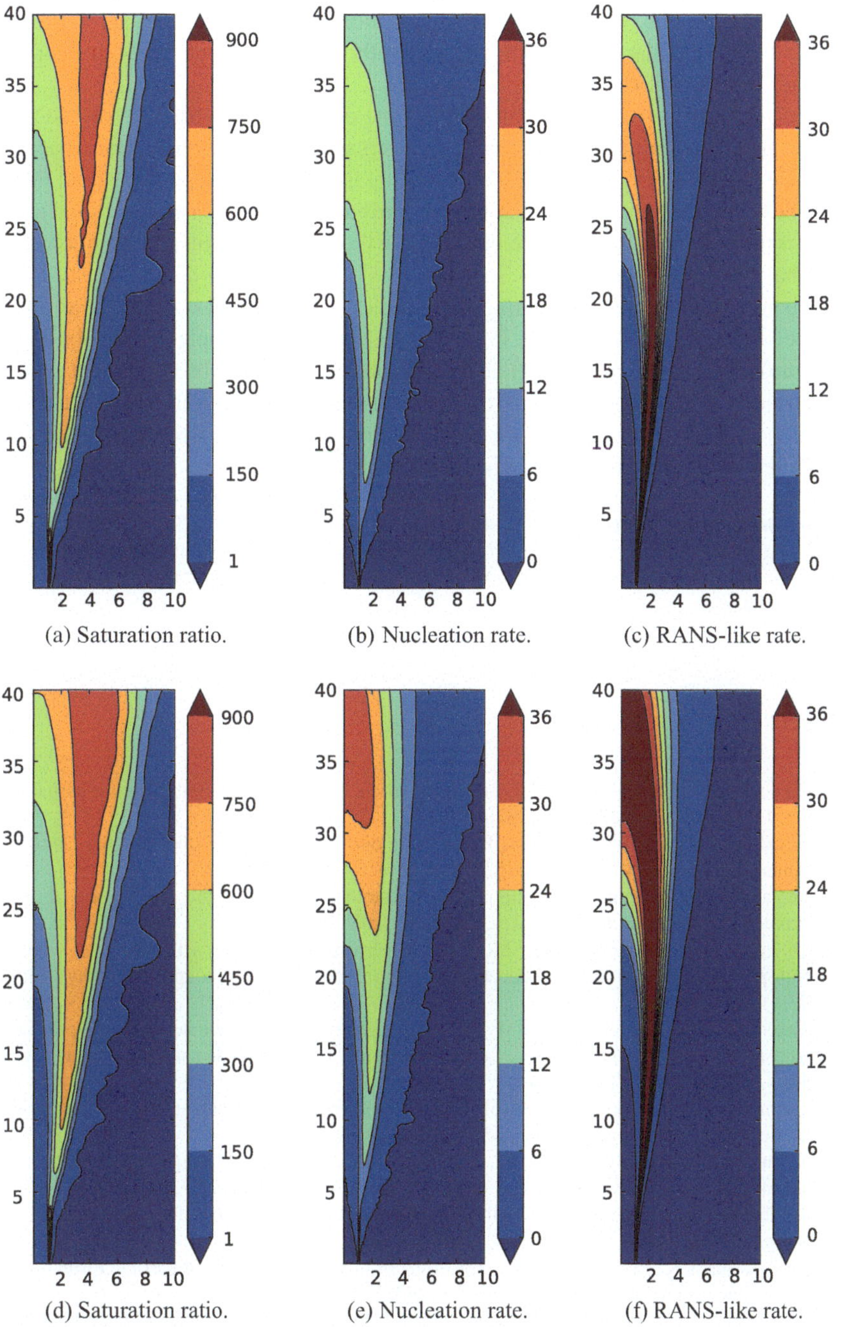

Fig. 19.1 Mean-fields comparison between two-way (top) and one-way (bottom) coupling regimes, for the same cases with $X_{jet} = 2.5 \times 10^{-4}$ and $\theta_{jet} = 140\,°\mathrm{C}$ as inlet conditions and ambient pressure at $\theta_{amb} = 25\,°\mathrm{C}$ in dry air, at rest. Environment conditions are taken as reference quantities

locally decrease the peak value of the nucleation rate, but overall to spread it to a wider region (Fig. 19.1b, c), with respect to the *RANS-like* one, in both coupling regimes (Fig. 19.1e, f).

The phase-change mainly contributes to the droplets back-reaction (on the thermodynamics) in a couple of ways: not only decreasing the amount of vapour, because of nucleation/condensation, but also increasing the droplets temperature (and therefore the fluid one), through the latent heat contribution of Eq. 19.5, proportional to the mass transfer (Eq. 19.7). Both effects decrease the saturation ratio and the nucleation rate that strongly depends on it.

We may conclude that neglecting the particles back-reaction is as wrong as it is to neglect the turbulent fluctuations. Probably, this is due to the droplets redistribution of vapour (condensing/evaporating) and temperature (releasing/absorbing heat), along their trajectory. It seems that the particles back-reaction further enhances the turbulent fluctuations. However, as far as we know, it is not possible to effectively model it in a simpler way, rather than to *simply* evolve each single droplet.

Acknowledgements We acknowledge the CINECA award under the ISCRA initiative, for the availability of HPC resources and support. The research leading to these results has received access to resource MARCONI, under the ISCRAC Project grant no. HP10CQBSIR from Cineca based in Bologna, Italy.

References

1. Balachandar, S.: A scaling analysis for point-particle approaches to turbulent multiphase flows. Int. J. Multiph. Flow. **35**(9), 801–810 (2009)
2. Becker, R., Döring, W.: Kinetische Behandlung der Keimbildung in Übersättigten Dämpfen. Annalen der Physik **416**(8), 719–752 (1935)
3. Dalla Barba, F., Picano, F.: Clustering and entrainment effects on the evaporation of dilute droplets in a turbulent jet. Phys. Rev. Fluids **3**(3), 034304 (2018)
4. Di Veroli, G.Y., Rigopoulos, S.: Modeling of aerosol formation in a turbulent jet with the transported population balance equation-probability density function approach. Phys. Fluids **23**(4), 043305 (2011)
5. Fager, A.J., Liu, J., Garrick, S.C.: Hybrid simulations of metal particle nucleation: a priori and a posteriori analyses of the effects of unresolved scalar interactions on nanoparticle nucleation. Phys. Fluids **24**(7), 075110 (2012)
6. Garmory, A., Mastorakos, E.: Aerosol nucleation and growth in a turbulent jet using the stochastic fields method. Chem. Eng. Sci. **63**(16), 4078–4089 (2008)
7. Gualtieri, P., Picano, F., Sardina, G., Casciola, C.M.: Exact regularized point particle method for multiphase flows in the two-way coupling regime. J. Fluid Mech. **773**, 520–561 (2015)
8. Jacobson, M.Z.: Fundamentals of Atmospheric Modelling. Cambridge University Press, Cambridge (2005)
9. Lesniewski, T.K., Friedlander, S.K.: The effect of turbulence on rates of particle formation by homogeneous nucleation. Aerosol Sci. Technol. **23**(2), 174–182 (1995)
10. Lesniewski, T.K., Friedlander, S.K.: Particle nucleation and growth in a free turbulent jet. Proc. Royal Soc. A **454**(1977), 2477–2504 (1998)
11. Lim, S., Cha, J., Lee, H., Kim, T., Shin, W.G.: Understanding the condensation process of turbulent steam jet using the PDPA system. Int. J. Multiph. Flow **98**, 168–181 (2018)

12. Maxey, M.R., Riley, J.J.: Equation of motion for a small rigid sphere in a nonuniform flow. Phys. Fluids **26**(4), 883–889 (1983)
13. Michaelides, E.E., Feng, Z.: Heat transfer from a rigid sphere in a nonuniform flow and temperature field. Int. J. Heat Mass Transf. **37**(14), 2069–2076 (1994)
14. Miller, R.S., Harstad, K., Bellan, J.: Evaluation of equilibrium and non-equilibrium evaporation models for many-droplet gas-liquid flow simulations. Int. J. Multiph. Flow **24**(6), 1025–1055 (1998)
15. Murfield, N.J., Garrick, S.C.: The effects of unresolved scalar fluctuations during homogeneous nucleation. Aerosol Sci. Tech. **47**(7), 806–817 (2013)
16. Oerlemans, S., Badie, R., Van Dongen, M.E.H.: An experimental and numerical study into turbulent condensing steam jets in air. Exp. Fluids **31**(1), 74–83 (2001)
17. Okong'o, N.A., Bellan, J.: Consistent large-eddy simulation of a temporal mixing layer laden with evaporating drops. Part 1. Direct numerical simulation, formulation and a priori analysis. J. Fluid Mech. **499**, 1–47 (2004)
18. Pesmazoglou, I., Kempf, A.M., Navarro-Martinez, S.: Aerosol nucleation in a turbulent jet using large eddy simulations. Chem. Eng. Sci. **116**, 383–397 (2014)
19. Vehring, R.: Pharmaceutical particle engineering via spray drying. Pharm. Res. **25**(5), 999–1022 (2008)
20. Zhou, K., Attili, A., Alshaarawi, A., Bisetti, F.: Simulation of aerosol nucleation and growth in a turbulent mixing layer. Phys. Fluids **26**(6), 065106 (2014)

Chapter 20
A-Priori Assessment of Interfacial Sub-grid Scale Closures in the Two-Phase Flow LES Context

J. Hasslberger, S. Ketterl and M. Klein

LES Formalism

Turbulent bubbly flows play an essential role in a large number of technical applications, e.g., for chemical reactors in the process industry. Since DNS is usually too expensive for the technically relevant Reynolds numbers, sub-grid closures for the computationally more efficient LES are urgently needed. Using the usual notation, the filtered LES equations [7] based on the one-fluid formulation read

$$\frac{\partial \overline{u}_i}{\partial x_i} = 0 \tag{20.1}$$

$$\frac{\partial \overline{\rho}\,\overline{u}_i}{\partial t} + \frac{\partial \overline{\rho}\,\overline{u}_i \overline{u}_j}{\partial x_j} = -\frac{\partial \overline{p}}{\partial x_i} + \frac{\partial}{\partial x_j}\left[\overline{\mu}\left(\frac{\partial \overline{u}_i}{\partial x_j} + \frac{\partial \overline{u}_j}{\partial x_i}\right)\right] + \overline{\rho} g_i + \sigma\, \overline{n}_i^s\, \overline{\kappa}^s\, \overline{\delta}_S$$
$$- \frac{\partial}{\partial x_j}\tau_{\rho uu,ij} + \frac{\partial}{\partial x_j}\tau_{\mu S,ij} + \tau_{nn,i} - \frac{\partial}{\partial t}\tau_{tt,i} \tag{20.2}$$

$$\frac{\partial \overline{\alpha}}{\partial t} + \frac{\partial}{\partial x_i}(\overline{\alpha}\,\overline{u}_i) = -\frac{\partial}{\partial x_i}\tau_{\alpha u,i} \tag{20.3}$$

J. Hasslberger (✉) · S. Ketterl · M. Klein
Institute of Applied Mathematics and Scientific Computing,
Bundeswehr University Munich, Munich, Germany
e-mail: josef.hasslberger@unibw.de

M. García-Villalba et al. (eds.), *Direct and Large Eddy Simulation XII*,
ERCOFTAC Series 27,
https://doi.org/10.1007/978-3-030-42822-8_20

with the corresponding sub-grid closure terms

$$\tau_{\rho uu,ij} = \overline{\rho\, u_i u_j} - \overline{\rho}\, \overline{u}_i\, \overline{u}_j \tag{20.4}$$

$$\tau_{tt,i} = \overline{\rho\, u_i} - \overline{\rho}\, \overline{u}_i \tag{20.5}$$

$$\tau_{\mu S,ij} = \overline{\mu \left(\frac{\partial\, u_i}{\partial x_j} + \frac{\partial\, u_j}{\partial x_i} \right)} - \overline{\mu} \left(\frac{\partial\, \overline{u}_i}{\partial x_j} + \frac{\partial\, \overline{u}_j}{\partial x_i} \right) \tag{20.6}$$

$$\tau_{nn,i} = \sigma\, \overline{n_i\, \kappa}^s\, \overline{\delta}_S - \sigma\, \overline{n}_i^s\, \overline{\kappa}^s\, \overline{\delta}_S \tag{20.7}$$

$$\tau_{\alpha u,i} = \overline{\alpha\, u_i} - \overline{\alpha}\, \overline{u}_i \tag{20.8}$$

In addition to volume filtering

$$\overline{\phi(x)} = \int \phi(y) G(x-y) dy \tag{20.9}$$

with a Gaussian filter kernel G, surface filtering is defined by $\overline{\phi}^s = \overline{\phi \delta_S} / \overline{\delta}_S$, which is used to reformulate the filtered surface tension term in this work:

$$\sigma\, \overline{n_i\, \kappa\, \delta_S} = \sigma\, \overline{n_i\, \kappa}^s\, \overline{\delta}_S. \tag{20.10}$$

Although the surface tension force is singular in theory, δ_S has to be approximated in numerical practice, usually by $\delta_S = |\nabla \alpha|$. It is known from literature [4] that the diffusive term closure (Eq. 20.6) is of inferior importance. The closure of momentum advection (Eq. 20.4) is of significant importance but a large variety of models is already available from single-phase flows that can be reasonably well modified for two-phase flows [3]. Thus, the present study focuses on the crucial gaps, namely the closures of volume fraction advection (Eq. 20.8) and the quasi-singular surface tension force (Eq. 20.7), which contribute only in the vicinity of the interface of both phases. It is worth noting that Eqs. 20.5 and 20.8 are of the same mathematical nature.

DNS Database

The DNS database has been generated by the state-of-the-art two-phase flow solver PARIS (unsteady incompressible Navier–Stokes equations, finite-volume discretization, QUICK scheme for momentum advection, geometrical interface reconstruction, curvature calculation by height function method, pressure projection method) [6], and explicitly filtered for varying filter width Δ. As indicated in Fig. 20.1, the DNS setup consists of a well-resolved (64 cells per spherical bubble diameter) single air bubble in a water-filled periodic box. A wobbling ellipsoidal bubble is observed due to the imposed fluid properties and a spherical diameter of 5 mm. The domain size is chosen such that the overall gas void fraction lies in the technically relevant

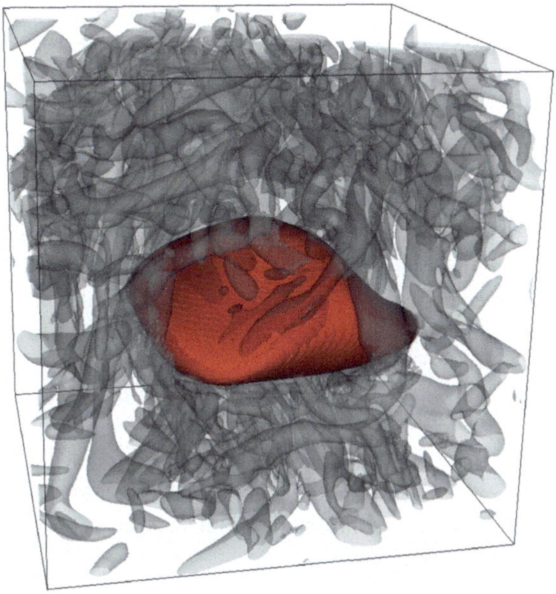

Fig. 20.1 The red surface represents the bubble interface in the turbulent flow and the semi-transparent gray $Q = 10^4/s^2$ iso-contour indicates dominant vortical structures in the analyzed snapshot. Gravitational acceleration is acting in vertical direction

range (6.6% here). Since the turbulence is induced by the buoyant bubble itself, the simulation is continued until a converged state of the domain-averaged enstrophy is reached. The chaotic flow structure resembles the behavior in a dense bubble swarm, characterized by a statistically significant number of turbulence–interface interaction events.

Models and Results

Besides potentially suitable existing models for $\tau_{\alpha u,i}$ (the gradient flux model GFM, Clark's tensor diffusivity model CTM, Bardina's scale similarity model SS, the two-phase version of the Bray–Moss–Libby model BML), several new models are tested. Depending on the model variant, the filter-scale velocity gradient tensor $\overline{A}_{ij} = \partial \overline{u}_i / \partial x_j$ in the original two-phase BML model (inspired by flamelet theory of turbulent combustion) [3]

$$\tau_{\alpha u,i}^{BML} = \overline{\alpha}(1 - \overline{\alpha}) \Delta \overline{A}_{ij} \overline{n}_j \tag{20.11}$$

is either replaced by the blending $(1 - F_{CS})\overline{S}_{ij} + (1 + F_{CS})\overline{W}_{ij}$ (BML-F), or by the blending $(1 - \overline{\alpha})\overline{S}_{ij} + \overline{\alpha}\,\overline{W}_{ij}$ (BML-SW), where $\overline{S}_{ij}$ and $\overline{W}_{ij}$ are the symmetric and anti-symmetric part of $\overline{A}_{ij}$, respectively. In the first formulation BML-F, the coherent structure function is defined as $F_{CS} = 2Q/(\overline{A}_{ij})^2$, where Q is the second invariant of $\overline{A}_{ij}$. The blending by means of $-1 \leq F_{CS} \leq 1$ takes into account whether the flow is locally strain- or vorticity-dominated on the resolved level. The same dominant flow behavior can then be assumed on sub-grid level because in LES, other than in (U)RANS, there is usually a sufficiently high correlation between the strain–vorticity relationship in the DNS and the resolved level [5]. It has been found in literature [2] that especially the regions of high interface curvature in bubbly flows are dominated by vortices. In regions characterized by $F_{CS} = 0$ (strain–vorticity equilibrium), the original formulation (Eq. 20.11) is locally recovered. The simplified second formulation BML-SW utilizes the fact that the bubble interior and exterior close to the interface are usually vorticity- and strain-dominated, respectively. Here, α is taken to be the volume fraction of the lighter phase, in which the vorticity tends to accumulate in unsteady turbulent flows [2, 8]. Finally, the scale similarity (SS) ansatz for this term reads

$$\tau_{\alpha u,i}^{SS} = \widehat{\overline{\alpha}\,\overline{u}_i} - \widehat{\overline{\alpha}}\,\widehat{\overline{u}}_i \tag{20.12}$$

where the secondary test filter $(\widehat{\phi})$ is implemented according to Anderson and Domaradzki [1]. A pragmatic approach (without stringent derivation) accounting for both the scale similarity principle as well as the bi-modal of α (not $\overline{\alpha}$) distribution can thus be constructed as

$$\tau_{\alpha u,i}^{SS-BML} = 4\,\overline{\alpha}(1-\overline{\alpha})\left(\widehat{\overline{\alpha}\,\overline{u}_i} - \widehat{\overline{\alpha}}\,\widehat{\overline{u}}_i\right) \tag{20.13}$$

which reduces to Eq. 20.12 at $\overline{\alpha} = 0.5$.

Some correlation results, before and after taking the divergence, are presented in Fig. 20.2. The Pearson correlation coefficient has a value between -1 and $+1$, where $+1$ is total positive linear correlation, 0 is no linear correlation, and -1 is total negative linear correlation. The bad performance of the GFM model is attributed to the fact that this model is the only one which cannot switch between gradient and counter-gradient transport depending on local conditions. It is interesting to note that Clark's model is obtained when $\overline{\alpha}(1-\overline{\alpha})$ in the BML model (Eq. 20.11) is replaced

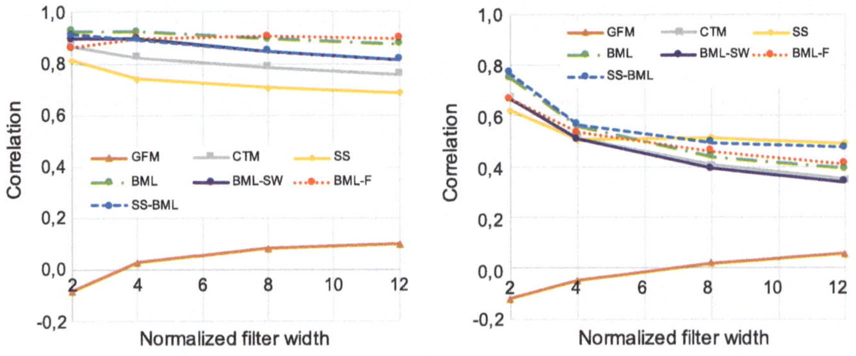

Fig. 20.2 Correlation of $\tau_{\alpha u,i}$ in vertical direction (left) and $\partial\tau_{\alpha u,i}/\partial x_i$ (right) with the corresponding models, depending on the normalized filter width Δ/Δ_{DNS}

by $|\nabla\overline{\alpha}|\ \Delta$. Whereas Clark's model is likely to perform better for a passive scalar, the $\overline{\alpha}(1-\overline{\alpha})$-proportionality accounts for the discontinuous two-phase interface prohibiting intermediates states of α (bi-modal distribution). This property is assumed to be the main reason for the superior performance of the BML-type models before taking the divergence. After taking the divergence, correlation values are generally lower due to additional differentiation errors on the LES grid and the SS-type models show the highest correlation results.

To the best of authors' knowledge, no well-established model exists in literature for $\tau_{nn,i}$ in the context of surface filtering. Multiplication of Shirani's original model in the context of volume filtering by the correction $(1-2\overline{\alpha})$ makes the model applicable in the surface filtering context:

$$\tau_{nn,i}^{SHIR-Corr} = C_{st}\,(1-2\overline{\alpha})\sqrt{|\overline{S}_{ij}|}\,\frac{\Delta}{\sqrt{\nu}}\,\sigma\,\overline{n}_i^s\,\overline{\kappa}^s\,\overline{\delta}_S \tag{20.14}$$

with $C_{st}=0.15$. According to the schematic in Fig. 20.3, it seems reasonable that the sub-grid surface tension force changes its sign on both sides of the resolved/filtered interface because it is the nature of (sub-grid) surface tension to minimize the interface surface area. The linear correction within the interface region (varies from $+1$ at $\overline{\alpha}=0$ to -1 at $\overline{\alpha}=1$) may be assumed as a first approximation. On average, the lowest sub-grid curvature occurs at $\overline{\alpha}=0.5$ and, hence, the sub-grid surface tension force is assumed to vanish in this region. An alternative, parameter-free, approach is again given by the scale similarity ansatz for this term:

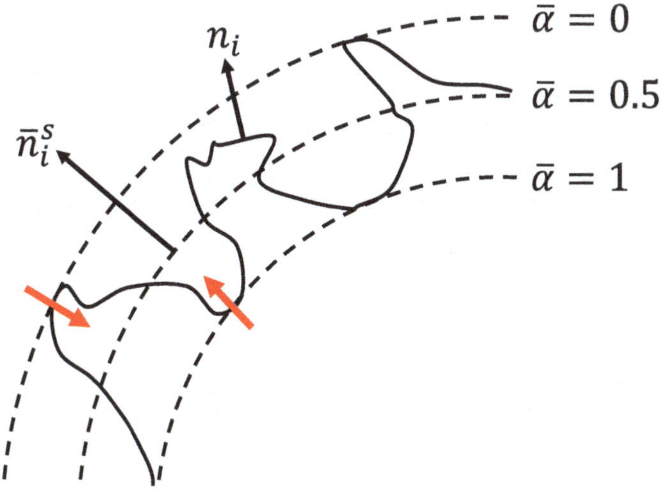

Fig. 20.3 Relationship between the real interface (continuous line) and the resolved/filtered interface (dashed lines). Red arrows indicate sub-grid surface tension forces

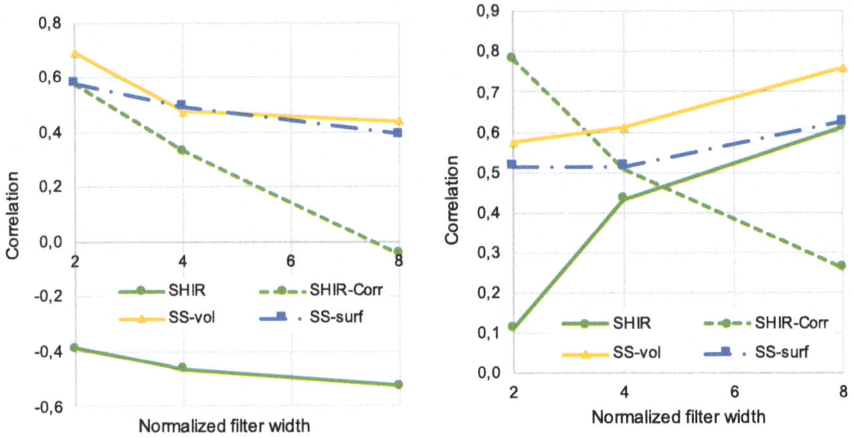

Fig. 20.4 Correlation of $\tau_{nn,i}$ and the corresponding models in vertical (left) and lateral (right) direction, depending on the normalized filter width Δ/Δ_{DNS}

$$\tau_{nn,i}^{SS-vol} = \sigma\left(\widehat{\overline{n}_i^s\,\overline{\kappa}^s} - \widehat{\overline{n}_i^s}\,\widehat{\overline{\kappa}^s}\right)\overline{\delta}_S \tag{20.15}$$

It can also be formulated with a surface test filter $\widehat{\phi}^s$ (SS-surf) instead of a volume test filter $\widehat{\phi}$ (SS-vol), which might be more consistent and robust in the surface filtering context.

The magnitude of this term can be of the same order as the magnitude of the momentum advection sub-grid term [4] and it is potentially important for topological changes, as well as vorticity production at the interface, cf. [2]. Considering the high sensitivity of this term, Fig. 20.4 attests a convincing performance of the SS-type models. The proposed correction leads to a significant improvement of existing models which are directly proportional to the resolved/surface-filtered curvature and normal vector.

Concluding Remarks

Using a-priori analysis, several promising candidates for closing the volume fraction advection equation have been identified and the impact of different model contributions is discussed. It seems to be important to account for the bi-modal character of the volume fraction. Regarding the closure of sub-grid surface tension, a linear correction is proposed to make available models applicable in the surface filtering context as investigated here. However, that modeling idea might be limited to wrinkled/corrugated interfaces and not be valid for highly fragmented interfaces. In addition, the scale similarity ansatz looks convincing.

The robustness of the tested/proposed models has to be checked in a-posteriori analysis—including additional challenges like the interaction with numerical schemes and sub-grid models of other terms, particularly the closure of momentum advection.

References

1. Anderson, B.W., Domaradzki, J.A.: A subgrid-scale model for large-eddy simulation based on the physics of interscale energy transfer in turbulence. PoF **24**(6), 065104 (2012)
2. Hasslberger, J., Klein, M., Chakraborty, N.: Flow topologies in bubble-induced turbulence: a direct numerical simulation analysis. J. Fluid Mech. **857**, 270–290 (2018)
3. Ketterl, S., Klein, M.: A-priori assessment of subgrid scale models for large-eddy simulation of multiphase primary breakup. Comput. Fluids **165**, 64–77 (2018)
4. Klein, M., Ketterl, S., Hasslberger, J.: Large eddy simulation of multiphase flows using the volume of fluid method: Part 1—Governing equations and a priori analysis. Exp. Comput. Multiph. Flow **1**(2), 130–144 (2019)
5. Kobayashi, H.: The subgrid-scale models based on coherent structures for rotating homogeneous turbulence and turbulent channel flow. Phys. Fluids **17**(4), 045104 (2005)
6. Ling, Y., Zaleski, S., Scardovelli, R.: Multiscale simulation of atomization with small droplets represented by a lagrangian point-particle model. Int. J. Multiph. Flow **76**, 122–143 (2015)
7. Toutant, A., Chandesris, M., Jamet, D., and Lebaigue, O.: Jump conditions for filtered quantities at an under-resolved discontinuous interface. Part 1: Theoretical development. International Journal of Multiphase Flow, **35(12)**, 1100–1118 (2009)
8. Tripathi, M.K., Sahu, K.C., Govindarajan, R.: Why a falling drop does not in general behave like a rising bubble. Scientific reports **4**, 4771 (2014)

Chapter 21
Analysis of Regularised Scale Similarity Type Models in the Context of Two-Phase Flow with Moving Boundaries

S. Ketterl, H. Kobayashi and M. Klein

Introduction

In contrast to Large-Eddy Simulation (LES) of single phase flows which is well established in computational fluid dynamics, the LES of two-phase flows with moving boundaries is at an early development stage. On the one hand, the presence of the phase interface results in additional subgrid scale (SGS) terms in the LES formalism which cannot be neglected [5]. On the other hand, existing modelling approaches for the SGS stress tensor cannot be simply transferred from single to multiphase flows due to the discontinuous character of the density across the phase interface [3]. The moving interface acts to some extent in analogy to a wall [2] and involves an anisotropic turbulence field. Eddy viscosity models showed to be less suitable in the vicinity of the interface and a damping of the turbulent viscosity towards the interface might be needed. Scale similarity models in contrast revealed reasonable accuracy in a-priori studies [3] but lack of sufficient dissipation in a-posteriori LES led to a destabilisation of the computation. Recently, Kobayashi [6] proposed a regularised scale similarity type model which prevents backscatter of energy from the small scales to the large scales. The modelling approach provides the accuracy of scale similarity type models, while the included regularisation guarantees stability of the simulation. The goal of this contribution is a combined a-priori and a-posteriori

S. Ketterl (✉) · M. Klein
Department of Aerospace Engineering, Universität der Bundeswehr München, Neubiberg, Germany
e-mail: sebastian.ketterl@unibw.de

M. Klein
e-mail: markus.klein@unibw.de

H. Kobayashi
Department of Physics & Research and Education Center for Natural Sciences, Hiyoshi Campus, Keio University, Yokohama, Japan
e-mail: hkobayas@keio.jp

M. García-Villalba et al. (eds.), *Direct and Large Eddy Simulation XII*, ERCOFTAC Series 27,
https://doi.org/10.1007/978-3-030-42822-8_21

analysis of a modified version of Kobayashi's model in the context of two-phase flow with moving boundaries. The model will be applied to the simulation of the primary atomization of a round liquid jet and LES results will be compared to a direct numerical simulation (DNS) database.

Two-Phase Flow Large-Eddy Simulation

Applying a filter operation to the incompressible Navier–Stokes equations in one-fluid formulation yields the Favre filtered one-fluid LES equations [8]

$$\frac{\partial \widetilde{u}_i}{\partial x_i} = \tau_{\rho u} \tag{21.1}$$

$$\begin{aligned}\frac{\partial \overline{\rho}\,\widetilde{u}_i}{\partial t} + \frac{\partial\,\overline{\rho}\,\widetilde{u}_i\widetilde{u}_j}{\partial x_j} &= -\frac{\partial \overline{p}}{\partial x_i} + \frac{\partial}{\partial x_j}\left[\overline{\mu}\left(\frac{\partial \widetilde{u}_i}{\partial x_j} + \frac{\partial \widetilde{u}_j}{\partial x_i}\right)\right] + \overline{\sigma}\,\overline{n}_i\,\overline{\kappa}\,\overline{\delta}_S \\ &+ \tau_{nn,i} - \frac{\partial}{\partial x_j}\rho\,\tau_{uu,ij} + \frac{\partial}{\partial x_j}\tau_{\mu S,ij}\end{aligned} \tag{21.2}$$

$$\frac{\partial \overline{\alpha}}{\partial t} + \frac{\partial}{\partial x_i}(\overline{\alpha}\,\widetilde{u}_i) = 0 \tag{21.3}$$

where $\widetilde{Q} = \overline{\rho\,Q}/\overline{\rho}$ denotes the Favre filtered value of a general quantity Q. Due to the convolution operation, four unclosed SGS terms $\tau_{uu,ij}$, $\tau_{\mu S,ij}$, $\tau_{nn,ij}$ and $\tau_{\rho u}$ emerge. These SGS terms contain the effects of small scale turbulence and interfacial motion which are unresolved by the LES mesh and need to be treated by modelling. The convective SGS term

$$\tau_{uu,ij} = (\widetilde{u_i u_j} - \widetilde{u}_i\,\widetilde{u}_j) \tag{21.4}$$

is known as SGS stress from single phase flow and is the focus of this work. In the context of multiphase flow, the well-known Clark's tensor diffusivity model [1]

$$\tau_{uu,\,ij}^{CTM} = \frac{\Delta^2}{12}\frac{\partial \widetilde{u}_i}{\partial x_k}\frac{\partial \widetilde{u}_j}{\partial x_k} \tag{21.5}$$

did show high correlations with the stresses evaluated from DNS [3]. For isotropic filters, the CTM model can be thoroughly derived from the model expression itself Eq. 21.4 assuming that the filter operation can be written as the Taylor series and truncating the expression after the second-order terms. A more general derivation can be found in [9]. It is known that such gradient type models often do not provide a sufficient amount of dissipation in real LES and very recently Kobayashi [6] suggested a model formulation to regularise scale similarity type models in order to make them usable in real LES. Following the ideas of Kobayashi [6], the regularised CTM model reads

$$\tau^{KC}_{uu,\,ij} = \frac{-\tau^{CTM}_{uu,\,ab}\widetilde{S}_{ab} + \left|-\tau^{CTM}_{uu,\,ab}\widetilde{S}_{ab}\right|}{-\tau^{CTM}_{uu,\,ab}\widetilde{S}_{ab}}\tau^{CTM}_{uu,\,ij} + \left(\tau^{CTM}_{uu,\,ij} + \frac{-\tau^{CTM}_{uu,\,ab}\widetilde{S}_{ab}}{\widetilde{S}_{ab}\widetilde{S}_{ab}}\widetilde{S}_{ij}\right) \tag{21.6}$$

where $\widetilde{S}_{ab}$ is the Favre filtered deformation rate. The model will henceforth be called KC model. In contrast to the suggestion in [6], Eq. 21.6 makes use of the tensor $\tau^{CTM}_{\rho uu,\,ij}$ rather than its deviatoric part and the multiplying factor of the expression in large parentheses is set to unity in this work. This is motivated by the fact that no separate transport equation for SGS kinetic energy is solved in the present work. The first term in Eq. 21.6 yields only forward scatter and is a scale similarity term, whereas the second term does not contribute to the SGS energy transfer but instead improves the SGS force [6]. It is worth mentioning that the same modelling philosophy can be applied to any other scale similarity type model and as well to the filtered (rather than Favre filtered) version of Eqs. 21.1–21.3 which have been analysed in detail [3].

A-Priori Analysis

By filtering DNS flow data [3], an a-priori analysis allows the evaluation of the exact SGS terms and a comparison with LES models. The a-priori study is based on DNS data of the primary atomization of a round Diesel jet characterised by $\mathrm{Re} = 5000$, $\mathrm{We} = 2000$ and a density and viscosity ratio of 40. The DNS database is generated using the *PARIS* simulator code [7]. Numerical methods including boundary conditions, domain size, etc., are explained in detail in [4, 5, 7]. The DNS is discretized with 128 cells per nozzle diameter which gives in total 906 million grid cells.

The models have been evaluated conditional on the flow region in the following manner: The two bulk flows, the areas of pure gas and pure liquid, are denoted ‘g’ and ‘l’, respectively. The interface is marked by ‘i’, the region towards the interface from the gas side is denoted by ‘g–i’ and the region approaching the interface from the liquid phase with ‘i–l’.

Figure 21.1 shows the local correlation between the individual contributions of the KC model and the baseline model CTM for a filter size of $\Delta = 4\Delta x$ and $\Delta = 12\Delta x$ for a selected tensor and vector component. The model and its individual contributions show the same tendencies for both filter widths. The complete (all individual terms) KC model correlates similar to original CTM model which has the highest correlation strength. Although backscatter occurs quite frequently according to our evaluation, the correlation of the first term which allows only forward scatter of energy is slightly lower but still in the typical range of scale similarity type models. The third term of the KC model Eq. 21.6 achieves low level of correlation with the true stresses similar to single phase flow. The deformation rate representing eddy viscosity type models (e.g. the Smagorinsky or Nicoud model, please refer to [3] for details) is not a suitable approximation of the subgrid stress. The correlation coefficient of the complete second term is higher than the first term and improves

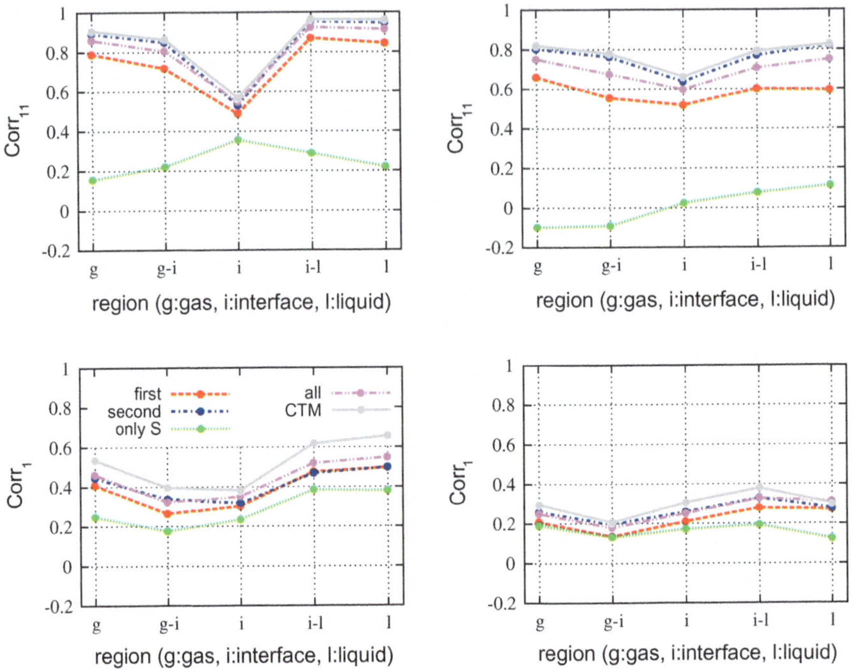

Fig. 21.1 Correlation coefficients of the individual contributions of the regularised scale similarity model for the SGS stress tensor $\tau_{uu,ij}$ depending on the flow region (gas and liquid flow 'g' and 'l', area adjacent to the interface 'g–i' and 'i–l', the interface 'i') plotted for filter width $\Delta = 4\,\Delta_{DNS}$ (left) and $\Delta = 12\,\Delta_{DNS}$ (right). Evaluated is the correlation for a single tensor component (top) and on vector level (bottom) after taking the divergence

the SGS force. The influence of the deformation rate in the second term, which is dominated by Clark's term $\tau^{CTM}_{uu,ij}$, is rather small.

A-Posteriori Analysis

The a-posteriori analysis is mainly intended to demonstrate that the model given by Eq. 21.6 runs stable which is not the case for a no model simulation and which is also not the case for the baseline expression $\tau^{CTM}_{uu,\,ij}$. However, the LES performance of the overall modelling approach will depend on a variety of terms [4] (see Eq. 21.2) and a final assessment based on a-posteriori analysis will hence not be possible. The current work focuses on the SGS stress modelling of $\tau_{uu,ij}$, while the SGS surface tension force $\tau_{nn,i}$ is modelled by a grid-based Weber number model [4]. The LES is only discretized with 16 cells per nozzle diameter resulting in 18 million cells which corresponds to a filter width $\Delta = 8\Delta x$ relative to the DNS. The computational effort

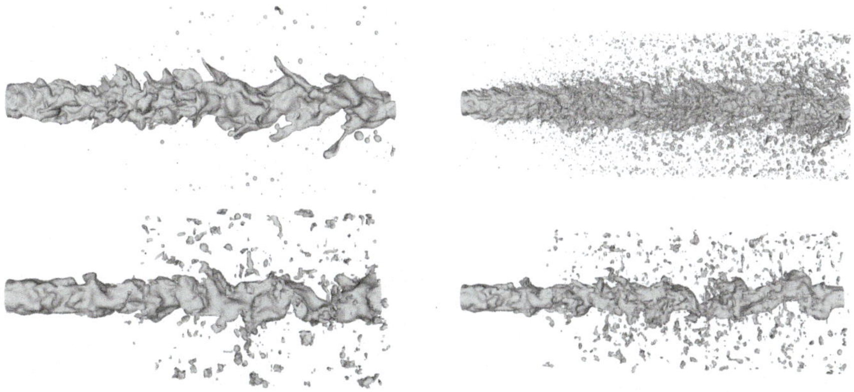

Fig. 21.2 Interface of a round liquid jet exhausting into a gaseous atmosphere for Re = 5000, We = 2000 (left) and Re = 8000, We = 5000 (right). The DNS result (top) is compared to the LES (bottom)

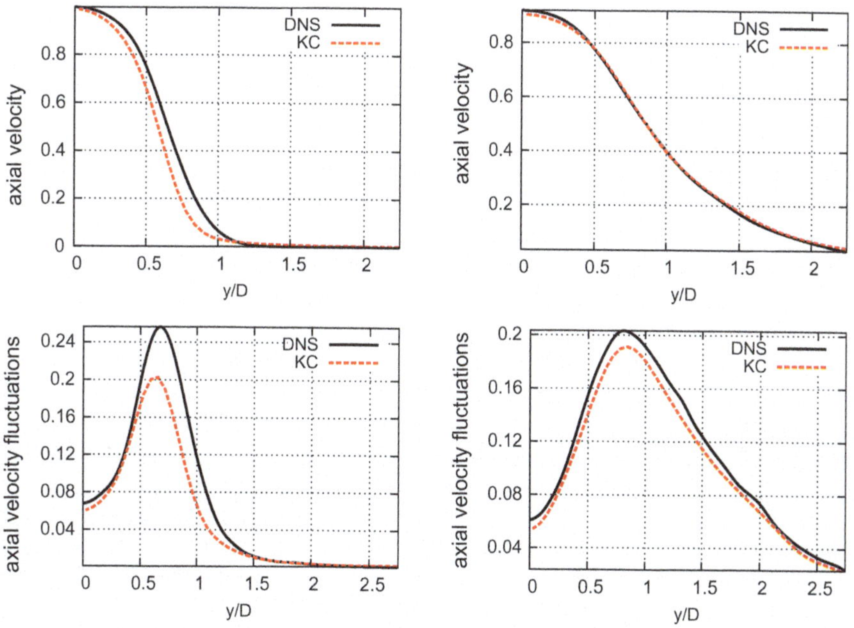

Fig. 21.3 Averaged axial velocity (top) and axial velocity fluctuations (bottom) for two axial positions $x/D = 3$ (left) and $x/D = 11$ (right) for a round jet characterised by Re = 5000, We = 2000. Compared is the DNS result to an LES with the regularised scale similarity model (KC)

is reduced by a factor of 4096 (assuming a complexity of the algorithm that scales with the number of unknowns and a timestep proportional to the grid spacing).

Figure 21.2 shows an atomizing round liquid jet for Re = 5000, We = 2000 (left) and Re = 8000, We = 5000 (right). It is important to note that the present LES model cannot predict droplets smaller than the filter width and a multiscale approach will be needed to describe such subgrid effects. A-posteriori results of the KC model (bottom) are compared to DNS data (top). Figure 21.3 shows the averaged axial velocity and the axial velocity fluctuations evaluated from the jet axis to the lateral boundaries for Re = 5000, We = 2000. Right after the injection of the jet, the LES depicts pronounced interface instabilities and the jet behaviour resembles the DNS results. Applying an eddy viscosity model would suppress this early formation of interface instabilities [4]. Further downstream strong corrugations at the surface emerge, ligaments are built and the formation of the Kelvin–Helmholtz instability in the shear layer and the Rayleigh–Taylor instability at the interface is supported. At this stage, an unphysical breakup of rather large liquid structures occurs which is to some extend counteracted by the model for $\tau_{nn,i}$ [4]. Although these results are encouraging, further work will be needed to refine the models and to analyse the effect of additional SGS terms ($\tau_{\mu S,ij}$ and $\tau_{\rho u}$, see Eq. 21.2) [4].

Conclusion

This contribution analysed a modified version of Kobayashi's regularised scale similarity model for the subgrid stress in the context of LES of two-phase flow with moving boundaries. The model has been evaluated by a combined a-priori and a-posteriori study based on a DNS database of primary atomization of a liquid jet. The a-priori study revealed that the included regularisation mechanism has only minor effects on the correlation strength which is nearly as high as for a pure scale similarity model. In the a-posteriori study, the regularisation procedure stabilises the gradient model as well as the overall simulation and promising results are obtained for the prediction of primary breakup by means of LES. Especially after jet injection, the regularised scale similarity model improves the formation of instabilities and the deformation of the interface.

Acknowledgements Support by the German Research Foundation (DFG, GS: KL1456/1-1) is gratefully acknowledged. Computer resources for this project have been provided by the Gauss Centre for Supercomputing/Leibniz Supercomputing Centre under grant: pr48no. The authors also express their gratitude to the developers of the PARIS Simulator for providing the source code.

References

1. Clark, R.A., Ferziger, J.H., Reynolds, W.C.: Evaluation of subgrid scale models using an accurately simulated turbulent flow. J. Fluid Mech. **91**(1), 1–16 (1979)
2. Fulgosi, M., Lakehal, D., Banerjee, S., De Angelis, V.: Direct numerical simulation of turbulence in a sheared air-water flow with a deformable interfac. J. Fluid Mech. **482**, 319–345 (2003)

3. Ketterl, S., Klein, M.: A-priori assessment of subgrid scale models for large-eddy simulation of multiphase primary breakup. Comput. Fluids **165**, 64–77 (2018)
4. Ketterl, S., Reissmann, M., Klein, M.: Towards large eddy simulation of multiphase flows using the volume of fluid method: Part 2 - a-posteriori analysis of liquid jet atomization. Exp. Comput. Multiph. Flow **1**(3), 201–211 (2019)
5. Klein, M., Ketterl, S., Hasslberger, J.: Towards large eddy simulation of multiphase flows using the volume of fluid method: Part 1 - governing equations and a-priori analysis. Exp. Comput. Multiph. Flow **1**(2), 130–144 (2019)
6. Kobayashi, H.: Improvement of the SGS model by using a scale-similarity model based on the analysis of SGS force and SGS energy transfer. Int. J. Heat Fluid Flow **72**, 329–336 (2018)
7. Ling, Y., Zaleski, S., Scardovelli, R.: Multiscale simulation of atomization with small droplets represented by a lagrangian point-particle model. Int. J. Multiph. Flow **76**, 122–143 (2015)
8. Liovic, P., Lakehal, D.: Subgrid scale modeling of surface tension within interface tracking-based Large-Eddy and Interface Simulation of 3D interfacial flows. Comput. Fluids **63**, 27–46 (2012)
9. Vreman, B., Geurts, B., Kuerten, H.: Large-eddy simulation of the temporal mixing layer using the clark model. Theoret. Comput. Fluid Dyn. **8**, 309–324 (1996)

Chapter 22
A-Posteriori Assessment of Sub-Filter Scale Models for Turbulence–Interface Interaction with the Two-Fluid Formulation Considering a Single Rising Gas Bubble in Liquid

R. Meller, M. Klein, D. Lucas and F. Schlegel

Introduction

Gas–liquid flows are of great importance for a large number of different industrial applications, e.g., in energy sector or metal processing industry. From there arises a strong need for numerical tools, which help to predict dynamics of two-phase flows in technical facilities with high predictive accuracy, reasonable computational expense, and without any need for initial knowledge of the present flow regime. A big challenge in investigations of such problems is the large range of interfacial and turbulent scales, which need to be accounted for in numerical simulations. A typical approach to capture all these dynamics and their interactions are multi-scale multi-regime methods, which make use of coupling of different individual modeling concepts.

The overall aim is to adopt the hybrid approach of Hänsch et al. [7], which combines an Eulerian–Eulerian approach with a volume-of-fluid(VOF)-like method in a two-fluid model in order to represent interfacial structures on subgrid-scale and on grid-scale, respectively. In situations, where transitions between different morphologies occur, interfacial structures appear to be too large for an Eulerian–Eulerian model approach and, at the same time, too small for the flow dynamics to be fully resolved in a VOF-like model. In other words, in the range of mesh scale

R. Meller (✉) · D. Lucas · F. Schlegel
Institute of Fluid Dynamics Helmholtz-Zentrum Dresden-Rossendorf,
Bautzner Landstrasse 400, 01328 Dresden, Germany
e-mail: r.meller@hzdr.de
URL: https://www.hzdr.de

M. Klein
Department of Aerospace Engineering, Bundeswehr University Munich,
Werner-Heisenberg-Weg 39, 85577 Neubiberg, Germany
URL: https://www.unibw.de

M. García-Villalba et al. (eds.), *Direct and Large Eddy Simulation XII*,
ERCOFTAC Series 27,
https://doi.org/10.1007/978-3-030-42822-8_22

and slightly above, interfacial and turbulent structures need to be simulated in an under-resolved manner. For this to deliver physically reasonable results, modeling of subgrid-scale (SGS) dynamics becomes necessary.

Numerical Method

Phase averaging is applied to the incompressible Navier–Stokes equations [6] regarding the phase volume fraction r_α of phase α. A filter operation $\overline{\cdot}$ is formally applied to the equations [12] to mimic implicit filtering due to the limited mesh resolution in a large-eddy simulation. Assuming commutativity between filter operation and derivatives, this results in the filtered, phase-averaged Navier–Stokes equations:

$$\frac{\partial \overline{r}_\alpha}{\partial t} + \frac{\partial \overline{r}_\alpha \, \overline{u}_{\alpha,i}}{\partial x_i} = -\frac{\partial \tau_{ru,\alpha,i}}{\partial x_i}, \tag{22.1}$$

$$\begin{aligned} \rho_\alpha \left(\frac{\partial \overline{r}_\alpha \, \overline{u}_{\alpha,i}}{\partial t} + \frac{\partial \overline{r}_\alpha \, \overline{u}_{\alpha,i} \, \overline{u}_{\alpha,j}}{\partial x_j} \right) &= -\overline{r}_\alpha \frac{\partial \overline{p}}{\partial x_i} + \mu_\alpha \frac{\partial}{\partial x_j} \left(2 \overline{r}_\alpha \, \overline{S}_{\alpha,ij} \right) \\ &+ \overline{K} (\overline{u}_{\beta,i} - \overline{u}_{\alpha,i}) + \sigma \, \overline{r}_\alpha \, \overline{n}_i \, \overline{\kappa} \, \overline{\delta}_S + \overline{f}_{\alpha,i} - \rho_\alpha \frac{\partial}{\partial t} \tau_{ru,\alpha,i} \\ &+ \frac{\partial}{\partial x_j} \left(-\rho_\alpha \, \tau_{ruu,\alpha,ij} + \mu_\alpha \, \tau_{rS,\alpha,ij} \right) - \tau_{rp,\alpha,i} + \tau_{K,\alpha,i} + \sigma \, \tau_{nn,\alpha,i}, \end{aligned} \tag{22.2}$$

with velocity u, density ρ, shared pressure p, dynamic viscosity μ, drag coefficient K, surface tension coefficient σ, vector normal to the interface n_i, interface curvature κ, interface indicator function δ_S, and volume forces f. The velocity strain rate tensor at grid-scale is $\overline{S}_{\alpha,ij} = 1/2 \left(\partial \overline{u}_{\alpha,i} / \partial x_j + \partial \overline{u}_{\alpha,j} / \partial x_i \right)$. Quantities, which are specific for phases α or β, are marked with corresponding subscripts. According to Ketterl and Klein [8], numerous unclosed SGS terms occur due to the filter operation. This formalism is adopted to the two-fluid model and results in SGS convective $\tau_{ruu,\alpha,ij} = \overline{r_\alpha \, u_{\alpha,i} \, u_{\alpha,j}} - \overline{r}_\alpha \, \overline{u}_{\alpha,i} \, \overline{u}_{\alpha,j}$, interfacial $\tau_{ru,\alpha,ij}$, diffusive $\tau_{rS,\alpha,ij}$, pressure $\tau_{rp,\alpha,i}$, drag $\tau_{K,\alpha,ij}$, and surface tension $\tau_{nn,\alpha,ij}$ terms. The momentum equations, specific for different phases α, are coupled to each other via the drag formulation by Štrubelj and Tiselj [15]. In this way, the homogeneous model is recovered within the given two-fluid formulation. Surface tension force is modeled as a volume force [3]. The system of equations is numerically solved with a finite volume method on an unstructured computational grid. Spacial discretization is realized with second-order accuracy. Time integration is treated explicitly for Eq. (22.1) and implicitly for Eq. (22.2), both with first-order accuracy. Pressure–velocity coupling is realized with a consistent momentum interpolation scheme [5]. Simulations are carried out with the OpenFOAM C++ Library.

Table 22.1 Closure models for SGS convective term

SM	$\tau^{\mathrm{SM}}_{ruu,\alpha,ij} = -2\overline{r}_\alpha \nu_{\mathrm{SGS}} \overline{S}_{\alpha,ij},\ \nu_{\mathrm{SGS}} = \left(C_s \overline{\Delta}\right)^2 \left\lvert\overline{S}_{\alpha,ij}\right\rvert,$ $\left\lvert\overline{S}_{\alpha,ij}\right\rvert = \sqrt{2\overline{S}_{\alpha,ab}\overline{S}_{\alpha,ab}},\ C_s = 0.18$
SigM	$\tau^{\mathrm{SigM}}_{ruu,\alpha,ij} = -2\overline{r}_\alpha \nu_{\mathrm{SGS}} \overline{S}_{\alpha,ij},\ \nu_{\mathrm{SGS},\alpha} = \left(C_\sigma \overline{\Delta}\right)^2 \frac{\sigma_3(\sigma_1-\sigma_2)(\sigma_2-\sigma_3)}{\sigma_1^2}$ $\sigma_1 \geq \sigma_2 \geq \sigma_3 = \sqrt{Eig\left(\overline{G}_{\alpha,ij}\right)},\ \overline{G}_{\alpha,ij} = \frac{\partial \overline{u}_{\alpha,k}}{\partial x_i}\frac{\partial \overline{u}_{\alpha,k}}{\partial x_j},\ C_\sigma = 1.5$
SS	$\tau^{\mathrm{SS}}_{ruu,\alpha,ij} = \left(\widehat{\overline{r}_\alpha \overline{u}_{\alpha,i}\overline{u}_{\alpha,j}}\right) - \widehat{\overline{r}}_\alpha \widehat{\overline{u}}_{\alpha,i}\widehat{\overline{u}}_{\alpha,j}$
CL	$\tau^{\mathrm{CL}}_{ruu,\alpha,ij} = \frac{\overline{\Delta}^2}{12}\overline{r}_\alpha \frac{\partial \overline{u}_{\alpha,i}}{\partial x_k}\frac{\partial \overline{u}_{\alpha,j}}{\partial x_k}$
M-SS-SM	$\tau^{\mathrm{M-SS-SM}}_{ruu,\alpha,ij} = \tau^{\mathrm{SS}}_{ruu,\alpha,ij} + \tau^{\mathrm{SM}}_{ruu,\alpha,ij}$
KO	$\tau^{\mathrm{KO}}_{ruu,\alpha,ij}(L_{\alpha,ij}) =$ $\frac{-L_{\alpha,ab}\overline{S}_{\alpha,ab} + \left\lvert -L_{\alpha,ab}\overline{S}_{\alpha,ab}\right\rvert}{-L_{\alpha,ab}\overline{S}_{\alpha,ab}} L_{\alpha,ij} + \left(L_{\alpha,ij} - \frac{-L_{\alpha,ab}\overline{S}_{\alpha,ab}}{\overline{S}_{\alpha,ab}\overline{S}_{\alpha,ab}}\overline{S}_{\alpha,ij}\right)$
KO-CL	$\tau^{\mathrm{KO-CL}}_{ruu,\alpha,ij} = \tau^{\mathrm{KO}}_{ruu,\alpha,ij}(L_{\alpha,ij} = \tau^{\mathrm{CL}}_{ruu,\alpha,ij})$
KO-SS	$\tau^{\mathrm{KO-SS}}_{ruu,\alpha,ij} = \tau^{\mathrm{KO}}_{ruu,\alpha,ij}(L_{\alpha,ij} = \tau^{\mathrm{SS}}_{ruu,\alpha,ij})$
KL-SS	$\tau^{\mathrm{KL-SS}}_{ruu,\alpha,ij} = L_{\alpha,ij} - 2\,\nu_{\mathrm{SGS},\alpha}\overline{S}_{\alpha,ij},$ $\nu_{\mathrm{SGS},\alpha} = max\left(\frac{L_{\alpha,ab}\overline{S}_{\alpha,ab}}{\overline{S}_{\alpha,ab}\overline{S}_{\alpha,ab}}, 0\right),\ L_{\alpha,ij} = \tau^{\mathrm{SS}}_{ruu,\alpha,ij}$

Subgrid-Scale Modeling

The modeling approaches applied to the unclosed SGS convective term are shown in Table 22.1.

The first two models are of eddy-viscosity type, namely the Smagorinsky model SM and the sigma model SigM, as proposed by Nicoud et al. [14], where the filter width of implicit filter operation is denoted as $\overline{\Delta}$. The remaining models assume similar structures at subgrid- and at grid-scales, so-called scale similarity [2]. Liu et al. [13] proposed a model that incorporates an explicit test filter operation $\widehat{\cdot}$ with filter size $\widehat{\Delta} > \overline{\Delta}$. Following Ketterl and Klein [8], this model is adapted to consider the structure of gas–liquid interfaces, $\overline{r}_\alpha$ in this case, and is referred to as SS. Scale similarity models are not strictly dissipative but may allow for backscatter in some situations [2], which is undesirable for a-posteriori investigations, as simulations may become unstable [1]. Bardina et al. [2] proposed a so-called mixed model M-SS-SM. In order to use scale similarity models in real LES simulations, Kobayashi proposed a regularized model variant KO for models of scale similarity type [11]. In Kobayashi's original model formulation, the second term consists of an additional factor of $2\,k_{\mathrm{SGS}}/L_{kk}$, which is set to unity here, following Ketterl et al. [9]. Additionally, Kobayashi proposed to use the traceless part of sub-model $L_{\alpha,ij}$, which is not done here [9]. Originally Kobayashi's model is applied to the gradient model CL by Clark et al. [4], which is referred to as KO-CL. In order to account for resolved

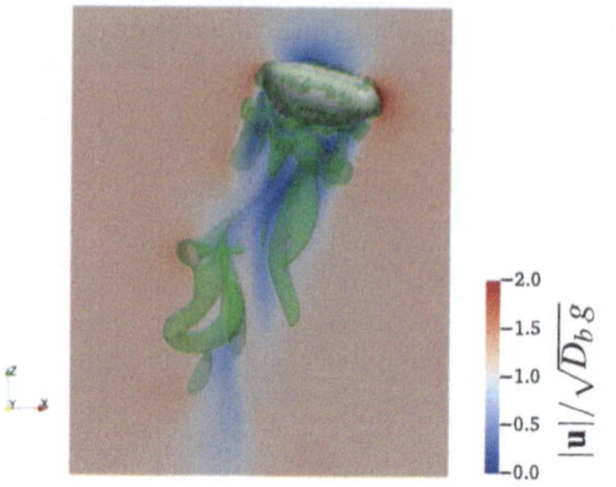

Fig. 22.1 Instant contours of velocity magnitude $|\mathbf{u}|$ (normalized with velocity scale $\sqrt{D_b\, g}$ with equivalent bubble diameter D_b and gravitational acceleration g), gas–liquid interface (white), and isosurfaces of Q-criterion (green)

interfacial structures ($\overline{r}_\alpha$), KO is applied to SS in addition, resulting in the KO-SS model. Klein et al. proposed a reformulated version KL of such regularized structural SGS models [10] by using an existing model, SS in this case, and adding just the right amount of viscosity in case of backscatter, which results in the model KL-SS.

Model Assessment

In order to assess the performance of different SGS models in an a-posteriori context, a single gas bubble rising in stagnant liquid under the influence of gravity is numerically investigated. A visualization is shown in Fig. 22.1. Material quantities are chosen for the bubble to be in a wobbling regime (Table 22.2). The cuboid domain is of dimensions $4D_b \times 4D_b \times 6D_b$ with equivalent bubble diameter D_b. At the top boundary, a liquid inflow with a controlled velocity is imposed, such that the bubble remains at nearly constant height. All sideward boundaries are set up to be periodic in x- and y- direction (see Fig. 22.1). An outflow boundary is placed at the bottom of the computational domain. A-posteriori investigations are carried out on an uniform hex-mesh of $60 \times 60 \times 90$ grid points, which results in $D_b/\Delta x = 15$, where Δx is the spacing of the grid cells. For the direct numerical simulation (DNS), double the number of grid points is used in each spacial direction.

Radial profiles of vertical mean velocity u_z and resolved turbulent kinetic energy $k = 0.5\,\overline{u}'^{\,2}_{\alpha,i}$ are shown in Fig. 22.2, where $\overline{u}'_{\alpha,i}$ denotes resolved velocity fluctuations. It is worth noting that both quantities are quasi identical for gas and liquid phases due to the drag formulation. As shown in Fig. 22.2a, u_z has a clear dependence on SGS modeling in the gas phase inside the bubble, while in the liquid phase, only a minor sensitivity is observed. All simulations have a positive peak value of u_z at $r/D_b = 0$ in common. In simulations without any SGS model (NoModel) and with SM and SigM, this center peak velocity is much higher compared to DNS. With KO-CL, the center peak velocity is slightly above the DNS value. All models, which

Table 22.2 Overview of flow regime in terms of dimensionless numbers, namely the Reynolds number Re_b, Morton number Mo, Eötvös number Eo and ratios of density Π_ρ and dynamic viscosity Π_μ; subscripts L and G denote liquid and gas quantities, respectively; the rising velocity of the bubble is referred to as U_b, gravitational acceleration as g, and density difference between liquid and gas as $\Delta\rho$

Symbol	Expression	Value
Re_b	$U_b D_b \rho_L/\mu_L$	367
Mo	$g\mu_L^4 \Delta\rho/(\rho_L^2\sigma^3)$	1.5×10^{-9}
Eo	$\rho_L g D_b^2/\sigma$	2.5
Π_ρ	ρ_L/ρ_G	100
Π_μ	μ_L/μ_G	10

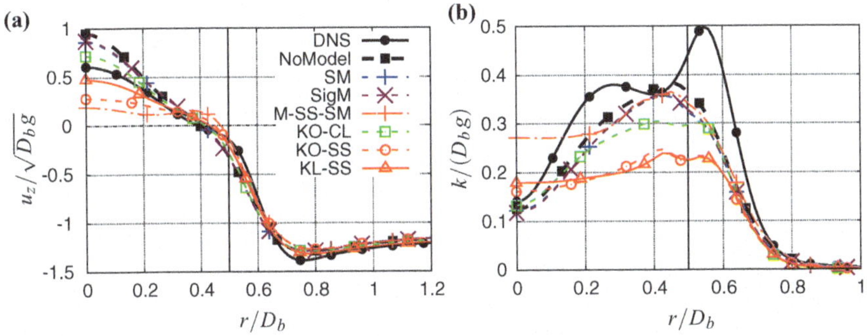

Fig. 22.2 Radial profiles of **a** vertical component of mean velocity u_z and **b** resolved turbulent kinetic energy k, normalized with $\sqrt{D_b\, g}$, averaged in time and circumferential direction; position of gas–liquid interface for sphere equivalent bubble radius is illustrated by vertical solid line

refer to SS and therefore take the interfacial structure ($\overline{r}_\alpha$) into account, show a value for the center peak velocity, which is below the one from DNS, with KL-SS resulting in the lowest deviation from DNS. DNS as well as all closure models, which concern SS, lead to a bump in the profile of u_z in the interface region at $r/D_b \approx 0.5$. For all remaining simulations, this velocity bump is nearly undetectable.

In terms of k (see Fig. 22.2b), radial profiles differ a lot between DNS and LES simulations with different SGS modeling strategies. DNS results in two peaks for k, one in the region of the interface and a second, smaller one inside the bubble at $r/D_b \approx 0.25$. As all other simulations were run on the coarse grid, the overall level of k is lower in the corresponding results. SM, SigM, M-SS-SM as well as NoModel result in a single peak for k in the interface region, where the values of k are nearly identical in all of these simulations. KO-CL, KO-SS, and KL-SS are able to predict these two peaks, even if they are much less pronounced than in DNS. Apart from that, all models incorporating the SS model predict a plateau in the profile of k in the center region $r/D_b < 0.2$, while all other simulations, including DNS, result in a local minimum of k right at the bubble center.

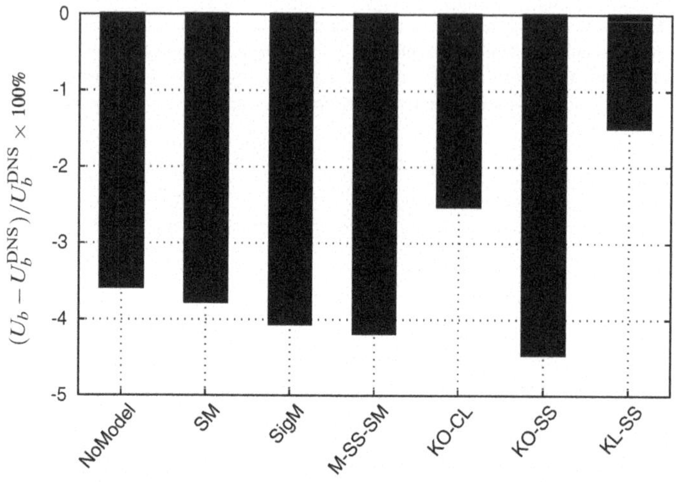

Fig. 22.3 Bubble rising velocity U_b, averaged over time for simulations with different SGS closure models; values are shown as percentage deviation from the value obtained with DNS

Figure 22.3 shows time averaged bubble rising velocity U_b for different SGS modeling strategies.

Computations on the coarse mesh generally result in a diminished value for U_b, compared to DNS. The authors assume that low spatial resolution of the interfacial region does not allow for a fully resolved interfacial boundary layer but instead the latter is artificially enlarged, which results in an increased friction drag of the bubble and therefore in a lower rising velocity U_b. The simulation without SGS model results in a value of U_b, which is about -3.5% below the DNS result, while values obtained with SM, SigM, M-SS-SM ,and KO-SS are even lower with up to -4.5% deviation from DNS. Solely KO-CM and KL-SS models result in bubble rising velocity values, which are much closer to DNS compared to NoModel, with KL-SS resulting in the lowest deviation of -1.5% from DNS. This indicates that modeling of the SGS convective term in a scale similarity fashion and under consideration of the interfacial structure $\overline{r}_\alpha$ might help to counteract effects of spatial under-resolution.

Conclusions

Different modeling strategies for the unclosed subgrid-scale convective term are assessed in the context of a two-fluid model considering a single gas bubble rising in stagnant liquid. The focus of this work is on modeling approaches for under-resolved gas–liquid interfaces and their interaction with flow dynamics. Closure models are compared in terms of mean velocity, resolved turbulent kinetic energy, and bubble rising velocity. In the present case, scale similarity models perform well and there is indication that these models are able to counteract effects of spatial under-resolution in interfacial boundary layers.

References

1. Anderson, B.W., Domaradzki, J.A.: A subgrid-scale model for large-eddy simulation based on the physics of interscale energy transfer in turbulence. Phy. Fluids **24**(6), 065104 (2012)
2. Bardina, J., Ferziger, J., Reynolds, W.C.: Improved subgrid-scale models for large-eddy simulation. In: 13th Fluid and PlasmaDynamics Conference, pp. 1357–1366 (1980)
3. Brackbill, J.U., Kothe, D.B., Zemach, C.: A continuum method for modeling surface tension. J. Comput. Phys. **100**, 335–354 (1992)
4. Clark, R.A., Ferziger, J.H., Reynolds, W.C.: Evaluation of subgrid-scale models using an accurately simulated turbulent flow. J. Fluid Mech. **91**, 1–16 (1979)
5. Cubero, A., Sánchez-Insa, A., Fueyo, N.: A consistent momentum interpolation method for steady and unsteady multiphase flows. Comp. Chem. Eng. **62**, 96–107 (2014)
6. Drew, D.A., Passman, S.L.: Theory of multicomponent fluids. Appl. Math. Sci. **135** (1999)
7. Hänsch, S., Lucas, D., Krepper, E., Höhne, T.: A multi-field two-fluid concept for transitions between different scales of interfacial structures. Int. J. Multiph. Flow **47**, 171–182 (2012)
8. Ketterl, S., Klein, M.: A-priori assessment of subgrid scale models for large-eddy simulation of multiphase primary breakup. Comput. Fluids **165**, 64–77 (2018)
9. Ketterl, S., Reißmann, M., Klein, M.: Large eddy simulation of multiphase flows using the volume of fluid method: Part 2 - a-posteriori analysis of liquid jet atomization. Exp. Comput. Multiph. Flow **1**(3), 201–211 (2019)
10. Klein, M., Ketterl, S., Kobayashi, H.: Regularised, parameter free scale similarity type LES models (Submitted)
11. Kobayashi, H.: Improvement of the SGS model by using a scale-similarity model based on the analysis of SGS force and SGS energy transfer. Int. J. Heat Fluid Flow **72**, 329–336 (2018)
12. Labourasse, E., Lacanette, D., Toutant, A., Lubin, P., Vincent, S., Lebaigue, O., Caltagirone, J.-P., Sagaut, P.: Towards large eddy simulation of isothermal two-phase flows: governing equations and a priori tests. Int. J. Multiph. Flow **33**, 1–39 (2007)
13. Liu, S., Meneveau, C., Katz, J.: On the properties of similarity subgrid-scale models as deduced from measurements in a turbulent jet. J. Fluid Mech. **275**, 83–119 (1994)
14. Nicoud, F., Toda, H.B., Cabrit, O., Bose, S., Lee, J.: Using singular values to build a subgrid-scale model for large eddy simulations. Phy. Fluids **23**, 085106 (2011)
15. Štrubelj, L., Tiselj, I.: Two-fluid model with interface sharpening. Int. J. Num. Methods Eng. **85**, 575–590 (2011)

Chapter 23
Ambient Flow Properties of Kolmogorov-Length-Scale Size Non-Spherical Particles in Isotropic Turbulence

K. Fröhlich, L. Schneiders, M. Meinke and W. Schröder

Introduction

Due to the stochastic nature of the fluid and the solid phase in particle-laden flows, a large number of phenomena can be observed, which highly depend on the flow configuration. Consequently, a generalized theory for particle-laden turbulent flows is missing and the analysis has often to be specialized to assess specific aspects of the multiphase dynamics.

Recently, direct particle-fluid simulations of Kolmogorov-length-scale size [4] and of finite size [7] ellipsoidal particles have been performed. These simulations fully resolve the particle-fluid interaction including all turbulent and all particle scales, which facilitates an assessment of Lagrangian point-particle models as in [3]. An important step for the validation of ellipsoidal Lagrangian point-particle models is a definition of the ambient flow field in fully resolved simulations, i.e., a characteristic flow field which enables the determination of the fluid velocity and its gradients encountered by the particles. This is precisely the motivation of this study. The ambient flow field is extensively used in Lagrangian point-particle models and if the self-induced disturbances of the particles are not significant, it can be determined directly at the particle position via an interpolation routine. In particle-resolved simulations, however, the ambient flow is not clearly defined and different methods have been proposed for finite-size particles to recover the flow field seen by particles. Most studies rely on a filtering approach due to the finite size of the particles, where the flow field is averaged in the vicinity of the particles such that the disturbances of

K. Fröhlich (✉) · L. Schneiders · M. Meinke · W. Schröder
RWTH Aachen University, Institute of Aerodynamics, Wüllnerstr. 5a, 52062 Aachen, Germany
e-mail: k.froehlich@aia.rwth-aachen.de

M. Meinke · W. Schröder
JARA Center for Simulation and Data Science, RWTH Aachen University, Seffenter Weg 23, 52074 Aachen, Germany

M. García-Villalba et al. (eds.), *Direct and Large Eddy Simulation XII*, ERCOFTAC Series 27,
https://doi.org/10.1007/978-3-030-42822-8_23

the particles are mostly excluded. Depending on the chosen configuration, various alternative methods have been proposed in, e.g., [1, 2, 6]. A unified approach is not available, which can partially be explained by the vast parameter space of particle-laden turbulent flows, where a single unique definition of the ambient flow field is hardly possible.

In this contribution, a recently developed method to estimate the ambient flow encountered by ellipsoidal particles [7] is further assessed for 60,000 fully resolved Kolmogorov-length-scale size ellipsoidal particles. A sensitivity analysis is conducted to provide confidence intervals of the estimation method and averaged near-particle flow patterns yield further information on the disturbances of the particles.

Numerical Method and Computational Setup

The conservation equations of mass, momentum, and energy are solved using a cell-centered finite-volume discretization on adaptively refined hierarchical Cartesian grids. The particles are tracked using a level-set method. A sharp resolution of the interfaces is provided by cut cells and a strictly conservative flux-redistribution technique is employed to stabilize arbitrary small cut cells. The computational requirements for the full resolution of Kolmogorov-length-scale size particles is mitigated by solution-adaptive refinement. A complete description of the mathematical model for the fluid phase, the solid phase, and their coupling is presented in [8] and references therein.

Isotropic decaying turbulence is established in a fully periodic cube with edge length L using the model energy spectrum $E(\mathbf{k}) = \left(3u_0^2/2\right)\left(\mathbf{k}/k_\mathrm{p}^2\right)\exp\left(-\mathbf{k}/k_\mathrm{p}\right)$ with the rms velocity u_0 and the peak wave number $k_\mathrm{p} = 8\pi/L$. At $t^* = t\varepsilon_0/u_0 = 0.28$, the turbulent flow is developed and the velocity-derivative skewness converges to -0.5, where ε_0 is the initial dissipation rate. At this time, 60,000 ellipsoidal particles are injected in the turbulent flow field and initialized with the local linear and angular fluid velocity. The initial Taylor length based Reynolds number is chosen $\mathrm{Re}_{\lambda,0} = 79.1$ and converges to $\mathrm{Re}_\lambda \approx 35$ at $t^* = 1.0$. The ambient turbulent scales are resolved via a uniform mesh using 512^3 cells. Four additional refinement levels are introduced in the vicinity of the particles to fully resolve the particle-fluid interaction. Four particle-laden cases are considered with identical density ratio $\rho_p/\rho_f = 400$ and volume loading $\phi = 0.47 \times 10^{-3}$. The volumetric diameter is chosen $d_p^{vol}/L = 0.0025$ which corresponds to $d_p^{vol}/\eta = 1.3$ at insertion with the Kolmogorov-length scale η. Only the particle aspect ratio is varied and ranges from 1/3 through 8. All instantaneous statistics in the subsequent section are presented for the time $t^* = 2.0$, when the auto-correlation of the particle angular velocity is significantly decreased (not shown here).

Results and Discussion

The model equations derived in [5] are the basis for ellipsoidal Lagrangian point-particle models and indicate ambient flow properties to have a significant effect on the particle dynamics, i.e., the ambient fluid velocity $\mathbf{U}_p$, the ambient fluid rotation rate $\mathbf{O}_p$, and the ambient fluid shear rate $\mathbf{S}_p$. Although the particle dynamics are fully resolved in direct particle-fluid simulations, these ambient flow properties are required for a direct comparison with ellipsoidal point-particle models and their estimation will subsequently be presented for Kolmogorov-length-scale size ellipsoidal particles. The results are quite similar for the analyzed particle shapes and hence will only be presented for ellipsoids with aspect ratio 4.

Figure 23.1 illustrates an upwind-biased estimation method, which has been previously used in [4, 7]. The ambient fluid velocity $\mathbf{U}_p$ seen by the particles is reconstructed using a distance-weighted filtering in the upwind region, i.e.,

$$\mathbf{U}_p = \left(\int_{\Sigma_p} \mathscr{C}_p^{up}(\mathbf{x}) \mathrm{d}V \right)^{-1} \int_{\Sigma_p} \mathscr{C}_p^{up}(\mathbf{x}) \mathbf{u}(\mathbf{x}) \mathrm{d}V, \tag{23.1}$$

where the first term normalizes the upwind-biased weights

$$\mathscr{C}_p^{up}(\mathbf{x}) = \begin{cases} \exp\left\{ -18 \left[\delta_p(\mathbf{x}) - \delta_0 \right]^2 / \sigma^2 \right\}, & \text{if } \angle(\mathbf{n}_p, \mathbf{F}_p) < \theta_{up} \\ 0, & \text{otherwise} \end{cases}, \tag{23.2}$$

with $\angle(\mathbf{n}_p, \mathbf{F}_p)$ the angle between the hydrodynamic force vector $\mathbf{F}_p$ and the normal vector $\mathbf{n}_p$ toward the closest particle surface point. The angle θ_{up} should be chosen such that the disturbances of the particle wakes are largely excluded in the upwind region. The reconstruction region Σ_p is further limited by a minimum and a maximum distance $\delta_{\min}$ and $\delta_{\max}$ with

$$\Sigma_p = \left\{ \mathbf{x} \mid \delta_{\min} < \delta_p(\mathbf{x}) < \delta_{\max} = \delta_{\min} + \sigma \right\} \tag{23.3}$$

and σ represents its width. The Gaussian weight is depicted in Fig. 23.1 and its definition in Eq. 23.2 is chosen, such that δ_0 denotes the peak of the Gaussian. In total, the three parameters θ_{up}, $\delta_{\min}$, and σ fully define the upwind-biased estimation. Figure 23.1 shows the absolute relative fluid velocity $< |\mathbf{u}(\mathbf{x}) - \mathbf{v}_p| >_{85}$ with the particle velocity $\mathbf{v}_p$ referenced by $|\mathbf{U}_p - \mathbf{v}_p|$, where $< \ >_N$ denotes the average for all particles inclined by $N \pm 2°$, and the upwind region defined by $\{\delta_{\min} = 1.5 d_p^{vol}, \theta_{up} = 90°, \sigma = d_p^{vol}\}$. A lower $\delta_{\min}$ may include the particle boundary layer in the upwind region. Otherwise, if $\delta_{\min}$ is chosen to high, the estimation procedure suffers from filtering effects and the estimated ambient flow properties are less correlated to the flow field seen by the particles. Although, the mean particle Reynolds number $\mathrm{Re}_p \sim 2$ is moderate, pronounced particle wakes can be observed which should be excluded by the estimation procedure. As recapitulated in the Intro-

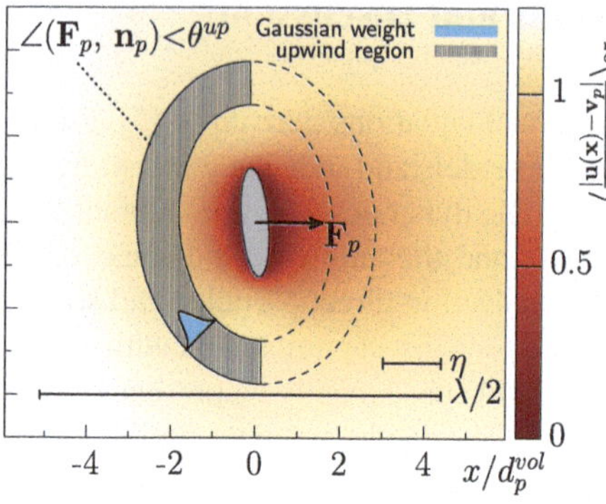

Fig. 23.1 Averaged absolute local near-particle relative fluid velocity of the particles with aspect ratio 4 inclined by $85 \pm 2°$. The upwind region is defined by $\left\{\delta_{\min} = 1.5 d_p^{vol}, \theta_{up} = 90°\right\}$

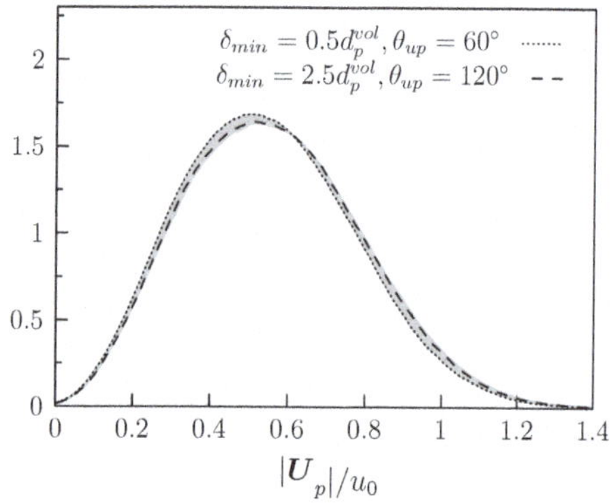

Fig. 23.2 pdf of the absolute ambient fluid velocity seen by the particles $|\mathbf{U}_p|$ referenced by the initial rms velocity u_0. The shaded area illustrates the estimation by Eq. 23.2 with $\delta_{\min} = \{0.5 d_p^{vol}, 1.5 d_p^{vol}, 2.5 d_p^{vol}\}$ and $\theta_{up} = \{60°, 90°, 120°\}$

duction, a unique definition for the ambient flow field is hardly possible and may depend on the aspect ratio, the particle orientation, and on the Reynolds number. Therefore, a sensitivity analysis is performed to estimate the impact of the method specifications on the results. As shown in [7], the parameter σ has a negligible impact on the results and is set to $\sigma = d_p^{vol} \sim \eta$. The sensitivity analysis will be based on 9 combinations of parameter specifications with $\delta_{\min} = \{0.5 d_p^{vol}, 1.5 d_p^{vol}, 2.5 d_p^{vol}\}$ and $\theta_{up} = \{60°, 90°, 120°\}$.

Figure 23.2 shows the probability density function (pdf) of the ambient fluid velocity $\mathbf{U}_p$ for the parameter specifications $\left\{\delta_{\min} = 0.5 d_p^{vol}, \theta_{up} = 60°\right\}$ and $\left\{\delta_{\min} = 2.5 d_p^{vol},\ \theta_{up} = 120°\right\}$, which represent the smallest and largest weighting volume. The shaded area illustrates the data of all other parameter specifications. That is, the estimation of the ambient velocity $\mathbf{U}_p$ is almost independent of the parameter specification. This can be explained by the small size of the particles compared to the ambient turbulent scales. The Kolmogorov length η and the Taylor length λ are depicted in Fig. 23.1. Note that the upwind region is significantly smaller than λ. Since the velocity fluctuations are mainly defined by larger scales, filter effects are negligible for the estimation of $\mathbf{U}_p$ in the case of Kolmogorov-length-scale size particles.

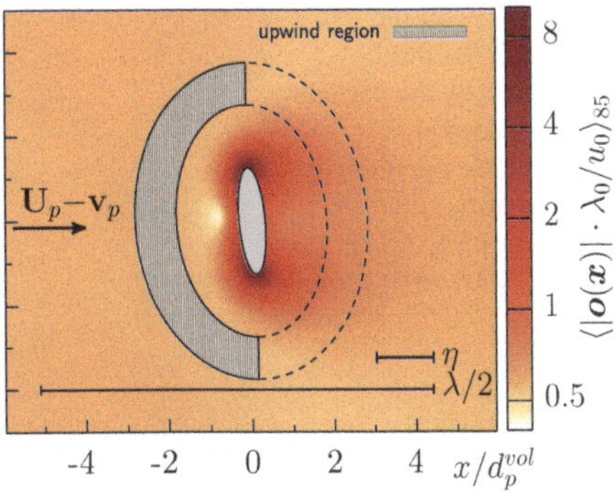

Fig. 23.3 Averaged absolute local near-particle angular fluid velocity of the particles with aspect ratio 4 inclined by $85 \pm 2°$, referenced by the initial Taylor-length scale λ_0 and the rms velocity u_0

Figure 23.3 shows the absolute angular fluid velocity $< |\mathbf{o}| >_{85} = < 0.5|\text{curl } \mathbf{u}(\mathbf{x})| >_{85}$. Due to the small scales and the complex shape of the particle, the vorticity is an order of magnitude higher near the particle surface compared to the background turbulent flow and a logarithmic scale has been used for the visualization. Clearly, the upwind region defined by $\{\delta_{\min} = 1.5d_p^{vol}, \theta_{up} = 90°\}$ does not include the disturbances of the particle. To estimate the ambient angular fluid velocity

$$\mathbf{O}_p = \frac{1}{2}\left(\frac{\partial u_z}{\partial y}_p - \frac{\partial u_y}{\partial z}_p, \frac{\partial u_x}{\partial z}_p - \frac{\partial u_z}{\partial x}_p, \frac{\partial u_y}{\partial x}_p - \frac{\partial u_x}{\partial y}_p\right)^T, \tag{23.4}$$

each of the nine components of the gradient tensor $\partial u_i/\partial x_j$ is individually filtered using Eq. 23.2 to obtain $\left(\partial u_i/\partial x_j\right)_p$. The *pdf* of the ambient angular fluid velocity $\mathbf{O}_p$ is presented in Fig. 23.4. As in Fig. 23.2, all combinations for $\delta_{\min}$ and θ_{up} are enclosed by the shaded area, and $\{\delta_{\min} = 0.5d_p^{vol}, \theta_{up} = 60°\}$ as well as $\{\delta_{\min} = 2.5d_p^{vol}, \theta_{up} = 120°\}$ represent the smallest and largest upwind region. Depending on the specification of $\delta_{\min}$ and θ_{up}, deflections of high vorticity are damped. Hence, Kolmogorov-length-scale size ellipsoidal particles are subject to small scale fluid vorticity fluctuations due to their finite length. This is consistent with the findings of [9] for inertialess fibers in forced isotropic turbulence, where filtering effects have been accounted for the attenuation of fiber dynamics.

The averaged absolute fluid shear rate $< |\mathbf{s}(\mathbf{x})| >_{85}$ with

$$\mathbf{s} = \frac{1}{2}\left(\frac{\partial u_z}{\partial y} + \frac{\partial u_y}{\partial z}, \frac{\partial u_x}{\partial z} + \frac{\partial u_z}{\partial x}, \frac{\partial u_y}{\partial x} + \frac{\partial u_x}{\partial y}\right)^T, \tag{23.5}$$

is shown in Fig. 23.5. Similar to the vorticity, the shear rate is an order of magnitude higher at the particle surface compared to the shear rate of the background turbulence. The disturbances of the particles are more expansive in the upwind region, while the depicted upwind region specified by $\{\delta_{\min} = 1.5d_p^{vol}, \theta_{up} = 90°\}$ still mostly avoids regions of extreme shear rate.

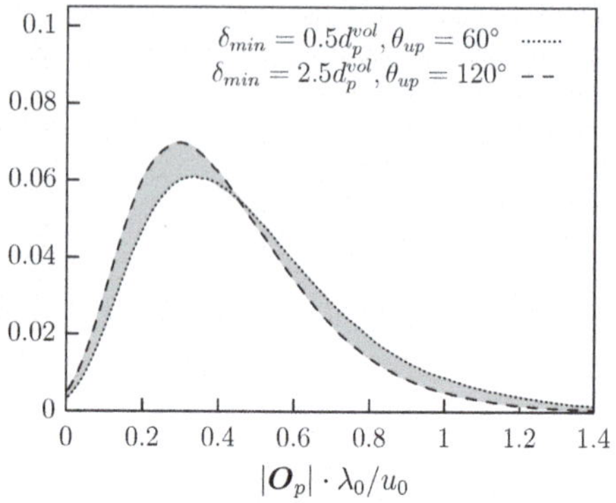

Fig. 23.4 pdf of the ambient angular fluid velocity seen by the particles $\mathbf{O}_p$. The shaded area illustrates the estimation by Eq. 23.2 with $\delta_{\min} = \{0.5d_p^{vol}, 1.5d_p^{vol}, 2.5d_p^{vol}\}$ and $\theta_{up} = \{60°, 90°, 120°\}$

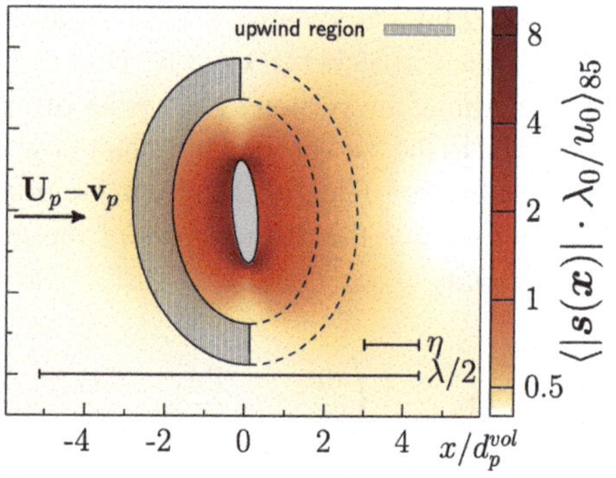

Fig. 23.5 Averaged absolute local near-particle shear rate of the particles with aspect ratio 4 inclined by 85 ± 2°, referenced by the initial Taylor-length scale λ_0 and the rms velocity u_0

Analogously to $\mathbf{O}_p$, the ambient shear rate $\mathbf{S}_p$ is defined by individually filtered gradients $\left(\partial u_i/\partial x_j\right)_p$ and the corresponding pdf is presented in Fig. 23.6. In contrast to the pdfs of $\mathbf{U}_p$ and $\mathbf{O}_p$, the shaded area includes only results of $\delta_{\min} = \{1.5d_p^{vol}, 2.5d_p^{vol}\}$ to avoid the disturbances of the particles. Compared to the pdf of $\mathbf{O}_p$, the sensitivity of the estimation is higher and filtering effects of the procedure are more pronounced. Additionally, the estimated ambient shear rate $\mathbf{S}_p$ is shown for $\left\{\delta_{\min} = 0.5d_p^{vol}, \theta_{up} = 60°\right\}$, which includes regions, where the shear rate is disturbed by the particle. This leads to a significantly higher estimation of the ambient shear rate $\mathbf{S}_p$.

It can be concluded that estimation procedures with distance-based filtering can accurately determine the ambient fluid velocity encountered by the particles $\mathbf{U}_p$, if applied for fully resolved Kolmogorov-length-scale size non-spherical particles. However, the ambient fluid rotation rate $\mathbf{O}_p$ and the shear rate $\mathbf{S}_p$ are partially filtered, which is related to the estimation procedure *and* to the finite size of the particles. The approximation of a velocity gradient field seen by the particles with a single rotation rate and a shear rate inherently introduces filtering. Therefore, estimations of gradient fields using filtering methods are sensitive to the specific method. Further statistics using $\mathbf{O}_p$ and $\mathbf{S}_p$ should include a confidence interval generated by sensitivity analysis.

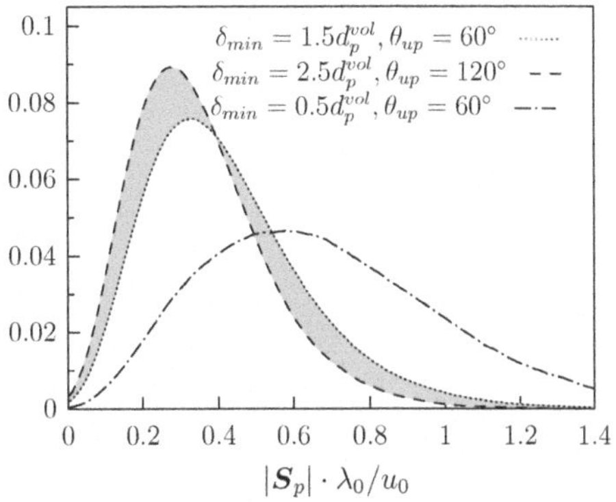

Fig. 23.6 pdf of the ambient shear rate seen by the particles $\mathbf{S}_p$. The shaded area illustrates the estimation by Eq. 23.2 with $\delta_{\min} = \{1.5d_p^{vol}, 2.5d_p^{vol}\}$ and $\theta_{up} = \{60°, 90°, 120°\}$

Acknowledgements This work has been financed by the German Research Foundation (DFG) within the framework of the SFB/Transregio 129 Oxyflame (subproject B2). The support is gratefully acknowledged. Computing resources were provided by the High Performance Computing Center Stuttgart (HLRS) and by the Jülich Supercomputing Center (JSC) within a Large-Scale Project of the Gauss Center for Supercomputing (GCS).

References

1. Cisse, M., Homann, H., Bec, J.: Slipping motion of large neutrally buoyant particles in turbulence. J. Fluid Mech. **735** R1 (2013)
2. Fornari, W., Picano, F., Brandt, L.: Sedimentation of finite-size spheres in quiescent and turbulent environments. J. Fluid Mech. **788**, 640–669 (2016)
3. Fröhlich, K., Schneiders, L., Meinke, M., Schröder, W.: Validation of Lagrangian two-way coupled point-particle models in Large-Eddy simulations. Flow Turb. Combust. **101**(2), 317–341 (2018)
4. Fröhlich, K., Schneiders, L., Meinke, M., Schröder, W.: Direct particle-fluid simulation of Kolmogorov-length-scale size ellipsoidal particles in isotropic decaying turbulence. In: Eleventh International Symposium on Turbulence and Shear Flow Phenomena (2019)
5. Jeffery, G.B.: The motion of ellipsoidal particles immersed in a viscous fluid. Proc. Royal Soc. Lond. **102**(715), 161–179 (1922)
6. Kidanemariam, A.G., Chan-Braun, C., Doychev, T., Uhlmann, M.: Direct numerical simulation of horizontal open channel flow with finite-size, heavy particles at low solid volume fraction. New J. Phys. **15**(2), 025,031 (2013)
7. Schneiders, L., Fröhlich, K., Meinke, M., Schröder, W.: The decay of isotropic turbulence carrying non-spherical finite-size particles. J. Fluid Mech. **875**, 520–542 (2019)
8. Schneiders, L., Günther, C., Meinke, M., Schröder, W.: An efficient conservative cut-cell method for rigid bodies interacting with viscous compressible flows. J. Comput. Phys. **311**, 62–86 (2016)
9. Shin, M., Koch, D.L.: Rotational and translational dispersion of fibres in isotropic turbulent flows. J. Fluid Mech. **540**, 143–173 (2005)

Chapter 24
Nanoparticle Behavior and Formation in Turbulent Spray Flames Investigated by DNS

A. Abdelsamie and D. Thévenin

Introduction

The behavior of nanoparticles in turbulent spray flames is a process of growing importance for practical purposes but involves many, strongly coupled physicochemical processes of high complexity. This work investigates the formation and behavior of nanoparticles in turbulent spray combustion with the help of direct numerical simulation (DNS). The simulations are carried out using the in-house DNS code DINO [1]. Detailed reaction schemes are taken into account to describe kinetics. The DNS code has been coupled with additional models to describe nucleation and production of nanoparticles. In this manner, it will be ultimately possible to describe with a high accuracy all the processes occurring after liquid jet atomization.

Numerical Approaches

In this study, three main numerical approaches are employed: (1) DNS to solve the gas phase relying on the low Mach number approach; (2) a Lagrangian description for tracking spray droplets, and (3) an Eulerian approach to model nanoparticle growth, aggregation, coagulation, etc. Droplets, being noticeably smaller than the

A. Abdelsamie (✉) · D. Thévenin
Laboratory of Fluid Dynamics and Technical Flows (LSS/ISUT),
University of Magdeburg "Otto von Guericke" Universitätsplatz 2,
39106 Magdeburg, Germany
e-mail: abouelmagd.abdelsamie@ovgu.de

D. Thévenin
e-mail: thevenin@ovgu.de

A. Abdelsamie
Laboratory of Fluid Mechanics, Facutly of Engnieering (Elmattaria),
Helwan University, Cairo, Egypt

M. García-Villalba et al. (eds.), *Direct and Large Eddy Simulation XII*,
ERCOFTAC Series 27,
https://doi.org/10.1007/978-3-030-42822-8_24

grid resolution, are modeled as point droplets with a variable diameter. To implement numerically these models, the in-house DNS code DINO has been used. DINO is a Fortran 2000 code. Spatial discretization relies on a sixth-order central finite-difference scheme. A semi-implicit third-order Runge–Kutta method is used for temporal integration. In DINO, the open-source library Cantera 2.4.0 is used to compute all chemical reaction rates, thermodynamic parameters, and molecular transport processes in the gas phase. More details about DINO can be found in Ref. [1]. The Lagrangian frame in DINO is relying on the discrete particles simulation approach, DPS. Hence, the resulting simulations can be categorized as DNS-DPS. A two-way coupling between both phases (gas and spray) is implemented via the exchange of mass, momentum, and energy. The droplet equations rely on the model first introduced by Abramzon and Sirignano [4], taking into account the modifications suggested by [5]. All details concerning the implemented equations describing droplet location, momentum, mass transfer, and heat transfer can be found in previous publications [2, 3].

For the nanoparticle simulations, additional models have been implemented based on the ideas proposed by [6] and modified by [8, 10]. This model works as follows:

$$\frac{\mathrm{d}N}{\mathrm{d}t} = -\frac{1}{2}\,\beta\,N^2 + I, \tag{24.1}$$

$$\frac{\mathrm{d}A}{\mathrm{d}t} = -\frac{1}{\tau_s}\,(A - N\,a_s) + I\,a_0, \tag{24.2}$$

$$\frac{\mathrm{d}V}{\mathrm{d}t} = I\,v_0. \tag{24.3}$$

In these equations (Eqs. 24.1–24.3), N is the nanoparticle number concentration, A is the total surface area concentration, V is the total volume concentration, v_0 is the monomer volume, a_0 is the monomer surface area, and I is the nucleation rate. The coagulation kernel, β is computed as

$$\beta = 4\,\pi\,d_c\,\mathscr{D}\left[\frac{0.5\,d_c}{d_c + g\,\sqrt{2}} + \frac{\mathscr{D}\,\sqrt{2}}{0.5\,c\,d_c}\right]^{-1}, \tag{24.4}$$

$$g = \frac{\left[(d_c + L)^3 - (d_c^2 + L^2)^{1.5}\right]}{3\,L\,d_c} - d_c, \tag{24.5}$$

$$L = \frac{8\,\mathscr{D}}{\pi\,c}, \tag{24.6}$$

$$c = \sqrt{\frac{8\,k_b\,T}{\pi\,\rho_n\,V}}, \tag{24.7}$$

$$\mathscr{D} = \frac{k_b\,T}{3\,\pi\,\mu\,d_c}. \tag{24.8}$$

where L is the mean free path, k_b is the Boltzmann constant, ρ_n is the particle density, T is the gas temperature, μ is the viscosity of the gas, c is the particle velocity, and $\mathscr{D}$ is the diffusion coefficient of the particles. The primary diameter d_p, aggregate diameter d_a, and collision diameter d_c are computed as follows:

$$d_p = \left(\frac{6\,V}{A}\right), \tag{24.9}$$

$$d_a = \left(\frac{6\,V}{\pi\,N}\right)^{1/3}, \tag{24.10}$$

$$d_c = d_p\,n_p^{1/1.8}, \tag{24.11}$$

$$n_p = \frac{6V}{\pi\,N\,d_p^3}. \tag{24.12}$$

In Eq. 24.2, the surface area of the completely fused particles is

$$a_s = \left(\frac{V}{N\,v_0}\right)^{2/3} a_0. \tag{24.13}$$

The characteristic sintering time τ_s for titanium dioxide particles TiO_2 (the material considered in this work) is computed as

$$\tau_s = 7.4\,\times\,10^{16}\,T\,d_a^4 \exp\left(\frac{-31000}{T}\right). \tag{24.14}$$

Numerical Configuration

The main purpose of this first study is to obtain conditions similar to that of the experimental investigations conducted at the University of Duisburg-Essen [9] within the DFG Priority programme 1980 "Nanoparticle Synthesis in Spray Flames SpraySyn: Measurement, Simulation, Processes". In these experiments, the main solvent is ethanol, which is mixed (in liquid state) with a precursor and then injected as a spray with a dispersion gas (O_2) through an injector. The liquid is evaporated by the pilot flame (CH_4/Air) as shown in Fig. 24.1.

In the current simulations, a reduced kinetic mechanism is used for ethanol oxidation; this mechanism consists of 36 species and 85 elementary reactions [7]. This mechanism is already quite large for a DNS. As a consequence, a planar jet is currently considered instead of a full three-dimensional domain. The computational domain with dimensions of 4.5 mm (transverse direction) $\times$ 9.0 mm (streamwise direction) $\times$ 0.28 mm (spanwise direction) is discretized over 2 million grid points. This domain is associated to inflow/outflow boundary conditions in streamwise direction, and to periodic boundary conditions in the other two directions.

Fig. 24.1 Schematic picture showing the boundary and inflow conditions

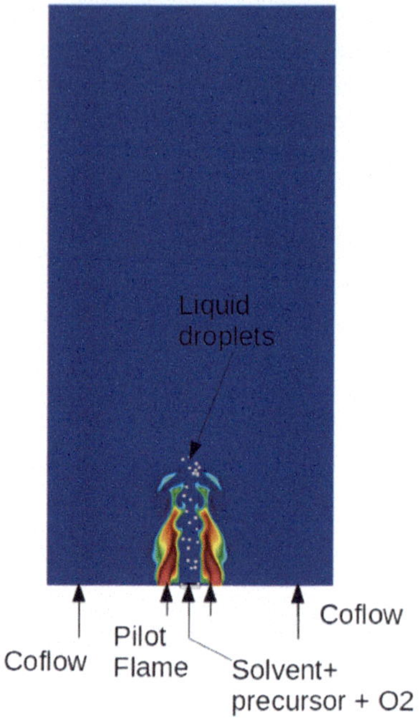

In all simulations, initially monodispersed liquid droplets (ethanol + $TiCL_4$), starting with a diameter of 11 µm and a temperature of 300 K, are injected through a planar nozzle with a diameter of $d_j = 0.3$ mm into a domain filled with air at a temperature of 300 K. The dispersion gas is injected with jet velocity, u_j, of 60 m/s. The gas driving the pilot flame enters the domain at 40 m/s, while the coflow is injected at 10 m/s. In order to trigger turbulence, fluctuation velocities with a turbulence intensity of 10% are added at the inlet boundary condition within the area of the injection nozzle.

Results

In this section, the results at three different times t, as a function of the jet timescale $\tau_j = d_j/u_j$, will be discussed.

Figure 24.2 shows the temporal evolution of the gas temperature. With time, the evaporation increases, therefore the temperature near the droplets (represented as white spheres in Fig. 24.2) decreases.

Following the evaporation of the liquid droplets (solvent + precursor), nanoparticles are produced as shown in Fig. 24.3. This figure shows the number concentration of nanoparticles N. At $t = 25\tau_j$, the nanoparticles first appear near the top-end of the

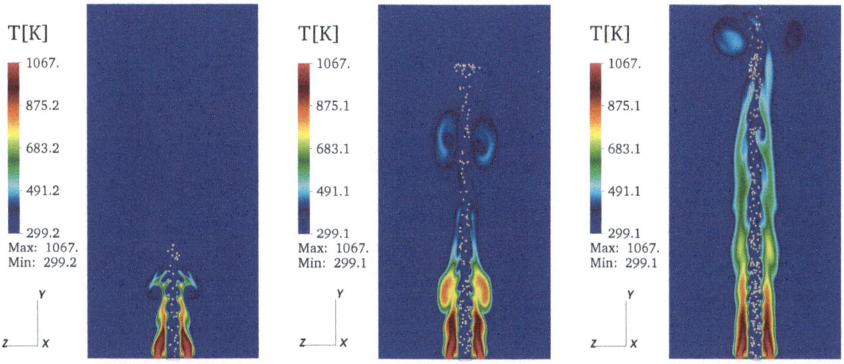

Fig. 24.2 Temporal evolution of gas temperature at three successive times, t. Left: $t = 25\,\tau_j$. Center: $t = 80\,\tau_j$. Right: $t = 160\,\tau_j$. In this figure, the white spheres represent the liquid droplets; $\tau_j = d_j/u_j$ is the jet timescale

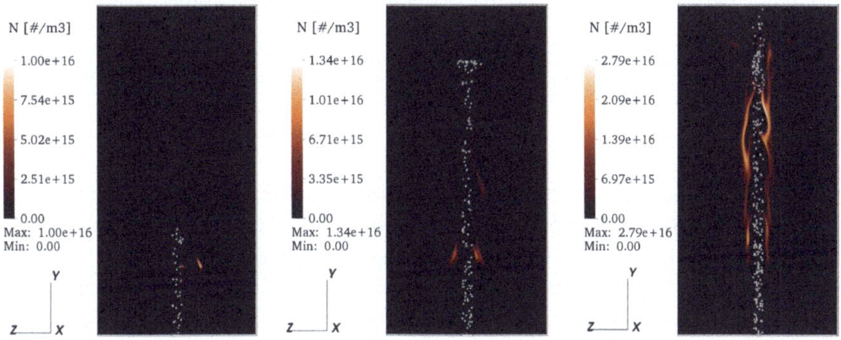

Fig. 24.3 Temporal evolution of number concentration, N of TiO_2 nanoparticles at three successive times, t. Left: $t = 25\,\tau_j$. Center: $t = 80\,\tau_j$. Right: $t = 160\,\tau_j$ (same as Fig. 24.2). The white spheres represent the liquid droplets

jet. Later on, the nanoparticles are produced and accumulate predominantly along the edges of the jet core, forming elongated filaments.

In order to investigate the local correlation between temperature and nanoparticle production, a scatter plot of the number concentration of nanoparticles versus gas temperature is presented in Fig. 24.4. The scenario dominating nanoparticles' production can be summarized as follows. From Fig. 24.4(Left), the nanoparticles are first produced at low temperature; this is the place corresponding to the final phase of intense evaporation. At a later time, $t = 80\tau_j$ the nanoparticle production switches to a higher temperature range ($300 < T < 850$ K), with a peak around 750 K as can be observed from Fig. 24.4(Center). At time $t = 160\,\tau_j$, corresponding to Fig. 24.4(Right), two peaks are found: (1) around $T = 580$ K and (2) around $T = 700$ K. This might correspond to two different physicochemical processes: (1)

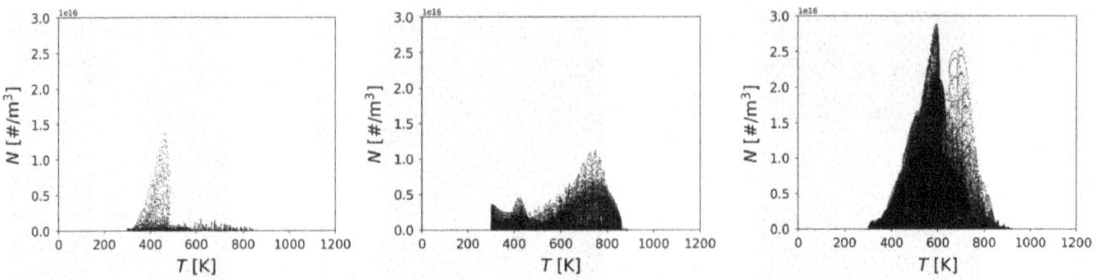

Fig. 24.4 Temporal evolution of scatter plot of N versus T at three successive times, t. Left: $t = 25\,\tau_j$. Center: $t = 80\,\tau_j$. Right: $t = 160\,\tau_j$ (same as Fig. 24.2)

dominated by concentration (abundance of reactants in the gas phase); (2) dominated by the exponential term in the kinetic rate.

The production rate and accumulation of nanoparticles depend on many further parameters: jet velocity, evaporation rate, flame structure, etc. The impact of these parameters will be investigated in more detail in our further studies.

Conclusions

Nanoparticle production from spray combustion has been investigated using DNS with the help of the discrete particle simulation approach to track the liquid droplets. A suitable model has been implemented to simulation nanoparticle production. It has been found that the nanoparticles are mainly produced along the sides of the jet core. Peak production is found at two different temperatures: a low temperature corresponding to large reactant concentration; a higher temperature favoring kinetic effects. Overall, the coupling processes appear to be very complex. It is clear that the production and accumulation of nanoparticles strongly depend on many parameters: jet velocity, evaporation rate, flame structure…Further systematic studies are needed to delineate the relative importance of these contributions.

Acknowledgements The financial support of the DFG (Deutsche Forschungsgemeinschaft) for Abouelmagd Abdelsamie within the project SPP1980 "Nanoparticle Synthesis in Spray Flames SpraySyn: Measurement, Simulation, Processes" is gratefully acknowledged. The computer resources provided by the Gauss Center for Supercomputing/Leibniz Supercomputing Center Munich under grant pr84qo have been essential to obtain the results presented in this work.

References

1. Abdelsamie, A., Fru, G., Oster, T., Dietzsch, F., Janiga, G., Thévenin, D.: Towards direct numerical simulations of low-Mach number turbulent reacting and two-phase flows using immersed boundaries. Comput. Fluids **131**, 123–141 (2016)

2. Abdelsamie, A., Thévenin, D.: Direct numerical simulation of spray evaporation and autoignition in a temporally-evolving jet. Proc. Combust. Inst. **36**(2), 2493–2502 (2017)
3. Abdelsamie, A., Thévenin, D.: On the behavior of spray combustion in a turbulent spatially-evolving jet investigated by direct numerical simulation. Proc. Combust. Inst. **37**(3), 3373–3382 (2019)
4. Abramzon, B., Sirignano, W.A.: Droplet vaporization model for spray combustion calculations. Int. J. Heat Mass Transf. **32**(9), 1605–1618 (1989)
5. Kitano, T., Nishio, J., Kurose, R., Komori, S.: Effects of ambient pressure, gas temperature and combustion reaction on droplet evaporation. Combust. Flame **161**, 551–564 (2014)
6. Kruis, F.E., Kusters, K.A., Pratsinis, S.E., Scarlett, B.: A simple model for the evolution of the characteristics of aggregate particles undergoing coagulation and sintering. Aerosol Sci. Tech. **19**(4), 514–526 (1993)
7. Marinov, N.M.: A detailed chemical kinetic model for high temperature ethanol oxidation. Int. J. Chem. Kinet. **31**(3), 183–220 (1999)
8. Panda, S., Pratsinis, S.E.: Modeling the synthesis of aluminum particles by evaporation-condensation in an aerosol. Nanostruct. Mater. **5**, 755–767 (1995)
9. Schneider, F., Suleiman, S., Menser, J., Dreier, T., Wlokas, I., Kempf, A., Shulz, C., Wiggers, H.: SpraySyn—A standardized burner configuration for nanoparticle synthesis in spray flames. Accpeted for publicatoin, Rev. Sci. Instrum. (2019)
10. Weise, C., Menser, J., Kaiser, S.A., Kempf, A., Wlokas, I.: Numerical investigation of the process steps in a spray flame reactor for nanoparticle synthesis. Proc. Combust. Inst. **35**, 2259–2266 (2015)

Chapter 25
Characterization of Dynamics in a Premixed Flame

Y. Shen, N. Jaouen and C. Duwig

Abstract The dynamics of a lean premixed flame is investigated using LES. The instantaneous fields show a V-shaped preheat zone while POD results exhibit the dominant single and double helix structures, which contribute to the rotational motion.

Introduction

Lean premixed combustion is widely used for industrial burners in order to achieve low emissions. Under this condition, the flame tends to exhibit combustion instabilities, thermoacoustic oscillations and lean blowout. The underlying mechanisms involving complex interactions of unsteady flow structure, mixing and flame are still not well enough understood today [1].

To understand the chaotic non-linear behaviour of the unsteady flame, we investigated the quasi-period dynamics. Being an important indicator of combustion regime, formaldehyde indicates the preheat zone whereas OH radicals indicate the oxidation and post-flame zone [2]. The present work contributes to the comprehension of the entangled dynamic of intermediate products distribution. To do this, reactive Large Eddy Simulation (LES) with finite rate chemistry is employed for computing the reactive flow-field. We also present a modal analysis technique called proper orthogonal decomposition to demonstrate its capabilities for understanding flame dynamics.

Y. Shen (✉) · N. Jaouen · C. Duwig
KTH Royal Institute of Technology, Stockholm, Sweden
e-mail: yashen@kth.se

N. Jaouen
e-mail: njaouen@kth.se

C. Duwig
e-mail: duwig@kth.se

M. García-Villalba et al. (eds.), *Direct and Large Eddy Simulation XII*,
ERCOFTAC Series 27,
https://doi.org/10.1007/978-3-030-42822-8_25

Computational Study

Computational Geometry

The study focuses on a model combustor designed by Turbomeca within the EU project PRECCINSTA (Prediction and Control of Combustion Instabilities in Industrial Gas Turbines) [3]. The burner operates with a perfectly premixed methane/air mixture corresponding to a lean equivalence ratio of 0.83. The air/gas mixture with a mass flow rate of 12.9 g/s is led into the radial swirler and combustion chamber consecutively.

Governing Equations

A low-Mach number LES formalism is used; as described by the following partial differential equations for the conservation of mass, momentum, the conservation of N_s chemical species and energy. The system is closed with the ideal gas law.

$$\partial_t \overline{\rho} + \nabla \cdot (\overline{\rho u}) = 0 \,, \tag{25.1}$$

$$\partial_t \overline{\rho} \tilde{u} + \nabla \overline{\rho} \tilde{u} \tilde{u} = -\nabla \tilde{p} + \nabla \cdot (\overline{\rho} \tilde{u} \tilde{u} - \overline{\rho u u} + \sigma) \,, \tag{25.2}$$

$$\partial_t \overline{\rho} \tilde{Y}_k + \nabla \overline{\rho} \tilde{u} \tilde{Y}_k = \nabla \cdot \left(\overline{\rho} \tilde{u} \tilde{Y}_k - \overline{\rho u Y_k} + \rho D_k \nabla \tilde{Y}_k \right) + \overline{\dot{\omega}}_k \,, \tag{25.3}$$

$$\partial_t \overline{\rho} \tilde{h} + \nabla \overline{\rho} \tilde{u} \tilde{h} = \nabla \cdot \left(\overline{\rho} \tilde{u} \tilde{h} - \overline{\rho u h} + \lambda \nabla \tilde{T} \right) + \overline{\dot{\omega}}_T \,, \tag{25.4}$$

$$\overline{\rho} = \overline{\left(\frac{P}{RT} \right)} \,, \tag{25.5}$$

where u is the velocity vector, P the pressure, T is the temperature of the mixture and ρ its density, σ the viscous contribution, λ is the thermal conductivity, R represents the ideal gas constant, D_k is the diffusivity of species k, Y_k is species k mass fraction and $\dot{\omega}_k$ chemical source term of the species k. The term h is the mass sensible enthalpy of the mixture and $\dot{\omega}_T$ stands for the combustion heat release. The overline corresponds to the averaging operation and the tilde denotes the density weighted averaging. The species diffusion is modelled with Fick's law. The viscosity of the mixture is computed with Sutherland's law and a similar description is employed for the thermal conduction. The Smagorinsky model is used for computing the sub-grid scale stresses.

Chemistry Modelling

The importance of the chemical description on the global structure of flame is assessed using the DRM19 combustion scheme [4]. This skeletal mechanism is designed for methane combustion into air and involves 19 species and 82 elementary reactions. It is validated with shock-tube experiments for a range of equivalence ratio ranging from 0.2 to 2.0, an initial pressure range of 0.1 to 50 atm and temperature from 1300 to 2500 K, as well as methane–air adiabatic flames from 1 to 20 atm.

The Thickened Flame Model (TFM) is used to describe the interaction between the turbulence and chemistry [5]. The flame front is thickened artificially by a factor F, so that the progress variable can be resolved on the LES mesh.

Numerical Methods

The CFD code employed for this simulation relies on the OpenFOAM object oriented library. Discretisation in space is based on an unstructured Finite Volume method. The convective fluxes are reconstructed by central differencing giving a second-order accuracy in space and a backward time-discretisation is employed providing a second-order accuracy as well. An implicit time step is employed and the coupling of the velocity with pressure is performed with the PISO method. The maximum Courant–Friedrichs–Lewy number is fixed to 0.4. The settings of the LES simulation are corresponding to earlier work proposed by Duwig et al. [6].

The heat losses at the wall of combustion chamber are unknown. Considering the overtemperature near wall caused by the adiabatic wall condition, a wall temperature Dirichlet boundary condition is adopted to reach the agreement with experimental temperature of recirculation zones.

The geometry is decomposed into 3,278,741 hexahedral elements. The basic grid size is 1 mm. A refined discretisation with a size of 0.5 mm is set in the nozzle and in the flame region to get a detailed description of the injection system and of the physics of the reaction.

Modal Analysis

Conventional post-processing strategies such as time-average and root mean square neglect the fluctuation part of the turbulent flow. In this study, Proper Orthogonal Decomposition (POD) is employed to reveal the dynamic (dominant features) of the flame through a set of statistical proportions [7]. In the following equation, the quantity Ω that is obtained from the LES is to be decomposed into a linear combination of the orthogonal vectors $\phi_k(x)$ and their corresponding temporal coefficients $a_k(t)$, where the zeroth eigenfunction, which is mode 0, represents the mean field, while subsequent modes contain the fluctuations:

$$\Omega(x,t) = a_0(t)\phi_0(x) + \sum_{k=1}^{M} a_k(t)\phi_k(x)\,, \tag{25.6}$$

The analysis is performed using Sirovich's [7] method of snapshots according to the development of Duwig et al. [6] In this study, the approach is applied on 880 snapshots.

Extended POD (EPOD) approach is employed for its good capabilities for correlating the flame/flow interaction based upon simultaneously measured quantities [8]. The sequence of eigenvectors obtained trough POD of the OH field are used to reconstruct the main structure of the velocity fluctuations.

Results

Figure 25.1 provides randomly chosen instantaneous LES fields of the axial velocity, temperature, CH_2O and OH mass fractions. The burner exhibits a large Central Recirculation Zone (CRZ) that brings hot combustion products (of about 1900 K) back to the area close to nozzle. These recirculating burnt gases help igniting the fresh gas and stabilising the flame at the lips of the nozzle. The Outer Recirculation Zones (ORZ) are also clearly identifiable close to the wall with a temperature of about 1800 K. The adopted skeletal reaction scheme allows to capture the intermediates behaviour.

The CH_2O distribution corresponds to the area with temperature close to 1300 K, indicating a V-shaped preheat zone. The addition of heat loss on the chamber wall eliminates the high temperature of ORZ and a further external flame front. The OH distribution replenishes the area where formaldehyde is completely consumed, which shows the wrinkled flame front caused by turbulent eddies.

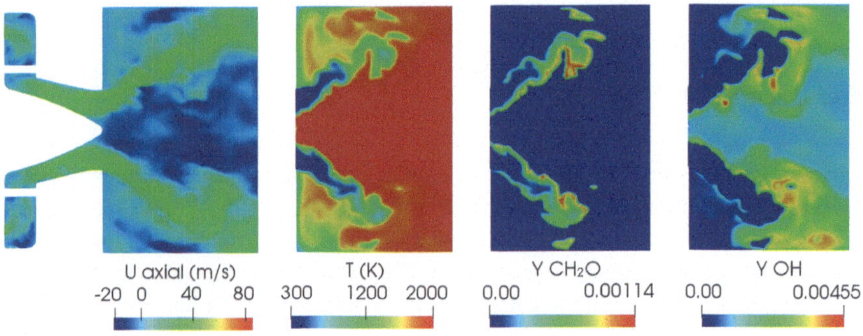

Fig. 25.1 Instantaneous LES fields of velocity, temperature, CH_2O and OH mass fraction

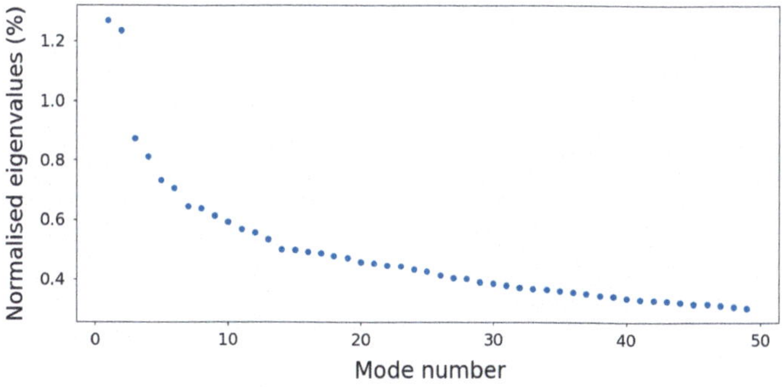

Fig. 25.2 Variance distribution of POD modes

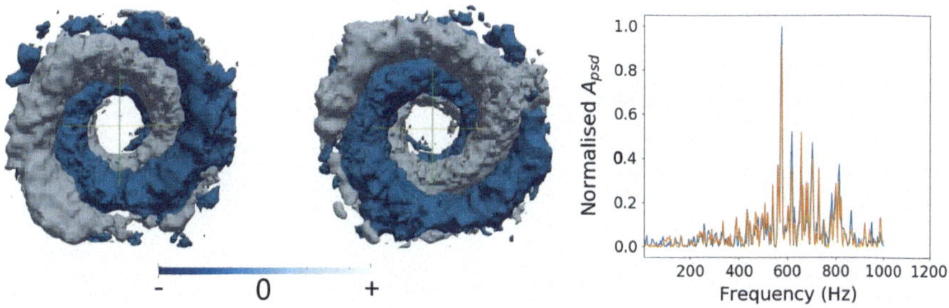

Fig. 25.3 POD modes 1–2 for OH field (left) and the Fourier analysis of their time coefficients (right)

Figure 25.2 shows the variance distribution of the first 50 modes. The studied flame does not exhibit any strongly dominant POD mode and the fluctuation is spread in the modes. Modes 1 and 2 each contain about 1.2% of the total variance, which makes it into a pair mode. Similarly, modes 3 and 4 with 0.8% variance are in pair. The relatively uniform variance distribution limits the possibility of lower order reconstruction of the turbulent field from the POD modes.

Figure 25.3 gives the first two POD modes of OH field. The first two modes show similar helical and axis-symmetry structure. The combination gives a rotating intermediate combustion products fluctuation. The overlaps of peaks in the power spectra density distribution also show these two modes are transverse complementary modes. Similar to modes 1 and 2, Fig. 25.4 shows modes 3 and 4 are in pair. Compared with the single helix shown in the first pair modes, modes 3 and 4 exhibit double helix structures. The first and second pairs of modes form into harmonic modes, which contribute to the main rotational flame dynamics simultaneously.

The EPOD modes, computed using the weights from the POD of OH and snapshots of velocity, are presented in Fig. 25.5. These two modes area are composed of a regular arrangement of vortices in the shear layer. The combination results in counter rotating flow vortices, which are generated and transported downstream.

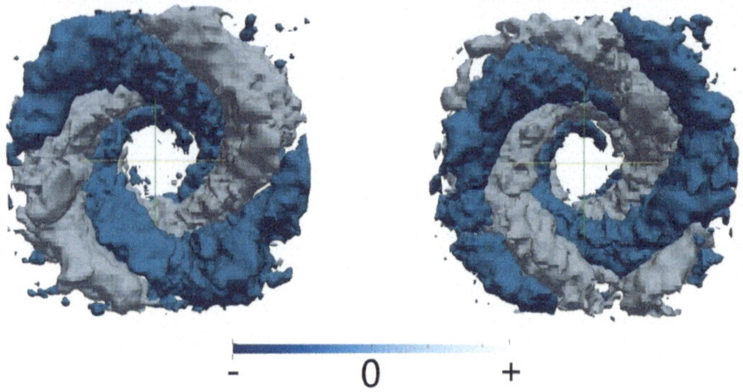

Fig. 25.4 POD modes 3–4 for OH field

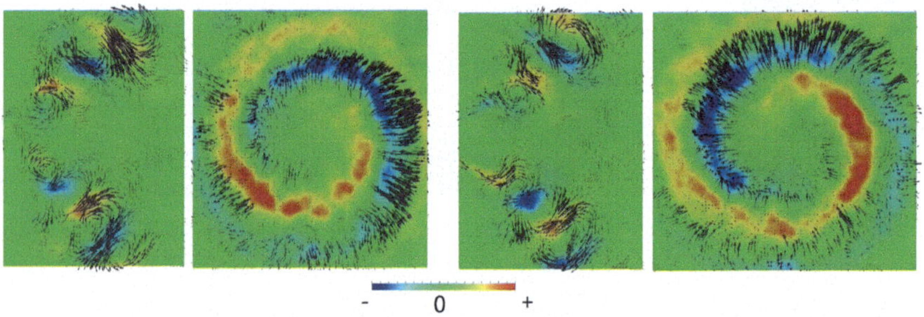

Fig. 25.5 EPOD modes 1–2: axial (left) and cross (right) section

Conclusion

In this work, the dynamics of a lean premixed flame has been investigated using a reactive Large Eddy Simulation. The instantaneous results show a detailed flow structure and different flame zones. The flame dynamics is analysed by the means of POD. Two pairs of POD modes with single and double helix structures contribute to the rotational trend of the flame dynamics. The POD and EPOD results provide detailed insights into the flow-flame interaction of this premixed turbulent flame.

References

1. Renard, P.H., Thevenin, D., Rolon, J.C., Candel, S.: Dynamics of flame/vortex interactions. Prog. Energy Combust. Sci. **26**(3), 225–282 (2000)
2. Sjöholm, J., Rosell, J., Li, B., Richter, M., Li, Z., Bai, X., Aldén, M.: Simultaneous visualization of OH, CH, CH_2O and toluene PLIF in a methane jet flame with varying degrees of turbulence. Proc. Combust. Inst. **34**(1), 1475–1482 (2013)

3. Meier, W., Weigand, P., Duan, X., Giezendanner-Thoben, R.: Detailed characterization of the dynamics of thermoacoustic pulsations in a lean premixed swirl flame. Combust. Flame **150** (1–2), 2–26 (2007)
4. Kazakov, A., Frenklach, M.: Drm19. http://www.me.berkeley.edu/drm/. Accessed June 18, 2019
5. Colin, O., Ducros, F., Veynante, D., Poinsot, T.: A thickened flame model for large eddy simulations of turbulent premixed combustion. Phys. Fluids **12**(7), 1843–1863 (2000)
6. Duwig, C., Nogenmyr, K.J., Chan, C.K., Dunn, M.J.: Large eddy simulations of a piloted lean premix jet flame using finite-rate chemistry. Combust. Theory Model. **15**(4), 537–568 (2011)
7. Sirovich, L.: Turbulence and the dynamics of coherent structures. i. coherent structures. Q. Appl. Math. **45**(3), 561–571 (1987)
8. Borée, J.: Extended proper orthogonal decomposition: a tool to analyse correlated events in turbulent flows. Exp. Fluids **35**(2), 188–192 (2003)

Chapter 26
LES of a Turbulent Partially-Premixed Flame Near Extinction Based on REDIM-PFDF Model

P. Wang, P. Shrotriya and T. Hou

Introduction

In recent years, turbulent partially-premixed flame with inhomogeneous inlets has drawn the attention of combustion community because of better flame stability near jet exit plane and its application in the practical combustors [1]. Large eddy simulation (LES) is a promising tool to study the turbulent partially-premixed flame but to balance accuracy and cost, it is often used in combination with reduced chemical kinetics. Reaction–Diffusion Manifold (REDIM) [2] is the well-established chemistry reduction technique which provides detailed chemistry in tabulated form. However, interaction of turbulence and chemistry in LES is still a challenging task, and to do so, sub-grid scale (SGS) combustion models are required. In the present paper, REDIM is combined with the Presumed Filtered Density Function (PFDF) [3] method to inherit the advantage of tabulated chemistry and solve the filtered equations that appear in the LES.

P. Wang (✉)
Institute for Energy Research, Jiangsu University, Xuefu Road 301,
212013 Zhenjiang, China
e-mail: pingwang@ujs.edu.cn

P. Shrotriya · T. Hou
School of Energy & Power Engineering, Jiangsu University,
Xuefu Road 301, Zhenjiang 212013, China
e-mail: shrotriya.prashant@yahoo.com

M. García-Villalba et al. (eds.), *Direct and Large Eddy Simulation XII*, ERCOFTAC Series 27,
https://doi.org/10.1007/978-3-030-42822-8_26

REDIM-PFDF Combustion Model

Thermochemical state vector (ψ) in any reacting system can be defined as

$$\psi = \left(h, p, \phi_1, \phi_2, \ldots, \phi_n\right)^T \tag{26.1}$$

where h assigns specific enthalpy, p pressure, and ϕ_i specific mass fraction of species i. In the REDIM method, this state is reduced to m-dimensional manifold ($m << n+2$) and can be written as $\psi = \psi(\theta)$, where $\theta = \theta_1, \theta_1, \ldots, \theta_m$ represents the local coordinates in the manifold.

In order to generate the look-up tables in the REDIM method, following evolution equation is solved [2]:

$$\frac{\partial \psi}{\partial t} = \left(I - \psi_\theta \psi_\theta^+\right)\left\{F(\psi) + \frac{d}{\rho}\chi^\circ \psi_{\theta\theta}^\circ \chi\right\} \tag{26.2}$$

Here, ρ is density, χ is the vector of spatial gradient estimates for θ, $F(\psi)$ is the vector of the chemical source terms, ψ_θ is the matrix of partial derivatives of ψ with respect to θ, ψ_θ^+ is its Moore–Penrose pseudo-inverse, and $\psi_{\theta\theta}$ is the Hessian matrix. The symbol '$^\circ$' in Eq. (26.2) is an abbreviation for the multiplication of two vectors with a tensor of third order.

Here, we use mass fractions of CO_2 (Y_{CO_2}) and N_2 (Y_{N_2}) as reduced coordinates for the REDIM to represent reaction progress and mixing process, respectively. All other quantities like species mass fraction, density, temperature as well as production rate of CO_2 (rCO_2) can be determined from the look-up tables. In LES of turbulent combustion, we solve the equation for the Favre filtered quantities. The Favre filtered equation for mass fraction is given as

$$\frac{\partial \bar{\rho}\widetilde{Y_k}}{\partial t} + \frac{\partial}{\partial x_i}(\bar{\rho}\widetilde{u_i}\widetilde{Y_k}) + \frac{\partial}{\partial x_i}(\overline{\rho u_i Y_k} - \bar{\rho}\widetilde{u_i}\widetilde{Y_k}) = \frac{\partial}{\partial x_i}(\bar{\rho}\bar{D}\frac{\partial \widetilde{Y_k}}{\partial x_i}) + \bar{\rho}\widetilde{\omega_k} \tag{26.3}$$

where the index $k = 1, 2$ denotes CO_2 and N_2, respectively. Note that for N_2, the source term in Eq. 26.3 is zero. The Favre filtered production rate is determined from the pre-calculated REDIM/FDF table by employing the PFDF method. In the PFDF method, the Favre filtered thermochemical quantities are calculated by

$$\tilde{f} = \int_{Y_{1_{\min}}}^{Y_{1\max}} \int_{Y_{2_{\min}}}^{Y_{2\max}} f(Y_1, Y_2)\tilde{P}(Y_1)\tilde{P}(Y_2)dY_1 dY_2 \tag{26.4}$$

where f is a general scalar variable. Under the assumption of statistical independence of Y_{CO_2} and Y_{N_2}, the joint FDF can be written as the product of the two marginal FDFs, i.e., $\tilde{P}(Y_{CO_2}, Y_{N_2}) = \tilde{P}(Y_{CO_2})\tilde{P}(Y_{N_2})$. In this work, $\tilde{P}(Y_{CO_2})$ is presumed to be clipped Gaussian FDF, while $\tilde{P}(Y_{N_2})$ is presumed to be top-hat distribution. The clipped Gaussian shape has been widely used with other joint FDFs because of its simplicity and good performance in reacting flows [3]. For multi-stream mixing problem, top-hat FDF is found to be more suitable choice compared to ubiquitous beta-FDF [4]. Mean and variance of the variable are used to parameterize the shape of the FDF. Mean values are calculated using Eq. (26.4) and variance is modeled algebraically [4].

Experimental and Numerical Setup

To validate the present REDIM-PFDF model for turbulent partially-premixed flame with inhomogeneous inlets, we chose experimental dataset provided from University of Sydney for piloted turbulent methane/air jet flames with variable inlet conditions [5]. It consists of two concentric pipes surrounded by the pilot stream. Inside diameter of the inner, outer, and pilot pipes are 4.0 mm, 7.5 mm, and 18 mm, respectively. Fuel is issued from the inner pipe and air is carried by the outer pipe. Inner pipe is recessed by length $Lr = 75$ mm so that inhomogeneous mixture of fuel and air is availed at the main jet exit. Five-gas pilot ($5GP$) mixture consisting of C_2H_2, H_2, CO_2, N_2, and air in proper proportion was used in experiments, while in this study we use stoichiometric mixture of CH_4/air whose adiabatic flame temperature and composition are the same as $5GP$ mixture. The whole burner assembly is centrally kept in the square cross sectioned wind tunnel of 25×25 cm^2 dimension, where coflow air flows at a fixed velocity $U_c = 15$ m/s. Here, we performed the simulation for FJ200-5GP-Lr75-103 flame case among all the available flame cases in the dataset. This flame case provides the inhomogeneous inlet conditions and more close to the blow off velocity. This case was run with $U_J = 103$ m/s (90% of blow off). For brevity, the name of the flame case is abbreviated to the U103 in this paper.

The in-house Finite-Volume code LESOCC2C is used to do the LES. This code solves the Low-Mach-number version of the compressible Navier–Stokes equations. The momentum equations are spatially discretized by a second-order central schemes and a three-step Runge–Kutta method of second order is used to advance in time. The mesh used here is the body-fitted curvilinear block-structured grids, in which whole computational domain is divided into 349 blocks and contains 5.25 million cells.

Results and Discussion

Flow and Mixing Pattern

Figure 26.1 shows instantaneous velocity field for the U103 case. In the instantaneous velocity field, the turbulence is fully developed by using periodic inlet flow condition. An accurate prediction of the turbulent flow inside the pipes is believed to be

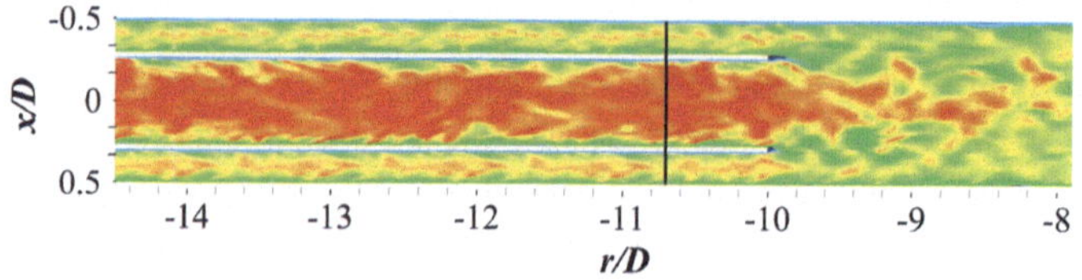

Fig. 26.1 Instantaneous velocity field

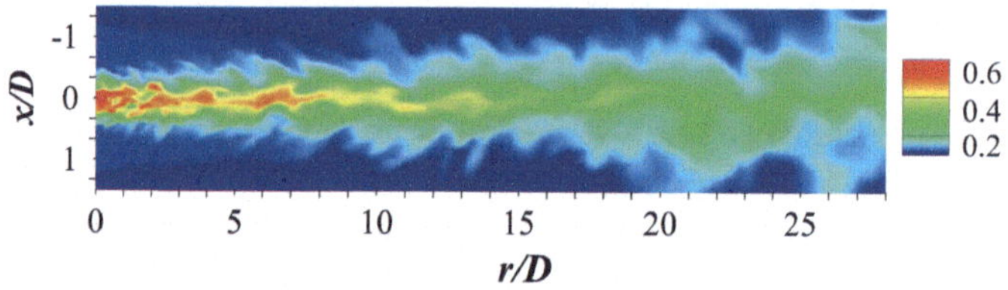

Fig. 26.2 Instantaneous mixture fraction field

important for a correct prediction of the mixing process of fuel and air, hence it plays an important role in the downstream flame propagation. The instantaneous mixture fraction field in Fig. 26.2 shows the inhomogeneous mixture fraction composition within the main jet, near jet exit plane. While at downstream locations, uniform mixture fraction field is obtained due to proper mixing of the fuel/air mixture with the coflow air. This inhomogeneous mixture fraction field has a specific advantage of the stratified-premixed combustion near the jet exit plane which augments the effect of the pilot flame, and this additional heat release by the pilot flame increases the flame stability.

Radial Statistics

Figures 26.3 and 26.4 show the radial profiles of the Favre mean and rms of the mixture fraction, respectively, at different axial locations. LES results are compared with the measured data, and the good performance of the REDIM-PFDF model is found. The REDIM-PFDF model predicted well the Favre mean and rms values of the mixture fraction at different axial locations. For $x/D \leq 10$, the rms values are under-predicted by the REDIM-PFDF model but the qualitative agreement is good. Near jet exit plane (at $x/D = 1$), the peak value of the Favre mean mixture is found slightly away from the center of the jet which is around 0.5. With increase in the downstream locations, stronger spreading in the Favre mean mixture fraction profile is found because of the higher Reynolds number of the current case (Re > 26000), and at $x/D = 15$, peak of the mean mixture fraction profile decays to 0.32. The rms profile at $x/D = 1$ shows the double peaks at the edges of the jet which represents the inhomogeneous behavior of the current flame case. At downstream location

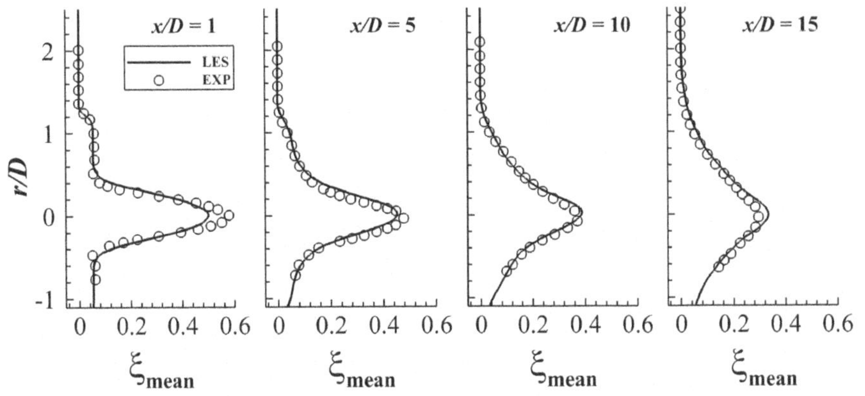

Fig. 26.3 The comparison of the measured and calculated radial statistics of the Favre mean mixture fraction

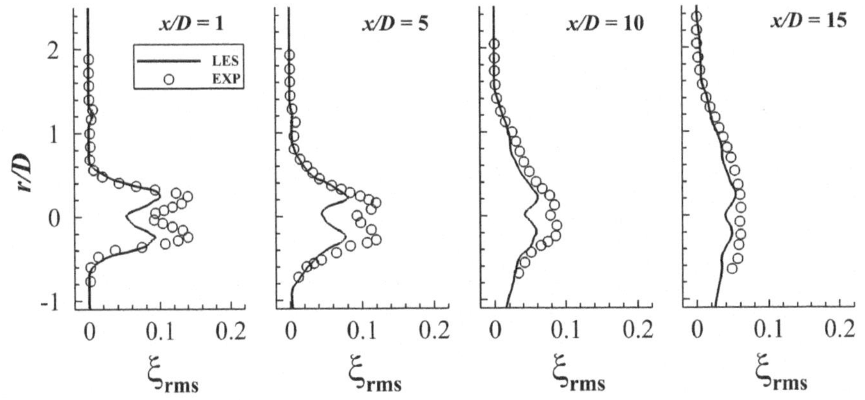

Fig. 26.4 The comparison of the measured and calculated radial statistics of rms mixture fraction

($x/D = 15$), this attains almost flat profile due to the uniform mixing of fuel and air mixture and advection of the coflow air in the main jet, which has the constant mixture fraction over the time.

The radial profiles of the Favre mean and rms temperature at different axial locations are shown in Figs. 26.5 and 26.6, respectively. For the Favre mean and rms profiles, agreement between the calculated and experimental value is good up to $x/D \leq 10$. But at $x/D = 15$, it is seen that both the Favre mean and rms values are over-predicted by the REDIM-PFDF model and it is thought to be caused by the neglected radiative heat losses in the present simulation. At $x/D = 1$, mean profile shows plateau in the pilot flow and double peaks in rms profile, corresponding to the

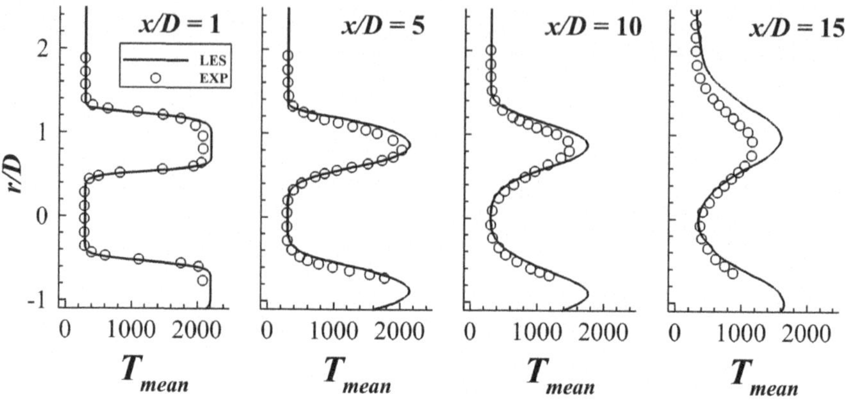

Fig. 26.5 The comparison of the measured and calculated radial statistics of the Favre mean temperature

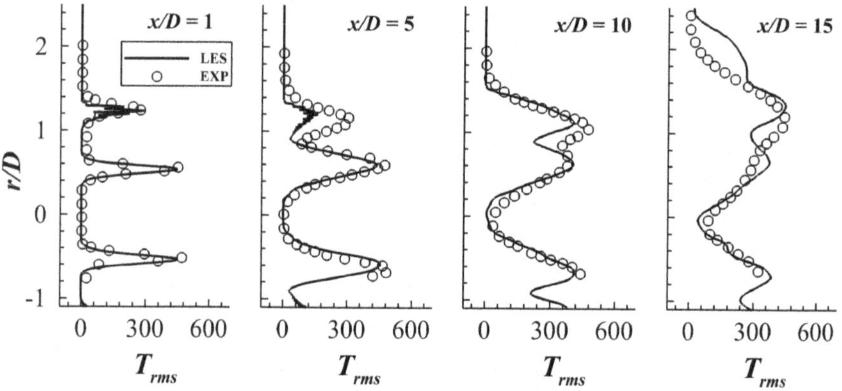

Fig. 26.6 The comparison of the measured and calculated radial statistics of rms temperature

inner reaction zone at the edge of the jet and the outer mixing layer between the pilot and coflow. A similar tendency is also observed for the Y_{CO}, which is not shown here due to the space limitation.

Conclusions

The REDIM-PFDF SGS model is used to perform the simulation for turbulent partially-premixed flame with inhomogeneous inlets and close to blowoff velocity. The predictions are in good agreement with the experimental results except at downstream locations, where jet spreading is slightly under-predicted by the REDIM-PFDF model. The inhomogeneity of the mixture near jet exit plane is clearly seen in the calculations. Existence of the stratified-premixed combustion is shown near jet exit plane by the flame index parameter, which further converts into non-premixed type of combustion at downstream locations.

References

1. Gicquel, L.Y.M., Staffelbach, G., Poinsot, T.: Large eddy simulations of gaseous flames in gas turbine combustion chambers. Prog. Energy Combust. Sci. **38**, 782–817 (2012)
2. Bykov, V., Maas, U.: Problem adapted reduced models based on reaction-diffusion manifolds (REDIMs). Proc. Combust. Inst. **32**(1), 561–568 (2009)
3. Wang, P., Platova, N.A., Fröhlich, J., Maas, U.: Large eddy simulation of the PRECCINSTA burner. Int. J. Heat Mass Transf. **70**, 486–495 (2014)
4. Floyd, J., Kempf, A.M., Kronenburg, A., Ram, R.H.: A simple model for the filtered density function for passive scalar combustion LES. Combust. Theory Model. **13**, 559–588 (2009)
5. Barlow, R.S., Meares, S., Magnotti, G., Cutcher, H., Masri, A.R.: Local extinction and near-field structure in piloted turbulent CH4/air jet flames with inhomogeneous inlets. Combust. Flame **162**, 3516–3540 (2015)

Chapter 27
Large-Eddy Simulation of Primary Atomization Using an Entropy Stable Conservative Level Set

O. Lehmkuhl, D. Mira, L. Gasparino, H. Owen and G. Houzeaux

Introduction

The atomization process in applications of propulsion and power is commonly characterized by high pressure injections in complex geometries. The liquid undertakes a large acceleration during the expansion at the nozzle and breaks down into small pieces or filaments that eventually form droplets [1]. Vallet et al. [1] proposed an Eulerian model based on the application of concepts for turbulent mixing to flow atomization in the context of RANS. Based on this work, the model was updated with new closures and different modelling strategies [2–4] and extended to LES [5, 6]. Other modelling strategies for multiphase flow in LES have also been presented in the literature [7–9]. The use of large-eddy simulation (LES) is mandatory when modelling applications of industrial interest, as DNS at engine-realistic conditions is out of the scope even with current HPC systems from today. Nevertheless, DNS conditions have also been addressed in the literature at moderate Reynolds numbers [8, 10, 11]. The numerical discretization of two-phase flows poses

O. Lehmkuhl · D. Mira (✉) · L. Gasparino · H. Owen · G. Houzeaux
Computer Applications of Science and Engineering Department,
Barcelona Supercomputing Centre (BSC), Barcelona, Spain
e-mail: daniel.mira@bsc.es

O. Lehmkuhl
e-mail: oriol.Lehmkuhl@bsc.es

L. Gasparino
e-mail: lucas.gasparino@bsc.es

H. Owen
e-mail: herbert.owen@bsc.es

G. Houzeaux
e-mail: guillaume.houzeaux@bsc.es

M. García-Villalba et al. (eds.), *Direct and Large Eddy Simulation XII*,
ERCOFTAC Series 27,
https://doi.org/10.1007/978-3-030-42822-8_27

several challenges in the modelling community, specially at high Reynolds and Weber numbers. Different discretization strategies can be found in the literature. Menard et al. [2] proposed a level set/VOF approach, Balcazar et al. [10] and Desjardins et al. [8] use a conservative level set with a re-initialization equation, while Vaudor et al. [11] introduce a consistent calculation of mass and momentum fluxes.

The present study proposes a numerical strategy for solving primary atomization using LES based on an entropy stable conservative level set method. The modelling approach includes an extension of the conservative level set equation stabilized with an entropy stable method [12] for the regions where the interface can be resolved, while it includes a subgrid contribution when the slip-velocity of the two-phase flow cannot be resolved with the given filter size [5]. This modelling approach has been coupled to the transport equation of subgrid interphase surface density in order to predict primary atomization in a pure Eulerian framework. In the present work, continuous finite elements are used in order to enable the modelling of realistic geometries.

Mathematics

The modelling approach presented here is developed to describe primary atomization at high Weber and Reynolds number flows [1]. The system of equations is given by the conservation of momentum, liquid volume fraction ϕ, and liquid/gas interface Σ in the incompressible limit, with $\nabla \cdot \mathbf{u} = 0$. The system is closed by the equation of state for a liquid/gas mixture given by $\rho = \phi\rho_l + (1-\phi)\,\rho_g$. The liquid/gas interface density Σ represents the quantity of liquid/gas interface per unit of volume and it is used to provide the characteristics and sizes of the liquid droplets in the dense part of the spray. The conservation equations will be presented below using spatially filtered quantities for LES. A decomposition of $\overline{\Sigma}$ following Chesnel et al. [5] is used and the subgrid surface density $\overline{\Sigma}'$ that evolves similar to $\overline{\Sigma}$ is solved instead. This separation is based on the existence of a minimum surface density $\overline{\Sigma}_{\min}$ due to the presence of liquid and a subgrid surface density $\overline{\Sigma}'$ that evolves similar to $\overline{\Sigma}$.

$$\overline{\Sigma} = \overline{\Sigma}_{\min} + \overline{\Sigma}' \tag{27.1}$$

where $\overline{\Sigma}_{\min}$ is obtained from empirical correlations based on DNS [5] and given by

$$\overline{\Sigma}_{\min} = \frac{\alpha}{\overline{\Delta}}\sqrt{\overline{\phi}\left(1-\overline{\phi}\right)} \tag{27.2}$$

with $\overline{\Delta}$ being a characteristic length scale (filter size in standard LES with implicit filtering) and $\alpha = 2.4$. Therefore, the transport equation for the liquid–gas interface can be expressed in terms of $\overline{\Sigma}'$ rather than $\overline{\Sigma}$.

The complete set of governing equations is presented here using filtered quantities and reads

$$\frac{\partial(\bar{\rho}\widetilde{\mathbf{u}})}{\partial t} + \nabla \cdot (\bar{\rho}\widetilde{\mathbf{u}}\widetilde{\mathbf{u}}) = \nabla \cdot (\mu + \mu_t)\nabla\widetilde{\mathbf{u}} + \nabla \cdot (\mu + \mu_t)(\nabla\widetilde{\mathbf{u}})^T - \nabla\bar{p} + \sigma k \mathbf{n}\delta_\Gamma \tag{27.3}$$

$$\frac{\partial\bar{\phi}}{\partial t} + \nabla \cdot \left(\widetilde{\mathbf{u}}\bar{\phi}\right) = -\nabla \cdot \tau_\phi \tag{27.4}$$

$$\frac{\partial\overline{\Sigma}'}{\partial t} + \nabla \cdot \left(\widetilde{\mathbf{u}}\overline{\Sigma}'\right) = \nabla \cdot \left(\overline{\Sigma}'\left(\widetilde{\mathbf{u}} - \widetilde{\mathbf{u}}_\Gamma\right)\right) + \dot{\overline{\Sigma}}_{\text{int}} \tag{27.5}$$

where the variables are expressed in standard notation.

The turbulent flux τ_ϕ and the slip-velocity $\overline{\Sigma}'\left(\widetilde{\mathbf{u}} - \widetilde{\mathbf{u}}_\Gamma\right)$ are both modelled using a gradient assumption closure [5, 6]. The source term $\dot{\overline{\Sigma}}_{\text{int}}$ represents the production/destruction of surface density by the mean shear, turbulence and liquid structure interactions [5] and is modelled by

$$\dot{\overline{\Sigma}}_{\text{int}} = \frac{\overline{\Sigma}}{\tau_b}\left(1 - \frac{\overline{\Sigma}}{\overline{\Sigma}_{\text{crit}}}\right) \tag{27.6}$$

The equilibrium value of surface density $\overline{\Sigma}_{\text{crit}}$ is obtained from a critical Weber number expressed in terms of balance of turbulent kinetic energy and takes the form of

$$\overline{\Sigma}_{\text{crit}} = \overline{\Sigma}_{\text{min}} + \overline{\phi}\left(1 - \overline{\phi}\right)\frac{\bar{\rho}\bar{k}}{\sigma \text{We}_{\text{crit}}} \tag{27.7}$$

The spatial discretization is based on a low dissipation finite element method with an explicit algorithm of the fractional step to solve the velocity/pressure coupling [14] and an entropy stable conservative level set [12] for the volume fraction equation. The main difference with Guermond work is the use of a cell-based viscosity instead of an edge-based version. The temporal discretization is based on a third-order energy preserving Runge–Kutta scheme.

Results

Validation

In order to validate the proposed interface tracking algorithm, the single vortex deformation of a circle is considered. Three regular $\mathbb{Q}_1$ meshes are considered in the validation (**m1**: 100×100, **m2**: 200×200 and **m3**: 400×400). Mass conservation is preserved in the three levels of mesh, being the L2 error for all cases below 10^{-6}.

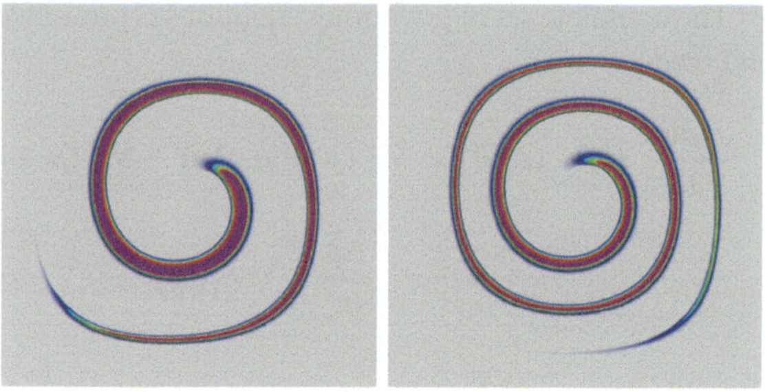

Fig. 27.1 Single vortex deformation of a circle using 400 × 400 nodes at t = 2 s (left) and t = 3 s (right)

In Fig. 27.1, snapshots for $t = 2$ and $t = 3$ and **m3** mesh are presented. Results are similar to the ones presented by Guermond et al. [12] using entropy stable edge-based viscosity instead of our proposed cell-based viscosity. Therefore, we can conclude that the proposed extension of the initial Guermond methodology to cell-based viscosity results in similar performance but on a more general numerical framework easier to merge with subgrid scale models for LES.

Turbulent Jet

The application considered to address the presented modelling approach is a test case representative of primary atomization. The configuration was proposed by Menard et al. [2] and has been the subject of some investigations [5, 6]. The test case corresponds to a liquid jet issuing into a quiescent air at moderate pressure. Two unstructured meshes composed by tetrahedrons with local refinement regions of 5 and 3 μm ending in 10 million and 60 million cells, respectively, are tested. A turbulent inlet velocity with a prescribed velocity profile [2] is imposed using fluctuations generated by diffusion [13]. A length scale of 10 and 20% of the jet diameter are used for the coarse and fine mesh, respectively.

The evolution of the flow can be distinguished in Fig. 27.2 by an iso-surface of the volume fraction and a contour plot of volume fraction along the spanwise direction. It is observed the atomization of jet with the main liquid core surrounded by a cloud of droplets and how the smallest droplets are formed at the front of the jet. After 510 diameters, the jet starts to break and filaments are generated.

The results are time-integrated and compared with the DNS data from Menard et al. [2]. The data for comparison includes the liquid volume fraction and surface density. As the surface density equation results are sensitive to the initial and boundary conditions, the solutions from DNS and the present LES are scaled by Σ_0 [5]. The liquid dispersion and surface density compared to the DNS data along the centreline is shown in Fig. 27.3a. The results indicate a good level of correlation with the DNS data for the two meshes tested. The penetration is well captured and the results show the influence of resolution on the jet decay. The results show a similar liquid core

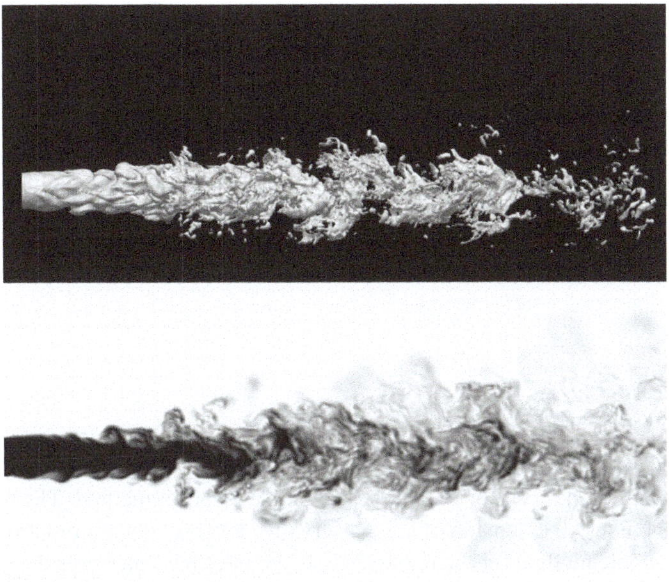

Fig. 27.2 Instantaneous snapshot of iso-contour (top) and contour plot over central plane (bottom) of volume fraction from a $\mathbb{P}_1$ grid with 60 M cells after 45 μs of injection

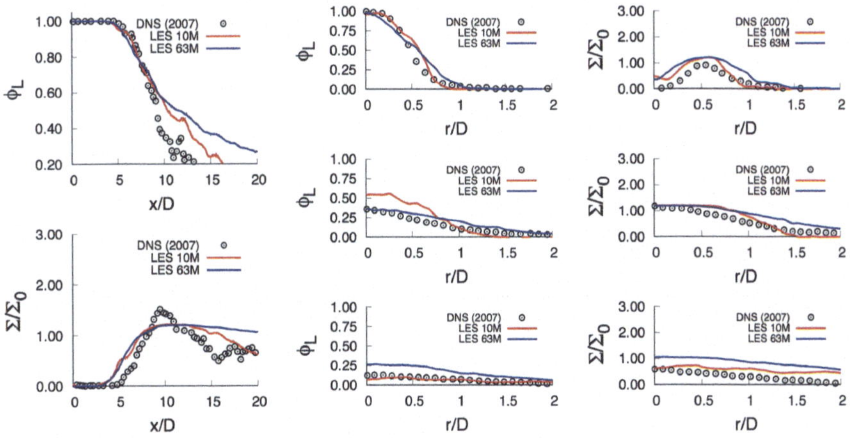

Fig. 27.3 Mean liquid volume fraction and surface density along the centreline (left) and radial distributions at three axial locations (middle and right): x/D = 5 (top), x/D = 10 (middle), and x/D = 20 (bottom)

length, but different dispersion. The surface density distribution is also rather similar along the centreline. The dispersion of the liquid can be distinguished in Fig. 27.3b. At the two upstream locations (x/D = 5, and x/D = 10), the results are in good agreement and the fine mesh shows better predictions than the coarse mesh in terms of liquid dispersion. The predictions of surface density show a similar behaviour for the two cases, which suggests a further revision of the critical Weber number would be recommended and is left for future work. At downstream conditions, the results on the finest mesh show high penetration and lower decay of the liquid core, that in turns affects the surface density.

Conclusions

The present study proposes a numerical strategy for solving primary atomization using LES based on an entropy stable conservative level set method. The modelling approach includes an extension of the conservative level set equation stabilized with an entropy stable method for the regions where the interface can be resolved, while it includes a subgrid contribution when the slip-velocity of the two-phase flow cannot be resolved with the given filter size. This strategy has been applied to the study of primary breakup of a turbulent jet at moderate Re and We numbers using two levels of resolution. The results indicate that the present approach is able te reproduce the liquid penetration and dispersion, and the comparison with the reference data from DNS is satisfactory.

Acknowledgements This work has been partially financially supported by the Ministerio de Economía y Competitividad, Secretaría de Estado de Investigación, Desarrollo e Innovación, Spain (Ref. TRA2017-89139-C2-2-R, CHEST) and by the Clean Sky Program from European Unions Horizon 2020 H2020-CS2-CFP06-2017-01 Fire Extinction ID: 785549. We also acknowledge Red Espanola de Surpercomputación (RES) for awarding us access to the MareNostrum IV (FI-2019-1-0019 and IM-2019-2-0012), and Juan de la Cierva personal grant IJCI-2015-26686.

References

1. Vallet, A., Burluka, A.A., Borghi, R.: Development of a Eulerian model for the atomization of a liquid jet. At. Sprays **11**, 619–642 (2001)
2. Menard, T., Tanguy, S., Berlemont, A.: Coupling level set/VOF/ghost fluid methods: validation and application to 3D simulation of the primary break-up of a liquid jet. Int. J. Multiph. Flow **33**(5), 510–524 (2007)
3. Lebas, R., Menard, T., Beau, P., Berlemont, A., Demoulin, F.: Numerical simulation of primary break-up and atomization: DNS and modelling study. Int. J. Multiph. Flow **35**(3), 247–260 (2009)
4. Garcia-Oliver, J.M., Pastor, J.M., Pandal, A., Trask, N., Baldwin, E., Schmidt, D.P.: Diesel spray CFD simulations based on the eulerian atomization model. At. Sprays **23**(1), 71–95 (2013)

5. Chesnel, J., Reveillon, J., Menard, T., Demoulin, F.-X.: Large eddy simulation of liquid jet atomization. At. Sprays **21**(9), 711–736 (2011)
6. Navarro-Martinez, S.: Large eddy simulation of spray atomization with a probability density function method. Int. J. Multiph. Flow **63**, 11–22 (2014)
7. Oefelein, J.: Large eddy simulation of turbulent combustion processes in propulsion and power systems. Prog. Aerosp. Sci. **42**, 2–37 (2006)
8. Desjardins, O., McCaslin, J., Owkes, M., Brady, P.: Direct numerical and large-eddy simulation of primary atomization in complex geometries. At. Sprays **23**(11), 1001–1048 (2013)
9. Pitsch, H., Desjardins, O., Balarac, G., Ihme, M.: Large-eddy simulation of turbulent reacting flows. Prog. Aerosp. Sci. **44**, 466–478 (2008)
10. Balcazar, N., Jofre, L., Lehmkuhl, O., Castro, J., Rigola, J.: A finite-volume/level-set method for simulating two-phase flows on unstructured grids. Int. J. Multiph. Flow **64**, 55–72 (2014)
11. Vaudor, G., Mnard, T., Aniszewski, W., Doring, M., Berlemont, A.: A consistent mass and momentum flux computation method for two phase flows. Application to atomization process. Comput. Fluids **152**, 204–216 (2017)
12. Guermond, J.-L., de Luna, M.Q., Thompson, T.: An conservative anti-diffusion technique for the level set method. J. Comput. Appl. Math. **321**, 448–468 (2017)
13. Kempf, A., Klein, M., Janicka, J.: Efficient generation of initial- and inflow-conditions for transient turbulent flows in arbitrary geometries. Flow, Turbul. Combust. **74**, 67–84 (2015)
14. Lehmkuhl, O., Houzeaux, G., Owen, H., Chrysokentis, G., Rodriguez, I.: A low-dissipation finite element scheme for scale resolving simulations of turbulent flows. J. Comput. Phys. **390**, 51–65 (2019)

Chapter 28
Toward DNS of the Ultimate Regime of Rayleigh–Bénard Convection

R. J. A. M. Stevens, D. Lohse and R. Verzicco

Introduction

Heat transfer mediated by a fluid is omnipresent in nature as well as in technical applications and it is always among the fundamental mechanisms of the phenomena. The performance of modern computer processors has reached a plateau owing to the inadequacy of the fluid-based cooling systems to get rid of the heat flux which increases with the operating frequency [1]. On much larger spatial scales, circulations in the atmosphere and oceans are driven by temperature differences whose strength is key for the evolution of the weather and the stability of regional and global climate [2].

The core of the problem, which is referred to as *natural convection*, is relatively simple since it reduces to determining the strength of the heat flux crossing the system for given flow conditions. Unfortunately, the governing equations (Navier–Stokes) are complex and non-linear, thus preventing the possibility to obtain analytical solutions. On the other hand laboratory experiments, aimed at tackling these phenomena, have to cope with the issue of how to make a setup of size $h_m = \mathcal{O}(\text{cm m})$ dynamically similar to a system of $h = \mathcal{O}(\text{m km})$ (Fig. 28.1). Indeed, this upscaling problem is common to many experiments in hydrodynamics; in thermal convection however it is exacerbated since, as we will see in the next section, the most relevant governing parameter, the Rayleigh number (Ra), depends on the *third* power of the leading spa-

R. J. A. M. Stevens · D. Lohse · R. Verzicco (✉)
PoF, UTwente, Enschede, The Netherlands
e-mail: r.verzicco@utwente.nl; verzicco@uniroma2.it

R. J. A. M. Stevens
e-mail: r.j.a.m.stevens@utwente.nl

D. Lohse
e-mail: d.lohse@utwente.nl

R. Verzicco
DII, Uniroma2, Rome, Italy

M. García-Villalba et al. (eds.), *Direct and Large Eddy Simulation XII*,
ERCOFTAC Series 27,
https://doi.org/10.1007/978-3-030-42822-8_28

Fig. 28.1 Cartoon of the scaling problem for a Rayleigh–Bénard flow: a model system of size h_m has to be operated in dynamic similarity with a real system of size h

tial scale. This implies that in real applications Ra easily attains huge values while it hardly hits its low-end in laboratory experiments.

Numerical simulations are subjected to similar limitations because of the spatial and temporal resolution requirements that become more severe as the Rayleigh number increases [3]. Only recently the former have become a viable alternative to experiments, thanks to the continuously growing power of supercomputers.

Indeed, if the dynamics of the system could be expressed by power laws of the form $\approx A\mathrm{Ra}^{\beta}$, experiments and numerical simulations, performed at moderate values of the driving parameters, could be scaled up to determine the response of the real systems at extreme driving values. Unfortunately, this strategy works only assuming that the coefficients of the power law (A and β) remain constant for every Ra and this could not be the case for thermal convection [4]. More in detail, Malkus [5] and Priestley [6] conjectured that all the mean temperature profile variations occur within the thermal boundary layers at the heated plates while the mean temperature in the bulk of the flow is essentially constant. Assuming also that the thermal boundary layers are far enough to evolve independently, one immediately obtains $\beta = 1/3$. A few years later, however, Kraichnan [7] noticed that as Ra increases, also the flow strengthens and the viscous boundary layers eventually must become turbulent. In this case, referred to as *ultimate regime*, velocity profiles are logarithmic with the wall normal distance, and it results $\beta = 1/2$ (times logarithmic corrections) which yields huge differences with respect to the previous theory.

This last observation and the fact that most of the practical applications evolve in the range of very high Rayleigh numbers motivate the effort to study turbulent thermal convection in the ultimate regime even if it requires the solution of formidable difficulties that we will detail in the next section.

A Trap Problem

One of the most appealing features of thermal convection is that the essence of the phenomenon can be reduced to a very simple model problem in which a fluid layer of thickness h, kinematic viscosity ν and thermal diffusivity κ is heated from below and cooled from above with a temperature difference $T_h - T_c = \Delta$. The temperature field, in a constant gravity field g produces a flow motion via the thermal expansion, parameterized by a constant coefficient α: this is the Rayleigh–Bénard flow. The heat flux $\dot{Q}$ between the plates will be a function of the form $\dot{Q} = f(h, \nu, \kappa, g, \alpha, \Delta)$ which involves $N = 7$ different quantities whose minimum number of independent dimensions is $K = 4$. The Buckingham Π-theorem assures that, in non-dimensional form, the above relation is equivalent to one with only $N - K = 3$ parameters that can be written as $\text{Nu} = F(\text{Ra}, \text{Pr})$ being $\text{Ra} = g\alpha\Delta h^3/(\nu\kappa)$, $\text{Pr} = \nu/\kappa$ and $\text{Nu} = \dot{Q}/\dot{Q}_{\text{diff}}$ with Q_{diff} the diffusive heat flux through the fluid in absence of motion (Figs. 28.2 and 28.3).

The above relation looks very attractive since it depends only on two independent parameters Ra and Pr, the latter being determined solely by the fluid properties. Many efforts have been made to study the function $F(\cdot)$ through laboratory experiments and, in the last two decades, also by numerical simulations as a viable alternative. Unfortunately, the practical realization of the RB convection is substantially different from the theoretical problem and many additional details come into play.

The first point is that while in the ideal flow the fluid layer is laterally unbounded, in real experiments, by necessity, it must be somehow confined thus introducing a second length d or equivalently an additional parameter $\Gamma = d/h$, the aspect-ratio.

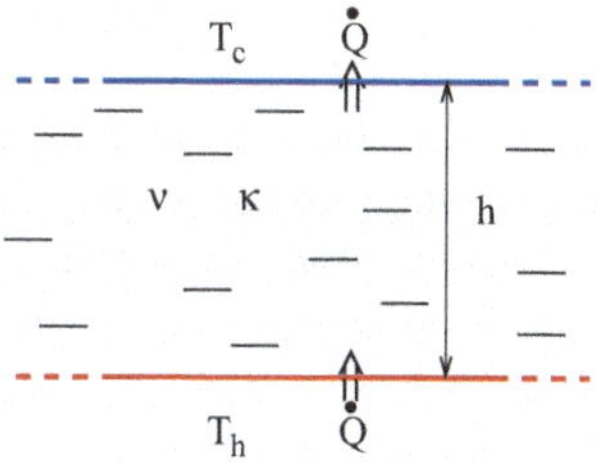

Fig. 28.2 Cartoon of the ideal Rayleigh–Bénard flow

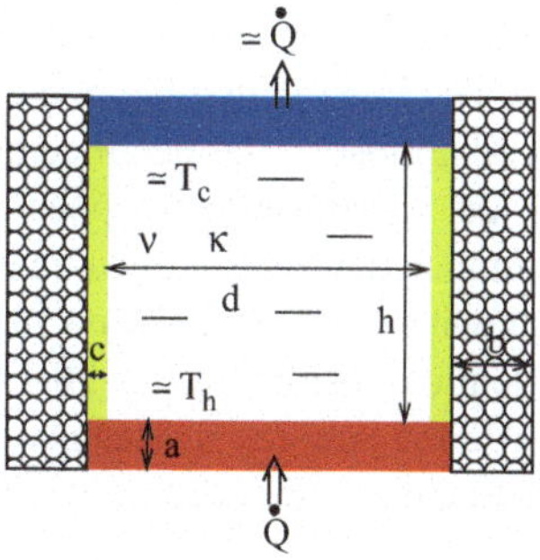

Fig. 28.3 Simplified sketch of a real experimental setup. Outside of the cell, only an insulating layer of foam has been reported while all the details about heating and cooling systems and thermal shields have been neglected

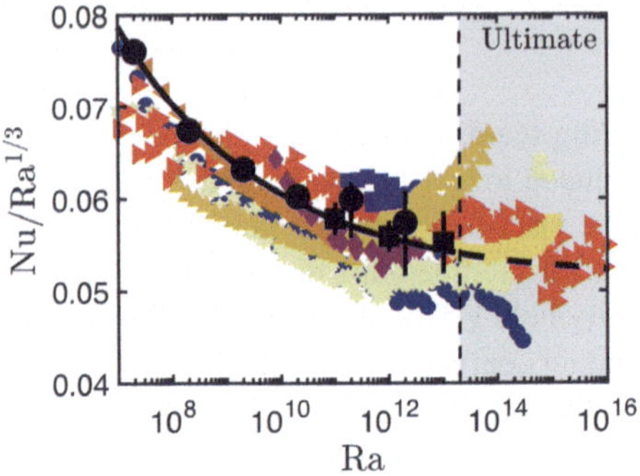

Fig. 28.4 Compilation of experimental and numerical data for the compensated Nusselt versus Rayleigh numbers: the solid black line is the theory by [4], blue bullets are the experiments by [8], purple squares [9], dark red diamonds [10], red right triangles [11], orange left triangles [12], yellow up triangles [13], yellow down triangles [14], yellow stars [15], black bullets, numerical simulations by [16], black squares, preliminary numerical simulations by Stevens (2019), (*Personal Communication*)

The parameters become more than one if the tank does not have a cylindrical or square cross section.

Several additional variables are introduced by the physical realization of the thermal boundary conditions; in fact, in the ideal problem, all the heat entering the fluid through the lower isothermal hot plate leaves the fluid only crossing the upper cold plate, without heat leakage across the side boundaries. In the real flow, the isothermal surfaces are obtained by thick metal plates ($a = \mathcal{O}(1-5)\,\text{cm}$) of high thermal diffusivity κ_{pl} (copper or aluminum) that provide stable temperature values at the fluid interface for every flow condition. The sidewall, in contrast, should minimize the heat transfer and it is therefore made of low thermal diffusivity κ_{sw} materials (steel or Plexiglas) and with reduced thickness ($e = \mathcal{O}(1-5)\,\text{mm}$). To further prevent parasite heat currents, insulation foam layers and even active thermal shields are installed outside of the cell, thus further increasing the number of input parameters. Within this scenario, the heat flux function looks like $\dot{Q} = f'(h, \nu, \kappa, g, \alpha, \Delta, d, e, \kappa_{sw}, a, \kappa_{pl}, b, \kappa_{foam}, ...)$ whose variable counting is $N > 14$ while it results always $K = 4$ thus implying that the non-dimensional counterpart ($\text{Nu} = F'(\text{Ra}, \text{Pr}, ...)$) involves more than $N - K > 10$ independent parameters.

The hope is that when the function $F'(\cdot)$ is explored, experimentally or numerically, most of the parameters introduced by the experimental technicalities do not affect significantly the phenomena and the relevant variables reduce to a tractable number. Unfortunately, some of the recent, and not so recent, experiments have shown that it is not always the case since dynamically equivalent flows do not yield identical results. In Fig. 28.4 we report a collection of Nusselt numbers taken from different sources showing some disagreement both at the low- and high-end of Ra. While the former differences have been attributed mainly to the heat leakage through

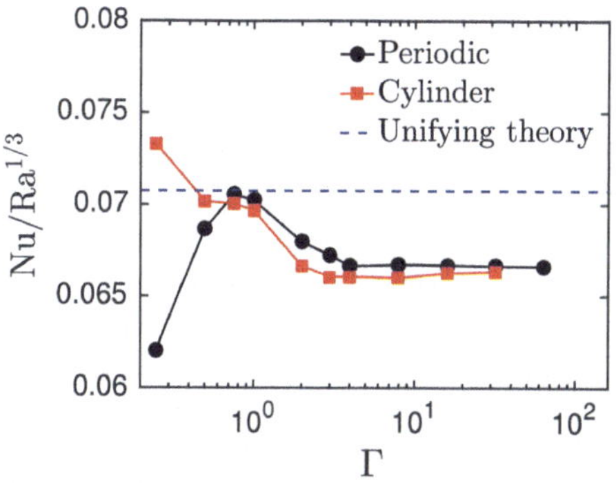

Fig. 28.5 Nusselt number versus domain aspect-ratio $\Gamma = d/h$ at $\mathrm{Ra} = 10^8$ and $\mathrm{Pr} = 1$. The dashed line for the "unifying theory" [4] is computed from experimental data at $\Gamma = 1$

the sidewall [17–20] and some of the latter to the non-perfect thermal sources [21], most of the discrepancies are still unexplained and they are the subject of intense investigation by many research groups worldwide. The situation is even more complex if one does not restrict the analysis only to the Nusselt number since higher order statistics show higher sensitivity to external perturbations.

In this context, numerical experiments can be particularly helpful since they can be used as ideal tests to isolate the different perturbations of the basic problem and assess their effectiveness. However, performing direct numerical simulation of Rayleigh–Bénard convection implies several non-obvious choices and requires huge computational resources that, for the parameter range of the ultimate regime, are not fully available yet. In the following, we will present estimates of the computational costs and discuss some of the possible open choices.

The first relevant point is whether the simulation should be aimed at the ideal RB flow or rather has to mimic a laboratory experiment. In the first case, the computational domain should be laterally unbounded which can be approximated by periodic boundary conditions applied to a rectangular domain of horizontal size d; this configuration results in a horizontally homogeneous flow that benefits from easy and efficient uniform spatial discretizations and fast converging statistics. On the other hand, in a real setup, the boundedness of the fluid layer generally results in a smaller fluid volume although the kinematic boundary layer at the sidewall has to be resolved by additional gridpoints.

Which of the two configurations is more advantageous is not obvious since the computational efficiency of the fully bounded flow solvers is smaller than those with homogeneous directions but the former setup involves smaller flow volumes.

Besides computational considerations, there are also physical issues since the presence of lateral boundaries, and even the shape of the container affect the flow dynamics [22]. In Fig. 28.5, we report the dependence of the Nusselt number on the domain aspect-ratio Γ for rectangular and cylindrical geometries. It is immediately evident that for slender domains Nu depends on Γ in the opposite way and only for $\Gamma \approx 1$ the two geometries yield similar values while for $\Gamma \geq 4$ the Nusselt number converges to the asymptotic value for unbounded domains. All the laboratory experiments reported in Fig. 28.4 have been performed in low aspect-ratio cylindrical cells and accordingly also the numerical simulations have been run in a cylinder at $\Gamma = 0.5$. In contrast, if a horizontally homogeneous flow has to be simulated, in order to get rid of the numerical confinement effect, it must be computed on domains at least of $\Gamma = 4$, even if spectra and higher order statistics indicate that $\Gamma = 8$ or $\Gamma = 16$ is needed to eliminate confinement effects [23].

In the following, we give an estimate of the computational resources needed for direct numerical simulation of turbulent RB convection in rectangular and cylindrical geometries by evaluating the number of nodes contained in the relative mesh. The basic assumption is that the flow can be divided into bulk and boundary layer regions, the former discretized by a mesh of the same size as the smallest between the Kolmogorov and Batchelor scales and the latter with the resolution criteria suggested by [3]. We further assume that the rectangular box has a size $d \times d \times h$ discretized in Cartesian coordinates, while the cylinder has a diameter d and a height h discretized in polar coordinates.

For the ease of discussion, we will restrict to $\mathrm{Pr} = 1$ keeping in mind that as the Prandtl number deviates substantially from unity the simulation becomes even more demanding either because the velocity field develops finer scales than the temperature ($\mathrm{Pr} \ll 1$) or vice versa ($\mathrm{Pr} \gg 1$).

For the mean Kolmogorov scale η, we can easily write $\eta/h \approx (\mathrm{RaNu})^{1/4}$ that with a fit $\mathrm{Nu} = A\mathrm{Ra}^{\beta}$ ($A \simeq 0.05$ and $\beta = 1/3$ from the high-end of Ra in Fig. 28.4) yields a number of nodes per unit length in the bulk $N_{bu} = 0.473\mathrm{Ra}^{1/3}$. For the resolution of each boundary layer, we rely on the correlation derived by [3] which suggest a number of nodes $N_{bl} \approx 0.35\mathrm{Ra}^{0.15}$. Within these figures, the total number of nodes for the rectangular domain reads $N_{\mathrm{Car}} = \Gamma^2(0.105\mathrm{Ra} + 0.156\mathrm{Ra}^{0.816})$.

We proceed along the same lines for the cylindrical domain keeping in mind that there is an extra boundary layer at the sidewall and that the polar coordinates have azimuthal isolines that diverge radially. Therefore the resolution requirements in this direction are dictated by the location farthest from the symmetry axis. Using the same correlations as above we obtain $N_{\mathrm{Cyl}} = 0.5\pi\Gamma^2(0.105\mathrm{Ra} + 0.156\mathrm{Ra}^{0.816}) + \pi\Gamma(0.223\mathrm{Ra}^{0.816} + 0.116\mathrm{Ra}^{0.633})$.

It is worth mentioning that these expressions have been obtained by simplifying assumptions; therefore their results should be taken as coarse estimates and not as precise measures. For example, A and β have been assumed constant and equal to the high-end Ra values of Fig. 28.4 and we have used $h - 2\delta_{bl} \approx h$ (with δ_{bl} the boundary layer thickness): all these positions concur to an overestimate of the number of nodes. On the other hand, the correlation $N_{bl} \approx 0.35\mathrm{Ra}^{0.15}$ of [3] was obtained for a Prandtl–Blasius laminar boundary layer that is expected to underestimates the resolution when the ultimate regime sets in and the boundary layers transition to turbulence. At the transitional Rayleigh number, the above factors might compensate each other and the estimates could give reasonable numbers.

A comparison of the two expressions immediately shows that the leading order term increases at the same rate with Ra and Γ although the cylindrical mesh has asymptotically 60% more nodes than the Cartesian counterpart. This is true although,

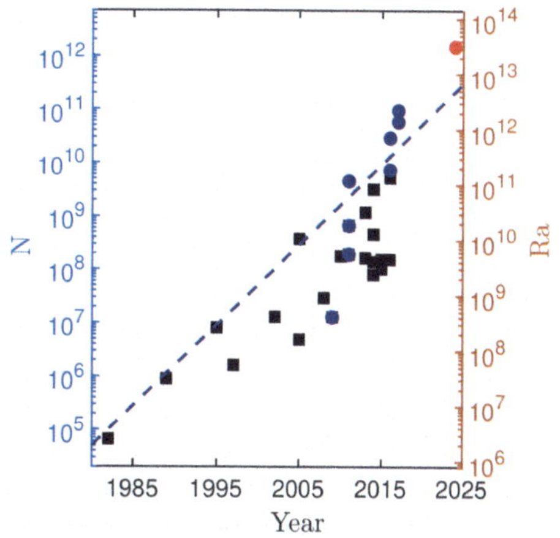

Fig. 28.6 Number of nodes N (in a cylindrical cell of $\Gamma = 0.5$) and achievable flow Rayleigh number Ra versus the years for direct numerical simulations of Rayleigh–Bénard convection. Black squares for various data from the literature, blue bullets for simulations from our research group, big red bullet final goal for the ultimate regime simulation

for the same aspect-ratio, the latter has a volume ($\Gamma^2 h^3$) which is more than 20% bigger than the former ($\pi \Gamma^2 h^3/4$).

If now we focus on the onset of the ultimate regime we have to determine the critical Rayleigh number at which the boundary layer undertakes the transition to the turbulent state. This is triggered by the large scales of convection that sweep the plates by the induced winds; according to Ref. [24] the boundary layer transition occurs for a shear Reynolds number of $\mathrm{Re}_S \approx 420$ that Grossmann and Lohse [4] have estimated to happen around $\mathrm{Ra}_C \approx 10^{14}$.

In a rectangular domain with $\Gamma = 4$ this Rayleigh number implies a mesh with $N_{\mathrm{Car}} > 10^{14}$ nodes that is clearly infeasible in the mid-term future. In a cylindrical cell of $\Gamma = 0.5$, however, it results $N_{\mathrm{Cyl}} \approx 10^{12}$ nodes that could be achieved within the next 5 years (see Fig. 28.6). Indeed, we are already running simulations at $\mathrm{Ra} = 10^{13}$ at $\Gamma = 0.5$ and even $\mathrm{Ra} = 10^{14}$ at $\Gamma = 0.25$ with meshes of the order of 10^{11} nodes (R. Stevens, *Personal Communication*) although we expect to tackle the ultimate regime only by the "next generation" simulations.

We wish to point out that if we compare the numbers coming from the present formulas with those currently used for the highest Rayleigh number simulations we find that the former produces a systematic overestimate of the required resolution. For example, for a cylindrical mesh of aspect-ratio $\Gamma = 0.5$ at $\mathrm{Ra} = 10^{13}$ our prediction yields a number of nodes $N_{\mathrm{Cyl}} \approx 4 \times 10^{11}$ while a simulation on a mesh $4608 \times 1400 \times 4480$ ($N_{\mathrm{Cyl}} \approx 2.9 \times 10^{10}$) yielded the same Nusselt number as another run on the finer grid $6144 \times 1536 \times 6144$ with $N_{\mathrm{Cyl}} \approx 5.8 \times 10^{10}$ nodes (R. Stevens, *Personal Communication*).

A possible explanation for this difference is that in our model we have assumed that in the bulk the mesh has to be as fine as the mean Kolmogorov scale η. However, looking at the dissipation spectra of turbulence [25] one finds its peak around 10η thus implying that also a mesh of size $1.5 - 2\eta$ already resolves most of the dissipation. In three-dimensions, this difference yields a factor $6.25 - 8$ less in the node counting that is about the mismatch between our prediction and the actual meshes.

For a while, we have been working at improving the simulation code [16, 26, 27] by more efficient implementations of the solution algorithms and of the parallelization strategies in order to reduce the time-to-solution. In addition we are also figuring out alternatives to achieve the ultimate regime in more affordable problems.

One possible way is to exploit the analogy between Rayleigh–Bénard and Taylor–Couette (TC) flow [28]. The latter is the flow developing in the gap between two coaxial cylinders rotating at different angular velocities and whose angular momentum flux across the cylinders behaves as the heat flux between the plates in a RB flow [29]. It turns out, however, that the mechanical forcing of the TC flow is more efficient in producing turbulent boundary layers than the thermal forcing of RB flows and the ultimate regime can be achieved for smaller values of the driving parameters that are affordable by numerical simulation [30]. In Ref. [31], thanks to the presence of baffled cylinders, which disrupted the logarithmic part of the turbulent boundary layer profiles, it has been possible to get rid of the logarithmic correction and obtain a pure $1/2$ power law in the analogous of the Nu versus Ra relationship.

Another possibility is to simulate a two-dimensional RB flow that allows, already now, to tackle Rayleigh numbers $> 10^{14}$; indeed in Ref. [32] (and successive developments) simulations have been run up to $\mathrm{Ra} \simeq 5 \times 10^{14}$ with the appearance of a transition already for $\mathrm{Ra} \geq 10^{13}$.

Closing Remarks

In this contribution, we have briefly introduced the problem of turbulent thermal convection with a particular look at its transition to the ultimate regime and the resolution requirements needed for the direct numerical simulation of this flow.

Leaving aside all the complications related to the spurious heat currents through the sidewall and the imperfect character of the thermal sources, already addressed in some of the referred papers, it appears that a preliminary fundamental question is whether the simulation should be aimed at replicating an experimental setup with a lateral confinement or to mimic the truly Rayleigh–Bénard flow that is virtually infinite in the horizontal directions.

We have shown that in the latter case a domain with aspect-ratio no smaller than $\Gamma = 4$ is required and this implies, at the estimated critical Rayleigh number $\mathrm{Ra}_C \approx 10^{14}$, a computational mesh with more than 10^{14} nodes that is not likely to be tractable within the next decade. On the other hand, although for a given Ra and Γ cylindrical, laterally confined geometries contain about 60% more nodes than the rectangular "unbounded" domains, when restricted to the existing, slender cylinders of the laboratory experiments the number of nodes become more feasible. In particular, for $\Gamma = 0.5$ and $\mathrm{Ra} = 10^{14}$ the present estimate gives a mesh slightly larger than a trillion of nodes. Even if this number might look impressive, it is "only" one order of magnitude bigger than the current state-of-the-art simulations and, according to Fig. 28.6, such meshes will become affordable within the next 5 years or so. It is also worthwhile mentioning that the present estimates assume a mesh in the bulk of the flow that is everywhere as fine as the mean Kolmogorov scale η while actual grid refinement checks performed on Rayleigh–Bénard turbulence have shown converged results already for meshes of size 2η. This implies that in three-dimensional flows the actual mesh sizes can be about one order of magnitude smaller a this is fully confirmed by our ongoing simulations.

Needless to say, once the ultimate regime will have been hit by numerical simulations also in three-dimensions, a *terra incognita* will be entered. Turbulent boundary layers have more severe resolution requirements than the laminar counterparts and once the ballistic plumes of Kraichnan [7], which can be thought of as pieces of detached thermal boundary layer, are shot into the bulk also the resolution of that flow region is likely to become more demanding. Clearly, attempting resolution estimates beyond the onset of the transition would be even more speculative than those of the present paper and only with the data of those simulations at hand, further reasonable projections can be made.

While waiting for adequate computational resources to tackle thermal convection in the ultimate regime we can nevertheless compute turbulent flows that exhibit similar dynamics or that can be reduced to a tractable size by simplifying assumptions. These include the Taylor–Couette flow, that can be rigorously shown to be analogous to Rayleigh-Bénard convection, or two-dimensional thermal convection that already now can be simulated well beyond $\mathrm{Ra} = 10^{14}$ and has indeed shown evidence of transition to the ultimate state.

Acknowledgements This work is supported by the Twente Max-Planck Center and the ERC (European Research Council) Starting Grant no. 804283 UltimateRB. The authors gratefully acknowledge the Gauss Centre for Supercomputing e.V. (https://www.gauss-centre.eu) for funding this project by providing computing time on the GCS Supercomputer SuperMUC-NG at Leibniz Supercomputing Centre (https://www.lrz.de).

References

1. Chu, R.C., Simons, R.E., Ellsworth, M.J., Schmidt, R.R., Cozzolino, V.: Review of cooling technologies for computer products. IEEE Trans. Dev. Mat. Reliab. **4**(4), 568–585 (2004)
2. Egan, P.J., Megan Mullin, M.: Recent improvement and projected worsening of weather in the United States. Nature **532**, 357–360 (2016)
3. Shishkina, O., Stevens, R.J.A.M., Grossmann, S., Lohse, D.: Boundary layer structure in turbulent thermal convection and its consequences for the required numerical resolution. New J. Phys. **12**, 075022 (2010)
4. Grossmann, S., Lohse, D.: Scaling in thermal convection: a unifying theory. J. Fluid Mech. **407**, 27–56 (2000)
5. Malkus, M.V.R.: Heat transport and spectrum of thermal turbulence. Proc. R. Soc. London, Ser. A **225**, 169 (1954)
6. Priestley, C.H.B.: Turbulent Transfer in the Lower Atmosphere. University of Chicago Press, Chicago (1959)
7. Kraichnan, R.H.: Turbulent thermal convection at arbitrary Prandtl number. Phys. Fluids **5**(11), 1374 (1962)

8. Castaing, B., Gunaratne, G., Heslot, F., Kadanoff., L., Libchaber, A., Thomae, S., Wu, X.Z., Zaleski, S., Zanetti, G.: Scaling of hard thermal turbulence in Rayleigh-Bénard convection. J. Fluid Mech. **204**, 1–30 (1989)
9. Fleisher, A.S., Goldstein, R.J.: High Rayleigh number convection of pressurized gases in a horizontal enclosure. J. Fluid Mech. **469**, 1–12 (2002)
10. Chaumat, S., Castaing, B., Chilla, F.: Rayleigh–Bénard cells: influence of plate properties. In: Castro, I.P., Hancock, P.E., Thomas, T.G. (eds.) Advances in Turbulence IX. International Center for Numerical Methods in Engineering, CIMNE
11. Niemela, J., Skrbek, L., Sreenivasan, K.R., Donnelly, R.: Turbulent convection at very high Rayleigh numbers. Nature **404**, 837–840 (2000)
12. Ahlers, G., He, X., Funfshilling, D., Bodenshatz, E.: Heat transport by turbulent Rayleigh–Bénard convection for $Pr \simeq 0.8$ and $3 \times 10^{12} \leq Ra \leq 10^{15}$: aspect ratio $\Gamma = 0.50$. New J. Phys. **14**, 103012 (2012)
13. Roche, P.E., Gauthier, F., Kaiser, R., Salort, J.: On the triggering of the ultimate regime of convection. New J. Phys. **12**, 085014 (2012)
14. He, X., Funfshilling, D., Bodenshatz, E., Ahlers, G.: Heat transport by turbulent Rayleigh–Bénard convection for $Pr \simeq 0.8$ and $4 \times 10^{11} \leq Ra \leq 2 \times 10^{14}$: aspect ratio $\Gamma = 1.00$. New J. Phys. **14**, 103012 (2012)
15. Urban, P., Musilovsá, V., Skrbek. L.: Efficiency of heat transfer in turbulent Rayleigh–Bénard convection. Phys. Rev. Lett. **107**, 014302 (2011)
16. Stevens, R.J.A.M., Lohse, D., Verzicco, R.: Prandtl and Rayleigh number dependence of heat transport in high Rayleigh number thermal convection. J. Fluid Mech. **688**, 31–43 (2011)
17. Roche, P.E., Castaing, B., Chabaud, B., Hebral, B., Sommeria, J.: Side wall effects in Rayleigh-Bénard experiments. Eur. Phys. J. B **24**, 405–408 (2001)
18. Ahlers, G.: Effect of sidewall conductance on heat-transport measurements for turbulent Rayleigh-Bénard convection. Phys. Rev. E **63**, 015303 (2000)
19. Verzicco, R.: Sidewall finite conductivity effects in confined turbulent thermal convection. J. Fluid Mech. **473**, 201–210 (2002)
20. Stevens, R.J.A.M., Lohse, D., Verzicco, R.: Sidewall effects in Rayleigh-Bénard convection. J. Fluid Mech. **741**, 1–27 (2014)
21. Verzicco, R.: Effect of non perfect thermal sources in turbulent thermal convection. Phys. Fluids **16**, 1965 (2004)
22. Daya, Z.A., Ecke, R.E.: Does turbulent convection feel the shape of the container? Phys. Rev. Lett. **89**(4501), U81–U83 (2001)
23. Stevens, R.J.A.M., Blass, A., Zhu, X., Verzicco, R., Lohse, D.: Turbulent thermal superstructures in Rayleigh-Bénard convection. Phys. Review Fluids **3**(4), 041501 (2018)
24. Landau, L.D., Lifshitz, E.M.: Fluid Mechanics. Pergamon Press, Oxford (1984)
25. Pope, S.B.: Turbulent Flows. Cambridge University Press, Cambridge (2000)
26. Verzicco, R., Orlandi, P.: A finite-difference scheme for three dimensional incompressible flows in cylindrical coordinates. J. Comp. Phys. **123**, 402 (1996)
27. Verzicco, R., Camussi, R.: Numerical experiments on strongly turbulent thermal convection in a slender cylindrical cell. J. Fluid Mech. **477**, 19–49 (2003)
28. Busse, F.H.: Viewpoint: the Twins of Turbulence Research. Physics **5**, 4 (2012)
29. Eckhardt, B., Grossmann, S., Lohse, D.: Torque scaling in turbulent Taylor-Couette flow between independently rotating cylinders. J. Fluid Mech. **581**, 221–250 (2007)
30. Ostilla-Monico, R., van der Poel, E., Verzicco, R., Grossmann, S., Sun, C., Lohse, D.: Exploring the phase diagram in the fully turbulent Taylor-Couette flow. J. Fluid Mech. **761**, 1–26 (2014)
31. Zhu, X., Verschoof, R.A., Bakhuis, D., Huisman, S.G., Verzicco, R., Sun, C., Lohse, D.: Wall roughness induces asymptotic ultimate turbulence. Nature Phys. **14**(4), 417–423 (2018)
32. Zhu, X., Mathai, V., Stevens, R.J.A.M., Verzicco, R., Lohse, D.: Transition to the ultimate regime in two-dimensional Rayleigh-Bénard convection. Phys. Review Lett. **120**(14), 144502 (2018)

Chapter 29
Thermal Structures in Turbulent Couette Flows by DNS

F. Alcántara-Ávila, S. Gandía-Barberá and S. Hoyas

Introduction

Turbulent channel flows confined between two parallel plates where the flow is driven by shear are called Couette flow (CF) [1–3]. The computational cost of simulating CFs is very high compared with Poiseuille flows (PFs). Therefore, relatively small Reynolds numbers are used when doing these simulations. The reason for this high computational cost is the existence of very large-scale roll-like motions that extend along the streamwise direction of the domain, that were first found experimentally [4, 5]. In order to properly represent these rolls in the simulation, the computational box must be long and wide. These velocity rolls are defined as coherent regions of either positive or negative streamwise velocity fluctuations. They appear in pairs, creating a *structure* in the velocity field composed by a couple of counter-rotating vortices aligned with the streamwise direction [1, 4]. For the thermal field, these structures have been observed for the first time here. Velocity and thermal structures are organized in a larger structure. The generic term Couette Thermal Flow Superstructure (CTFS) will be used to identify this composition of velocity and thermal structures [6].

For plane CF, the highest Reynolds number simulated is $Re_\tau = 1000$ by Kraheberger et al. [7]. When the thermal field is included the highest Reynolds number simulated was $Re_\tau = 400$ by Kawamura et al. [8]. However, the computational box

F. Alcántara-Ávila (✉) · S. Gandía-Barberá · S. Hoyas
Instituto Universitario de Matemática Pura y Aplicada,
Universitat Politècnica de València, 46022 Valencia, Spain
e-mail: fraalvi@alumni.upv.es

S. Gandía-Barberá
e-mail: serganba@etsid.upv.es

S. Hoyas
e-mail: serhocal@mot.upv.es

M. García-Villalba et al. (eds.), *Direct and Large Eddy Simulation XII*,
ERCOFTAC Series 27,
https://doi.org/10.1007/978-3-030-42822-8_29

used was too narrow and short to properly describe the CTFS [6]. In the present work, up to the knowledge of the authors, we present the results for $\mathrm{Re}_\tau = 500$, the highest simulated.

The paper is structured as follows. In the second section, the numerical setup is presented. In the third section, an analysis of the CTFS is developed. Finally, fourth section contains the conclusions.

Numerical Setup

In this work, several DNS of a passive thermal flow in a turbulent CF have been performed. Periodic boundary conditions are applied in the streamwise and spanwise directions. x, y, and z represent the streamwise, wall-normal, and spanwise coordinates, respectively. The corresponding velocity components are U, V and W or, using index notation, U_i. The average operator $\langle\cdot\rangle_{x_i}$ is defined as in [6]. Statistically averaged quantities in x and z are denoted by an overbar, $\overline{\phi} = \langle\phi\rangle_{xz}$, whereas fluctuating quantities are denoted by lowercase letters, i.e., $U = \langle U\rangle_{xz} + u = \overline{U} + u$. The superscript (+) indicates that the quantities have been normalized either by the friction velocity u_τ, or by the friction temperature θ_τ, where

$$u_\tau = \sqrt{\tau_w/\rho} \quad \text{and} \quad \theta_\tau = q_w/\left(\rho c_p u_\tau\right), \tag{29.1}$$

and τ_w is the statistically averaged wall shear stress and q_w is the normal heat flux to the walls.

Temperature is denoted by T. The boundary condition for the thermal field is the Mixed Boundary Condition (MBC) [9], where both walls of the computational box are heated with a constant heat flux q_w. Therefore, T increases linearly in the streamwise direction. In order to make the temperature field periodic, the transformed temperature is used, which is defined as, $\Theta = T - T_w$, where T_w is the temperature at the wall. This transformation allows the use of highly efficient Fourier methods in the streamwise direction.

The equations of the problem are the Navier–Stokes equations, together with the energy equation. The code used to solve them is the LISO code [1, 10–15]. The grid size in the wall-normal direction is $\Delta y_w^+ = 0.83$ and $\Delta y_c^+ = 2.3$ in the walls and center of the channel, respectively. The wall-parallel resolution for x and z is $\Delta x^+ \simeq 8.4$ and $\Delta z^+ \simeq 4.3$.

The simulations that have been performed are shown in Table 29.1. All cases are run using the Prandtl number of air, $\mathrm{Pr} = 0.71$. Three different friction Reynolds numbers have been used ($\mathrm{Re}_\tau = 180$, 250 and 500). The friction Reynolds number is defined as $\mathrm{Re}_\tau = u_\tau h/\nu$, where h is the semi-height of the channel and ν is the kinematic viscosity. The computational box dimensions are $16\pi h \times 6\pi h$ in x and z, respectively. A comparison with PFs will be done during the analysis of the results.

Table 29.1 Parameters of the simulation. Friction and bulk Reynolds numbers are shown in columns three and four, respectively. N_x, N_y, and N_z are the number of collocation points in each direction. The last column contains the time spent to compile statistics in terms of eddy turnovers

Cases	Re_τ	Re	N_x	N_y	N_z	Tu_τ/h
I	178	3500	1536	251	1152	59.1
II	262	5400	1536	251	1152	88.9
III	476	10500	3072	251	2304	58.1

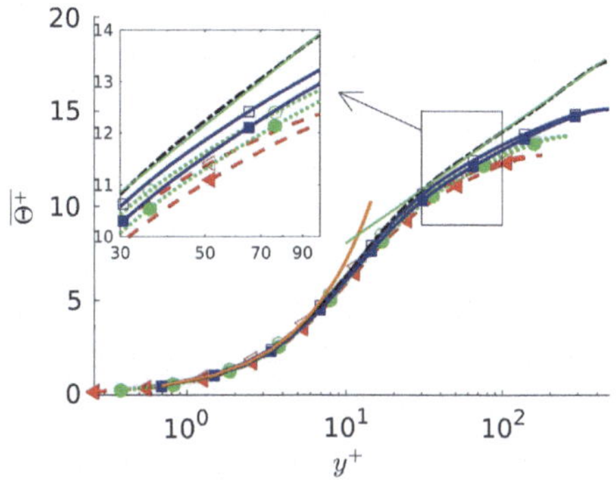

Fig. 29.1 $\overline{\Theta^+}$ for cases all cases (I: triangle red dashed; II: circle green dotted; and III: square blue solid). Closed symbols, bottom half. Empty symbols, top half. Black dash-dotted line represents a PF at $Re_\tau = 497$. Orange solid line: thermal law of the wall, $\overline{\Theta^+} = Pr \cdot y^+$. Green solid line represents the logarithmic behavior

Results and Discussion

Figure 29.1 shows the mean temperature, $\overline{\Theta^+}$, for all cases. Also, the mean temperature profile for a PF at $Re_\tau = 500$ is represented [14]. One of the first important results is that the temperature profile is no longer symmetric in both halves of the channel, in contrast with a PF. It can be seen in the zoom of Fig. 29.1 that the temperature in the upper half of the channel (empty marks) is slightly above the temperature in the lower half of the channel (closed marks). A second difference in $\overline{\Theta^+}$ between CFs and PFs is the absence of a logarithmic behavior in the former. To check if this logarithmic layer is formed in a CF for a higher Reynold number, a very high computational power is required. This will be a challenge for the next decade.

In Fig. 29.2a, b, the temperature fluctuation, θ'^+, and the streamwise heat flux, $\overline{u\theta}$, respectively, are represented. As a general result, all turbulent intensities increase and move toward the wall with an increase of the Reynolds number. Also, as it was stated for the mean temperature, all statistics, where the temperature is involved, are no totally symmetric or antisymmetric, in contrast with the PF. This asymmetry is due to the use of boundary conditions for the velocity and the thermal field that are not equivalent: a CF imposes different velocity on each wall and MBC imposes the

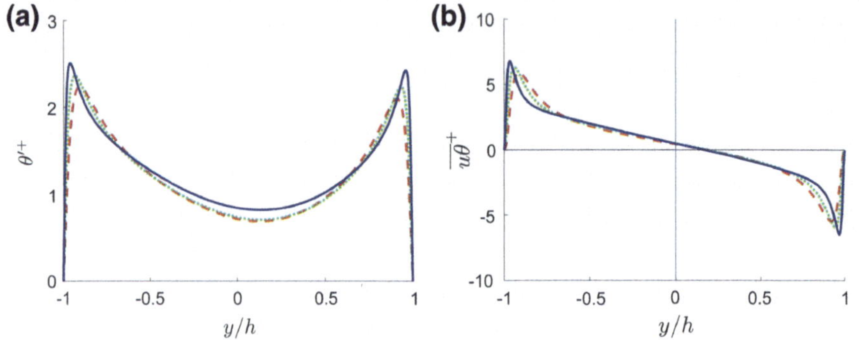

Fig. 29.2 **a** θ'^{+}. Case I: red dashed; Case II: green dotted; and Case III: blue solid. **b** $\overline{u\theta}$. Case I: red dashed; Case II: green dotted; and Case III: blue solid

same temperature in both walls. It is unclear if this will happen with even larger computational boxes and longer integration times. Another difference between CFs and PFs is seen in the, almost, antisymmetric profile of $\overline{u\theta}$ (Fig. 29.2b). For PF, this $\overline{u\theta}$ is perfectly symmetric.

One of the distinctive features of CF is the presence of coherent structures in the velocity field. Similar structures are also present in the thermal field. These thermal structures can be observed in Fig. 29.3. In the center, instantaneous images of θ' are represented. The structures appear in a strip-like pattern in $x - z$ planes. Isocontours of θ' show how these structures expand along the entire domain in the streamwise direction and are repeated along the spanwise direction. The CTFS are a consequence of this particular distribution of the values of u' and θ', as they reflect the effect of these instantaneous structures in the time-averaged flow and thermal field. A first proof of the existence of the CTFS can be observed in the top image of Fig. 29.3, where contours of θ' averaged for the entire database are represented. When the average is done in the streamwise direction, the values of $\langle u \rangle_x$, $\langle v \rangle_x$, $\langle w \rangle_x$, and $\langle \theta \rangle_x$ are obtained. These values are shown in the bottom image of Fig. 29.3. On one hand, u-velocity structures expand across the whole height of the channel, and v and w-velocity structures form pairs of counter-rotating vortices. On the other hand, a pair of hot-cold thermal structures appears for every velocity structure. While the velocity structures look more or less symmetric, thermal structures on the lower half look more intense than the ones on the upper half.

Inside the black box of the bottom image of Fig. 29.3, two velocity structures and four thermal ones are captured. This composition of structures is what is called Couette Thermal Flow Superstructures (CTFS). They are organized in such a way that the thermal flux, $\overline{u\theta}$, is negative in the upper half of the channel and positive in the bottom one. In the center of the CTFS appears a large vortex created by the streamwise velocities $\langle v' \rangle$ and $\langle w' \rangle$.

Two non-trivial symmetries, never described before, can be observed in the CTFS. First, there is a translation symmetry of period L_R. This length is the distance between

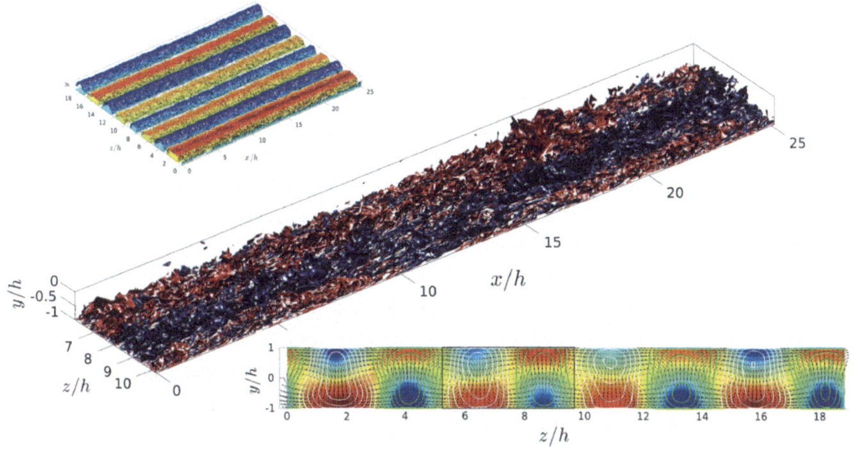

Fig. 29.3 Velocity and thermal structures for $\mathrm{Re}_\tau = 500$. Top: Isocontours of cold (blue) and hot (red) $\langle\theta'\rangle$. The values of the isocontours are $0.4\langle\theta'\rangle_{\min}$ and $0.4\langle\theta'\rangle_{\max}$ for the cold and hot $\langle\theta'\rangle$, respectively. Middle: isocontours of θ'. Red, hotter than mean; blue, colder than mean. Bottom: Contours of $\langle\theta'\rangle_x$. White and green lines represent contours of positive and negative $\langle u'\rangle_x$, respectively. $(\langle v'\rangle_x, \langle w'\rangle_x)$ vector field is represented by arrows

the centers of two consecutive rolls with the same sign. For the specific case where the computational box width is $6\pi h$, the value of L_R is $6\pi/4 \approx 4.7$. The second symmetry appears for $\langle\theta\rangle_x$, with respect to the origin of every (v', w') vortex.

Finally, a spectral analysis for u^+ and θ^+ has been performed to see how the CTFS affects to the energy spectrum. Contour plots of the 1D spectral density of the two intensities are shown in Fig. 29.4a, b to visualize the y dependencies of the spectrum. First, on the left part of both figures, there is a concentration of the energy, corresponding to turbulent structures intrinsic to every turbulent flow. They are small in length and are close to the wall, $y^+ = 10$. The new feature of the CFs appears at $\lambda_z^+ \approx 4.7$, which corresponds to the length of the CTFS. Here is where most of the energy is concentrated, which corresponds to the coherent structures formed for CFs. It can be seen how the velocity rolls are perfectly symmetric, while the temperature ones are more intense in the lower half of the channel, as mentioned before.

Conclusions

As a conclusion, three simulations of thermal flows at larger numbers of Reynolds and larger computational boxes than those previously simulated have been described. Thermal quantities appear to be not totally symmetric or antisymmetric, in contrast to PFs. The perturbations of the flow field present a well-organized structure in the form of positive and negative velocity rolls. The superstructure made of two velocity

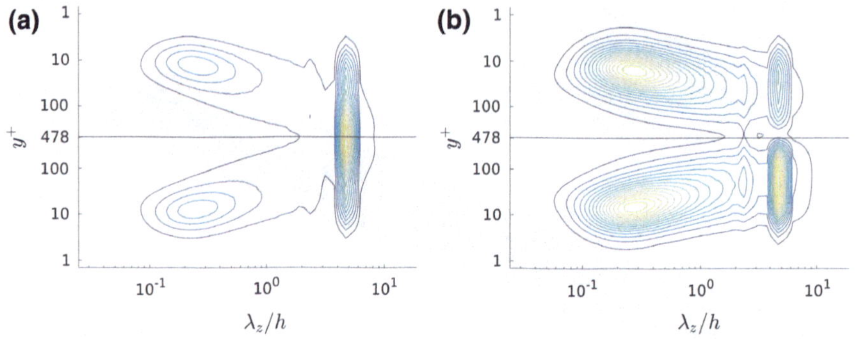

Fig. 29.4 **a** Contours of u^+ spectral density for $\mathrm{Re}_\tau = 500$ in the $Y - Z$ plane. **b** Contours of θ^+ spectral density for $\mathrm{Re}_\tau = 500$ in the $Y - Z$ plane

rolls and four thermal ones in the time-averaged field has been defined as a Couette Thermal Flow Superstructure. The influence of the CTFS on the energy spectrum has also been studied. The statistics of all simulations can be downloaded from the web page of our group: http://personales.upv.es/serhocal/.

Acknowledgements This work was supported by the MINECO/FEDER, under project RTI2018-102256-B-I00. The computations of the new simulations were made possible by a generous grant of computing time from the Barcelona Supercomputing Centre, reference FI-2018-1-0037. FAA is partially funded by GVA/FEDER project ACIF2018.

References

1. Avsarkisov, V., Hoyas, S., Oberlack, M., García-Galache, J.P.: Turbulent plane Couette flow at moderately high Reynolds number. JFM **751**, R1 (2014)
2. Pirozzoli, S., Bernardini, M., Orlandi, P.: Turbulence statistics in Couette flow at high Reynolds number. JFM **758**, 323–343 (2014)
3. Lee, M., Moser, R.: Extreme-scale motions in turbulent plane Couette flows. JFM **842**, 128–145 (2018)
4. Tillmark, N.: Experiments on transition and turbulence in plane Couette flow. KTH, Royal Institute of Technology (1995)
5. Kitoh, O., Nakabyashi, K., Nishimura, F.: Experimental study on mean velocity and turbulence characteristics of plane Couette flow: low-Reynolds-number effects and large longitudinal vortical structure. JFM **539**, 199–227 (2005)
6. Alcántara-Ávila, F., Gandía, B., Hoyas, S.: Evidences of persisting thermal structures in Couette flows. IJHFF **76**, 287–295 (2019)
7. Kraheberger, S., Hoyas, S., Oberlack, M.: DNS of a turbulent Couette flow at constant wall transpiration up to $Re_\tau = 1000$. JFM **835**, 421–443 (2018)
8. Kawamura, H., Abe, H., Shingai, K.: DNS of turbulence and heat transport in a channel flow with different Reynolds and Prandtl numbers and boundary conditions. Turbul Heat Mass Transf **3** (Proceedings of the 3rd International Symposium on Turbulence, Heat and Mass Transfer) (2000)

9. Kasagi, N., Tomita, Y., Kuroda, A.: Direct numerical simulation of passive scalar field in a turbulent channel flow. JHT **114**, 598–606 (1992)
10. Alcántara-Ávila, F., Hoyas, S., Pérez-Quiles, M.J.: DNS of thermal channel flow up to $Re_\tau =$ 2000 for medium to low Prandtl numbers. IJHMT **127**, 349–361 (2018)
11. Avsarkisov, V., Oberlack, M., Hoyas, S.: New scaling laws for turbulent Poiseuille flow with wall transpiration. JFM **746**, 99–122 (2014)
12. Gandía-Barberá, S., Hoyas, S., Oberlack, M., Kraheberger, S.: The link between the Reynolds shear stress and the large structures of turbulent Couette-Poiseuille flow. PF **30**, 041702 (2018)
13. Hoyas, S., Jiménez, J.: Scaling of the velocity fluctuations in turbulent channels up to $Re_\tau =$ 2003. PF **18**, 011702 (2006)
14. Lluesma-Rodríguez, F., Hoyas, S., Peréz-Quiles, M.J.: Influence of the computational domain on DNS of turbulent heat transfer up to $Re_\tau = 2000$ for $Pr = 0.71$. IJHMT **122**, 983–992 (2018)
15. Hoyas, S., Jiménez, J.: Reynolds number effects on the Reynolds-stress budgets in turbulent channels. PF **20**, 101511 (2008)

Chapter 30
DNS of Turbulent Pipe Flow with Temperature-Dependent Fluid Properties Subject to Non-uniform External Heating

A. Antoranz, O. Flores and M. García-Villalba

Introduction

In the present article, we report results of Direct Numerical Simulations (DNS) of a pressure-driven fully developed turbulent flow in a pipe, with sinusoidal heat flux boundary conditions, and with temperature-dependent viscosity and thermal diffusivity, representative of the conditions in the tubes of heat receivers in Solar Power Tower plants. The main objective is to study the influence of variable fluid properties on the mean values and turbulence statistics in the heat transfer fluid: a molten salt. To that end, three different conditions are considered: a case with constant fluid properties and two cases with temperature-dependent fluid properties.

Numerical studies of heat transfer in a pipe with *homogeneous* heating can be found in Refs. [1, 2] while the experimental study of *circumferentially-varying* heat flux is treated in Refs. [3, 4]. In previous works [5, 6], the present authors reported and analyzed DNS of pipe flow with *circumferentially-varying* heat flux with constant fluid properties.

Regarding the studies on temperature-dependent fluid properties, the works in Refs. [7, 8], despite not being done for a pipe flow, are relevant for the present study.

A. Antoranz · O. Flores · M. García-Villalba (✉)
Bioengineering and Aerospace Engineering Department,
Universidad Carlos III de Madrid, Leganés, Spain
e-mail: mgarciav@ing.uc3m.es

A. Antoranz
e-mail: antonio.antoranz@uc3m.es

A. Antoranz
ITP Aero, Alcobendas, Spain

M. García-Villalba et al. (eds.), *Direct and Large Eddy Simulation XII*,
ERCOFTAC Series 27,
https://doi.org/10.1007/978-3-030-42822-8_30

Computational Setup

The equations solved are the Navier–Stokes equations for the incompressible flow (continuity and momentum), together with an advection–diffusion equation for the temperature (neglecting viscous energy dissipation). Energy and momentum equations are coupled by the viscosity variation with temperature. The system of equations is completed with appropriate boundary conditions on the pipe surface: velocity is zero at the wall, and the wall-normal temperature gradient is given by an imposed sinusoidal heat flux distribution

$$q''_w(\theta) = \pi \overline{q}''_w \sin\theta = -\rho C_p \alpha(T) \left.\frac{\partial T}{\partial r}\right|_{r=R} . \tag{30.1}$$

Since the net heat flux to the domain is null, the axial direction, z, is homogeneous and we can apply periodic boundary conditions at the inlet and outlet of the pipe.

The variation of the kinematic viscosity and the thermal diffusivity with temperature, $\nu(T)$ and $\alpha(T)$, are prescribed by the following power-laws:

$$\frac{\nu}{\nu_0} = \left(\frac{T}{T_0}\right)^{e_\nu}, \quad \frac{\alpha}{\alpha_0} = \left(\frac{T}{T_0}\right)^{e_\alpha}, \tag{30.2}$$

The exponents $e_\nu = -3$ and $e_\alpha = 0.3$ are selected to represent the behavior of typical molten salts encountered in solar central receivers. Apart from these exponents, the problem is governed by three non-dimensional parameters: the global Reynolds number $\mathrm{Re}_{\tau 0} = u_{\tau 0} R/\nu_0$, the global Prandtl number $\mathrm{Pr}_0 = \nu_0/\alpha_0$, and the normalized temperature $T_0/T_{\tau 0}$, where $u_{\tau 0}$ and $T_{\tau 0}$ are, respectively, the friction velocity and friction temperature for the case with constant fluid properties. Note that T_0 is not defined for the constant fluid property case. Values for the cases studied are provided in Table 30.1.

The simulations are performed using the code Nek5000 [9], which uses a spectral element method, solving the equations on Gauss–Lobatto–Legendre nodes. The computational domain consists of a straight circular pipe of length $25R$, discretized with 55440 spectral elements of polynomial order $N = 7$, with 105 elements in the stream-wise direction and 528 elements in the cross-plane.

Table 30.1 Parameters of the simulations. Case 1: constant fluid properties, Case 2: low sensitivity to temperature, Case 3: high sensitivity to temperature

Case	$\mathrm{Re}_{\tau 0}$	Pr_0	$T_0/T_{\tau 0}$	Line style
1	180	0.7	–	Solid (Black)
2	180	0.7	1000	Dashed (Red)
3	180	0.7	500	Dashed-dotted (Blue)

Results and Discussion

For illustration, a snapshot of the instantaneous temperature $T/T_{\tau 0}$ for Case 3 is shown in Fig. 30.1. The circumferential distribution of the averaged temperature on the pipe wall is presented in Fig. 30.2a for the three cases as the change from their corresponding bulk temperatures, $(\langle T_w \rangle - T_b)/T_{\tau 0}$. Case 1 shows an anti-symmetrical temperature distribution with peak amplitude $(\langle T_w \rangle - T_b)/T_{\tau 0} \sim 65$, with Cases 2 and 3, where ν and α are allowed to vary, being slightly shifted to lower temperatures.

As opposed to cases with iso-thermal or mixed boundary conditions [1], a pure heat flux boundary condition allows temperature fluctuations at the wall. Figure 30.2b shows the circumferential variation of the RMS temperature fluctuations on the wall for the three cases of study. Values range from $T'_{\rm rms}/T_{\tau 0} \sim 5.5$ to 7.7 and are maximum in the regions where the heat flux is more intense, $\theta = \pi/2$ and $\theta = 3\pi/2$. These fluctuations are found to be largely influenced by the variable fluid properties. As $T_0/T_{\tau 0}$ decrease, the fluctuations at the heated side are mitigated, while they are exacerbated at the cooled side. For Case 3, we have a +12% increment in $T'_{\rm rms}/T_{\tau 0}$ peak in the cold region compared to the case of constant fluid properties, while fluctuations are reduced by −7% at the hot top.

Variation of the flow properties in the crossplane are shown through the use of contour plots, where the left half of the pipe corresponds to the case with constant properties (Case 1) and the right half belongs to the more sensitive case (Case 3), facilitating the comparison between the two cases. The turbulent kinetic energy, $k/u_{\tau 0}^2$, and the RMS temperature fluctuations, $T'_{\rm rms}/T_{\tau 0}$, contours in the cross-plane are compared in Fig. 30.3. We observe that, despite the wall mean temperature profile on the pipe does not change significantly, both temperature and velocity fluctuations are increased on the cooled wall but damped on the heated wall for the cases with variable fluid properties, in agreement with previous observations [7, 8].

The introduction of variable fluid properties in the computation leads to the existence of small but discernible mean velocities in the radial and circumferential direc-

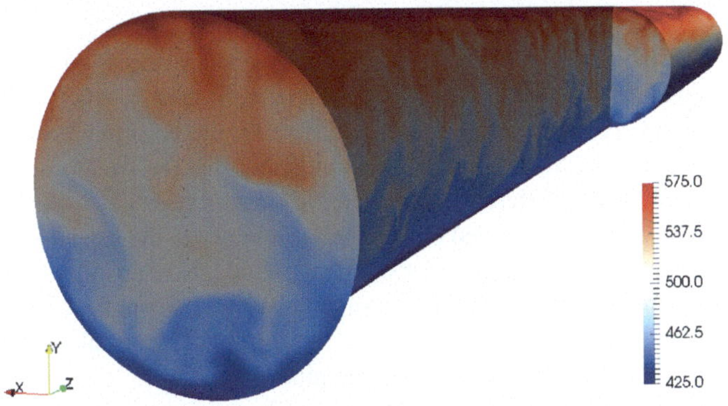

Fig. 30.1 Instantaneous temperature field $T/T_{\tau 0}$ for Case 3

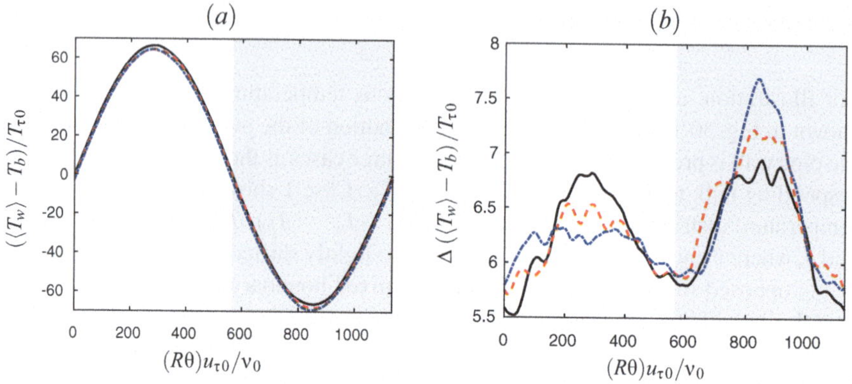

Fig. 30.2 Circumferential variation of **a** wall temperature minus bulk temperature and **b** RMS temperature fluctuation on the pipe wall. Shaded area indicates cooled half of the pipe. Lines are as described in Table 30.1

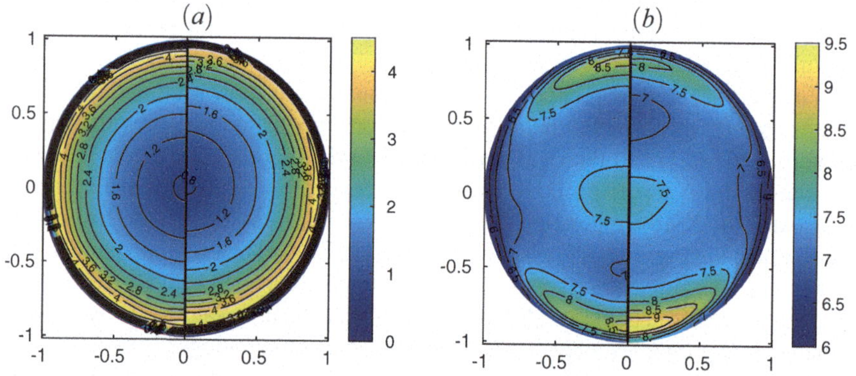

Fig. 30.3 **a** Turbulent kinetic energy contours, $k/u_{\tau 0}^2$. **b** Root mean square temperature fluctuation contours, $T'_{\mathrm{rms}}/T_{\tau 0}$. Left half of each subfigure: case 1, right half: Case 3

tions caused by the variations in temperature in the cross-plane. Figure 30.4a shows the magnitude of the secondary flows for Case 3 to be of $O(u_\tau) \sim 0.1$, equivalent to approximately a 0.7% of the bulk velocity. A detailed study of the terms contributing to the generation of the secondary flows indicates that the stream-wise vorticity is mainly driven by the difference in Reynolds stresses near the walls. Thus, this secondary flow can be categorized as secondary flow of second kind.

These secondary flows, despite their small magnitude, have an important practical contribution to the heat fluxes in the cross-plane. We can distinguish in Case 3 two anti-symmetrical vortex moving hotter fluid from top to bottom along the vertical diameter and returning colder fluid along the pipe walls.

In order to assess the influence of the secondary flows in the heat fluxes, let consider the energy equation:

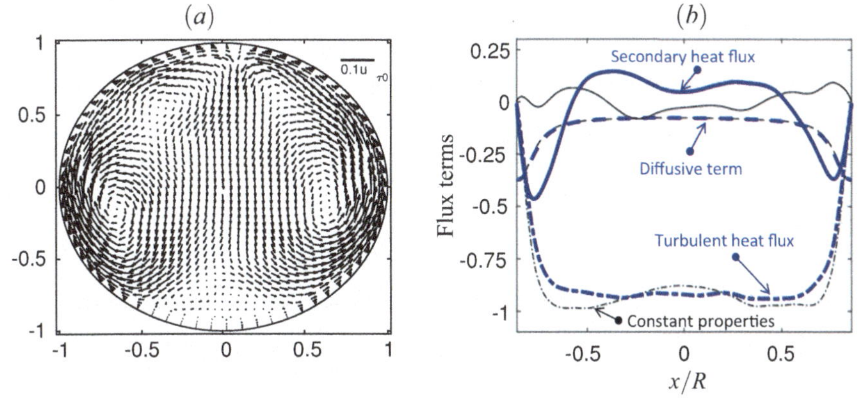

Fig. 30.4 **a** Secondary flows for Case 3. **b** Vertical heat fluxes for Cases 1 and 3 at plane $y/R = -0.5$

$$\frac{\partial \langle u_i T \rangle}{\partial x_i} = \frac{\partial}{\partial x_i} \left\langle \alpha(T) \frac{\partial T}{\partial x_i} \right\rangle. \tag{30.3}$$

We define the total heat flux, Φ, as

$$\Phi = \langle u_i \rangle (\langle T \rangle - T_b) + \langle u_i' T' \rangle - \langle \alpha(T) \rangle \left\langle \frac{\partial T}{\partial x_i} \right\rangle. \tag{30.4}$$

where we have removed the contribution of the solenoidal component of the flux $\langle u_i \rangle T_b$ and we have neglected the term $\langle \alpha' \partial x_i T' \rangle$ (note that $T'/T_0 \ll 1$, and then $\alpha'/\alpha_0 \ll 1$).

In Fig. 30.4b, we plot the three terms of the vertical component of the heat flux Φ, the secondary flow heat flux, $\langle u_y \rangle (\langle T \rangle - T_b)$, the turbulent heat flux, $\langle u_y' T' \rangle$, and the diffusive heat flux, $-\langle \alpha \rangle \partial \langle T \rangle / \partial y$, normalized with $u_{\tau 0} T_{\tau 0}$, for Case 3, compared with the uniform fluid properties case. Heat flux distributions are provided on the horizontal plane $y/R = -0.5$, within the cooled lower part of the pipe. The sign criteria we follow is that negative values indicate heat flux going from top to bottom.

As shown in the figure, the normalized turbulent heat fluxes $\langle u_y' T' \rangle / (u_{\tau 0} T_{\tau 0})$ are dominant in most of the flow domain. Starting from zero on the pipe surface, the profiles reach a maximum near the wall and then decreases slightly toward the mid-pipe diameter.

For the case of the heat flux contribution of the secondary flows, $\langle u_y \rangle (\langle T \rangle - T_b)/(u_{\tau 0} T_{\tau 0})$, we observe in Fig. 30.4b that for Case 3, the flux is strong and negative near the walls, indicating heat flux going down created by a cold stream moving upwards. These results are consistent with the vectors plot in Fig. 30.4a. Their contribution is of the same order as the contribution of the diffusive terms, accounting for up to one-fourth of the total heat flux crossing the pipe. Note that the heat flux contribution of the secondary flows should be zero for Case 1 and that deviations from zero give an indication of the statistical error due to lack of convergence.

Finally, we have found that the overall friction coefficient increases by up to 2.9% from Case 1 to Case 3, whilst the heat transfer on the wall, characterized by the Nusselt number, varies only within 1% in all three cases at the top and bottom locations.

Conclusions

In this study, we compare and analyze the influence of variable fluid properties (temperature-dependent viscosity and diffusivity) in the kinematic and thermal characteristics of a turbulent pipe flow, subjected to circumferentially varying heat transfer boundary conditions.

We observe that, as mentioned in previous studies, the turbulent fluctuations of velocities and temperature are enhanced near the cold wall, while are damped close to the hot wall.

The temperature gradient in the pipe cross-plane, together with the variations of the fluid properties, produces a non-uniform shear stress on the pipe wall. This promotes the appearance of discernible secondary flows, which enhances the flow mixing in the pipe. We have quantified the effect of the secondary flows in the heat transfer from top to bottom, finding that they are responsible for up to one-fourth of the total vertical heat flux.

References

1. Piller, M.: Direct numerical simulation of turbulent forced convection in a pipe. Int. J. Numer. Meth. Fluids **49**, 583–602 (2005)
2. Redjem-Saad, L., Ould-Rouiss, M., Lauriat, D.: Direct numerical simulation of turbulent heat transfer in pipe flows: effect of Prandtl number. Int. J. Heat Fluid Flow **28**, 847–861 (2007)
3. Black, A.W., Sparrow, E.M.: Experiments on turbulent heat transfer in a tube with circumferentially varying thermal boundary conditions. J. Heat Transfer **89**, 258–268 (1967)
4. Quarmby, A., Quirk, R.: Measurements of the radial and tangential eddy diffusivities of heat and mass in turbulent flow in a plain tube. Int. J. Heat Mass Transfer **15**, 2309–2327 (1972)
5. Antoranz, A., Gonzalo, A., Flores, O., Garcia-Villalba, M.: Numerical simulation of heat transfer in a pipe with non-homogeneous thermal boundary conditions. Int. J. Heat Fluid Flow **55**, 45–51 (2015)
6. Antoranz, A., Ianiro, A., Flores, O., Garcia-Villalba, M.: Extended proper orthogonal decomposition of non-homogeneous thermal fields in a turbulent pipe flow. Int. J. Heat Mass Transf. **118**, 1264–1275 (2018)
7. Zonta, F., Marchioli, C., Soldati, A.: Modulation of turbulence in forced convection by temperature-dependent viscosity. J. Fluid Mech **697**, 150–174 (2012)
8. Lee, J., Jung, S.Y., Sung, H.J., Zaki, T.A.: Effect of wall heating on turbulent boundary layers with temperature-dependent viscosity. J. Fluid Mech **726**, 196–225 (2013)
9. Fischer, P.F., Lottes, J.W., Kerkemeier, S.G.: nek5000 web page (2008)

Chapter 31
DNS of Variable-Property Rayleigh–Bénard Convection in a 3D Cavity Filled with Water

A. D. Demou and D. G. E. Grigoriadis

Introduction

So far, 3D simulations of Rayleigh–Bénard (RB) convection in fluids with variable properties are relatively rare. For water the only two studies in the literature are those of Horn and Shishkina [3] who studied the non-Oberbeck–Boussinesq (NOB) RB convection in a cylindrical domain and of Demou et al. [2] who considered a cuboid domain. The reason behind the lack of 3D variable-property studies is the significant increase in the computational cost. More specifically, the inclusion of density variations in the governing equations leads to the emergence of a Poisson equation for the pressure with variable coefficients. This characteristic prohibits the use efficient numerical tools such as Fast Direct Solvers (FDS).

Nevertheless, 2D simulations cannot always replace their 3D counterparts. For example, Schmalzl et al. [4] compared 2D with 3D Oberbeck–Boussinesq (OB) solutions for the case of RB convection with fluids of different Prandtl number (Pr) values at a Rayleigh number of $\mathrm{Ra} = 10^6$. Even at this relatively low Rayleigh number value, only high Prandtl number fluids ($\mathrm{Pr} > 100$) produce comparable results in 2D compared to 3D simulations. For lower values of the Prandtl number, the Nusselt (Nu) and Reynolds (Re) numbers exhibit a large deviation when the 2D and 3D predictions are compared. Demou et al. [2] conducted a similar comparison for the RB convection in water ($\mathrm{Pr} = 4.38$) under NOB conditions and reported a deviation in Nusselt numbers in the order of 15–20%.

A. D. Demou (✉) · D. G. E. Grigoriadis
Mechanical and Manufacturing Engineering Department,
University of Cyprus, Nicosia, Cyprus
e-mail: ademou01@ucy.ac.cy

D. G. E. Grigoriadis
e-mail: grigoria@ucy.ac.cy

M. García-Villalba et al. (eds.), *Direct and Large Eddy Simulation XII*,
ERCOFTAC Series 27,
https://doi.org/10.1007/978-3-030-42822-8_31

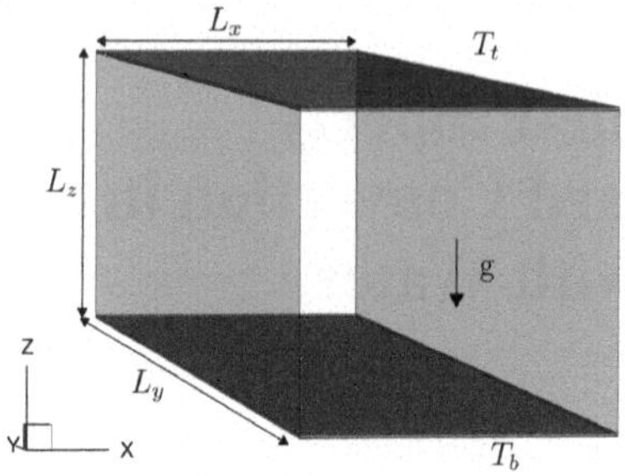

Fig. 31.1 Schematic representation of the cuboid Rayleigh–Bénard cavity used in the simulations, with $(L_x, L_y, L_z) = (1, \pi, 1)$. Solid walls confine the x and z directions, while the spanwise y direction is periodic. The cavity is heated and cooled by the bottom and top walls, respectively, and the sidewalls are considered adiabatic

In the present study, an efficient numerical methodology is used to conduct 3D simulations of RB convection in water under NOB conditions. The domain geometry is shown in Fig. 31.1. Several quantities that relate to temperature statistics are investigated at $\mathrm{Ra} = 10^9$ and under a temperature difference in the range of 0–60 K.

Mathematical Formulation

To model flows of water with variable properties, the continuity equation (31.1), Navier–Stokes equations (31.2) and the energy equation (31.3) take the following non-dimensional form:

$$\frac{\partial u_j}{\partial x_j} = 0, \quad (31.1)$$

$$\frac{\partial u_i}{\partial t} + \frac{\partial u_i u_j}{\partial x_j} = -\frac{1}{\rho}\frac{\partial P}{\partial x_i} + \frac{1}{\rho}\frac{\mathrm{Pr}}{\sqrt{\mathrm{Ra}}}\frac{\partial}{\partial x_j}\left[\mu\left(\frac{\partial u_i}{\partial x_j} + \frac{\partial u_j}{\partial x_i}\right)\right] + \frac{1}{\mathrm{Fr}^2}\delta_{i3}, \quad (31.2)$$

$$\frac{\partial \Theta}{\partial t} + \frac{\partial u_j \Theta}{\partial x_j} = \frac{1}{\rho c_p}\frac{1}{\sqrt{\mathrm{Ra}}}\frac{\partial}{\partial x_j}\left(k\frac{\partial \Theta}{\partial x_j}\right). \quad (31.3)$$

where $i = 1, 2, 3$, x_i is the non-dimensional Cartesian position vector, also denoted as (x, y, z) and u_i represents the non-dimensional velocity vector. The gravitational acceleration g acts along the z-direction. Moreover, t is the non-dimensional time, P the non-dimensional pressure and Θ the non-dimensional temperature. The scales used to non-dimensionalize these variables are the height of the cavity $L_0 = L_z$ as the length scale, $V_0 = \bar{\alpha}(T_0)\mathrm{Ra}^{1/2}/L_0$ as the velocity scale, $t_0 = L_0/V_0$ as the time scale, and $P_0 = \bar{\rho}(T_0)V_0{}^2$ as the pressure scale. Temperature is made non-dimensional as $\Theta = (T - T_0)/\Delta T$, where $\Delta T = T_b - T_t$ is the temperature difference between the heated (T_b) and cooled (T_t) boundaries of the domain. The reference temperature is denoted by $T_0 = (T_b + T_t)/2$. Using these scales, the characteristic dimensionless groups emerging are the Rayleigh number $\mathrm{Ra} = g\bar{\beta}(T_0)L^3\Delta T/(\bar{\nu}(T_0)\bar{\alpha}(T_0))$, the Prandtl number $\mathrm{Pr} = \bar{\nu}(T_0)/\bar{\alpha}(T_0) = 4.38$ for water and the Froude number $\mathrm{Fr} = \bar{\alpha}(T_0)\sqrt{\mathrm{Ra}/(gL_0{}^3)}$, which expresses the ratio of the flow inertia to the gravitational field. To distinguish between the dimensional and non-dimensional form of the fluid properties, an overbar is used to denote the dimensional form, i.e., the dimensional form of the fluid density, thermal expansion coefficient, kinematic viscosity, and thermal diffusivity are denoted as $\bar{\rho}$, $\bar{\beta}$, $\bar{\nu}$ and $\bar{\alpha}$, respectively. Since the medium under investigation is water, its fluid properties are effectively independent on pressure and can be approximated by the temperature dependent polynomials reported by Ahlers et al. [1]. These fluid properties were non-dimensionalized using the value of each property at the reference temperature, e.g., $k(T) = \bar{k}(T)/\bar{k}(T_0)$.

Numerical Methodology

Equations (31.1)–(31.3) were solved following the fractional-step method, using second-order central differences for space discretization and a fully explicit, second-order Adams-Bashforth scheme to advance the solution in time. More specifically, a pressure-splitting technique was utilized for the transformation of the derived variable coefficient Poisson equation for the pressure into a constant coefficient equation. The Poisson equation is solved using a FDS based on the public domain package FISHPACK [5]. The overall NOB methodology has a relatively low performance overhead ($\sim$20%) compared to the OB case. A more comprehensive presentation of the numerical implementation used here and its validation can be found in [2].

Computational Parameters and Results

The simulations were conducted inside a cuboid cavity (Fig. 31.1) with $L_x = L_z = 1$ at $\mathrm{Ra} = 10^9$, $\mathrm{Pr} = 4.38$ and $\Delta T < 60$ K. The spanwise extend of the cavity was chosen as $L_y = \pi$ so that the structures of the flow were not suppressed by the imposition of the periodic boundary condition in the spanwise direction. Moreover, a no-slip boundary condition was applied on the solid walls. The bottom and the top walls were uniformly heated ($\Theta_b = +0.5$) and cooled ($\Theta_t = -0.5$), respectively, while the side walls were considered adiabatic. Preliminary simulations were used to determine the proper resolution, which was verified by an a posteriori analysis in-line with the criteria set by Shiskina et al. [6]. The grid resolution consisted of $604 \times 576 \times 604$ cells, amounting to approximately 210 million nodes. After an initial time-interval of approximately 100 time-units, a statistically stationary state was developed and statistical sampling commenced for an additional 200 time-units.

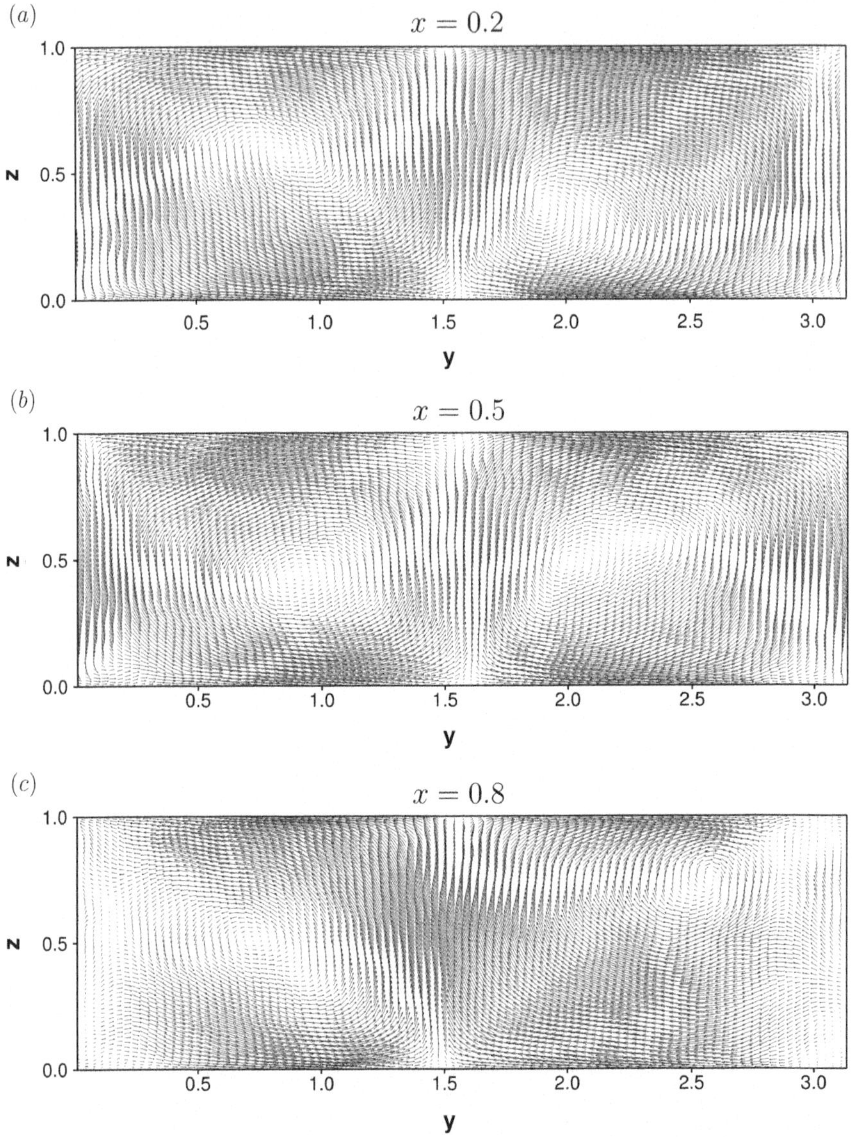

Fig. 31.2 Time-averaged velocity vectors at different $y - z$ planes, for $\Delta T = 40$ K. **a** $x = 0.2$, **b** $x = 0.5$ and **c** $x = 0.8$. Two LSC structures are clearly visible

The hot and cold plumes ejecting from the bottom and top boundary layers respectively are organized into large-scale circulation (LSC) structures. The time-averaged velocity vectors, shown in Fig. 31.2 for $\Delta T = 40$ K, reveal the presence of two LSC structures along the y-direction. More specifically, these structures are aligned with each $y - z$ plane and extend along the x-direction, swiping almost the entirety of the top and bottom boundary layers.

The effects of the property variations on the time- and $x - y$ plane-averaged temperature distributions are shown in Fig. 31.3a. In all cases, the bulk of the cavity is approximately isothermal, at a temperature that increases with ΔT. These findings are in-line with the experimental work of Ahlers et al. [1] and the 2D DNS results of Sugiyama et al. [7]. Additionally, due to the increased temperature at the central region of the cavity, the thermal boundary layers become thicker at the top compared to the bottom of the cavity. Furthermore, Fig. 31.4 shows the $x - y$ plane-averaged temperature r.m.s. distributions next to the top and bottom walls. Two maxima are revealed, one next to the heated and one next to the cooled wall. The top-bottom asymmetry of the maximum temperature r.m.s. values is enhanced as ΔT increases, with higher values at the top and lower values at the bottom of the cavity. This asymmetry results from the increased thermal diffusivity at the bottom of the cavity (shown in Fig. 31.3b) where temperature inhomogeneities are redistributed more effectively, reducing the fluctuations of the temperature field in this region. Moreover, Fig. 31.5 shows the power spectra density of temperature at the location of the maximum temperature r.m.s. next to the bottom and the top wall. The increased viscosity in the vicinity of the top wall (shown in Fig. 31.3b) causes a more efficient dissipation of the dominant frequencies compared to the bottom wall.

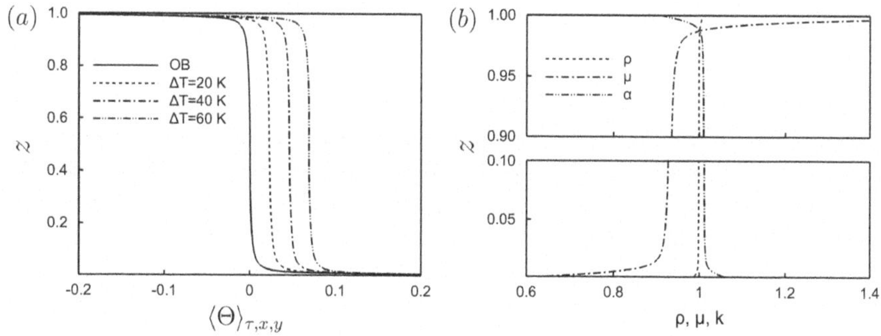

Fig. 31.3 **a** Time and $x - y$ plane-averaged temperature as a function of height z. Solid line, OB results; dashed line, $\Delta T = 20$ K; dashed-dotted line, $\Delta T = 40$ K; dashed-dotted-dotted line, $\Delta T = 60$ K. **b** Time and $x - y$ plane-averaged non-dimensionalised fluid properties as a function of height, considering $\Delta T = 60$ K. Dashed line, ρ; dashed-dotted line, μ; dashed-dotted-dotted line, α

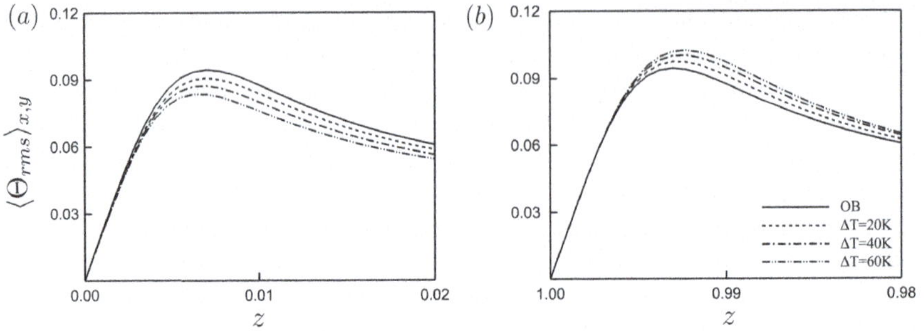

Fig. 31.4 Time and $x - y$ plane-averaged temperature r.m.s. distributions, in the vicinity of **a** the bottom wall and **b** the top wall. Solid line, OB results; dashed line, $\Delta T = 20$ K; dashed-dotted line, $\Delta T = 40$ K; dashed-dotted-dotted line, $\Delta T = 60$ K

Fig. 31.5 Power spectra density Φ of temperature at the locations of the maximum temperature r.m.s.: bottom location $x = 0.5$, $y = \pi/2$, $z = 6.96 \times 10^{-3}$ (red); top location $x = 0.5$, $y = \pi/2$, $z = 9.938 \times 10^{-1}$ (blue)

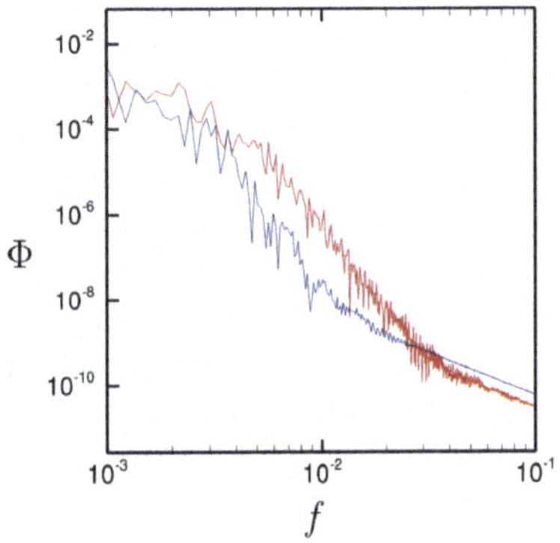

Conclusions

The 3D DNS results for the Rayleigh–Bénard convection in water with variable properties revealed an increased top-bottom asymmetry for higher temperature differences. Most importantly, the temperature at the bulk of the cavity is increased while temperature r.m.s. values increase at the top and decrease at the bottom of the cavity. In addition, the dissipation of the dominant frequencies was found to increase at the top compared to the bottom of the cavity.

References

1. Ahlers, G., Brown, E., Araujo, F.F., Funfschilling, D., Grossmann, S., Lohse, D.: Non-Oberbeck-Boussinesq effects in strongly turbulent Rayleigh-Bénard convection. J. Fluid Mech. **569**, 409–445 (2006)
2. Demou, A.D., Frantzis, C., Grigoriadis, D.G.E.: A numerical methodology for efficient simulations of non-Oberbeck-Boussinesq flows. Int. J. Heat Mass Transf. **125**, 1156–1168 (2018)

3. Horn, S., Shishkina, O.: Rotating non-Oberbeck-Boussinesq Rayleigh-Bénard convection in water. Phys. Fluids **26**(5), 055111 (2014)
4. Schmalzl, J., Breuer, M., Hansen, U.: On the validity of two-dimensional numerical approaches to time-dependent thermal convection. EPL (Eur. Lett.) **67**(3), 390 (2004)
5. Schumann, U., Sweet, R.A.: A direct method for the solution of Poisson's equation with Neumann boundary conditions on a staggered grid of arbitrary size. J. Comp. Phys. **20**(2), 171–182 (1976)
6. Shishkina, O., Stevens, R.J., Grossmann, S., Lohse, D.: Boundary layer structure in turbulent thermal convection and its consequences for the required numerical resolution. New J. Phys. **12**(7), 075022 (2010)
7. Sugiyama, K., Calzavarini, E., Grossmann, S., Lohse, D.: Flow organization in two-dimensional non-Oberbeck-Boussinesq Rayleigh-Bénard convection in water. J. Fluid Mech. **637**, 105–135 (2009)

Chapter 32
Study of Heat Transfer Characteristics in a Wall-Bounded Plane-Jet Using Large-Eddy Simulation

P. Kakka and K. Anupindi

Introduction

Turbulent jets find applications in the film cooling of gas turbine blades, gas turbine combustor liners, automobile defroster, and cooling of electronic components. With the advent of new technologies like 3D printing, complex shapes of the gas turbine blades, and liners can be manufactured to optimize heat transfer. An understanding of the thermal characteristics of wall-bounded jets helps in the better design of these devices. Wall jets have been extensively studied for flow characteristics, as they contain turbulent boundary layer and a free shear layer and this makes their interaction interesting. Few experiments were done for studying heat transfer in wall jets. Akfirat et al. [2] studied heat transfer on an isothermal flat plate with a fully developed turbulent flow at different Reynolds numbers and obtained a relation with Nusselt number as $\text{Nu} = 0.097(\text{Re})^{0.8}\ (x/h)^{-0.6}$ for $x/h > 30$ where h is the inlet height of the jet. Mabuchi and Kumada [6] obtained a similar relation as $\text{Nu} = 0.102(\text{Re})^{0.8}$ $(x/h)^{-0.6}$. AbdulNour et al. [1] experimentally studied heat transfer characteristics of a wall jet for different heat flux conditions on the wall near the inlet for $0 < x/h < 13$. Rathore and Das [8] evaluated several Reynolds-averaged Navier–Stokes (RANS) models and studied its heat transfer characteristics of the wall-bounded jets. Dejoan and Leschziner [3] performed LES to study the flow characteristics. Recently, Naqavi et al. [7] performed a direct numerical simulation (DNS) of a plane-jet over a heated isothermal wall and presented the heat transfer characteristics and the turbulent budgets. RANS simulations are relatively inexpensive, however, some simple models may not accurately capture the variation of temperature in the

P. Kakka · K. Anupindi (✉)
Indian Institute of Technology Madras, Chennai 600036, India
e-mail: kanupindi@iitm.ac.in

P. Kakka
e-mail: kakkapriyesh@hotmail.com

M. García-Villalba et al. (eds.), *Direct and Large Eddy Simulation XII*,
ERCOFTAC Series 27,
https://doi.org/10.1007/978-3-030-42822-8_32

thermal boundary layer [8] and the second-order turbulence statistics. On the other hand, DNS is very accurate, yet is computationally expensive. Therefore, there is a need to evaluate the capabilities of LES in accurately predicting the heat transfer characteristics.

Numerical Methodology

The governing equations are the unsteady, three-dimensional incompressible continuity, and Navier–Stokes equations. These are integrated using a finite-volume solver available in the open-source fluid dynamics software OpenFOAM [10]. Second-order accurate central-difference schemes are used for spatial discretization, and the Crank–Nicolson scheme is used for the time integration. The LES models tested are two sub-grid scale (SGS) models, namely, the Smagorinsky model with constant-coefficient C_s as 0.168 and the dynamic one-equation model. An implicit LES is also tested to check the influence of dissipation due to the numerical truncation errors. For the implicit LES, first-order upwind scheme is used for the spatial discretization to enable a stable and converged solution. At the inlet to the domain, a steady mean velocity profile extracted from the DNS studies of Naqavi et al. [7] is used, and the Reynolds number is set to 7500. The inlet velocity profile is superimposed with random isotropic fluctuations with a turbulence intensity of 1%. A small co-flow with 4% of the inlet velocity is induced from the top-boundary to aid the entrainment, and on the outlet, an advective boundary condition is applied. The lower wall and the wall above the inlet are impermeable and have a no-slip boundary condition. The lower wall is isothermal with the temperature higher than the ambient, and the wall above the inlet is kept as adiabatic whereas, the jet inlet is retained at the ambient temperature. Periodic boundary conditions are applied in the span-wise direction. The Prandtl number is set to 0.71 while the turbulent Prandtl number is assumed constant with 0.85 [5]. The simulation domain extends $30h \times 10h \times 5.5h$, where h is the jet inlet height as shown in Fig. 32.1. From the energy spectrum (Fig. 32.2), we can conclude that mesh resolves the inertial scales well for the mesh of around 4

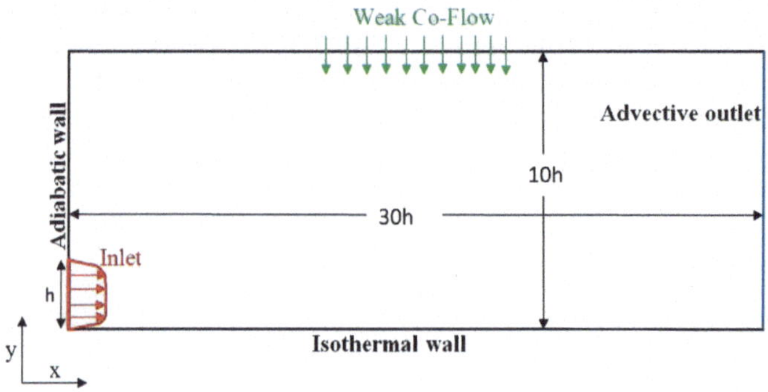

Fig. 32.1 Geometry of the flow-domain

million with 360 × 222 × 50 nodes in the stream-wise, wall-normal, and span-wise directions, respectively. A non-uniform structured grid is generated with a near-wall resolution that satisfies $y^+ \leq 1$.

Results

In this section, the results obtained are discussed. The wall jet approaches self-similarity around $x/h = 15$ if non-dimensionalized using different variables for the inner and outer scales [9]. The outer scaling is understood from the schematic in Fig. 32.3 where the y-axis is scaled with $y/y_{0.5}$, the results after scaling with a good match are shown in Figs. 32.4 and 32.5 where the profiles by sub-grid models at $x/h = 25$ collapse with the experimental profile [4] which is at $x/h = 20$. It can be noted that the implicit LES performs poorly on the same grid. For the inner scaling, velocity terms (Fig. 32.6) are scaled with friction velocity u_τ and the temperature

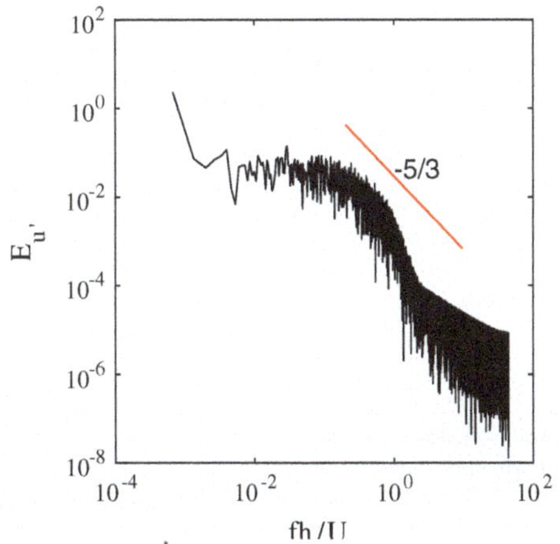

Fig. 32.2 The energy spectrum of stream-wise velocity fluctuations

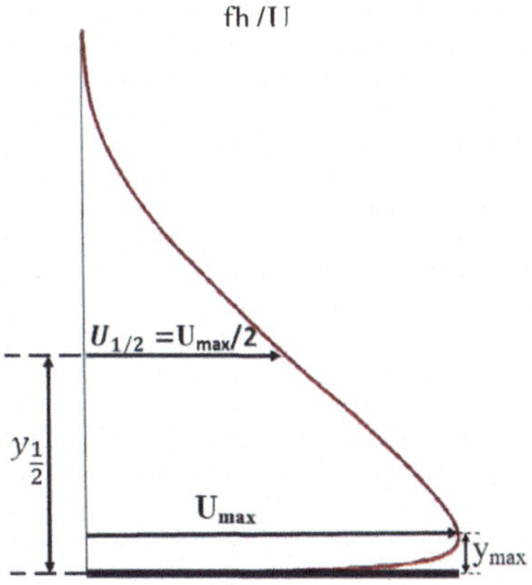

Fig. 32.3 Schematic of outer-scaling

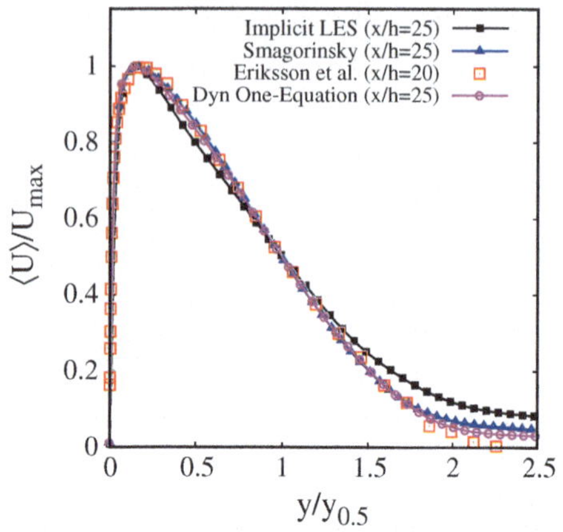

Fig. 32.4 Outer-scaled stream-wise velocity profile

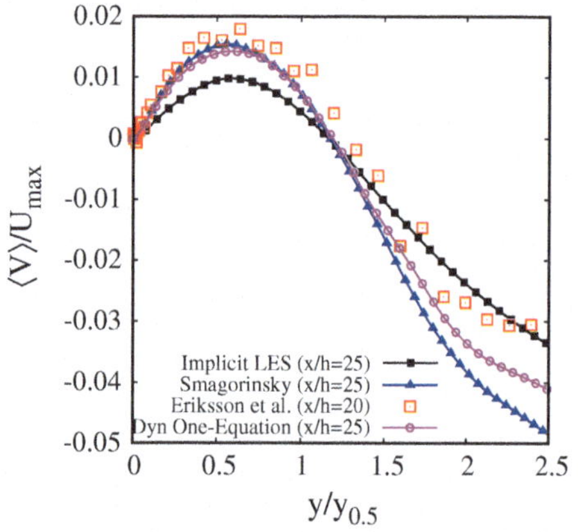

Fig. 32.5 Outer-scaled wall-normal velocity profile

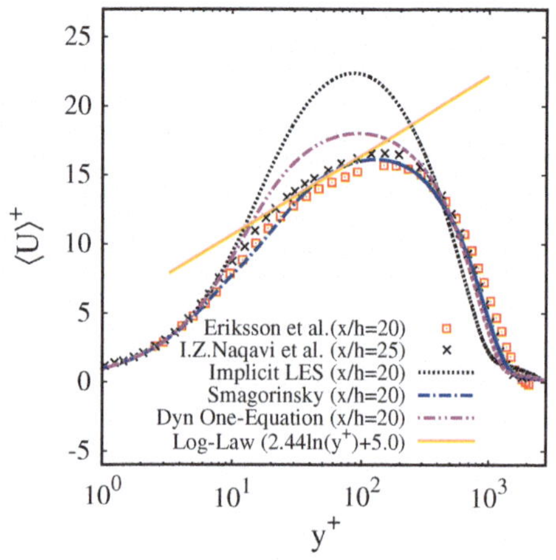

Fig. 32.6 Inner-scaled stream-wise velocity profile

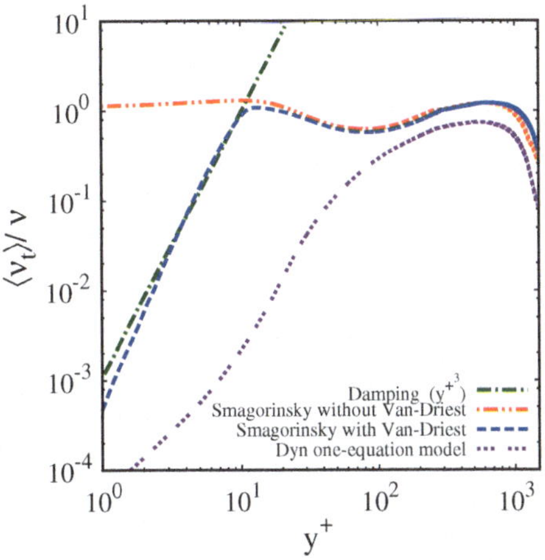

Fig. 32.7 Damping of turbulent viscosity

parameters with $T_\tau = \frac{\overline{q_w}}{\rho u_\tau C_p}$ with, $\overline{q_w} = -k_f \left.\frac{d\overline{T}}{d\eta}\right|_{wall}$. Here, the SGS models particularly Smagorinsky shows a better match with the DNS results of Naqavi et al. [7] and with the log-law $<u>^+ = \ln(y^+)/0.4 + 5.2$ for y^+ in the range of 28–55. The reason for the better performance is the Van-Driest damping used in the Smagorinsky model which enables the damping of turbulent viscosity near the wall to approach zero which is not the case for constant-coefficient SGS model without damping as can be seen in Figs. 32.7 and 32.8. For the given inlet velocity, the friction velocity u_τ calculated by Eriksson [4] at $x/h = 20$ is 0.546. The implicit LES over-predicts the profile as it under-estimates the friction velocity as 0.427 and therefore, the velocity gradient at the wall. The reason for this behavior can be attributed to excessive artificial dissipation near the wall. Among the SGS models, Smagorinsky model predicts the friction velocity better at 0.515 than the dynamic models at 0.473. The reason for the dynamic one-equation model to under-estimates the wall friction compared to Smagorinsky is associated with its failure to capture the damping of turbulent viscosity in the order of y^{+3} near the wall as can be seen from Fig. 32.7. The wall-normal Reynolds stress calculated by the dynamic one-equation model is quite close to the experiment, as shown in Fig. 32.9.

The mean temperature profiles shown in Figs. 32.10 and 32.11 are scaled with the outer and the inner variables, respectively. From Fig. 32.11 it can be noted that the Smagorinsky model shows an excellent agreement with the DNS and with the log-law, $<T>^+ = \ln(y^+)/0.48 + 3.5$ in the range of y^+ from 50 to 90. Implicit LES fails to predict the temperature parameters in terms of quality as well as the magnitude. The variation in root-mean-square values of the temperature with y is shown in Fig. 32.12, a slight overshoot can be observed at the peak for both LES models when compared with the DNS. The stream-wise variation of Nusselt number is shown in Fig. 32.13. Both the SGS models capture the minima of the heat transfer at $x/h = 5$ whereas, for the implicit LES, no minima is obtained. For the region near

Fig. 32.8 Effect of damping on velocity profile

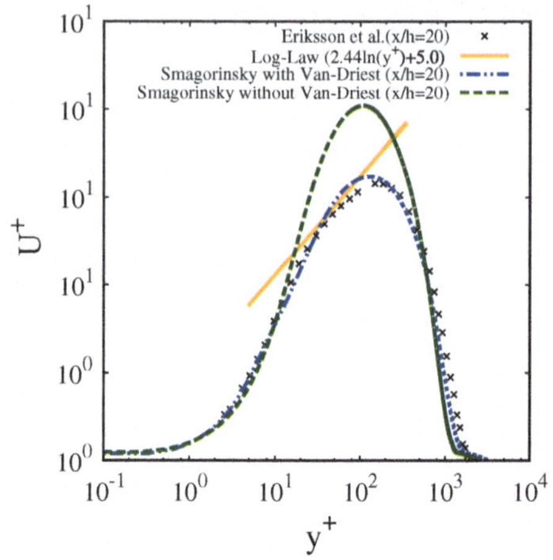

Fig. 32.9 Outer-scaled wall-normal stress

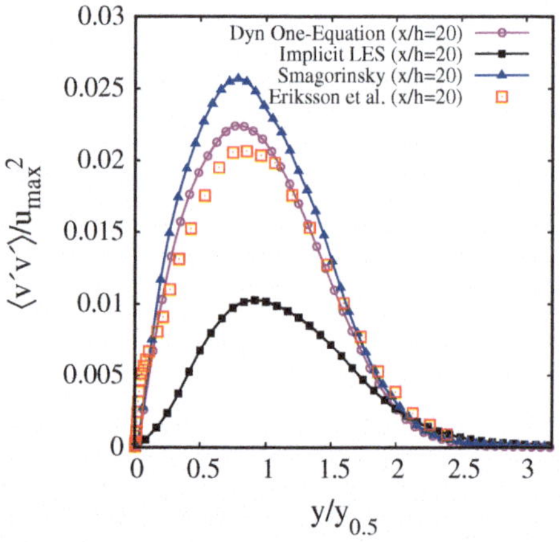

Fig. 32.10 Outer-scaled mean temperature profiles

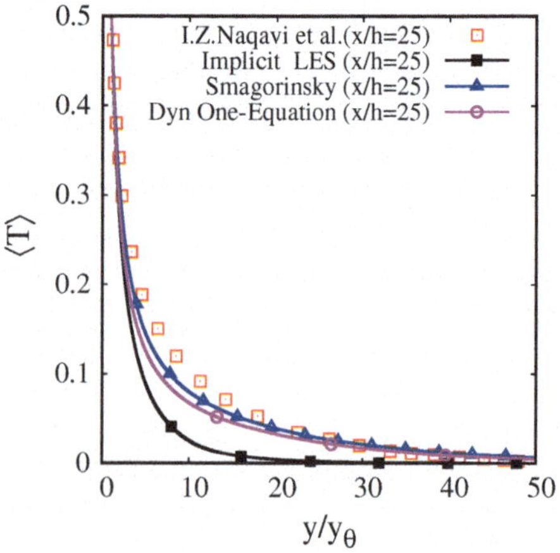

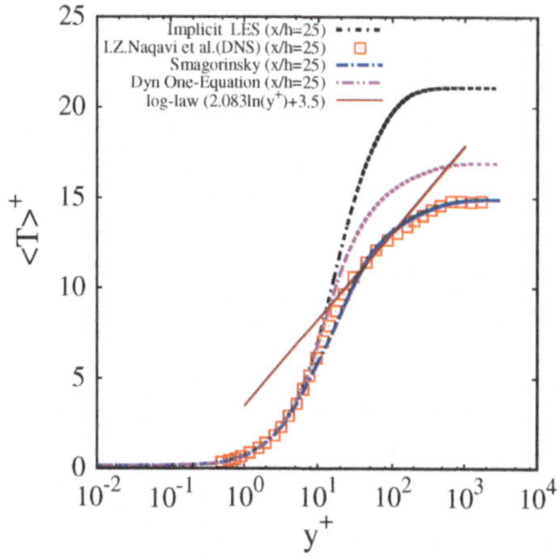

Fig. 32.11 Inner-scaled mean temperature profiles

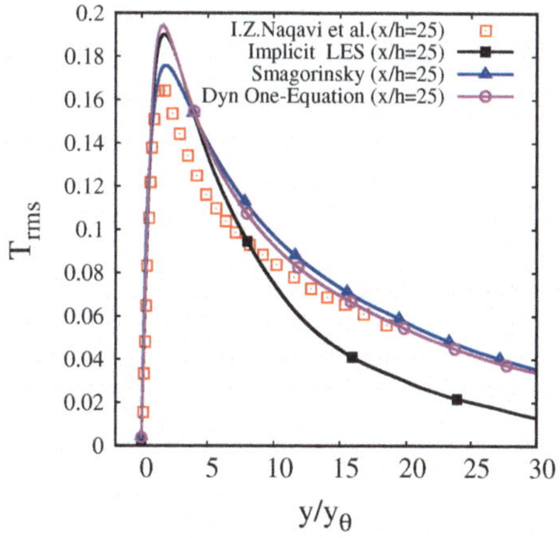

Fig. 32.12 Outer-scaled profiles of T_{rms}

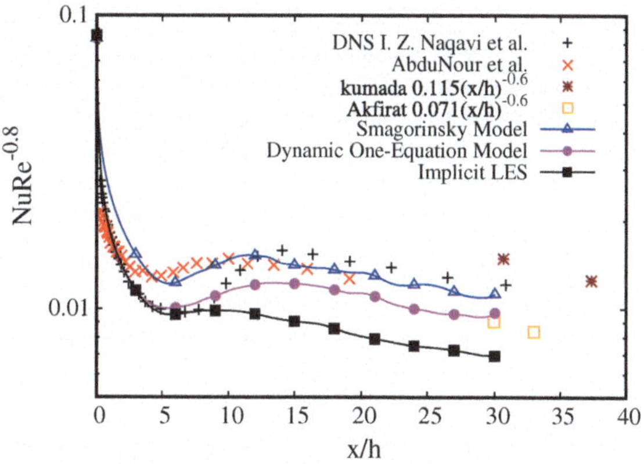

Fig. 32.13 Nusselt number variation in the stream-wise direction

the inlet, the Smagorinsky model performs well with an accurate prediction of the local minima and the maxima peak till $x/h = 13$ when compared to the experiment of AbdulNour et al. [1]. Further downstream both the SGS models seem to fall between the correlation provided by the experiments.

Conclusions

Two sub-grid scale LES models and an implicit LES are compared to test their capabilities in the prediction of heat transfer parameters in the wall jet. The implicit LES with first-order spatial discretization fails to predict the velocity as well as the temperature parameters qualitatively. The reason can be attributed to large artificial dissipation due to the lower order of spatial discretization. Among the SGS models, the dynamic one-equation model under-predicts the friction velocity due to improper damping of turbulent viscosity near the wall. Whereas, the Smagorinsky model with Van-Driest damping produces comparable results against the DNS to predict the flow as well as thermal characteristics of the planer jet on a heated isothermal surface. Therefore, Smagorinsky model can be used to study heat transfer characteristics of the wall jet using a mesh count that is lower by an order of magnitude when compared to DNS.

Acknowledgements The authors are grateful for the computational resources provided by P.G. Senapathy Center for Computer Resources on the VIRGO computing cluster at Indian Institute of Technology (IIT) Madras. The second author would like to acknowledge the funding received through the new-faculty initiation-grant given by IIT Madras.

References

1. AbdulNour, R.S., Willenborg, K., McGrath, J.J., Foss, J.F., AbdulNour, B.S.: Measurements of the convection heat transfer coefficient for a planar wall jet: uniform temperature and uniform heat flux boundary conditions. Exp. Thermal Fluid Sci. **22**(3–4), 123–131 (2000)
2. Akfirat, J.C.: Transfer of heat from an isothermal flat plate to a two-dimensional wall jet. In: International Heat Transfer Conference Digital Library. Begel House Inc. (1966)
3. Dejoan, A., Leschziner, M.A.: Large eddy simulation of a plane turbulent wall jet. Phys. Fluids. **17**(2), 025102 (2005)
4. Eriksson, J.G., Karlsson, R.I., Persson, J.: An experimental study of a two-dimensional plane turbulent wall jet. Exp. Fluids. **25**(50) (1998)
5. Kader, B.A., Yaglom, A.M.: Heat and mass transfer laws for fully turbulent wall flows. Int. J. Heat Mass Transf. **15**(12), 2329–2351 (1972)
6. Mabuchi, I., Kumada, M.: Studies on heat transfer to turbulent jets with adjacent boundaries. Bull. JSME **15**(88), 1236–1245 (1972)
7. Naqavi, I.Z., Tyacke, J.C., Tucker, P.G.: A numerical study of a plane wall jet with heat transfer. Int. J. Heat Fluid Flow **63**, 99–107 (2016)

8. Rathore, S., Das, M.: A comparative study of heat transfer characteristics of wall-bounded jets using different turbulence models. Int. J. Therm Sci. **89**, 337–356 (2014)
9. Wygnanski, I., Katz, Y., Horev, E.: On the applicability of various scaling laws to the turbulent wall jet. J. Fluid Mech. **234**, 669–690 (1992)
10. Weller, H.G., Tabor, G., Jasak, H., Fureby, C.: A tensorial approach to computational continuum mechanics using object-oriented techniques. Comput. Phys. **12**(6), 620–631 (1998)

Chapter 33
Influence of Buoyant Forces on Transition: Application to a Heated Cylinder

S. Rolfo, D. R. Emerson and C. Moulinec

Introduction

Overhead Lines (OHL) are a key component to transmit and distribute electrical energy. The aero-thermal design of OHL is generally carried out considering the cable as a perfect heated cylinder immersed in a uniform cross flow [1]. Moreover, the design also assumes that only two regimes can occur, depending on the Reynolds number (based on the line diameter and the incoming velocity): (a) natural convection for low velocities, e.g., below a wind speed of 0.6 m/s which roughly corresponds to a Reynolds number of $Re = 850$; (b) forced convection at higher Reynolds numbers. In both of these cases analytical correlations exist to compute the Nusselt number.

Fully turbulent flow is generally established at Reynolds number $Re = 3900$, which correspond to a wind speed of about 2.8 m/s, with full transition located in the shear layer [9]. However, transitional structures in cold flow have already been observed for Reynolds numbers $180 \leq Re \leq 250$ [7, 8]. Two modes for the transition have then been identified. Mode A, which occurs up to $Re \approx 200$ is linked to instabilities of the primary core vortex of the von Kármán's vortex street. Mode B occurs at $Re > 200$ and is related to instabilities of the shear layer. However, in the case of a heated cylinder a new mode has been also observed [5], which is known as mode E, where transition occurs at lower Reynolds numbers (i.e., from $Re \approx 80$). In our earlier work [6], we have shown that mode E transition is not only influenced by the Reynolds and Richardson numbers, but important roles are also played by the

S. Rolfo (✉) · D. R. Emerson · C. Moulinec
Scientific Computing Department, STFC Daresbury Laboratory, Warrington, UK
e-mail: stefano.rolfo@stfc.ac.uk

D. R. Emerson
e-mail: david.emerson@stfc.ac.uk

C. Moulinec
e-mail: charles.moulinec@stfc.ac.uk

M. García-Villalba et al. (eds.), *Direct and Large Eddy Simulation XII*,
ERCOFTAC Series 27,
https://doi.org/10.1007/978-3-030-42822-8_33

Prandtl number and the angle of attack of the incoming flow. More in details, the Prandtl number has a large influence on the Reynolds and Richardson numbers at which the transition occurs, whereas the angle of attack impairs, when positive, or enhances, when negative the chaotic behaviour of the 3-D flow.

The main focus of this work is to understand the evolution of mode B transition when affected by buoyancy forces.

Methodology

The simulations have been performed using the open source CFD software *Code_Saturne* [2, 3] using the incompressible solver. The Navier–Stokes–Fourier equations are discretised using a cell centred co-located finite volume approach. Velocity/pressure coupling is ensured through a prediction/correction method similar to the SIMPLEC algorithm and the Poisson's equation is solved using an algebraic multigrid solver. The code handles fully unstructured grids, including polyhedral cells and embedded refinements [3]. Second-order accurate schemes are used for both time and space discretisations.

The buoyancy term in the momentum equations is evaluated using the Boussinesq approximation, which is given by $-\rho \mathbf{g} = -\rho_\infty \mathbf{g} \beta (T - T_W)$ with ρ being the density, ρ_∞ being the undisturbed density, β the coefficient of thermal expansion, T the fluid temperature and T_W the temperature at the cylinder surface.

Test Case Definition and Mesh Generation

The study under consideration is the flow around a heated cylinder characterised by a temperature T_W, a diameter D, set in a uniform flow with a velocity U_∞. The temperature T_W is such that $T_W > T_\infty$, where T_∞ is the temperature of the undisturbed incoming flow. The Reynolds number is set to $\mathrm{Re} = U_\infty D/\nu = 250$, with ν being the kinematic viscosity and the temperature at the wall is varied to have the Richardson number $\mathrm{Ri} = Gr/\mathrm{Re}^2$ as $0 \leq \mathrm{Ri} \leq 2$. The Grashof number is defined as $Gr = g\beta (T_W - T_\infty) D^3/\nu^2$. The computational domain is presented in Fig. 33.1. It follows a C-shape configuration and has been meshed using a block structured approach. The mesh used is the same presented in [6] with 106 k cells in the 2-D cross section and 192 extrusion layers in the span-wise direction for a total mesh count of about 20 million cells.

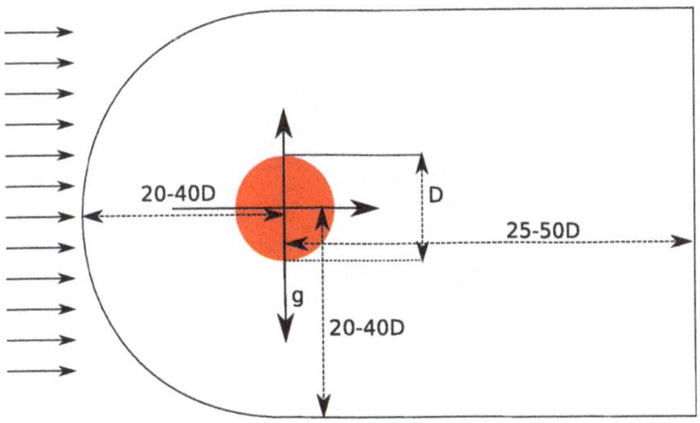

Fig. 33.1 Sketch of the computational domain and definition of the domain dimensions

Results

Figure 33.2 shows the wake behind the cylinder for four cases at different Reynolds and Richardson numbers. Case A is from [6] and depicts the typical wake structure in mode E transition at Re $=$ 150 and Ri $=$ 1.0. As described in [5] the transition mode is triggered by span-wise temperature gradient, that in turn generated stream-wise vorticity due to baroclinic vorticity production. The structures that are generated by the process have a very clear λ-shaped structure and are formed in the lower part of the cylinder at about $y/D \approx -0.5$ and assuming the the origin is placed in the centre of the cylinder. These structures evolve as they are convected downstream and by the fourth row the head has assumed a mushroom-like shape. These 3-D structures have a very orderly pattern with a span-wise wavelength of $2D$. In Fig. 33.2 Case B shows instead the typical mode B transition, (Re = 250 and Ri = 0), where some 3-D structures are formed as a result of instability of the shear layer. No clear pattern is present in this case and structures tend to break very quickly. Case C instead shows that even a small buoyancy force (i.e., Ri $=$ 0.1) has a large impact on the flow field. The generation mechanism of the 3-D transition switches to the one typical of mode E, which is driven by baroclinic vorticity production. However, the structures in the first row behind the cylinder do not have the clear λ-shaped pattern of Case A and again tends to break down very rapidly. Indeed, no clear span-wise wavelength is visible. Case D shows that at higher, but still moderate, Richardson numbers (i.e., Ri $=$ 0.5) the near wake behind the cylinder is characterised by structures very similar to the one observed for the mode E transition with a very elongated length in the vertical direction. However, the pattern tends to be more chaotic with well defined span-wise wavelength. Two rows of these λ-shaped structures form in the near wake of the cylinder and in contrast to case A they break down at $x/D \approx 8$ forming a turbulent-like wake. Figure 33.3 shows the time spectral analysis for the lift and drag coefficients, together with the Nusselt number for the heat transfer for different Richardson numbers at Re $=$ 250.

For the passive scalar case, (i.e., Ri $=$ 0) spectra are as expected with a dominant Strouhal number of $St = fD/U_B = 0.202$ for lift coefficient and $St = 0.404$ for both drag coefficient and Nusselt number. Higher harmonics are also present but

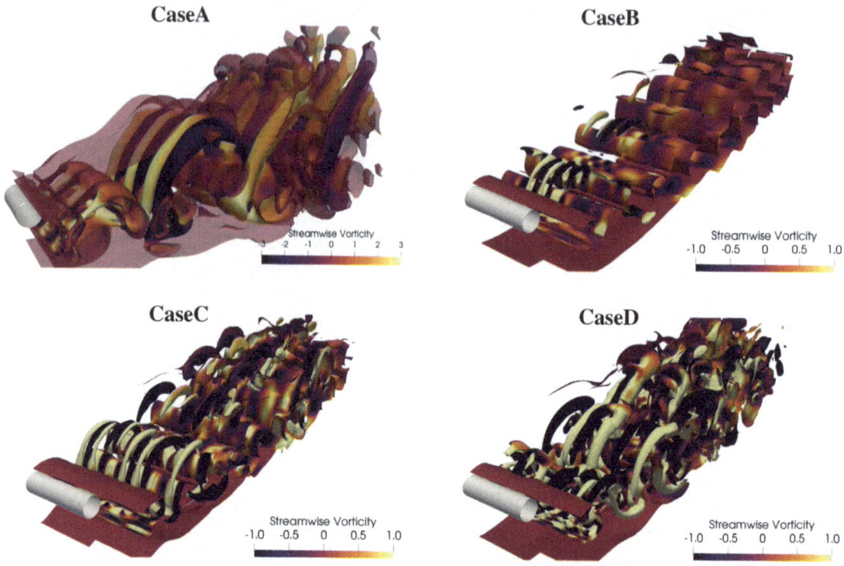

Fig. 33.2 Visualisation of the wake behind the cylinder at different Reynolds and Richardson numbers. CaseA: Re = 150, Ri = 1.0; CaseB: Re = 250, Ri = 0.0; CaseC: Re = 250, Ri = 0.1; CaseD: Re = 250, Ri = 0.5

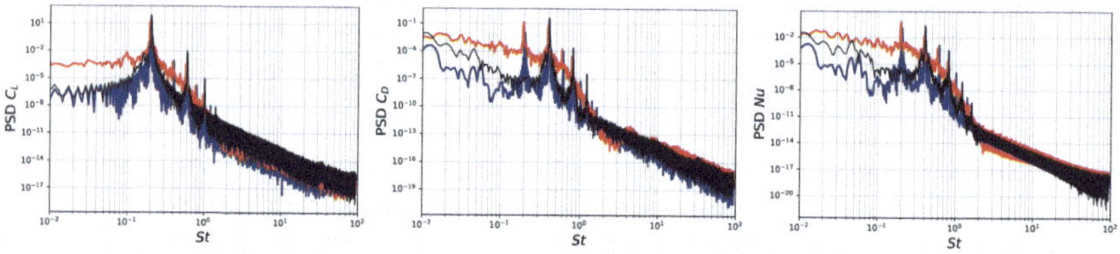

Fig. 33.3 Power spectral density for lift coefficient, drag coefficient, and Nusselt number at Re = 250 and for different Richardson numbers. Black line Ri = 0, blue line Ri = 0.1, red line Ri = 0.5

their relative strength is very weak in comparison with the main frequency. Spectra for Ri = 0.0 have a very banded energy content with only few energy containing frequencies. This is typical of a laminar-like unsteady flow where only relatively well-defined structures are present. The other interesting point to notice is the substantial spectral correspondence between the heat transfer (i.e., Nusselt number) and the drag coefficient meaning that the unsteady heat transfer is dominated by the unsteady variation of the forces in the horizontal direction.

Ri = 0.1 is still characterised by similar spectra as the pure force convection case, however, drag coefficient and Nusselt number spectra have now an additional dominant mode at $St = 0.202$, which corresponds to the main dominant frequency for the lift coefficient. This is not entirely unexpected since the main effect of moving toward mixed convection regime is that the averaged wake is no more symmetric with

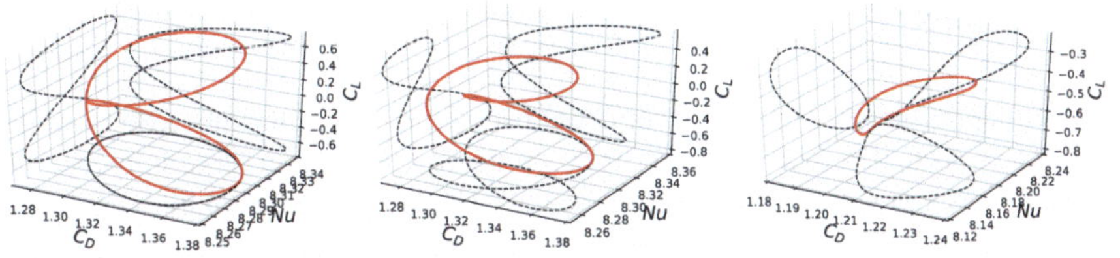

Fig. 33.4 3-D phase diagrams for lift coefficient, drag coefficient, and Nusselt number at Re = 250 and for different Richardson numbers. Left Ri = 0, middle Ri = 0.1, right Ri = 0.5

respect to the horizontal plane (i.e., $y/D = 0$), but rotates anticlockwise. However, it is interesting to notice that the effect is already quite noticeable for such low Richardson number, just above the force convection regime.

The more interesting case is Ri = 0.5, where spectra start to have a continuous distribution of the energy content among the different frequencies, with a cascade from low frequencies (i.e., low Strouhal numbers) to high frequencies. In other words, the spectra now resemble the typical one of a fully developed turbulent flow. Dominant frequencies are still present with a pattern similar to the Ri = 0.1 case, with two main dominant modes for both drag coefficient and Nusselt number, with the lowest of the two in-phase with the dominant mode for the lift coefficient. However, the dominant frequencies have a slight shift toward lower values and precisely the first frequency is at $St = 0.192$ and the second is at $St = 0.384$.

The effect of the modified spectra for the mixed convection regime is also clear in Fig. 33.4, where 3-D phase diagrams at different Richardson numbers are drawn. The 3-D diagrams are obtained putting together the phase average for lift coefficient, drag coefficient, and Nusselt number. The 3-D graphs show also the 2-D projections on the different planes (i.e., $C_L - C_D$, $C_L - NU$ and $C_D - Nu$) of the 3-D curve in order to show the more common 2-D phase diagrams (see [4] for an example). The 3-D curves for both Ri = 0 and Ri = 0.1 are relatively similar with an exhibiting double-crossing spiral pattern. However, for Ri = 0 the curve has a large degree of symmetry and the projection of the $C_L - C_D$ and $C_L - Nu$ curves produce the well known double knot shapes (see [4]) with both legs being symmetric. The $C_D - Nu$ curve is instead an ellipsoid with a very small eccentricity. This is a further evidence of the in-phase behaviour of the unsteady heat transfer with the unsteady horizontal forces. At Ri = 0.1 the 3-D curve has lost its symmetry and this affects the projections on the different planes. The double knot shapes for $C_L - C_D$ and for $C_L - Nu$ have one branch larger than the other and the $C_D - Nu$ has a very complex Venn diagram-like shape with three intersecting ellipses.

The 3-D curve for Ri = 0.5 has instead a simpler shape. This gives a projection for $C_L - Nu$ and $C_D - Nu$ diagrams of a very distorted ellipsoid shape. The $C_L - C_D$ phase diagram has still a double knot shape; however, the two branches are now largely different in size and are also anti-symmetric.

Conclusions

This work presents the study of the effect of buoyant forces on the mode B transition (i.e., Re > 200). The baroclinic vorticity production has still a large influence on the flow development generating large-scale λ-shaped structures in the vicinity of the cylinder like in the mode E transition. However, these structures tend to break rapidly and at Ri = 0.5 an almost full transition to turbulence is visible in the wake. This is also confirmed by spectral analysis on the averaged force and heat transfer coefficient. Spectral analysis and phase diagrams show also a different behaviour of the flow in the mixed convection regime with respect to the force convection case, even when the Richardson number is very low.

Acknowledgements The authors are grateful for financial support by the Hartree Centre and the UK Engineering and Physical Sciences Research Council (EPSRC) under the grants EP/P022286/1 and EP/L000261/1.

References

1. IEEE Standard for Calculating the Current-Temperature Relationship of Bare Overhead Conductors. IEEE Std 738-2012 (Revision of IEEE Std 738-2006 - Incorporates IEEE Std 738-2012 Cor 1-2013), 1–72 (2013)
2. Archambeau, F., Mechitoua, N., Sakiz, M.: Code_Saturne: a finite volume code for the computation of turbulent incompressible flows - Industrial Applications. Int. J. Finite **1**(1), (2004)
3. Fournier, Y., Bonelle, J., Moulinec, C., Shang, Z., Sunderland, A., Uribe, J.: Optimizing Code_Saturne computations on Petascale systems. Comput. Fluids **45**(1), 103–108 (2011)
4. Mittal, S., Kumar, B.: Flow past a rotating cylinder. J. Fluid Mech. **476**, 303–334 (2003)
5. Ren, M., Rindt, C.C.M., Van Steenhoven, A.A.: Three-dimensional transition of a water flow around a heated cylinder at $Re = 85$ and $Ri = 1.0$. J. Fluid Mech. 195–224, (2006)
6. Rolfo, S., Kopsidas, K., Rahman, S.A., Moulinec, C., Emerson, D.R.: Effect of large scale 3-d structures on the flow around a heated cylinder at low reynolds number. Flow, Turbul. Combust. **101**(2), 553–577 (2018). Sep
7. Williamson, C.H.K.: Three-dimensional wake transition. J. Fluid Mech. **328**, 345–407 (1996)
8. Williamson, C.H.K.: Vortex dynamics in the cylinder wake. Annu. Rev. Fluid Mech. **28**(1), 477–539 (1996)
9. Zdravkovich, M.: Flow Around Circular Cylinders. Fundamentals, vol. 1. Oxford University Press, New York, (1997)

Chapter 34
Assessment of Numerical Dissipation in Implicit LES of Turbulent Rayleigh-Bénard Convection

S. Yigit, J. Hasslberger and M. Klein

Abstract The methodology to quantify the numerical dissipation in implicit/explicit LES and under-resolved DNS (UDNS) based on the balance of the kinetic energy equation by Cadieux et al. [1] has been examined for Rayleigh-Bénard convection (RBC) in this study. To this end, this approach has been extended here for scalar transport (i.e., thermal variance) in addition to kinetic energy since momentum and thermal transport are coupled due to buoyancy forces in the RBC. Accordingly, the effect of grid resolution on numerical dissipation has been analysed in the context of implicit LES/UDNS in order to evaluate the accuracy of the simulations. It has been found that the aforementioned method works sufficiently well for quality assessment of UDNS of RBC.

Introduction

Direct Numerical Simulation (DNS) of Rayleigh-Bénard convection is not feasible especially at high Rayleigh and Prandtl numbers due to high-resolution requirements. Therefore, Large-Eddy Simulation (LES) can be used as a reliable tool for investigating turbulent Rayleigh-Bénard convection at high Rayleigh and Prandtl numbers for near future applications. However, LES quality assessment is immensely important in order to obtain predictive LES simulations. Recently, Cadieux et al. [1] evaluated

S. Yigit (✉) · J. Hasslberger · M. Klein
Institute of Applied Mathematics and Scientific Computing, Bundeswehr University Munich, Neubiberg, Germany
e-mail: sahin.yigit@unibw.de

J. Hasslberger
e-mail: josef.hasslberger@unibw.de

M. Klein
e-mail: markus.klein@unibw.de

M. García-Villalba et al. (eds.), *Direct and Large Eddy Simulation XII*, ERCOFTAC Series 27,
https://doi.org/10.1007/978-3-030-42822-8_34

numerical dissipation based on the balance of kinetic energy and used this information to assess the quality of implicit LES/Under-resolved DNS (UDNS) for flow around a sphere. This methodology has not been tested so far for non-isothermal flows when there is a two-way coupling between fluid flow and heat transfer. Thus, the aforementioned approach has been extended for scalar transport (i.e., thermal variance) in addition to kinetic energy since momentum and thermal transport are coupled due to buoyancy forces. The effect of grid resolution on numerical dissipation has been analyzed in the context of implicit LES/UDNS in order to evaluate the accuracy of the simulations.

Methodology

The balances of both kinetic energy ($e_{kin} = u_i^2/2$) and thermal variance ($e_t = T^2/2$) have been taken into account as shown in Eqs. 34.1 and 34.2 for Rayleigh-Bénard convection of incompressible Newtonian fluids:

$$\varepsilon_n^k = -\iiint\limits_V \Bigg[\frac{\partial e_{kin}}{\partial t} + \underbrace{\frac{\partial (e_{kin} u_i)}{\partial x_i}}_{F_{e_{kin}}} + \underbrace{\frac{1}{\rho}\frac{\partial (p u_i)}{\partial x_i}}_{F_p} \\ - \underbrace{\frac{\partial}{\partial x_i}\left(\nu \frac{\partial e_{kin}}{\partial x_i}\right)}_{F_\nu} + \underbrace{\nu \frac{\partial u_j}{\partial x_i}\frac{\partial u_j}{\partial x_i}}_{\varepsilon_\nu} - \underbrace{u_i g_i (1 - \beta(T - T_{ref}))}_{F_s} \Bigg] \mathrm{d}V, \tag{34.1}$$

$$\varepsilon_n^t = -\iiint\limits_V \Bigg[\frac{\partial e_t}{\partial t} + \underbrace{\frac{\partial (e_t u_i)}{\partial x_i}}_{F_{e_t}} - \underbrace{\frac{\partial}{\partial x_i}\left(\alpha \frac{\partial e_t}{\partial x_i}\right)}_{F_\alpha} + \underbrace{\alpha \frac{\partial T}{\partial x_i}\frac{\partial T}{\partial x_i}}_{\varepsilon_\alpha} \Bigg] \mathrm{d}V. \tag{34.2}$$

Conservative transport of momentum (temperature) does not necessarily imply conservation of kinetic energy (thermal variance). Hence, the residuals of Eqs. 34.1 and 34.2 represent according to [1] the numerical dissipation of kinetic energy (i.e., ε_n^k); numerical dissipation of thermal variance (i.e., ε_n^t) and ε_ν, ε_α denote the physical viscous dissipation and physical thermal dissipation, respectively. These equations are passively evaluated on the fly based on resolved velocity and temperature fields for different grid resolutions (see section "Numerical Implementation"). All the terms of Eqs. 34.1 and 34.2 were computed for a cubic domain from sequential snapshots of the velocity and temperature fields (see Fig. 34.1). Accordingly, the rate of the change in the kinetic energy and thermal variance is computed using a second-order central difference (i.e., $\partial e_\phi \,/\, \partial t = [e_\phi(t_{n+1}) - e_\phi(t_{n-1})] \,/\, (2\Delta t)$ where $e_\phi = e_{kin}$ or $e_\phi = e_t$).

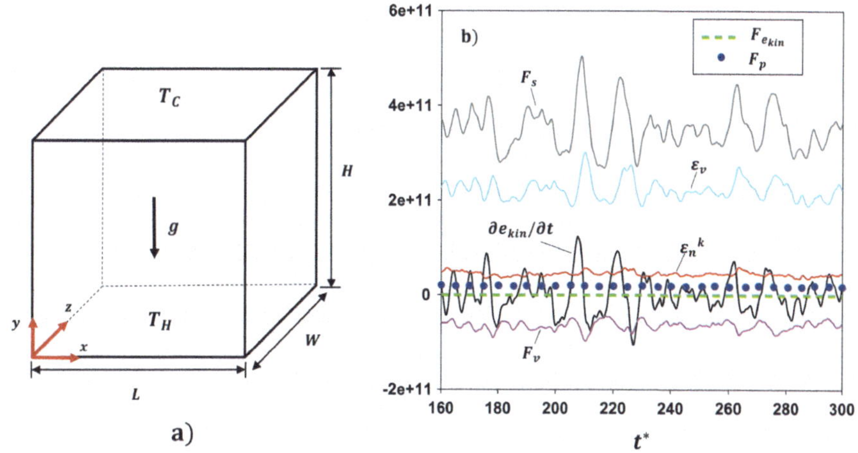

Fig. 34.1 **a** Schematic diagram of the simulation setup, **b** temporal evolution ($t^* = t\alpha\sqrt{\text{RaPr}}/H^2$) of volume averaged terms of Eq. 34.1 at Ra = 10^7 and Pr = 1 for a 40^3 uniform grid

Non-dimensional Numbers

In the Rayleigh-Bénard convection, the mean Nusselt number (Nu $= hH/k$) characterizes the heat transfer rate at the walls which depends on the non-dimensional parameters as Rayleigh number (Ra $= g\beta\Delta T H^3/(\nu\alpha)$), Prandtl number (Pr $= \nu/\alpha$), and the aspect ratios of the enclosure. Calzavarini et al. [2] suggested two alternative definitions of the mean Nusselt number based on volume and time averaged viscous (i.e., $\varepsilon_\nu = \nu(\partial_i u_j)^2$) and thermal (i.e., $\varepsilon_\alpha = \alpha(\partial_i T)^2$) dissipation rates in the following manner:

$$\overline{\text{Nu}}_{\varepsilon_\nu} = [(\langle\varepsilon_\nu\rangle\, H^4\,\text{Pr}^2)/(Ra\,\nu^3)] + 1\,,\ \overline{\text{Nu}}_{\varepsilon_\alpha} = (\langle\varepsilon_\alpha\rangle\, H^2)/(\alpha\,(\Delta T)^2) \tag{34.3}$$

where the average $\langle ... \rangle$ is over volume and over time.

Cadieux et al. [1] reported that the ratio of numerical and physical dissipation could be a good indicator of the overall accuracy of a given simulation (i.e., $\langle R_k\rangle = \langle\varepsilon_n^k\rangle/\langle\varepsilon_\nu\rangle$ and $\langle R_t\rangle = \langle\varepsilon_n^t\rangle/\langle\varepsilon_\alpha\rangle$ for kinetic energy and thermal variance, respectively). Based on this assumption for Nusselt numbers sufficiently larger than unity, Eq. 34.3 suggests that numerical dissipation errors directly result in mean Nusselt numbers errors (i.e., $\langle R_k\rangle = (\langle\varepsilon_n^k\rangle/\langle\varepsilon_\nu\rangle) \approx (\overline{\text{Nu}}_{\varepsilon_n^k}\,/\,\overline{\text{Nu}}_{\varepsilon_\nu})$ and $\langle R_t\rangle = (\langle\varepsilon_n^t\rangle/\langle\varepsilon_\alpha\rangle) \approx (\overline{\text{Nu}}_{\varepsilon_n^t}\,/\,\overline{\text{Nu}}_{\varepsilon_\alpha})$ where $\overline{\text{Nu}}_{\varepsilon_n^k}, \overline{\text{Nu}}_{\varepsilon_n^t}$ characterize changes in the mean Nusselt number due to numerical dissipation). Thus, the variation of $\langle R_k\rangle$ and $\langle R_t\rangle$ versus $\overline{\text{Nu}}_{\varepsilon_\nu}$ and $\overline{\text{Nu}}_{\varepsilon_\alpha}$ will be analyzed in this study in order to asses the quality of the UDNS.

Numerical Implementation

DNS have been performed for Rayleigh-Bénard convection for Ra $= 10^7$ and Pr $= 1$ in a three-dimensional cubical enclosure by using the finite volume code OpenFOAM (i.e., buoyantBoussinesqPimpleFoam) in order to validate UDNS results. The pressure-velocity coupling is addressed by the use of the PIMPLE algorithm. Convective and diffusive fluxes are evaluated by second-order central differencing schemes. Temporal advancement is achieved by the Crank–Nicolson scheme with adaptive time-stepping. It has been ensured that the CFL number is always sufficiently below unity. It is worth mentioning that the correction term in the pressure equation (i.e., fvc::ddtCorr(U, phi)) is not implemented in the current solver since it is extremely diffusive, as reported by Vuorinen et al. [4].

Results

One-dimensional (i.e., linear convection–diffusion equation) and two-dimensional (i.e., Rayleigh-Bénard convection) analysis have been carried out for better understanding of numerical dissipation in the context of this study. Results indicate that numerical dissipation is well correlated with the relative error and the value and sign of the numerical dissipation is a combined effect of temporal and spatial discretization schemes and grid spacing. In addition, the two-dimensional analysis demonstrated that convergence of numerical dissipation occurs on very fine grids. However, only the three-dimensional numerical dissipation results will be presented here for the sake of brevity.

The variations of absolute values of $\langle R_k \rangle$, $\langle R_t \rangle$ are shown in Fig. 34.2a with the percentage error in the mean Nusselt numbers based on Eq. 34.3 between UDNS and DNS versus non-dimensional grid size (i.e., $\Delta x/H$). The values of $\langle R_k \rangle$, $\langle R_t \rangle$ and $\overline{\mathrm{Nu}}_{\varepsilon_\nu}$ and $\overline{\mathrm{Nu}}_{\varepsilon_\alpha}$ can also be seen in Fig. 34.2b. The Nusselt number calculation based on Eq. 34.3 contains either the numerical dissipation error of kinetic energy or thermal variance, but as the temperature and momentum equation are coupled the prediction depends on both quantities simultaneously. For this reason Fig. 34.2a shows the error based on $0.5(|\langle R_k \rangle| + |\langle R_t \rangle|)$ besides the errors estimated on the individual quantities. It should be noted that averaging $|\langle R_k \rangle|$ and $|\langle R_t \rangle|$ is an acceptable approach since numerical dissipation from velocity and temperature fields are comparable for Pr $= 1$. It can be clearly seen from Fig. 34.2a that the error estimate remains of the same order of magnitude than the percentage error between UDNS and DNS of the mean Nusselt numbers (i.e., $e = 100 \cdot (|\overline{\mathrm{Nu}}_{DNS} - \overline{\mathrm{Nu}}_{UDNS}|)/\overline{\mathrm{Nu}}_{DNS}$). Figure 34.2a further demonstrates that the value of $0.5(|\langle R_k \rangle| + |\langle R_t \rangle|)$ decreases with decreasing non-dimensional grid size while the percentage error between UDNS and DNS of the mean Nusselt numbers also decreases in a similar manner and has a similar magnitude.

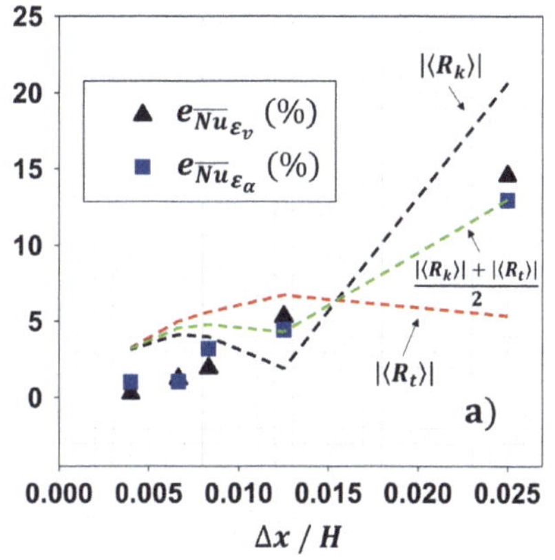

b)

Grid	$\langle R_k\rangle$	$\langle R_t\rangle$	$\overline{Nu}_{\varepsilon_v}$	$\overline{Nu}_{\varepsilon_\alpha}$
40^3	0.206	-0.0535	13.902	14.162
80^3	-0.0189	-0.067	15.412	15.55
120^3	-0.0396	-0.0561	15.96	15.75
150^3	-0.0412	-0.0499	16.073	16.101
250^3	-0.0312	-0.0324	16.229	16.106

Fig. 34.2 **a** The variations of absolute values of $|\langle R_k\rangle|$ (%), $|\langle R_t\rangle|$ (%) and percentage error in the mean Nusselt numbers based on Eq. 34.3 between UDNS and DNS versus non-dimensional grid size (i.e., $\Delta x/H$) for a central differencing scheme, **b** the results obtained from three-dimensional analysis for central differencing schemes at Ra $= 10^7$ and Pr $= 1$

The values of $|\langle R_k\rangle|$ and $|\langle R_t\rangle|$ can be converted to another useful quality assessment criterion by estimating effective viscosity and thermal diffusivity here. The effective viscosity ($\nu_{eff} = \nu + \nu_n$) of numerical simulations can be determined as the sum of the physical viscosity (ν is the nominal value used in the calculation) and the numerical viscosity (ν_n) due to numerical dissipation. Similarly, the effective thermal diffusivity ($\alpha_{eff} = \alpha + \alpha_n$) can also be determined as the sum of physical and numerical thermal diffusivity. This allows to define an effective Rayleigh ($\text{Ra}_{eff} = (g\beta\Delta T H^3)/(\nu_{eff}\alpha_{eff})$) and Prandtl ($\text{Pr}_{eff} = (\nu_{eff}/\alpha_{eff})$) number in terms of their nominal values (Ra, Pr) as $\text{Ra}_{eff} = \text{Ra}/[(1 + \nu_n/\nu)\,(1 + \alpha_n/\alpha)]$ and $\text{Pr}_{eff} = [\text{Pr}(1 + \nu_n/\nu)]/\,(1 + \alpha_n/\alpha)$.

The numerical viscosity and thermal diffusivity can be calculated as $\nu_n = \nu\langle\varepsilon_n^k\rangle/\langle\varepsilon_v\rangle = \nu\langle R_k\rangle$ and $\alpha_n = \alpha\langle\varepsilon_n^t\rangle/\langle\varepsilon_\alpha\rangle = \alpha\langle R_t\rangle$ respectively. Accordingly, Ra_{eff} and Pr_{eff} can be written in terms of $|\langle R_k\rangle|$ and $|\langle R_t\rangle|$ in the following manner:

$$\text{Ra}_{eff} = \text{Ra}/[(1 + \langle R_k\rangle)(1 + \langle R_t\rangle)]\,,\ \ \text{Pr}_{eff} = \text{Pr}(1 + \langle R_k\rangle)/(1 + \langle R_t\rangle). \quad (34.4)$$

Figure 34.3a shows the variations of Ra_{eff} and Pr_{eff} values calculated from Eq. 34.4 using absolute values of $|\langle R_k\rangle|$, $|\langle R_t\rangle|$ (given in Fig. 34.2b) for different non-dimensional grid sizes. It can be clearly seen from Fig. 34.3a that Ra_{eff} values are considerably smaller than their nominal value (i.e., Ra) for the coarsest grid configuration where numerical dissipation values are also high (i.e., $|\langle R_k\rangle| = 20.63\%$ and $|\langle R_t\rangle| = 5.35\%$). Also, Ra_{eff} values increase with decreasing non-dimensional grid size and approach the nominal value where numerical dissipation remains much smaller (i.e., $|\langle R_k\rangle| = 3.126\%$ and $|\langle R_t\rangle| = 3.24\%$). By contrast, Pr_{eff} decreases with decreasing grid size and approaches its nominal value (i.e., Pr) at the same set of parameters.

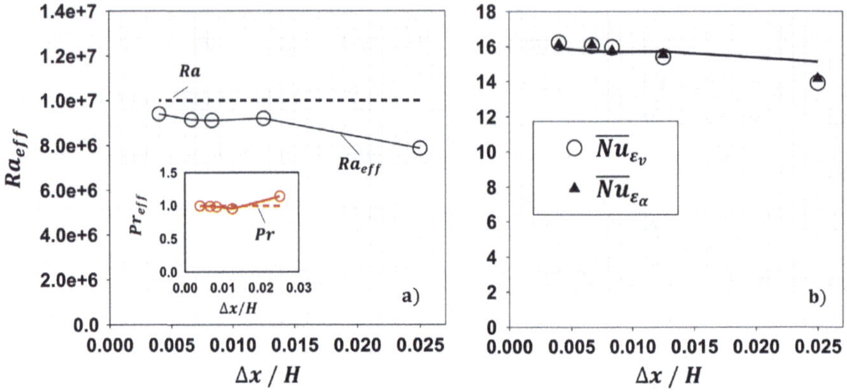

Fig. 34.3 **a** Variations of Ra_{eff} and Pr_{eff} values calculated from Eq. 34.4. **b** Variations of $\overline{\mathrm{Nu}}_{\varepsilon_v}$ and $\overline{\mathrm{Nu}}_{\varepsilon_\alpha}$ (shown in Fig. 34.2b) along with prediction of Eq. 34.5 (i.e., solid line) versus non-dimensional grid size (i.e., $\Delta x/H$)

Moreover, Fig. 34.3b exhibits that the values of $\overline{\mathrm{Nu}}_{\varepsilon_v}$ and $\overline{\mathrm{Nu}}_{\varepsilon_\alpha}$ increase with decreasing non-dimensional grid size. The mean Nusselt number is expected to scale with a power law of Ra and Pr (i.e., $\mathrm{Nu} \sim \mathrm{Ra}^m Pr^n$) [3].

Dropkin and Somerscales [3] experimentally analyzed turbulent Rayleigh-Bénard convection of Newtonian fluids in cubic enclosure for $5 \times 10^4 \leqslant Ra \leqslant 7.17 \times 10^8$ and $0.02 \leqslant \mathrm{Pr} \leqslant 11560$. They proposed a mean Nusselt number correlation as $\overline{\mathrm{Nu}} = 0.069\mathrm{Ra}^{1/3}\mathrm{Pr}^{0.074}$. However, the correlation by Dropkin and Somerscales [3] underestimates the mean Nusselt number at $\mathrm{Ra} = 10^7$ and $\mathrm{Pr} = 1$ (i.e., $\overline{\mathrm{Nu}}$ is predicted to be 14.86) since it is a general mean Nusselt number correlation for a quite wide range of parameters.

Neverthless, the above correlation can be revised in the following form $\overline{\mathrm{Nu}} = 0.07533\mathrm{Ra}^{1/3}\mathrm{Pr}^{0.074}$ to match the present DNS data. Using Ra_{eff} and Pr_{eff} values calculated from Eq. 34.4, this correlation can provide an estimate of the effective mean Nusselt number as follows:

$$\overline{\mathrm{Nu}}_{eff,model} = 0.07533\,\mathrm{Ra}_{eff}^{1/3}\,\mathrm{Pr}_{eff}^{0.074}\,. \tag{34.5}$$

Equation 34.5 can be seen as model for effective Nusselt number in order to estimate $\overline{\mathrm{Nu}}_{\varepsilon_v}$ and $\overline{\mathrm{Nu}}_{\varepsilon_\alpha}$ values based on Ra_{eff} and Pr_{eff} values calculated from Eq. 34.4 and $|\langle R_k\rangle|$, $|\langle R_t\rangle|$ values given in Fig. 34.2b. The predictions of $\overline{\mathrm{Nu}}_{\varepsilon_v}$ and $\overline{\mathrm{Nu}}_{\varepsilon_\alpha}$ based on Eq. 34.5 show quite well agreement with the current results (e.g., $\overline{\mathrm{Nu}}_{\varepsilon_v} = 14.981$ and $\overline{\mathrm{Nu}}_{\varepsilon_v} = 15.89$ for $\Delta x/H = 0.025$ and $\Delta x/H = 0.004$ respectively). The values of $\overline{\mathrm{Nu}}_{\varepsilon_v}$ and $\overline{\mathrm{Nu}}_{\varepsilon_\alpha}$ can be obtained based on Eq. 34.5 with percentage error of 7% and 2% in comparison to the values obtained from the simulation for $\Delta x/H = 0.025$ and $\Delta x/H = 0.004$, respectively. It is worth noting that the 7% difference is obtained only for the coarsest grid configuration and this value decreases significantly for the finer grids. Figure 34.3b suggests that effective Nusselt number can be satisfactorily predicted based on $|\langle R_k \rangle|$, $|\langle R_t \rangle|$. Further, by comparing the prediction of Eq. 34.5 with the one obtained based on nominal Ra and Pr numbers, an error estimate can be obtained.

Conclusions

The assessment of numerical dissipation has been analyzed for UDNS of Rayleigh-Bénard convection in cubical enclosure based on balances of kinetic energy and thermal variance in the OpenFOAM solver by extending the method used by Cadieux et al. [1]. It is found that the aforementioned method works sufficiently well for quality assessment of UDNS of Rayleigh-Bénard convection. This framework will be extended to explicit LES in the future which will give rise to an additional dissipation provided by the turbulence model.

References

1. Cadieux, F., Sun, G., Domaradzki, J.A.: Effects of numerical dissipation on the interpretation of simulation results in computational fluid dynamics. Comput. Fluids **154**, 256–272 (2017)
2. Calzavarini, E., Lohse, D., Toschi, F., Tripiccione, R.: Rayleigh and Prandtl number scaling in the bulk of Rayleigh-Bénard turbulence. Phys. Fluids **17**(5), 055107 (2015)
3. Dropkin, D., Somerscales, E.: Heat transfer by natural convection in liquids confined by two parallel plates which are inclined at various angles with respect to the horizontal. J. Heat Transf. **87**(1), 77–82 (1965)
4. Vuorinen, V., Keskinen, J.P., Duwig, C., Boersma, B.J.: On the implementation of low-dissipative Runge-Kutta projection methods for time dependent flows using OpenFOAM. Comput. Fluids **93**, 153–163 (2014)

Chapter 35
Shockwave/Boundary-Layer Interactions in Transitional Rectangular Duct Flows

D. J. Lusher and N. D. Sandham

Introduction

Shockwave/boundary-layer interactions (SBLI) are an important design consideration for many aeronautical applications. Supersonic engine intakes aim to efficiently decelerate incoming flow to subsonic conditions before the entrance to the compressor. Typically this is achieved by creating a sequence of oblique shocks terminated by a weak normal shock to minimise losses within the duct. The shock-induced adverse pressure gradient causes unsteadiness, a thickening of the boundary layer, and for sufficiently strong SBLI flow separation can occur. This reduction in flow uniformity can have a detrimental effect on engine performance and reliability.

In this work, the reflection of an oblique shockwave impinging on transitional boundary layers is investigated via Implicit Large-Eddy Simulations (ILES) for a full rectangular duct at Mach 2. In many studies of SBLI, a simplified periodic (quasi-2D) span is assumed to reduce computational cost, neglecting complex flow features that arise for internal confined flows. When flow confinement is included the oblique shock interacts with both the sidewall and bottom wall boundary layers, causing multiple regions of flow-reversal not present in quasi-2D simulations. Examples of previous numerical work on confined oblique SBLI include the turbulent SBLI of [1, 2], in which wall-modelled and wall-resolved LES were used, respectively. In both cases, the SBLI was highly three-dimensional and showed a strong dependence on the geometry of the duct. The swept-SBLI of the incident shock with sidewall boundary-layers led to a strengthened interaction on the centreline and regions of thin recirculation in the corners.

D. J. Lusher (✉) · N. D. Sandham
Aerodynamics and Flight Mechanics Group, University of Southampton,
Highfield Campus, Southampton SO17 1BJ, UK
e-mail: D.Lusher@soton.ac.uk

M. García-Villalba et al. (eds.), *Direct and Large Eddy Simulation XII*,
ERCOFTAC Series 27,
https://doi.org/10.1007/978-3-030-42822-8_35

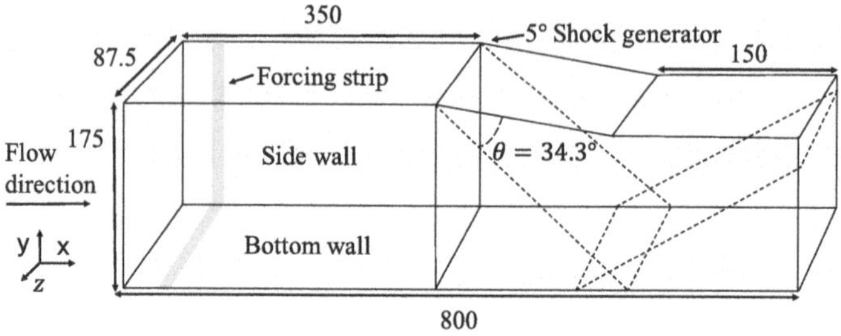

Fig. 35.1 Non-dimensional computational domain for the half aspect-ratio duct

Wind tunnel experiments are inherently three-dimensional due to the sidewalls of the test chamber, with large aspect ratios required to make comparison to span-periodic simulations. Some experiments [3] have proposed that the modified interaction could be predicted by the crossing location of compression waves emitted from corner separations. While turbulent duct SBLI has received some attention in recent years, laminar and transitional interactions in ducts are not as well understood. The work of [4] noted experimental difficulties when seeding boundary layers in a laminar case. Numerical simulations are well placed to investigate SBLI for transitional interactions and complement existing experimental literature.

Problem Specification

The fully-3D unsteady compressible Navier–Stokes equations for an ideal gas are solved directly in conservative form. The computational domain consists of a half aspect-ratio rectangular duct with a finite-length $\theta_{sg} = 5°$ internal shock generator as shown in Fig. 35.1. A Mach 2 freestream is initialised with a similarity solution laminar boundary layer [5] and reference temperature $T_{\text{ref}} = 288.0\,\text{K}$. The equations are solved in non-dimensional form; with the inlet displacement thickness δ^* taken as the characteristic length. Pressure is normalised by $\rho_\infty^* {U_\infty^*}^2$, to give $(U_\infty, \rho_\infty, T_\infty) = 1$ in the freestream. The boundary layers are initialised on the bottom and both sidewalls of the domain, with freestream conditions maintained on the top portion of the domain upstream of the shock generator. The Reynolds number based on the inlet displacement thickness is $\text{Re}_{\delta^*} = 1500$, with Prandtl number $\text{Pr} = 0.72$ and dynamic viscosity from Sutherland's law with $T_{\text{Suth}} = 110.4\,\text{K}$. No-slip isothermal wall conditions are enforced on the bottom wall, sidewalls, and on the shock generator for a non-dimensional wall temperature obtained from the similarity solution of $T_w = 1.676$ (4 significant figures). A pressure extrapolation is applied at the inlet, with zero gradient conditions enforced on the outlet and upstream of the shock generator.

To induce transition upstream of the shock reflection disturbances are added to the laminar boundary layer via wall-normal blowing/suction. The v velocity component is forced using a modified version of the forcing presented in [6] such that

$$v\left(x,z,t\right)=\begin{cases}A_0 f(x)g(z)h(t) & \text{if } x_a < x < x_b \\ 0 & \text{otherwise,}\end{cases} \tag{35.1}$$

for streamwise $f\left(x\right)$, spanwise $g\left(z\right)$ and time $h\left(t\right)$ dependence

$$f(x)=\exp\left(-\left(\frac{(x-(x_b-x_a))^2}{16}\right)\right), \tag{35.2}$$

$$g(z)=\sum_{l=1}^{l_{\max}} Z_l \sin\left(2\pi l\left(z/L_z+\phi_l\right)\right),\quad h(t)=\sum_{m=1}^{m_{\max}} T_m \sin\left(\omega_m t+2\pi\phi_m\right), \tag{35.3}$$

$$\sum_{l=1}^{l_{\max}} Z_l=1,\quad Z_l=1.25Z_{l+1},\quad \sum_{m=1}^{m_{\max}} T_m=1,\quad T_m=1.25T_{m+1}. \tag{35.4}$$

Random phases ϕ_l, ϕ_m distributed between [0, 1] are added for each of the $l, m =$ 20 terms. The same phases are fixed between different simulations but differ on each of the three forcing strips to add asymmetry to the breakdown. Forcing is applied for a range of evenly spaced frequencies ω_m in the interval [0.04, 0.12], taken from linear stability of compressible boundary layers [7]. The forcing strip is located between $x_a = 10.0, x_b = 30.0$ with $A_0 = 0.2$. A high amplitude is selected to trigger transition upstream of the shock reflection. On both sidewalls of the domain, the same forcing is applied for the wall-normal velocity w, where the spatial variation $g(z)$ is instead taken over $g(y)$. Two cases are considered in this work: one with disturbances added to both the bottom and sidewalls of the domain, and one where the bottom wall is kept laminar to observe how the transition develops over the span.

All computations were performed in the OpenSBLI code, validated for laminar SBLI in [8]. OpenSBLI uses high-order finite differencing on structured meshes with explicit time stepping. Convective terms of the governing equations are discretised by a 6th order low-dissipative Targeted Essentially Non-Oscillatory (TENO) scheme designed for ILES of shock-turbulence interactions [9]. A comparison of TENO methods for transitional SBLI in OpenSBLI was given in [10]. Viscous, heat-flux, and metric terms are discretized with a 4th order central scheme. Time stepping is performed by an explicit 3rd order low-storage Runge–Kutta scheme with $\Delta t = 1 \times 10^{-2}$. The grid is stretched toward each of the four walls of the duct and uniform in the streamwise direction with $\left(N_x, N_y, N_z\right) = (1050, 375, 325)$ points. Three flow-through times of the domain were simulated to clear the initial transient and let the shock reflection form, before time averaging was applied for a further four flow-through times.

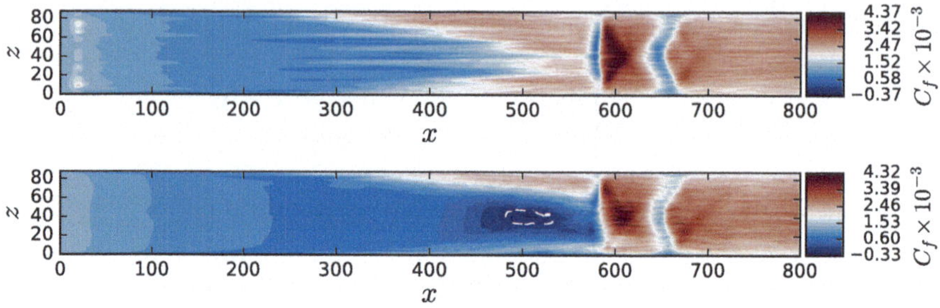

Fig. 35.2 Time-averaged skin friction distributions on the bottom wall. For the cases of (top) forced sidewalls and bottom wall and (bottom) forced sidewalls only. The dashed white line encloses regions of flow recirculation

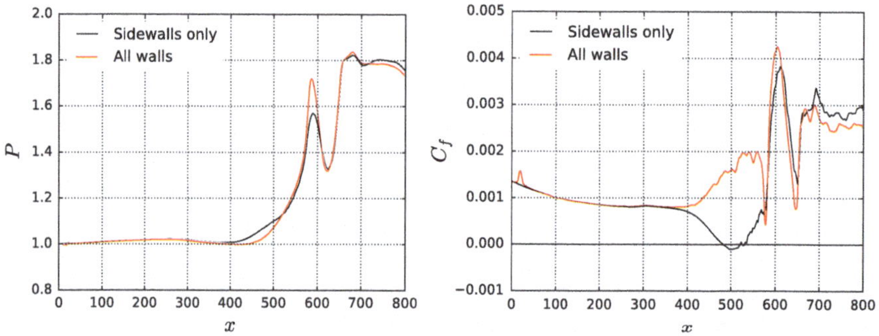

Fig. 35.3 Time-averaged centreline normalised pressure (left) and skin friction (right) on the bottom wall of the duct. For the cases of forced sidewalls and bottom wall (red) and forced sidewalls only (black)

Results

Time-averaged skin friction on the bottom wall is shown in Fig. 35.2 for the fully forced (top) and laminar bottom wall cases (bottom). The incident shock impinges on the bottom wall boundary layer at $x = 580$, which is curved as a result of the swept sidewall SBLI. For the fully forced case there is no mean flow separation at the centreline, the transition develops first in the corners and spreads out in a wedge across the span. In contrast, the laminar bottom wall case shows an oval region of mean flow separation bordered by the turbulence generated in the corners. The latter case also exhibits an asymmetry in the development of the transition, likely caused by the asymmetry in the random phases of the forcing and a short averaging time.

A curved secondary shock is observed at $x = 650$, caused by transverse reflections of the incident conical shocks between the sidewalls of the duct. This conical shock is seen to reflect as an expansion wave which crosses the centreline at $x = 700$. These complex shock structures are a result of the narrow duct and would not be present in infinite span simulations. Figure 35.3 shows average centreline wall distributions of normalised pressure and skin friction. The fully forced result has a stronger initial

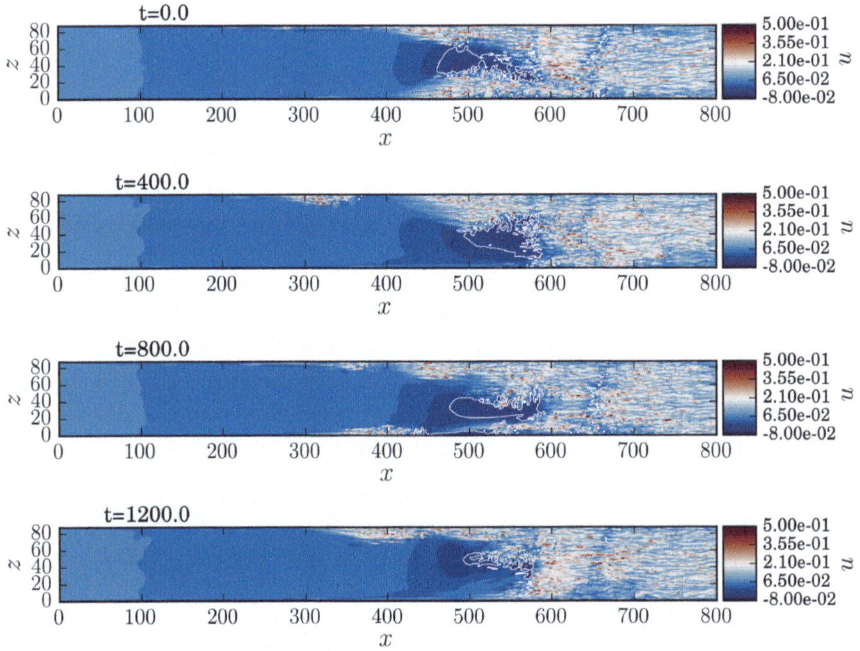

Fig. 35.4 Instantaneous streamwise velocity snapshots above the bottom wall. The solid white line represents the $u = 0$ contour enclosing regions of recirculation. The central recirculation bubble shifts laterally due to oncoming intermittent turbulent spots

pressure rise from the incident shock, but leads to a similar outlet pressure to the laminar bottom wall case. In the skin friction curve, the effect of the more developed transition is clear for the fully forced case, the rise in skin friction demonstrates the transition has reached the centreline and wards off mean flow separation. When only the sidewalls are tripped, incipient centreline separation is reached. The final level of skin friction is about 10% higher in this case, indicating a more energetic breakdown process.

To further elucidate the response of the central separation bubble to intermittent turbulence, a selection of instantaneous streamwise velocity snapshots are shown in Fig. 35.4. The case shown is the one with the unforced bottom wall, four figures are spaced at intervals of $\Delta t = 400$ which each corresponds to half a flow-through of the freestream. In the top image, the transition is relatively symmetric about the centreline and a small region of recirculation is present. At $t = 400$ the transition is stronger on the upper $z = 87.5$ sidewall, pushing the recirculation bubble down towards the $x = 0$ sidewall. By $t = 800$ a corner separation has formed on the lower sidewall at $x = 500$, which is amplifying initial disturbances and causing a breakdown near the corner reattachment point of $x = 525$. At $x = 400$ a turbulent spot has just reached the front of the corner separation and starts interacting with the bubble. Finally by $t = 1200$ the corner separation has been fully removed. The increased mixing rates inside the turbulent spot transfers low-momentum fluid away from the wall. The central separation has contracted and shifts back towards the upper sidewall.

Conclusions

Oblique shockwave/boundary-layer interactions (SBLI) have been simulated for a half aspect-ratio rectangular duct at Mach 2. Disturbances added to a laminar flow via blowing/suction caused intermittent turbulent spots that originated in the low-momentum corners of the duct. The transition spread out in a wedge shape across the width of the span. A pair of conical shockwaves form at the swept sidewall SBLIs and reflect laterally within the duct. For a case with disturbances added to both the side and bottom walls incipient mean separation was not reached for a $\theta_{sg} = 5.0°$ shock generator. When the bottom wall was kept initially laminar a region of shock-induced mean flow recirculation was observed on the centreline. Instantaneous snapshots of the near-wall streamwise velocity showed the recirculation bubble reacts dynamically to oncoming turbulent spots. Temporary corner separations were seen to develop that caused an earlier streamwise destabilisation of the downstream flow.

Acknowledgements DJL is funded by an EPSRC Centre for Doctoral Training grant (EP/L015382/1). Compute resources were provided by the 'University of Cambridge Service for Data Driven Discovery' (CSD3), EPSRC Tier-2 capital grant EP/P020259/1.

References

1. Bermejo-Moreno, I., Campo, L., Larsson, J., Bodart, J., Helmer, D., Eaton, J.K.: Confinement effects in shock wave/turbulent boundary layer interactions through wall-modelled large-eddy simulations. J. Fluid Mech. **758**, 5–62 (2014)
2. Wang, B., Sandham, N., Hu, Z., Liu, W.: Numerical study of oblique shock-wave/boundary-layer interaction considering sidewall effects. J. Fluid Mech. **767**, 526–561 (2015)
3. Xiang, X., Babinsky, H.: Corner effects for oblique shock wave/turbulent boundary layer interactions in rectangular channels. J. Fluid Mech. **862**, 1060–1083 (2019)
4. Giepman, R.H.M., Schrijer, F.F.J., van Oudheusden, B.W.: A parametric study of laminar and transitional oblique shock wave reflections. J. Fluid Mech. **844**, 187–215 (2018)
5. White, F.: Viscous Fluid Flow. McGraw-Hill, New York, NY (2006)
6. Pirozzoli, S., Grasso, F., Gatski, T.B.: Direct numerical simulation and analysis of a spatially evolving supersonic turbulent boundary layer at M=2.25. Phys. Fluids **16**, 530–545 (2004). Jan
7. Sansica, A.: Stability and unsteadiness of transitional shock-wave/boundary-layer interactions in supersonic flows. Ph.D. thesis, University of Southampton (2015)
8. Lusher, D.J., Jammy, S.P., Sandham, N.D.: Shock-wave/boundary-layer interactions in the automatic source-code generation framework OpenSBLI. Comput. Fluids **173**, 17–21 (2018)
9. Fu, L., Hu, X.Y., Adams, N.A.: A family of high-order targeted ENO schemes for compressible-fluid simulations. J. Comput. Phys. **305**, 333–359 (2016)
10. Lusher, D.J., Sandham, N.: Assessment of low-dissipative shock-capturing schemes for transitional and turbulent shock interactions. In: AIAA Aviation 2019 Forum. American Institute of Aeronautics and Astronautics, June (2019)

Chapter 36
Numerical Investigation of Hypersonic Boundary Layers of Perfect and Dense Gases

L. Sciacovelli, X. Gloerfelt, P. Cinnella and F. Grasso

Introduction

High-speed flows of gases with complex thermodynamic behavior (often referred to as "real gases") have been paid growing interest from the scientific community due to the manifold applications in aerospace and power generation systems. In this work, we focus on so-called dense-gas flows found in several engineering applications, ranging from energy production to high-speed wind tunnels. Dense gases are single-phase fluids of complex molecules, at pressure and temperature conditions of the same order of their thermodynamic critical point. These fluids, in particular those belonging to the Bethe–Zel'dovich–Thompson family, may exhibit nonclassical phenomena in the transonic and supersonic regimes such as expansion shocks or mixed waves; these nonclassical effects may be characterized by means of the fundamental derivative of gas dynamics Γ [8].

The influence of dense-gas effects on compressible turbulent flows has been previously investigated in isotropic turbulence [5, 6] and channel flow [7] configurations. Numerical simulations of hypersonic turbulent boundary layers (HTBL) enable the investigation of physical mechanisms characterizing a wall-bounded, unconfined configuration. A fundamental question arising in high-Mach number flows concerns

L. Sciacovelli (✉) · X. Gloerfelt · P. Cinnella
Arts et Métiers ParisTech, DynFluid Laboratory, Paris, France
e-mail: luca.sciacovelli@ensam.eu

X. Gloerfelt
e-mail: xavier.gloerfelt@ensam.eu

P. Cinnella
e-mail: paola.cinnella@ensam.eu

F. Grasso
Conservatoire National des Arts et Métiers, DynFluid Laboratory, Paris, France
e-mail: francesco.grasso@cnam.fr

M. García-Villalba et al. (eds.), *Direct and Large Eddy Simulation XII*,
ERCOFTAC Series 27,
https://doi.org/10.1007/978-3-030-42822-8_36

the laminar to turbulent transition. At hypersonic conditions higher instability modes of acoustic nature exist for air flows, as pointed out by Mack [3], leading to different mechanisms for instability growth and breakdown to turbulence, that must be understood in order to predict the transition location. Moreover, at these high flow speeds, the intricate coupling between thermal and dynamic effects, largely unexplored for the dense gas, affects the turbulent flow behavior and its contribution to skin friction and heat transfer.

In this work, we investigate HTBL at external Mach number $M = 6$ of a dense (DG) and a perfect gas (PFG) by means of Direct Numerical Simulations (DNS), from the laminar to fully turbulent state. The reference DNS considered for the purpose of comparison and validation for air is the HTBL investigated by Franko and Lele [2], who focused on the analysis of the transition mechanisms. A DG simulation is also conducted at the same Mach number. Although much higher than Mach numbers typically encountered in practical dense-gas applications, the chosen operating conditions are well suited to highlight dense-gas effects.

Direct Numerical Simulation Details

The selected working fluid for the dense-gas case is a heavy fluorocarbon, namely, PP11 ($C_{14}F_{24}$), previously considered for analyzing dense-gas effects in different configurations [6, 7]. As in [7], the thermodynamic behavior of PP11 is modeled by the Martin–Hou equation of state and the variation of the transport properties with temperature and density is described by means of the Chung–Lee–Starling dense-gas model. DNS simulations are carried out by means of a ninth-order accurate numerical scheme (see [5]), supplemented by a non-linear artificial viscosity term equipped with a shock-capturing sensor described in [7].

The computational domain is defined from the plate leading edge and transition to turbulent is obtained by forcing the most unstable waves by means of suction and blowing. For the PFG boundary layer, we simulate an oblique breakdown scenario (modal transition), whereby the forcing excites the three-dimensional Tollmien–Schlichting mode (or first mode), as in the reference DNS of [2]. However, since our aim is to achieve fully turbulent conditions, a much longer computational domain and integration time are used with respect to [2].

For the DG case, a preliminary linear stability study was carried out to determine the 3D unstable spatial modes [4]. Contrary to PFG, DG velocity profiles are largely insensitive to the Mach number. Additionally, the temperature rise at the wall is less than 3% in this high-specific-heat fluid, whereas it is about 600% in the PFG. As a consequence, friction heating is almost suppressed and the velocity profile remains close to the incompressible one. Despite the apparently incompressible behavior, acoustic waves are present and deeply altered by the complex thermodynamics, which turns into dramatically different stability properties. In DG, the Tollmien–Schlichting mode disappears for Mach above $\approx$1.5, due to the absence of a generalized inflection point in the boundary layer profiles. This is unlike the classical behavior for air, whereby the appearance of a generalized inflection point tends to counteract the stabilizing effect of compressibility and sustains the first mode. An acoustic mode (or "Mack mode") appears for Mach numbers of 3 or higher. This mode becomes supersonic, thus capable of radiating acoustic waves, and it is characterized by a much higher frequency than the corresponding mode in PFG. Based on the stability study, we chose to excite the 2D acoustic mode maximizing the N factor at $\mathrm{Re}_{\delta^*_{ref}} = 1200$, with $\omega\delta^*_{ref}/U_\infty = 0.6$ and an amplitude $\varepsilon = 0.02U_\infty$ to trigger transition in the dense-gas boundary layer.

Table 36.1 provides a summary of the reference conditions and numerical parameters for the perfect and dense-gas DNS. In order to enhance dense-gas effects, the thermodynamic conditions for PP11 are chosen such that Γ is negative outside of the boundary layer, and remains lower than unity close to the wall. Moreover, for the chosen values of density and temperature, the speed of sound for PP11 is equal to $c_\infty = 33.1$ m/s, so that hypersonic conditions are achieved for much smaller velocities with respect to air, as shown in table 36.1. An isothermal condition is applied at the wall, with $T_w = 422.5$ K and 663.2 K for air and PP11, respectively, corresponding to the wall temperature for the laminar boundary layer (pseudo-adiabatic condition). The streamwise, wall-normal, and spanwise resolutions, as well as properties of the boundary layer near the end of the computational domain, are summarized in Table 36.1. The total number of grid points is of the order of $\sim 10^9$, with the dense-gas case having twice the number of points in the streamwise direction with respect to the air one.

Numerical Results

PFG results are presented first. Comparisons are conducted in the transitional region, where DNS results for perfect gas are available (Fig. 36.1). The distribution of C_f matches the laminar correlation until $x/\delta^*_{ref} = 400$; then it rapidly departs from it and a significant overshoot is visible. The transition path and intensity of the overshoot are in very good agreement with Ref. [2]. Excellent agreement is also observed for the profiles of streamwise velocity and temperature. The mean temperature profiles closely follow Walz's law, confirming that pseudo-adiabatic conditions are obtained. A global view of the flow field for PFG is given in Fig. 36.2, showing an iso-surface

Table 36.1 Boundary layer properties and parameters

Fluid	U_∞ (m/s)	ρ_∞ (kg/m^3)	T_∞ (K)	Re_θ	δ (mm)	θ (mm)	C_f ($\times 10^3$)	Δx^+	Δy_w^+	Δz^+	$N_x \times N_y \times N_z$
Air	969.7	0.13	65	5720	5.82	0.22	0.87	3.76	0.26	2.09	7700 × 300 × 400
PP11	198.8	348.411	646.8	4402	0.032	0.0029	2.13	8.12	0.65	9.77	14336 × 320 × 280

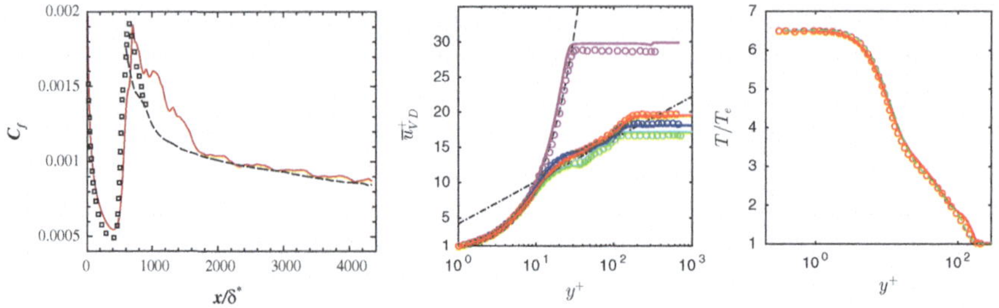

Fig. 36.1 Left, skin friction coefficient: (red,—) Present DNS; (square) Franko and Lele; (black, - - -) White's correlation. Center: Van Driest-scaled streamwise velocity profiles for the present DNS (solid lines) and Franko and Lele (circles) at locations $x/\delta^*_{ref} = 400$ (magenta), 650 (green), 800 (blue), 950 (red). Right, mean temperature profile at $x/\delta^*_{ref} = 950$: (red,—) present DNS; (red, circles) Franko and Lele; (green, - - -) Walz's law

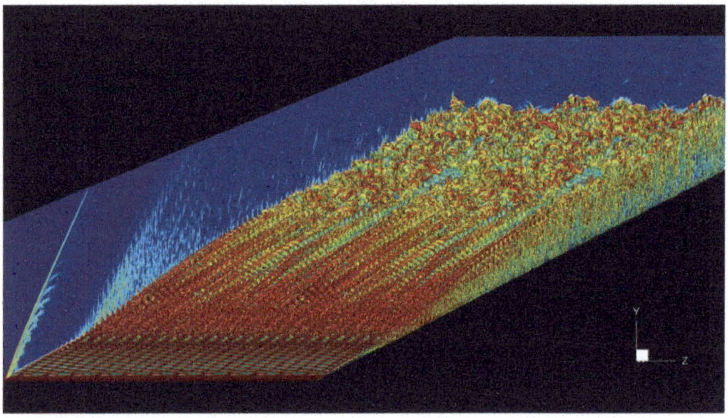

Fig. 36.2 Instantaneous views of the Q-criterion colored by the magnitude density gradient over the computational domain for the air HTBL

of the Q-criterion colored by the magnitude of the density gradient. The leading edge shock and the development of the unstable oblique waves downstream of the forcing strip are clearly visible, leading to the formation of an organized transition front and a transitional region characterized by long persistent streaky structures. A fully turbulent state is reached in the last third of the computational domain. Results for DG are reported in Fig. 36.3 and compared with PFG. The DG skin friction distribution remains close to the incompressible one; the transition behavior is deeply modified, and no overshoot is observed. In this case, the transition region is governed by streaks that eventually break leading to turbulent spots instead of an organized front. It can also be noticed that, while the local momentum-thickness-based Reynolds numbers (Re_θ) are reasonably similar for both flows, their incompressible counterparts obtained by scaling with the ratio of the freestream to wall viscosity ($\mathrm{Re}_{\theta,inc} = \mu_\infty/\overline{\mu}_w \mathrm{Re}_\theta$) are very different due to the negligible friction heating in the DG. Figure 36.4 shows selected first-order statistics profiles against the standard or

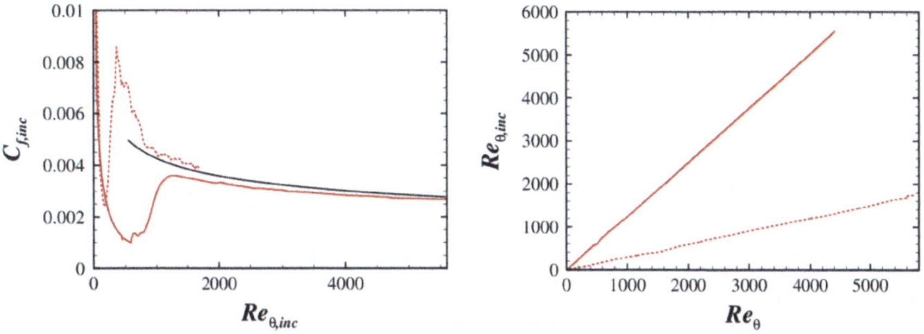

Fig. 36.3 Skin friction and Reynolds number distributions for DG (—) and PFG (– –)

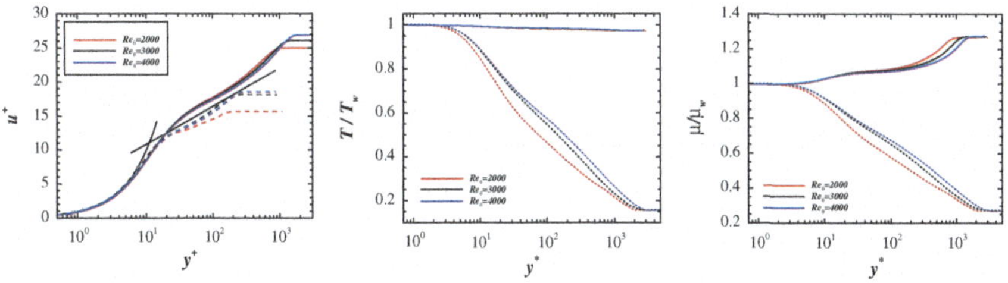

Fig. 36.4 Averaged profiles for PFG (– –) and DG (—) at various Reynolds numbers

semi-locally scaled wall coordinate at three stations ($Re_\theta = 2000$, 3000 and 4000) in the turbulent region. PFG is still transitional at the first station. For the DG velocity profile, the slope of the logarithmic region remains rather close to the incompressible case. Once again, the temperature variation across the boundary layer is very small. The density variation is also much smaller than in PFG (outer density is less than 1.25 times the wall density, instead of more than 6). Also note the peculiar variation of the viscosity that follows the density variation and decreases when approaching the wall (liquid-like behavior).

Selected second-order statistics are reported in Fig. 36.5. DG flow exhibits higher turbulent intensities, due to the higher local Reynolds number, and a dramatically different rms density profile. Compared to PFG, relative density fluctuations are much smaller and the fluctuation peak is located close to the wall and not in the outer region, as also observed in turbulent channel flow [7]. DG also reaches higher turbulent Mach numbers than PFG (not reported), with values reached in the production peak region ($y^* \approx 12$) sufficiently high ($M_{t,\max} > 0.3$) for the potential formation of eddy shocklets. Even when turbulent Mach number is lower than such a threshold, the existence of eddy shocklets is not excluded [1], since instantaneous Mach number values can be much higher. In PFG, high rms Mach number values are encountered in the outer region, and indeed sheet-like structures corresponding to high-density gradients and (mostly) strong compressions are observed in instantaneous Schlieren-like contours (Fig. 36.6), while in DG the peak rms Mach number is located in the

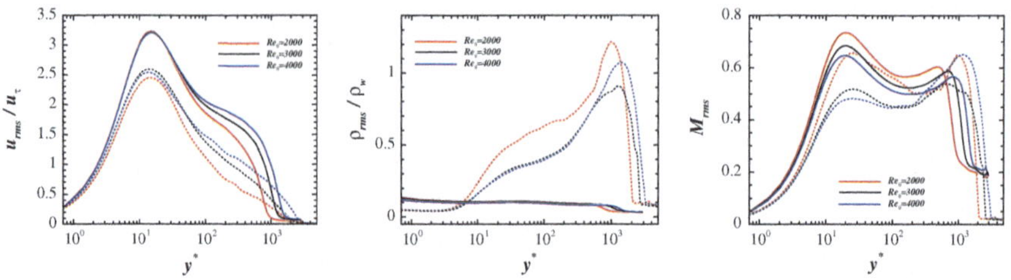

Fig. 36.5 Streamwise turbulent intensity (left), r.m.s. density (center) and r.m.s. Mach number (right) for PFG (– –) and DG (—) at various Reynolds numbers

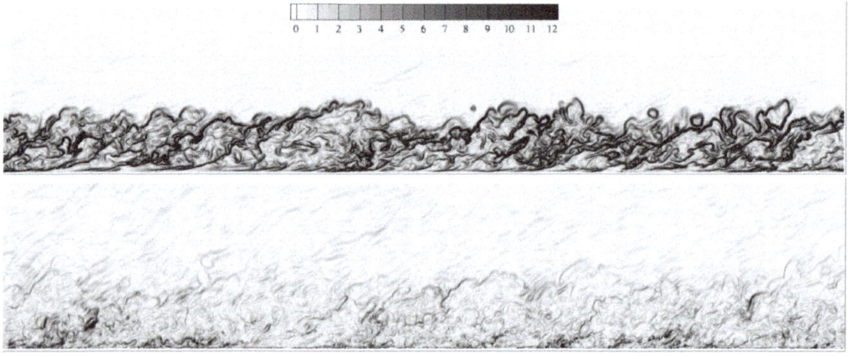

Fig. 36.6 Instantaneous snapshots of normalized numerical Schlieren $|\nabla\rho|/(\overline{\rho}/\delta_{99,\mathrm{end}})$ for PFG (top) and DG (bottom)

inner region. In this case, numerical Schlieren exhibit very high-density gradient spots (corresponding to both strong compressions and expansions) located near the wall (in a region where the rms density is also at its maximum), whereas sheet-like structures in the outer region are more sporadic.

Conclusions and Future Work

Direct numerical simulations of hypersonic boundary layers including the laminar, transitional, and fully turbulent regions were presented for perfect and dense gas at $M = 6$. For the dense gas, the complex thermodynamics and transport property variations play a crucial role at all flow regimes. Temperature variations are negligibly small throughout the flow field and friction heating is very small even at hypersonic conditions: as a consequence, both the laminar and turbulent velocity profiles remain very close to the incompressible ones. However, strong compressibility effects do exist and, due to the peculiar variation of the sound speed in dense gases, they deeply affect the boundary layer stability properties: a supersonic radiating mode appears, while the Tollmien–Schlichting mode is stable. The DG turbulent region is characterized by higher local incompressible Reynolds numbers, due to the liquid-like behavior of the viscosity. This leads to higher turbulent intensities and higher turbulent and fluctuating Mach numbers than in PFG. Despite this, sheet-like compression structures characterizing the external region of the PFG boundary layer are much rarer in DG, while intense density gradient spots appear in a region close to the turbulent kinetic energy peak. Further work will focus on better characterizing the flow structures and analyzing noise generation. Assessment of turbulence models is also planned.

Acknowledgements This work was granted access to the HPC resources of IDRIS and TGCC under the allocation 2018-7332 made by Grand Equipement National de Calcul Intensif (GENCI). We also acknowledge TGCC for awarding access to the Joliot-Curie supercomputer under the allocation "Grands Challenges" gch032.

References

1. Duan, L., Beekman, I., Martin, M.: Direct numerical simulation of hypersonic turbulent boundary layers. J. Fluid Mech. **672**, 245–267 (2011)
2. Franko, K.J., Lele, S.K.: Breakdown mechanisms and heat transfer overshoot in hypersonic zero pressure gradient boundary layers. J. Fluid Mech. **730**, 491–532 (2013)
3. Mack, L.M.: Boundary-layer linear stability theory. AGARD TR **709** (1984)
4. Sciacovelli L., Passiatore D., Cinnella P., Gloerfelt X., Grasso F.: Numerical investigation of supersonic dense-gas boundary layers. In: 2nd NICFD Conference, Bochum (DE), Oct (2018)
5. Sciacovelli, L., Cinnella, P., Grasso, F.: Dense gas effects in inviscid homogeneous isotropic turbulence. J. Fluid Mech. **800**, 140–179 (2016)
6. Sciacovelli, L., Cinnella, P., Grasso, F.: Small-scale dynamics of dense gas compressible homogeneous isotropic turbulence. J. Fluid Mech. **825**, 515–549 (2017)
7. Sciacovelli, L., Cinnella, P., Gloerfelt, X.: Direct numerical simulations of supersonic turbulent channel flows of dense gases. J. Fluid Mech. **821**, 153–199 (2017)
8. Thompson, P.A.: A fundamental derivative in gasdynamics. Phys. Fluids **14**(9), 1843 (1971)

Chapter 37
Towards Wavelet-Based Intelligent Simulation of Wall-Bounded Turbulent Compressible Flows

G. De Stefano and O. V. Vasilyev

Abstract The progress in the development of a wavelet-based intelligent modelling and simulation framework, capable of performing adaptive simulations of wall-bounded turbulent compressible flows, is discussed. A novel split mesh adaptation strategy based on mean and fluctuating components is introduced and tested. The preliminary results that are achieved for the transitional supersonic channel with isothermal walls are very promising. It is demonstrated that the new strategy is very efficient when dealing with inhomogeneous wall-bounded flows, in terms of resolving coherent flow structures, while substantially reducing the number of degrees of freedom, without loss of the solution fidelity.

Keywords Wavelet threshold filtering · Compressible turbulence · Split mesh adaptation

Introduction

The latest advancements in the wavelet-based methodology for the numerical solution of the Navier–Stokes equations, combined with the unique property of wavelet analysis to both unambiguously identify localised dynamically dominant flow structures and track them on adaptive computational meshes, make it feasible to develop intelligent methods for simulating turbulent flows. These numerical procedures effectively exploit the turbulence intermittency, namely, the fact that active turbulent flow

G. De Stefano (✉)
Engineering Department, University of Campania Luigi Vanvitelli, 81031Aversa, Italy
e-mail: giuliano.destefano@unicampania.it

O. V. Vasilyev
Keldysh Institute of Applied Mathematics, Russian Academy of Sciences, Moscow 125047, Russia
e-mail: oleg.v.vasilyev@gmail.com

Adaptive Wavelet Technologies LLC, Superior, CO80027, USA

M. García-Villalba et al. (eds.), *Direct and Large Eddy Simulation XII*,
ERCOFTAC Series 27,
https://doi.org/10.1007/978-3-030-42822-8_37

regions are distributed non-homogeneously in space and time, while extending over a broad range of scales.

In the wavelet-based approach to the numerical simulation of turbulent flows, the actual separation between resolved eddies and background unresolved motions is achieved by means of non-linear wavelet threshold filtering, where the level of thresholding controls the relative importance of filtered and residual fields and, thus, the actual fidelity of the solution. By modifying the wavelet threshold, a unified hierarchy of different fidelity methods is defined, where coherent flow structures are either totally or partially resolved on self-adaptive computational grids. For very low thresholds, the wavelet-based discretization of the governing equations is used to dynamically adapt the local mesh resolution, following the evolution of intermittent important flow structures. In this case, the effect of the background motions is completely neglected and the procedure is referred to as wavelet-based adaptive direct numerical simulation (WA-DNS), e.g., [1]. The use of relatively higher thresholds leads to the introduction of the wavelet-based adaptive large-eddy simulation (WA-LES) method that has the ability to resolve and dynamically track the most energetic coherent eddies, while modelling the significant effect of the less energetic ones. Originally developed for homogeneous isotropic turbulence, the method has been successfully extended to incompressible wall-bounded flows, e.g., [2]. Recently, the range of applicability of the above methods has been greatly extended due to the development of the adaptive-anisotropic wavelet collocation method on general curvilinear coordinate systems [3].

Traditionally, the wavelet-based adaptation is based on the analysis of instantaneous flow fields. This strategy, while being effective for homogeneous turbulence [4, 5], is not optimal for inhomogeneous flows, where the spatial variation of mean flow quantities is substantial. The situation is further complicated by the fact that the characteristic scales for fluctuating components could be substantially smaller than the scales associated with the mean flow. Thus, the main objective of this work is the preliminary development and testing of a new mesh adaptation strategy that fully exploits the spatial and temporal intermittency of turbulence. The method is capable of performing adaptive simulations of wall-bounded transitional and turbulent compressible flows on optimal computational meshes, with a more precise control of the relative resolution of mean and fluctuating components by using more physically relevant scales.

Wavelet-Based Intelligent Simulation

The key-role for wavelet-based methodologies is played by the thresholding level that controls the numerical accuracy of the solution. The built-in filtering effect of the wavelet collocation method is also exploited to determine the turbulence resolution in WA-LES, even if the superposition of an additional explicit wavelet filtering operation could be considered in order to separate numerical and filtering issues [6]. The method has been improved and made more flexible by developing the spatio-temporally variable wavelet thresholding procedure, which ensures that only a specified fraction of turbulence dissipation is actually resolved in the simulation [7]. By adopting such a strategy, the transition between the two different regimes becomes natural because WA-DNS switches to WA-LES as the amount of modelled turbulence dissipation becomes significant with respect to the resolved viscous dissipation.

For statistically steady turbulent flows, the spatial variability of a turbulent field variable is accounted for by the time average, while the fluctuation from the mean has a more or less uniform distribution. Based on this argument, rather than adapting on the instantaneous variable, the adaptation can be separately performed on mean and fluctuating components, by using two different thresholding levels. The adaptation on the time-averaged variable makes use of a smaller threshold, which ensures the adequate and accurate resolution of the mean flow structures. Conversely, the adaptation on the fluctuating component can be performed with a higher threshold, which actually reflects the WA-LES approximation.

The novel split mesh adaptation strategy is particularly useful and effective for wall-bounded flows, where, in order to resolve the coherent structures in the wall region, a relatively low thresholding level should be prescribed [2]. Without splitting the mesh adaptation, the use of a low threshold for the instantaneous field would lead to retaining an excessive number of collocation points throughout the flow domain, which substantially reduces the grid compression and highly deteriorates the efficiency of computations. On the other hand, the use of a unique relatively high threshold would result in aggressive filtering close to the wall, with the subsequent reduction of the accuracy of mean flow calculations. This unavoidably alters the turbulence production, along with introducing large numerical errors that greatly reduce the quality of the simulation. Thus, the main objective of split mesh adaptation is to achieve a more precise control of the relative resolution of fluctuating components, using more physically relevant scales based on the turbulence intensity, instead of relying on instantaneous or mean flow scales. Furthermore, the new strategy can be further extended to statistically unsteady or statistically quasi-periodic flows, where the split mesh adaptation can be based on moving time-averaged fields.

For compressible flows, the self-regulating continuous switch between the different fidelity regimes is accomplished through a two-way multi-objective feedback mechanism between the modelled sub-filter scale stress and heat flux terms and the resolution of the adapted multi-scale wavelet collocation grid. Actually, the resulting approach not only adapts the local mesh resolution in the physical domain, but also adjusts the turbulence modelling procedure. This intelligent method provides a unique systematic approach that allows for the synergistic transition among various levels of user-defined turbulence resolution. As to closure model for the compressible WA-LES regime, the minimum-dissipation model for anisotropic grids developed in [8] is used. This model appropriately switches off in flow regions where sub-filter energy is not being created.

Numerical Experiments

In this study, some preliminary experiments are conducted for the supersonic plane channel flow with isothermal walls, which represents a classical benchmark for wall-bounded turbulent compressible flows, e.g., [9]. Starting from the initial laminar flow conditions, with superimposed random disturbances in each space direction, the transitional flow in a bi-periodic channel is simulated. In order to achieve homogeneous flow conditions in the streamwise direction, appropriately, a forcing term is added to the streamwise momentum and energy equations to compensate for the effect of wall-friction forces. The ultimate goal of the calculation is the simulation of the transition to fully developed turbulent flow. The prescribed Reynolds and Mach numbers, based on bulk mass flux, bulk density, channel half height and wall temperature, are $\mathrm{Re} = 3000$ and $\mathrm{Ma} = 1.5$, respectively. The streamwise and spanwise extents of the channel are $2\pi H$ and $2\pi H/3$, respectively, where H stands for the half channel height. A suitable mapping function is employed between the computational space, where the wavelet collocation points are uniformly distributed on nested grids at seven levels of resolution, and the physical space, where the corresponding rectilinear meshes are stretched in the wall-normal direction following a hyperbolic tangent profile. At the finest level of resolution, the minimum mesh spacing in the wall-normal direction is equal to $5.7 \times 10^{-4} H$, which is close to the value adopted in the reference DNS solution [9]. Two different calculations are conducted, namely, with and without split mesh adaptation. In the first case, the levels of thresholding used for means and fluctuations are $\epsilon_1 = 2.5 \times 10^{-3}$ and $\epsilon_2 = 7.5 \times 10^{-3}$, respectively. The simulation without splitting is performed with a unique threshold equal to ϵ_1. The computational cost is much higher in the latter case, as illustrated in Fig. 37.1, where the number of retained wavelets is reported for some flow through times. However, the results are practically not affected by the use of less points, as demonstrated in Fig. 37.2, where the time histories of the wall-averaged density and the friction Reynolds number are depicted.

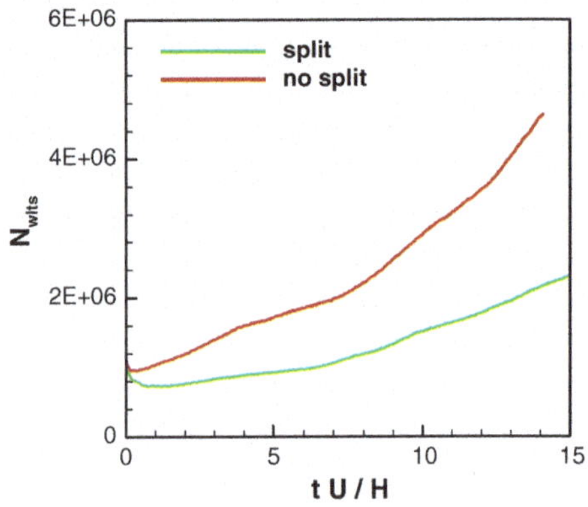

Fig. 37.1 Number of retained wavelets for simulations with and without split mesh adaptation

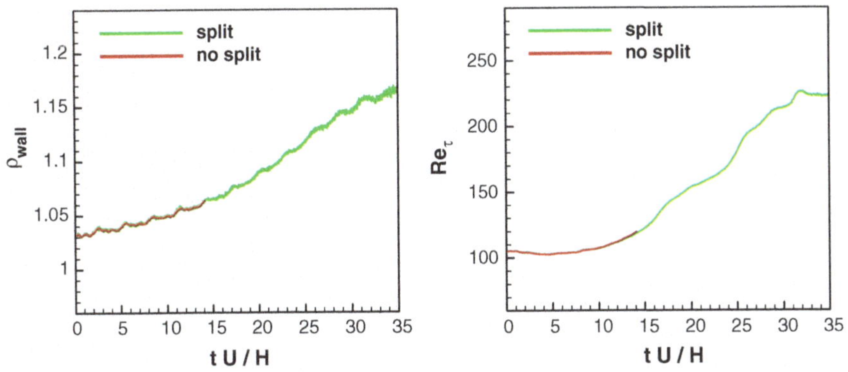

Fig. 37.2 Time histories of wall-averaged density (left) and friction Reynolds number (right) for simulations with and without split mesh adaptation

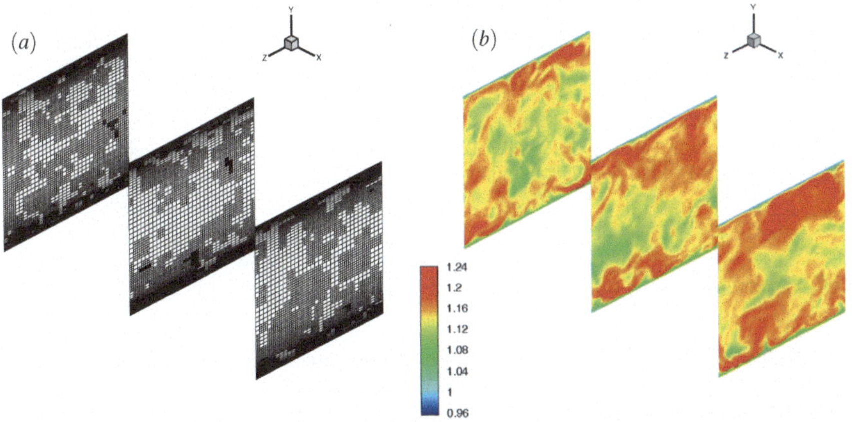

Fig. 37.3 Adapted mesh **a** and corresponding temperature contours **b** at three different channel sections along the streamwise direction

The simulation with split mesh adaptation is further conducted to simulate the transitional flow conditions. The adapted stretched mesh at a given time instant is illustrated in Fig. 37.3a, where three different channel sections along the streamwise direction are shown. The mesh is adapted to the local flow conditions as demonstrated, for instance, by looking at the corresponding temperature contours reported in Fig. 37.3b. The number of wavelets retained in the calculation is less than 2% of the total number of available wavelets corresponding to the finest grid with $384 \times 513 \times 256$ mesh points. The transition of the flow is particularly evident in the wall region, where elongated streak-like flow structures are apparently developing, as confirmed by inspection of Fig. 37.4, where the streamwise velocity contours for three sections aligned with the coordinate planes are shown.

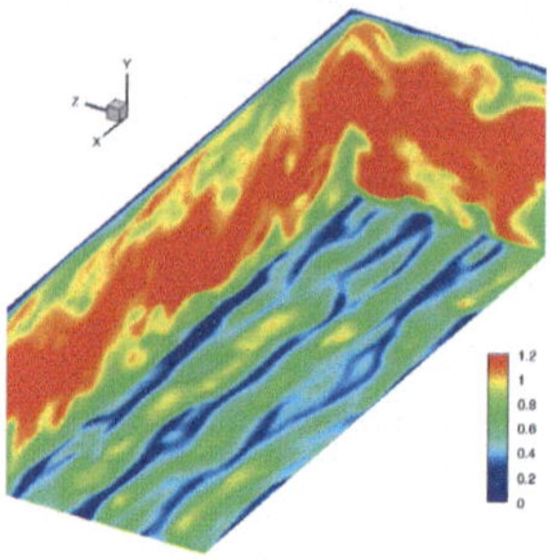

Fig. 37.4 Streamwise velocity contours for sections aligned with the coordinate planes

The new strategy is very efficient in terms of resolving coherent flow structures, while substantially reducing the number of degrees of freedom, without loss of the solution fidelity. These preliminary results are very encouraging and call for further development of wavelet-based intelligent modelling and simulation framework, capable of performing adaptive simulations of wall-bounded turbulent compressible flows.

Acknowledgements This work has been supported by the Russian Science Foundation (Project 16-11-10350). Authors are also thankful for high-performance computing resources provided by the CINECA award under the ISCRA initiative (Project IsC65).

References

1. De Stefano, G., Vasilyev, O.V.: Wavelet-based adaptive simulations of three-dimensional flow past a square cylinder. J. Fluid Mech. **748**, 433–456 (2014)
2. De Stefano, G., Nejadmalayeri, A., Vasilyev, O.V.: Wall-resolved wavelet-based adaptive large-eddy simulation of bluff-body flows with variable thresholding. J. Fluid Mech. **788**, 303–336 (2016)
3. Brown-Dymkoski, E., Vasilyev, O.V.: Adaptive-anisotropic wavelet collocation method on general curvilinear coordinate systems. J. Comput. Phys. **333**, 414–426 (2017)
4. Goldstein, D.E., Vasilyev, O.V., Kevlahan, N.K.-R.: CVS and SCALES simulation of 3-D isotropic turbulence. J. Turbul. **6**(37), 1–20 (2005)
5. De Stefano, G., Vasilyev, O.V.: Stochastic coherent adaptive large eddy simulation of forced isotropic turbulence. J. Fluid Mech. **646**, 453–470 (2010)
6. De Stefano, G., Vasilyev, O.V.: Wavelet-based adaptive large eddy simulation with explicit filtering. J. Comput. Phys. **238**, 240–254 (2013)
7. Nejadmalayeri, A., Vezolainen, A., De Stefano, G., Vasilyev, O.V.: Fully adaptive turbulence simulations based on Lagrangian spatio-temporally varying wavelet thresholding. J. Fluid. Mech. **749**, 794–817 (2014)
8. Rozema, W., Bae, H.J., Moin, P., Verstappen, R.: Minimum-dissipation models for large-eddy simulation. Phys. Fluids **27**, 085107 (2015)
9. Coleman, G.N., Kim, J., Moser, R.D.: A numerical study of turbulent supersonic isothermal-wall channel flow. J. Fluid Mech. **305**, 159–183 (1995)

Chapter 38
An Immersed Boundary Method for Moving Objects in Compressible Flows

F. De Vanna, F. Picano and E. Benini

Introduction

Numerical simulations of compressible flows around moving solids are a crucial target in many engineering applications, e.g. the aerodynamic flutter, rocket engines and landing devices. Because of the availability of a large variety of high-resolution schemes developed for Cartesian structured meshes in a finite-difference framework, the Immersed Boundary Method (IBM) appears to be one of the most efficient and reliable strategies to deal with the problem. Originally designed for biological incompressible flows, the method has been widely applied in every kind of fluid dynamic problems both laminar and turbulent, see, e.g. Iaccarino and Verzicco [1] for a review, Picano et al. [2] for a recent application to particulate flows and Boukharfane et al. [3] for compressible flows. Among the different IBM strategies [1], the most common are the Direct Forcing Method (DFM) and the Ghost Point Forcing Method (GPFM). The DFM consists in directly forcing the Navier–Stokes system of equations adding a source term that indirectly imposes the boundary conditions. In the DFM framework, while it is straightforward to impose Dirichlet boundary conditions, imposition of Neumann boundary conditions is more complex. On the other hand, the GPFM is able to deal with both kinds of boundary conditions, so the method appears more suitable in thermal and compressible flows where the Neumann boundary conditions are usually imposed. Despite its greater flexibility GPFM needs a more sophisticated data management and more coding effort. In addition, it is well known that IBM for moving objects in the compressible regime may suffer from several critical issues like spurious oscillations and instability. In this paper, we present results obtained by a recent developed solver based on GPFM in combination with a high-order method for compressible flow based on finite differences (see Pirozzoli [4]). The present method is able to accurately simulate moving objects in compressible

F. De Vanna (✉) · F. Picano · E. Benini
Department of Industrial Engineering,University of Padova,via Venezia 1,Padova, Italy
e-mail: francesco.devanna@phd.unipd.it

M. García-Villalba et al. (eds.), *Direct and Large Eddy Simulation XII*,
ERCOFTAC Series 27,
https://doi.org/10.1007/978-3-030-42822-8_38

subsonic and supersonic conditions without filtering and/or adding some artificial viscosity in order to remove numerical noise. The method has been applied to a tests campaign of both steady and moving objects in a wide range of both Reynolds and Mach number that will be presented in the following.

Governing Equations and Numerical Method

The present study was carried out using Unsteady Robust All-around Navier–Stokes Solver (URANOS) a Navier–Stokes solver recently developed at the Industrial Engineering Department of the University of Padua. The solver implements the compressible 3D Navier–Stokes equations in conservative formulation and in the hypothesis of an ideal gas. The solver relies on a co-located finite-difference conservative approach. In particular, the inviscid fluxes have been discretised using the sixth-order split-convective energy-preserving scheme proposed by Kennedy-Grüber-Pirozzoli (KGP) [4] hybridised with the fifth-order weighted essentially non-oscillatory scheme (WENO5) by Jiang and Shu [5]. The hybridisation strategy between the two schemes makes use of a Ducros sensor. The viscous terms, casted in Laplacian formulation, have been treated with a standard sixth-order central scheme. The time is advanced using the third-order Total Variation Diminishing (TVD) low-storage Runge–Kutta scheme by Gottlied and Shu [6]. The solver is fully parallelised with a hybrid MPI+OpenMP approach making use of a 3D domain decomposition.

Concerning IBM, we employed the Ghost-Point-Forcing-Method, here described in two-dimensional but general for three-dimensional objects. In order to discretise a 2D-geometry the `.msh` format has been employed describing the solid by a series of closed points and edges on a Cartesian mesh. Following O'Rourke [7] every Cartesian grid point is flagged with respect of the geometrical position, determining if lies inside/outside the polygon and the information is stored in a marker variable ζ. Once ζ is determined, a bilinear interpolation strategy was employed in order to couple the fluid and the solid domain similarly to Piquet et al. [8]. Dealing with moving objects, some points emerge in the fluid domain without previous history because they were belonging to the solid as ghost nodes, *fluid emerging points*. Being $\mathbf{x}_g(t)$ a ghost point at time level $t = t^n$ the point is flag as *fluid emerging point* if at time t^{n+1} is a fluid point. Because the fluid emerging points do not have a previous Runge–Kutta history an interpolation strategy must be carried out in order to assign them a more suitable value in respect of their neighbours. For every $\phi(\mathbf{x})$ a scalar field the smoothest Dirac delta proposed by Yang [9] was used. Another issue must be taken into account while a solid is moving in compressible flows. In particular, projecting the momentum equation along the outer normal direction of the solid boundary and dropping the viscous therms we conclude that the normal pressure gradient is coupled with the boundary acceleration via the approximated relation

$$\left(\frac{\partial p}{\partial \mathbf{n}}\right)_b \simeq -\left(\rho \frac{D\mathbf{u}}{Dt} \cdot \mathbf{n}\right)_b \tag{38.1}$$

thus, ensuring the boundary pressure gradient a correction of the equation proposed by Piquet et al. [8] must be fixed, so that

$$p_g = p_f + \Delta h \left(\frac{\partial p}{\partial \mathbf{n}}\right)_b + \mathcal{O}(\Delta h^2) \tag{38.2}$$

Results

The present section aims at validating the method in a wide range of well-documented test cases. Results regarding both steady and moving objects in compressible flow condition will be presented.

Subsonic Flow Past a Fixed Circular Cylinder

We simulated the flow past a steady cylinder at $\mathrm{Re} = u_\infty D/\nu = 10, 20, 40, 70$ at a fixed Mach number $\mathrm{M} = u_\infty/c_\infty = 0.05$. We denote with u_∞ the uniform velocity profile at the inlet, with D is the cylinder diameter and with ν the kinematic viscosity of the fluid. The computational mesh consists of a structured non-uniform grid, stretched around the position of the cylinder and featuring $N_x \times N_y = 800 \times 320 = 256000$ grid points. Grid stretching was applied along both x and y directions but, near the cylinder, a uniform grid spacing was employed ensuring 50 grid points per diameter. The CFL number was set equals to 0.5. In Fig. 38.1 the values of the drag coefficient are compared with those in literature showing a very good agreement (Fig. 38.1).

Subsonic Flow Past a Moving Circular Cylinder

To validate the IBM for moving objects we simulated a transversely oscillating cylinder in a free-stream configuration. Here the same domain configuration as the previous paragraph was employed. Following the available data in literature we simulated the flow at $\mathrm{Re} = 185$; the value of u_∞ was selected in such a way that a Mach number $\mathrm{M}_\infty = 0.25$ matching the conditions used in the computations performed by Khalili et al. [15]. The computational mesh, stretched around the cylinder features $N_x \times N_y = [800 \times 640]$ grid points granting a resolution of $\Delta x = \Delta y = [D/50]$ around the cylinder. The initial values for the non-dimensional pressure and temperature was set to $p/p_0 = T/T_0 = 1$, respectively, while the initial

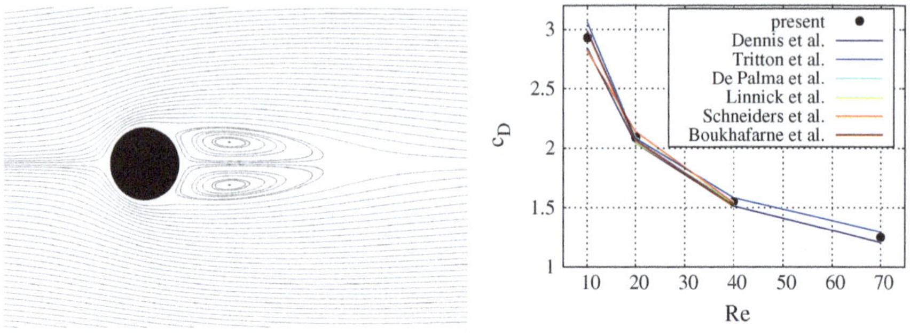

(a) (b)

Fig. 38.1 Wake recirculation **a** and drag coefficient, **b** of a steady cylinder in subsonic regime as a function of the Reynolds number with M_∞ 0.05 at various Reynolds number. The results are compared with Dennis and Chang [10], Tritton [11], De Palma et al. [12], Linnick and Fasel [13], Schneiders et al. [14] and Boukharfane et al. [3]

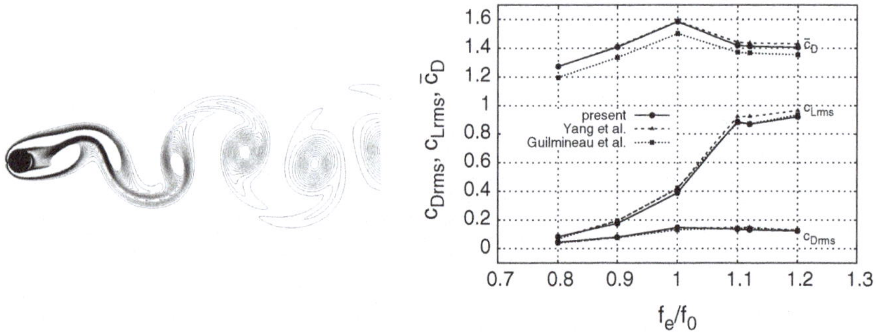

(a) (b)

Fig. 38.2 Vorticity contours **a** and force coefficients statistics, **b** for a transversely oscillating cylinder at Re = 185 and M = 0.25 as a function of the reduced frequency f_e/f_0. In the figure our results are compared with Guilmineau and Queutey [16] and Yang and Balaras [17]

velocity components was set equal to the free stream speed in all the domain. The CFL number was set equals to 0.5 and the cylinder was forced to obey to the a sinusoidal movement law $y(t) = A\sin(\omega t)$. Here $A = 0.2D$ and $\omega = 2\pi f_e$. f_e was expressed as a function of the cylinder natural shedding frequency $f_0 = St_0 D/u_\infty$ in the same steady conditions and a spectrum of six frequencies was considered, i.e. $f_e/f_0 = \{0.80, 0.90, 1.00, 1.10, 1.12, 1.20\}$. Previous simulations suggest $St_0 \simeq 0.195$ [15]. In Fig. 38.2 we present results concerning the force coefficient statistics in term of mean drag, root mean square of the drag, and root mean square of the lift as a function of the reduced frequency f_e/f_0.

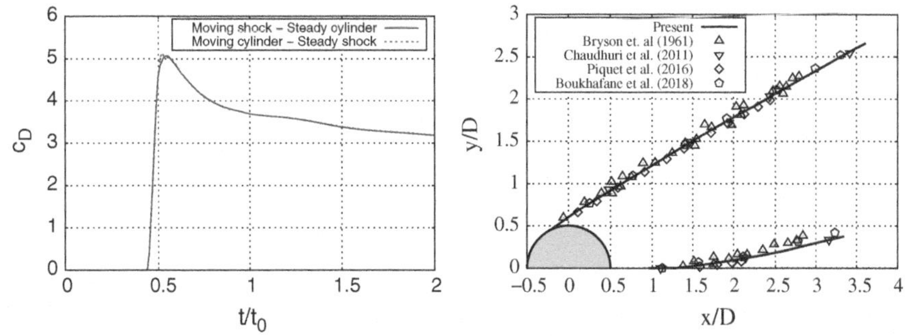

(a) (b)

Fig. 38.3 Drag coefficient history **a** and triple points trajectory, **b** of a shock wave diffraction against a cylinder in two reference frames. For data comparison see Boukharfane et al. [3]

Moving Cylinder in Supersonic Flow Conditions

As we mentioned in the introduction our method was especially developed to simulate moving objects also in supersonic compressible flow conditions. To the best of author knowledge, in this field it does not exist a well known benchmark test available in literature. Thus we decided to repeat the same physical case—consisting in the shock wave diffraction against a cylinder—in two different reference frames. In the primary, we have a fixed cylinder and a moving shock wave, while in the secondary reference frame the unperturbed shock is fixed, while the cylinder is moving. As far as the first reference frame is concerned, abundant literature data are available for the case at $\mathrm{Ma}_s = 2.81$ and $\mathrm{Re}_s = 1000$. Here the dimensionless groups are referred to the shock (the cylinder) speed. Extrapolation boundary condition was ensured in all the four edges of the domain. A uniform Cartesian mesh featuring $N_x \times N_y = 700 \times 400$ grid points discretising a domain of $L_x \times L_y = 14 \times 8$ has been used. The flow was initialised placing a shock wave two diameters upstream the cylinder and at the desired Mach equals to Ma_s. The time is advanced up to $t/t_0 = 2$ and the CFL number was set equal to 0.5. In Fig. 38.3 the results in terms of drag coefficient history and triple points trajectory are reported comparing the two reference frames.

Final Remarks

In the present manuscript, a method for moving objects in compressible flows has been presented and validated. The method combines high-order high-resolution numerical schemes available in literature with an improved immersed boundary technique granting a good prediction of the dynamics of moving objects in compressible conditions. The method has been validated showing a good agreement against well-documented test cases at different Mach and Reynolds numbers both in fixed and moving conditions.

References

1. Iaccarino, G., Verzicco, R.: Immersed boundary technique for turbulent flow simulations. Appl. Mech. Rev. (2003)
2. Picano, F., Breugem, W.P., Brandt, L.: Turbulent channel flow of dense suspensions of neutrally buoyant spheres. J. Fluid Mech. **764**, 463–487 (2015)
3. Boukharfane, R., Ribeiro, F.H.E., Bouali, Z., Mura, A.: A combined ghost-point-forcing/direct-forcing immersed boundary method (IBM) for compressible flow simulations. Comput. Fluids **162**, 91–112 (2018)
4. Pirozzoli, S.: Numerical methods for high-speed flows. Annu. Rev. Fluid Mech. **43**(1), 163–194 (2011)
5. Jiang, G.S., Shu, C.W.: Efficient implementation of weighted WENO schemes. J. Comput. Phys. **126**(126), 202–228 (1996)
6. Gottlieb, S., Shu, C.-W.: Total variation diminishing Runge-Kutta schemes. Math. Comput. Am. Math. Soc. **67**(221), 73–85 (1998)
7. O'Rourke, J.: Computational Geometry in C., vol. 64. Cambridge University Press (1998)
8. Piquet, A., Roussel, O., Hadjadj, A.: A comparative study of Brinkman penalization and direct-forcing immersed boundary methods for compressible viscous flows. Comput. Fluids **136**, 272–284 (2016)
9. Yang, X., Zhang, X., Li, Z., He, G.W.: A smoothing technique for discrete delta functions with application to immersed boundary method in moving boundary simulations. J. Comput. Phys. **228**(20), 7821–7836 (2009)
10. Dennis, S.C.R., Chang, G.-Z.: Numerical solutions for steady flow past a circular cylinder at Reynolds numbers up to 100. J. Fluid Mech. **42**, 471–489 (1970). Jul
11. Tritton, D.J.: Experiments on the flow past a circular cylinder at low Reynolds numbers. J. Fluid Mech. **6**(04), 547 (2006)
12. De Palma, P., de Tullio, M.D., Pascazio, G., Napolitano, M.: An immersed-boundary method for compressible viscous flows. Comput. Fluids **35**(7), 693–702 (2006)
13. Linnick, M., Fasel, H.: A high-order immersed boundary method for unsteady incompressible flow calculations, pp. 1–17 (2012)
14. Schneiders, L., Hartmann, D., Meinke, M., Schröder, W.: An accurate moving boundary formulation in cut-cell methods. J. Comput. Phys. **235**, 786–809 (2013)
15. Khalili, M.E., Larsson, M., Müller, B.: Immersed boundary method for viscous compressible flows around moving bodies. Comput. Fluids **170**, 77–92 (2018)
16. Guilmineau, E., Queutey, P.: A numerical simulation of vortex shedding from an oscillating circular cylinder. J. Fluids Struct. **16**(6), 773–794 (2002)
17. Yang, J., Balaras, E.: An embedded-boundary formulation for large-eddy simulation of turbulent flows interacting with moving boundaries. J. Comput. Phys. **215**(1), 12–40 (2006)

Chapter 39
Study of the Efficiency of Flapping and Heaving Wings in Tandem Configuration

G. Arranz, O. Flores and M. García-Villalba

Introduction

Contrary to birds, bats and most insects, dragonflies have two pairs of wings arranged in a tandem configuration which can move independently [1, 2]. By modifying the kinematics and the phase relationship between the fore and hind pair of wings, dragonflies are capable of performing rapid manoeuvres and cruising efficiently [2].

Several studies have analysed the kinematics of tandem flight that provides an optimal configuration in terms of propulsive efficiency and/or thrust [3–5]. However, the augmented parametric space (compared to the flight of a single pair of wings) restricts these studies to two dimensional (2D) configurations, in which the problem is simplified to that of two airfoils performing a heaving and/or pitching motion. Consequently, the applicability of such optimal kinematics is not straightforward in a three-dimensional (3D) configuration, since the wing shape and actual motion of the wing (dragonflies' wings do not heave but flap about an axis) are not defined in a 2D configuration.

Therefore, the aim of this work is to investigate the changes in the efficiency when a 2D optimal kinematics is used in wings of finite span. In particular, the effect of the wings' aspect ratio and the kinematics will be studied.

G. Arranz (✉) · O. Flores · M. García-Villalba
Universidad Carlos III de Madrid, Getafe, Spain
e-mail: garranz@ing.uc3m.es

M. García-Villalba et al. (eds.), *Direct and Large Eddy Simulation XII*,
ERCOFTAC Series 27,
https://doi.org/10.1007/978-3-030-42822-8_39

Methodology

We consider two identical wings arranged in a tandem configuration immersed in a constant, uniform free-stream of speed U. The wings are flat and have a rectangular planform of chord c and span b, with a constant thickness equal to $e = 0.01c$. Two different aspect ratios, $AR = b/c = [2, 4]$, are considered.

The motion of the wings is based on a 2D optimal kinematics which combines heaving and pitching about the mid chord [6] (see Fig. 39.1). These motions are assumed to be sinusoidal, namely,

$$h_i(t) = h_0 \cos(2\pi f t + \phi_i), \tag{39.1}$$

$$\theta_i(t) = \theta_0 \cos(2\pi f t + \xi_i), \tag{39.2}$$

where h_0 is the heaving amplitude, θ_0 is the flapping amplitude, f is the frequency, and ϕ_i and ξ_i are the heaving and pitching phase lags, respectively. In Eqs. (39.1) and (39.2), the subscript i refers to the fore-airfoil ($i = f$) or the hind-airfoil ($i = h$). In particular, in the present configuration, the fore- and the hind-airfoils are in counterphase, thus, $\phi_f = 0$ and $\phi_h = \pi$. The pitching phase lags are $\xi_f = 3\pi/4$ and $\xi_h = -\pi/4$, so both wings have and advanced pitching motion.

The spacing between the wings (distance from the trailing edge of the fore wing to the leading edge of the hind wing when the wings lay onto the horizontal plane) is constant for all the cases and equal to $S = 0.5c$. The maximum pitching amplitude is $\theta_0 = 25°$, the reduced frequency is $k = \pi f c/U \approx 2.2$ and the Strouhal number is $St = 2h_0 f/c = 0.34$.

We consider two kinds of motion for the wings: heaving (all chordwise sections of a wing move with the same velocity); and flapping (rotation of the wing about a fixed axis) about an axis parallel to the free-stream and located at a distance c from the inboard wing tips. For the latter, we choose the flapping amplitude so that the vertical motion of the section at $0.65b$ from the inboard wing tip equals the heaving motion of the 2D optimal configuration (see Fig. 39.2).

Finally, a Reynolds number based on c and U equal to Re = 1000 is considered. Consequently, the present study consists of a total of four simulations (i.e. one Reynolds number, two different ARs and heaving/flapping motions).

To perform the simulations, an in-house code, TUCAN, is used. TUCAN solves the Navier–Stokes equations of the incompressible flow using a fractional step method on a staggered grid. The wings are modelled using the immersed boundary

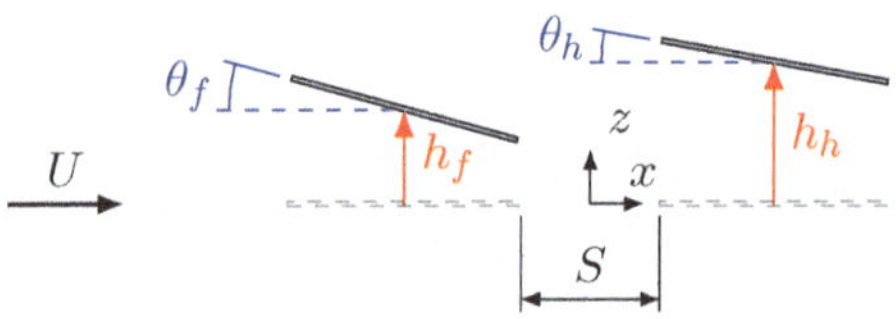

Fig. 39.1 Sketch of the 2D reference motion. θ_f and θ_h are the pitching angle of the fore- and hind wing, respectively

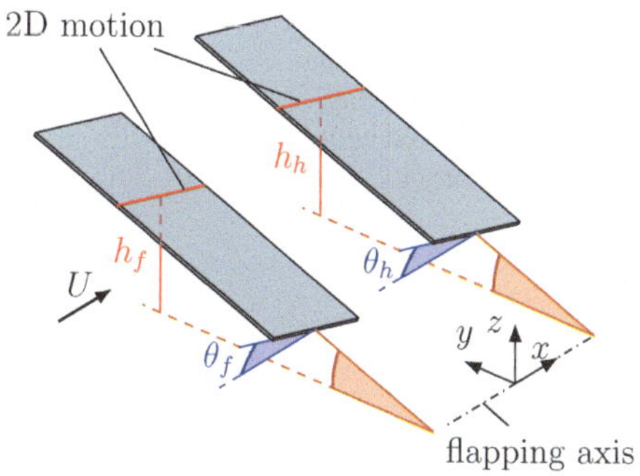

Fig. 39.2 Sketch of flapping motion. The chordwise section highlighted in red corresponds to the wing section that moves with the 2D optimal kinematics

method proposed by Uhlmann [7]. The code TUCAN has been extensively validated and employed for 2D and 3D simulations [8–10].

The fluid domain is a rectangular prism whose dimensions depend on the aspect ratio and the motion of the wings. The largest domain ($AR = 4$ in heaving motion) is $14c \times 13c \times 7.6c$. A uniform mesh is used in the region containing the wings, with a resolution of $96\,c/\Delta x$, where Δx is the cell size. Outside this region, the grid is stretched with a constant stretching factor of 1%.

Results

First, the effect of the aspect ratio is analysed. To that end, the 2D reference configuration is compared to the configuration of the wings in heaving motion for $AR = 2$ and 4. Figure 39.3 depicts the flow of the 2D configuration. It is observed that the fore-foil sheds a trailing edge vortex (TV) during the upstroke whose shear layer induces a leading edge vortex (iLEV) on the hind-foil. This iLEV and the TV shed by the fore-foil during the next downstroke interacts forming a dipole. The same interaction is observed, qualitatively, on wings of finite aspect ratio. However, the formation of the iLEV in the 3D configuration is more complex due to the interaction of the hind wing with the wing tip vortices (WTVs) shed by the fore wing, as shown in

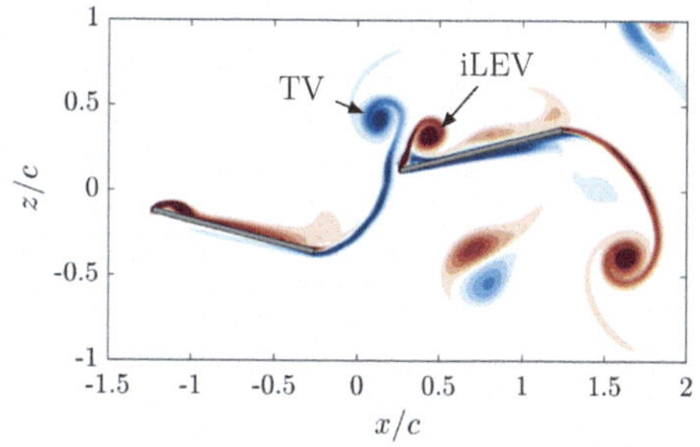

Fig. 39.3 Spanwise vorticity, $\omega_y c/U$, of the 2D, reference configuration. Contour limits range from −50 to 50

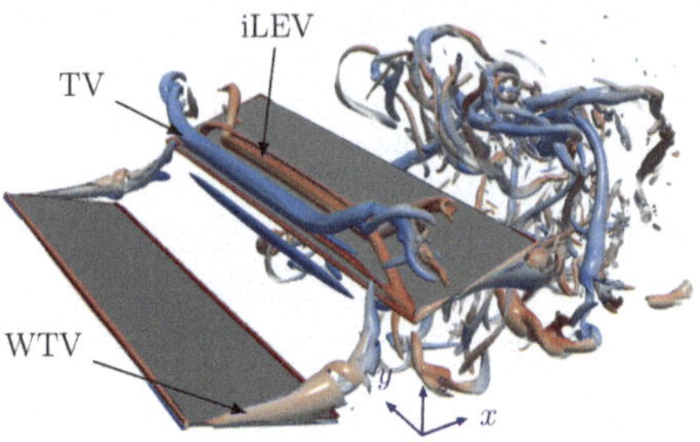

Fig. 39.4 Flow visualisation using the Q-criterion of the $AR = 4$ wings in heaving. Vortices are coloured with the spanwise vorticity, as in Fig. 39.3

Fig. 39.4 for $AR = 4$ (note that, although not shown, the flow is qualitatively similar for $AR = 2$). In particular, the interaction of the TV with the iLEV is not uniform all over the span of the hind wing, namely, both structures break at the tips and this vortex breakdown progresses towards the mid-span.

Regarding the effect of AR on the aerodynamic forces, Fig. 39.5 displays the thrust developed by the fore- and hind wing over a cycle for the 2D case and for both AR wings in heaving motion. Note that, the thrust is normalised with the free-stream dynamic pressure ($0.5\rho U^2$) times the wing surface (the chord, for the 2D case). It is observed that the same thrust is developed during the upstroke and downstroke of the finite wings, but the same is not true for the 2D case. This is due to a deflection of the 2D wake, which is not present for the 3D cases under consideration. A similar observation was made by Dong et al. [11] for single plunging and pitching wings.

Overall, it can be observed in Fig. 39.5 that the thrust increases (for both wings) as AR increases, being the largest for the 2D case. An analysis of the forces per chordwise section for the finite wings (not shown) reveals that sectional forces are the same for both AR near the tips (specifically for distances <0.75c from the tips). On the contrary, at mid-span, forces are larger for $AR = 4$ than for $AR = 2$ wings. This suggests that wing tip effects (i.e. downwash induced velocity) decrease the forces on both the fore- and hind wing. A more detailed explanation can be found in [12].

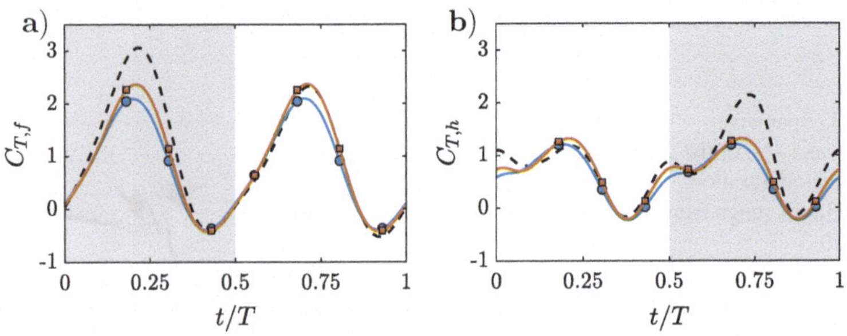

Fig. 39.5 Temporal evolution over a cycle ($T = 1/f$) of the thrust coefficient. **a** Fore wing; and **b** hind wing. (dashed line) 2D; (blue circles) $AR = 2$; and (orange squares) $AR = 4$. Shaded region correspond to the downstroke of each wing

Table 39.1 Average thrust, $\langle C_T \rangle$, and propulsive efficiency, η_p, for all the configurations under study

AR	2D	Heaving		Flapping	
	–	2	4	2	4
$\langle C_T \rangle$	0.92	0.69	0.78	0.54	0.58
η_p	0.23	0.23	0.23	0.21	0.20

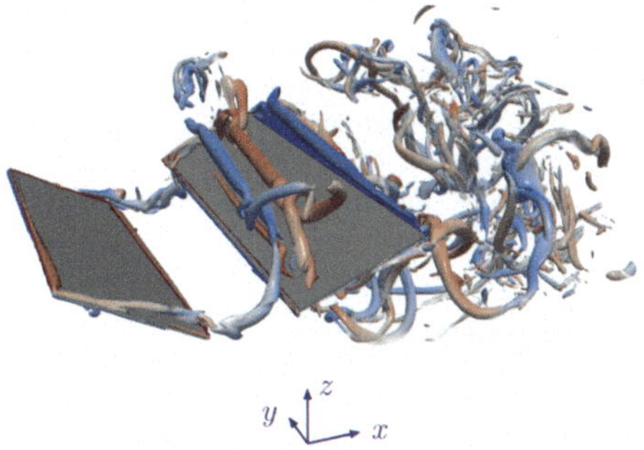

Fig. 39.6 Flow visualisation using the Q-criterion of the $AR = 4$ wings in flapping. Vortices are coloured with the spanwise vorticity, as in Fig. 39.3

Despite the aforementioned thrust reduction with decreasing AR, it is interesting to note that the global propulsive efficiency (η_p) remains constant for both AR and equal to the 2D case (see Table 39.1). Note that this entails that the power required decreases accordingly with the developed thrust.

From a mechanical perspective, heaving is not a realistic configuration. Instead, dragonfly and bio-inspired flapping micro air-vehicles flap their wings with respect to a fixed axis. Thus, the case in which the wing is in flapping motion (as described in section "Methodology") is compared to that in heaving motion. Figure 39.6 displays the vortical structures developed by the $AR = 4$ in flapping motion. Qualitatively, these structures are similar to that during heaving (Fig. 39.4). However, it is observed that the vortical structures are no longer symmetric with respect to the mid-span as in the heaving case.

From Table 39.1 it can be observed that the mean thrust decreases, for a given AR, from heaving to flapping. This behaviour is analogous to the one observed in [10] for a single pair of flapping wings. An explanation can be found by computing the sectional thrust coefficient (C_T') at different chordwise sections along the span. Figure 39.7 shows that, at the flapping section whose heaving amplitude is the same as the 2D case (located at $0.65b$ from the inboard wing tip), the sectional thrust is the same than in the same section in heaving. However, at inboard sections, the sectional force largely decreases (for both wings), leading to a decrease of the mean thrust. This poorer performance of inboard sections has been found to be linked to a shedding vorticity due to larger local angles of attack [12]. It should be noted that, as a consequence of this detriment in the performance, the propulsive efficiency decreases with respect to the heaving cases (Table 39.1).

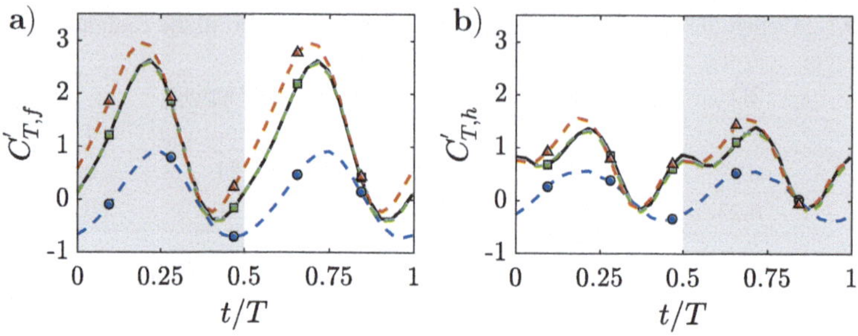

Fig. 39.7 Temporal evolution of the sectional thrust coefficient at different chordwise sections. (solid black) $0.65b$ section for heaving; (green squares) $0.65b$ section for flapping; (orange triangles) outboard wing section; and (blue circles) inner wing section for flapping. **a** fore wing; and **b** hind wing. Shaded region correspond to the downstroke of each wing

Conclusion

Direct numerical simulations of tandem wings undergoing a 2D optimal motion are performed. The objective of the study is to assess the effect of the aspect ratio and the kinematics (i.e. heaving vs. flapping) on the propulsive efficiency of tandem wings. The results showed that, when the wings are in heaving motion, the aerodynamic forces decrease with decreasing aspect ratio due to wing tip effects. Despite this, the propulsive efficiency remains constant and similar to the one obtained in the 2D case. On the contrary, when the wings are in flapping, the aerodynamic forces decrease when compared to the same wings in heaving. Additionally, the propulsive efficiency also decreases. It has been observed that this is due to the inboard sections of the wings, which undergo a sub-optimal motion.

References

1. Norberg, R.: Hovering flight of the dragonfly *Aeschna juncea L.*, kinematics and aerodynamics. In: Swimming and Flying in Nature, pp. 763–781 (1975)
2. Thomas, A., Taylor, G., Srygley, R., Nudds, R., Bomphrey, R.: Dragonfly flight: free-flight and tethered flow visualizations reveal a diverse array of unsteady lift-generating mechanisms, controlled primarily via angle of attack. J. Exp. Biol. **207**(24), 4299–4323 (2004)
3. Boschitsch, B., Dewey, P., Smits, A.: Propulsive performance of unsteady tandem hydrofoils in an in-line configuration. Phys. Fluids **26**(5), 051901 (2014)
4. Muscutt, L., Weymouth, G., Ganapathisubramani, B.: Performance augmentation mechanism of in-line tandem flapping foils. J. Fluid Mech. **827** (2017)
5. Kurt, M., Moored, K.: Flow interactions of two-and three-dimensional networked bio-inspired control elements in an in-line arrangement. Bioinspir. Biomim. **13**(6) (2018)
6. Ortega-Casanova, J., Fernández-Feria, R.: Analysis of the aerodynamic interaction between two plunging plates in tandem at low Reynolds number for maximum propulsive efficiency. J. Fluids Struct. **63**, 351–373 (2016)

7. Uhlmann, M.: An immersed boundary method with direct forcing for the simulation of particulate flows. J. Comput. Phys. **209**(2), 448–476 (2005)
8. Moriche, M., Flores, O., García-Villalba, M.: Three dimensional instabilities in the wake of a flapping wing at low Reynolds number. Int. J. Heat Fluid Flow **62**, 44–55 (2016)
9. Moriche, M., Flores, O., García-Villalba, M.: On the aerodynamic forces on heaving and pitching airfoils at low Reynolds number. J. Fluid Mech. **828**, 395–423 (2017)
10. Gonzalo, A., Arranz, G., Moriche, M., García-Villalba, M., Flores, O.: From flapping to heaving: a numerical study of wings in forward flight. J. Fluids Struct. **83**, 293–309 (2018)
11. Dong, H., Mittal, R., Najjar, F.M.: Wake topology and hydrodynamic performance of low-aspect-ratio flapping foils. J. Fluid Mech. **566**, 309–343 (2006)
12. Arranz, G., Flores, O., García-Villalba, M.: Three-dimensional effects on the aerodynamic performance of flapping wings in tandem configuration. J. Fluid Struct. **94**, 102893 (2020)

Chapter 40
Effects of Different Friction Control Techniques on Turbulence Developing Around Wings

M. Atzori, R. Vinuesa, D. Gatti, A. Stroh, B. Frohnapfel and P. Schlatter

Introduction

Developing efficient flow control techniques remain a challenging task due to the complexity of turbulent flows in industrial applications, a relevant example of which are turbulent boundary layers (TBL) subjected to pressure gradients. In the present study, we employ high-fidelity numerical simulations to assess the impact of different control strategies on the flow around a NACA4412 airfoil at a Reynolds number $Re_c = 200,000$ based on the chord length c and the inflow velocity U_∞. The choice of this specific study case is motivated by the relatively weak dependence of the pressure distribution around the airfoil on the Reynolds number [6], which allows distinguishing the effects of increasing Reynolds number and those of the non-uniform adverse pressure gradient (APG).

M. Atzori (✉) · R. Vinuesa · P. Schlatter
Linné FLOW Centre, KTH Mechanics and Swedish e-Science Research Centre (SeRC), Stockholm, Sweden
e-mail: atzori@mech.kth.se

R. Vinuesa
e-mail: rvinuesa@mech.kth.se

P. Schlatter
e-mail: pschlatt@mech.kth.se

D. Gatti · A. Stroh · B. Frohnapfel
Institute of Fluid Mechanics, Karlsruhe Institute of Technology (KIT), Karlsruhe, Germany
e-mail: davide.gatti@kit.edu

A. Stroh
e-mail: alexander.stroh@kit.edu

B. Frohnapfel
e-mail: bettina.frohnapfel@kit.edu

M. García-Villalba et al. (eds.), *Direct and Large Eddy Simulation XII*, ERCOFTAC Series 27,
https://doi.org/10.1007/978-3-030-42822-8_40

We consider three different control strategies: uniform blowing, uniform suction and body-force damping. The effectiveness of uniform blowing in reducing the skin friction in canonical cases, such as channel flow and developing TBL, has been widely documented [5]. Uniform suction, which is classically used for laminar flow control [1], increases the skin-friction drag in turbulent flows, but it is examined in the present case with turbulent boundary layers because it reduces the boundary-layer thickness and thereby potentially the pressure drag on the airfoil. Body-force damping is employed as a model for all the control strategies which aim at the reduction of the near-wall turbulent activity [7].

Note that we employ for the first time high-fidelity large-scale simulations to study friction control on finite aerodynamic bodies. The scope of the project is twofold: on the one hand, we assess how they affect the aerodynamic properties of the entire airfoil. On the other hand, we aim at characterising in detail how the different control strategies interact with the non-uniform adverse pressure gradient.

Methodology

We perform well-resolved large-eddy simulations (LES) with the code *Nek5000* [4], which integrates in time the incompressible Navier–Stokes equations using a spectral-element method (SEM). The computational domain is periodic in the z-direction and its extension is $L_x = 6c$, $L_y = 2c$ and $L_z = 0.2c$ (where x and y are the horizontal and vertical directions, respectively). A portion of the computational domain and vortex clusters detected with the λ_2 criterion are shown in Fig. 40.1.

Within the boundary layer, the grid spacing in viscous units in the tangential, wall-normal and spanwise directions are $\Delta x_t^+ = 18$, $\Delta y_n^+ = (0.64, 11)$ and $\Delta z^+ = 9$, respectively. The viscous length is $l^* = \nu/u_\tau$, where ν is the kinematic viscosity,

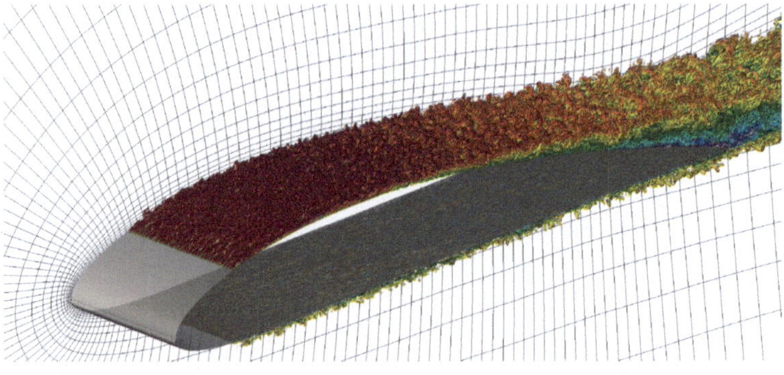

Fig. 40.1 Spectral elements in a portion of the computational domain and vortex clusters for the reference simulation at $\mathrm{Re}_c = 200{,}000$, coloured with the streamwise velocity: dark red is $1.6U_\infty$ and dark blue $-0.18U_\infty$

$u_\tau = \sqrt{\tau_w/\rho}$ the friction velocity, $\tau_w = \rho\nu(\mathrm{d}U_t/\mathrm{d}y_n)_{y_n=0}$ the wall-shear stress and ρ the fluid density. The LES subgrid-scale model is based on the approach introduced in [8] and the flow is tripped at $x/c = 0.1$. This resolution is obtained by employing about 125,000 spectral elements and the 11th polynomial order, leading to a total of 216 million grid points, and each of the cases discussed here required approximately 1.5 million CPU-hours in order to reach a proper statistical convergence for the physical quantities considered here. A more detailed description of the numerical setup is provided in [10].

The uniform blowing/suction is implemented as a boundary condition at the wall between approximately $0.25 < x/c < 0.86$, with the spirit of studying the best-case scenario: it does neither take into account the limitations of the actuators nor the mass source/sink required to provide/remove the fluid. For both blowing and suction, we consider two intensities: 0.1% and 0.2% of the inflow velocity U_∞.

The body-force damping, as in [7], is designed to suppress the component of the velocity normal to the airfoil surface in the same streamwise region where blowing and suction are applied, and in the wall-normal region below $y_n^+ = 20$, which is estimated using the local mean u_τ of the uncontrolled case. It is implemented as a volume force with expression $g = -\alpha(v_{n,x}, v_{n,y}, 0)$, where that $v_{n,x}$ and $v_{n,y}$ are the projections on the horizontal and vertical directions, respectively, of the instantaneous wall-normal component of the velocity.

In order to compare body-force and blowing effects at similar conditions, we calibrated the body force so that its integrated effect on the skin friction is similar to that of uniform blowing with intensity $0.1\%\,U_\infty$. This requires to prescribe an appropriate value of α, which was obtained via a parametric analysis using simulations with a shorter averaging time, sufficient for obtaining an approximated value of the skin-friction coefficient integrated over the suction side of the airfoil, C_f. Note that the local skin-friction coefficient is defined as $c_f = 2\tau_w/(\rho U_e^2)$, where U_e is the mean velocity at the boundary-layer edge (U_e is evaluated using the diagnostic method [11]), while C_f is the integrated value of the wall-shear stress projected along the direction of the incoming flow, normalised with the dynamic pressure of the incoming flow $\frac{1}{2}\rho U_\infty^2$. Figure 40.2 (left) shows c_f for the simulations with short time average performed to calibrate α. The effects of the body-force damping are non-uniform over the control region, as discussed in the next section, and not linear in α. This is made evident by comparing the effects of increasing α over the integrated C_f on the suction side, denoted $C_{f,s}$. For the uncontrolled cases, $C_{f,s} = 7.5 \times 10^{-3}$. For the values of $\alpha = 5, 25$ and 50, the corresponding values of $C_{f,s}$ are 7.4×10^{-3}, 7.0×10^{-3} and 6.7×10^{-3}, respectively. For much higher values, such as $\alpha = 500$, the skin friction is significantly lower than in the reference case ($C_{f,s} = 5.4 \times 10^{-3}$) and it is possible to observe a region of mean separation in the proximity of the trailing edge. These facts required a relatively high number of simulations in the calibration procedure to obtain a good matching with the uniform blowing with intensity $0.1\%U_\infty$. At the selected value of $\alpha = 32.5$, $C_{f,s} = 6.94 \times 10^{-3}$ for the body force and $C_{f,s} = 6.89 \times 10^{-3}$ for uniform blowing, which we considered sufficient for the scope of the present work.

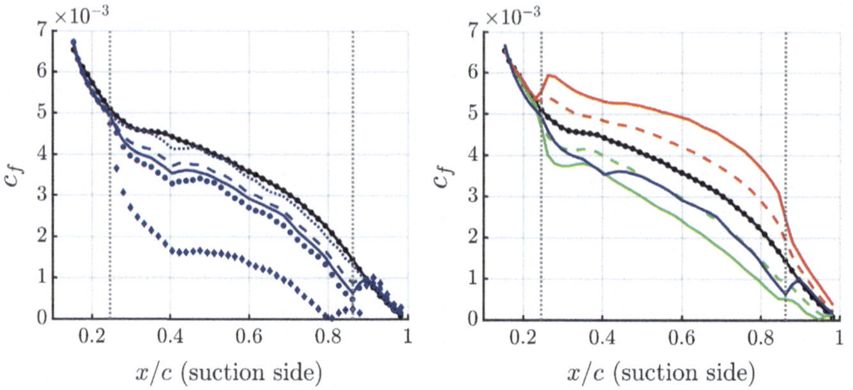

Fig. 40.2 Streamwise evolution of the local skin-friction coefficient c_f of the reference case compared with (left) the short-averaged cases employed to calibrate the body-force damping and (right) the long-averaged cases employed to study the control effects. The black solid line with symbols is the result for the suction side of the reference case. The body-force damping for different values of α is represented with blue lines. The dotted, dashed, solid, circles and diamonds lines are for $\alpha = 5, 25, 32.5, 50$ and 500, respectively. Green and red lines are the results of uniform blowing and suction, respectively, the solid lines for intensity $0.2\%U_\infty$ and the dashed lines for $0.1\%U_\infty$. The vertical dotted lines denote the streamwise extension of the control region, $0.25 < x/c < 0.86$

Results and Discussion

Figure 40.2 (right) shows c_f for all the cases considered here to study the interaction between the control strategies and the TBL. It is possible to observe that the effects of uniform blowing and uniform suction, which are, respectively, to decrease and increase the skin friction, are stronger within the control region but they persist downstream as well. On the other hand, the TBL returns to its natural state almost immediately as the body-force damping ceases, in agreement with the findings in [7]. It is interesting to note that the effect of the body-force damping exhibits a non-trivial trend, being more significant for $x/c \lesssim 0.45$, i.e. before the region of strong APG where the Clauser pressure-gradient parameter [2] β is larger than 1, and for $x/c \gtrsim 0.80$, where the APG is even stronger ($\beta > 6$). In particular, at $x/c = 0.4$, where the local skin-friction reduction achieved for $\alpha = 32.5$ is the highest, it is twice as that of uniform blowing with intensity $0.1\%U_\infty$ and it matches that of the blowing with intensity $0.2\%U_\infty$. Similarly, at $x/c = 0.86$, which is the location of highest β and lower c_f within the control region, the absolute skin-friction reduction of the uniform blowing is 60% of that of the body force. As a consequence of this trend, the agreement between body-force and uniform blowing could be achieved only in an integrated sense with the present formulation of the body-force damping, while to reach a local agreement would require to prescribe α as a function of x/c.

The considered control techniques have an impact not solely on the skin friction but on the thickness of the boundary layer as well, which can be evaluated, for

Table 40.1 Lift (C_l) and drag (C_d) coefficients, the latter decomposed into skin-friction (C_f) and pressure (C_p) contributions, and aerodynamic efficiencies for the considered cases. The values in coloured bold font are the relative differences compared with the reference case

Case	C_l	C_f	C_p	$C_d = C_f + C_p$	C_l/C_d
Ref.	0.871	0.0125	0.0071	0.0196	44
B. 1	0.842 (**−3.4%**)	0.0120 (**−4.4%**)	0.0082 (**+16%**)	0.0202 (**+3.2%**)	42 (**−6.3%**)
B. 2	0.816 (**−6.3%**)	0.0114 (**−8.4%**)	0.0091 (**+28%**)	0.0205 (**+4.8%**)	40 (**−11%**)
S. 1	0.894 (**+2.6%**)	0.0131 (**+4.6%**)	0.0063 (**−10%**)	0.0194 (**−0.8%**)	46 (**+2.3%**)
S. 2	0.912 (**+4.6%**)	0.0137 (**+9.8%**)	0.0058 (**−18%**)	0.0195 (**−0.4%**)	47 (**+4.6%**)
B. F.	0.878 (**+0.8%**)	0.0118 (**−5.2%**)	0.0078 (**+9.2%**)	0.0196 (**0%**)	45 (**+0.8%**)

instance, by considering the streamwise development of the momentum thickness θ (not shown here). Note that the momentum thickness θ is computed by integrating along the wall-normal direction y_n up to δ_{99}. Blowing and suction increases and decreases θ, respectively, and, as for the modification of the skin friction, this effect is also observed downstream the control region, eventually affecting the development of the wake. The body-force slightly reduces θ, but such reduction is relatively small compared with the corresponding reduction of the skin friction.

The integrated effects of the control on the lift and drag coefficients, defined as $C_l = 2f_l/(\rho U_\infty^2 c)$ and $C_d = 2f_d/(\rho U_\infty^2 c)$ (where f_l and f_d are the lift and drag forces per spanwise length-units, respectively) are reported in Table 40.1, together with the aerodynamics efficiency C_l/C_d. The drag coefficient is decomposed into the skin-friction contribution, C_f, and the pressure contribution, denoted C_p, which is the value of the pressure force projected along the direction of the incoming flow.

The comparison reveals that, at the considered Reynolds number, uniform blowing over the suction side is actually detrimental, which is due to the increase of the pressure drag and reduction of lift. On the other hand, uniform suction is beneficial for the opposite reason: it increases the skin friction but reduces the pressure drag and increases lift. Interestingly, the body-force damping reduces the skin friction as the uniform blowing, but has an almost equal and negative effect on the pressure drag, resulting in an almost negligible effect on the total C_d. However, it also has a positive effect on lift, which results in a mild improvement of the C_l/C_d.

To better characterise the effects of the control on the properties of the TBL, we examine the inner-scaled tangential and wall-normal mean velocity profiles as functions of the inner-scaled wall-normal distance from the wall, and we perform a comparison with the same quantities for zero-pressure gradient (ZPG) TBL at similar friction Reynolds number [3]. We consider in particular the profiles at $x/c = 0.8$ (Fig. 40.3). At this locations $\beta = 6.6$, which represents a relatively strong APG, and local effect of the body-force damping is the same as that of uniform blowing with intensity $0.1\% U_\infty$.

In the uncontrolled case, the tangential component of the velocity over the suction side is higher in the wake region than for the ZPG reference because of the stronger wall-normal convection associated with APG. From this perspective, the

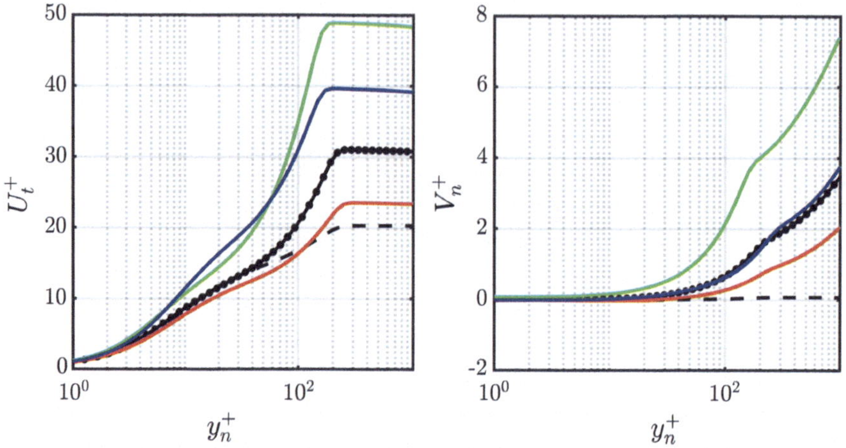

Fig. 40.3 Inner-scaled mean profile of (left) tangential and (right) wall-normal velocity at $x/c = 0.8$. Colour code as in Fig. 40.2. The dashed black lines are data from zero-pressure gradient TBL reported as Ref. [3]

effects of uniform blowing and uniform suction are similar to those of an adverse and a favourable pressure gradient, respectively. Uniform blowing promotes the wall-normal convection and it also results in higher inner-scaled velocity in the wake region, whilst the opposite behaviour is observed for uniform suction. On the other hand, the application of the body-force damping suppresses the turbulent activity in the near-wall region and results in a structural change in the profile of the tangential velocity component, which deviates from the canonical law of the wall. The body force has a comparably smaller effect on the wall-normal convection.

Acknowledgements This study was funded by the Swedish Foundation for Strategic Research, project "In-Situ Big Data Analysis for Flow and Climate Simulations" (Ref. number BD15-0082), by the Knut and Alice Wallenberg Foundation and by the Swedish Research Council (VR). The simulations were performed on resources provided by the Swedish National Infrastructure for Computing (SNIC) and within the project CWING on the national supercomputer Cray XC40 Hazel Hen at the High Performance Computing Center Stuttgart (HLRS).

References

1. Beck, N., Landa, T., Seitz, A., Boermans, L., Yaolong, L., Radespiel, R.: Drag reduction by laminar flow control. Energies **11**, 252 (2018)
2. Clauser, F.H.: Turbulent boundary layers in adverse pressure gradient. J. Aeronaut. Sci. **21**, 91–108 (1954)
3. Eitel-Amor, G., Örlü, R., Schlatter, P.: Simulation and validation of a spatially evolving turbulent boundary layer up to $Re_\theta = 8300$ Int. J. Heat Fluid Flow **47**, 57–69 (2014)
4. Fischer, P.F., Lottes, J.W., Kerkemeier, S.G.: NEK5000: Open source spectral element CFD solver. Available at: http://nek5000.mcs.anl.gov, (2008)

5. Kametani, Y., Fukagata, K., Örlü, R., Schlatter, P.: Effect of uniform blowing/suction in a turbulent boundary layer at moderate Reynolds number. Int. J. Heat Fluid Flow **55**, 132–142 (2015)
6. Pinkerton, R.M.: The variation with Reynolds number of pressure distribution over an airfoil section. NACA Ann. Rep. **24**, 65–84 (1938)
7. Stroh, A., Hasegawa, Y., Schlatter, P., Frohnapfel, B.: Global effect of local skin friction drag reduction in spatially developing turbulent boundary layer. J. Fluid Mech. **805**, 303–321 (2016)
8. Schlatter, P., Stolz, S., Kleiser, L.: LES of transitional flows using the approximate deconvolution model. Int. J. Heat Fluid Flow **25**, 549–558 (2004)
9. Schlatter, P., Örlü, R.: Turbulent boundary layers at moderate Reynolds numbers. Inflow length and tripping effects. J. Fluid Mech. **710**, 5–34 (2012)
10. Vinuesa, R., Negi, P.S., Atzori, M., Hanifi, A., Henningson, D.S., Schlatter, P.: Turbulent boundary layers around wing sections up to $Re_c = 1,000,000$. Int. J. Heat Fluid Flow **72**, 86–99 (2018)
11. Vinuesa, R., Bobke, A., Örlü, R., Schlatter, P.: On determining characteristic length scales in pressure-gradient turbulent boundary layers. Phys. Fluids **28**, 055101 (2016)

Chapter 41
Effect of the Actuation on the Boundary Layer of an Airfoil at Moderate Reynolds Number

O. Lehmkuhl, I. Rodriguez and R. Borrell

Introduction

Synthetic (zero net mass flux) jets are an active flow control technique to manipulate the flow field in wall-bounded and free-shear flows. The fluid necessary to actuate on the boundary layer is intermittently injected through an orifice and is driven by the motion of a diaphragm located on a sealed cavity below the surface [1]. Comprehensive reviews on active flow control techniques can be found in [1, 2]. Periodic excitation introduced at the surface has been shown as an efficient and practical means of flow control with the potential to significantly change the lift and drag of an airfoil and the separation of the boundary layer. In the particular case of active flow control of airfoils, there are different parameters that control the actuation such as the jet momentum coefficient, the slot position, the jet frequency (see for instance [3–5]). From a numerical point of view, the accurate prediction of the flow over an airfoil is a challenging task as is dominated by separation, transition to turbulence and, at some angles of attack, reattachment of the flow, not to mention the flow control mechanism. In this sense, large-eddy simulations (LES) offer a good compromise between accuracy and computational time required for performing the simulations. For instance, You and Moin [6] proved that LES techniques can be successfully applied for capturing the flow physics in these configurations. The present paper focuses on the particular range of moderate Reynolds numbers for

O. Lehmkuhl (✉) · R. Borrell
Barcelona Supercomputing Center, Barcelona, Spain
e-mail: oriol.lehmkuhl@bsc.es

R. Borrell
e-mail: ricard.borrell@bsc.es

I. Rodriguez
Universitat Politècnica de Catalunya, Barcelona, Spain
e-mail: ivette.rodriguez@upc.edu

M. García-Villalba et al. (eds.), *Direct and Large Eddy Simulation XII*,
ERCOFTAC Series 27,
https://doi.org/10.1007/978-3-030-42822-8_41

which the development of Micro Air Vehicles (MAV) is of interest. In this sense, the role of different actuation mechanisms on the boundary layer of a SD7003 airfoil at $\mathrm{Re} = U_{\mathrm{ref}}C/\nu = 6 \times 10^4$ is studied. Here, Reynolds number is defined in terms of the free-stream velocity U_{ref} and the airfoil chord C. At this Reynolds number, the formation of a laminar separation bubble (LSB) may have a dominant effect on the flow field and thus it is interesting to understand the role of the actuation on the reduction or suppression of the LSB as well as on post-stall behavior. To do so, angles of attack (AoA) of $\mathrm{AoA} = 4,\ 11,$ and $14°$ are considered. The first one corresponds with an AoA with a large laminar separation bubble, the second one is close to the maximum lift, whereas the last one corresponds with the flow in full stall.

Mathematical and Numerical Method

In this work, large-eddy simulations (LES) of the flow are performed by means of a low-dissipation finite-element method implemented into the code Alya [7]. The convective term is discretized using a Galerkin FEM scheme recently proposed by Charnyi et al. [8], which conserves linear and angular momentum, and kinetic energy at the discrete level. Neither upwinding nor any equivalent momentum stabilization is employed. In order to use equal-order elements, numerical dissipation is introduced only for the pressure stabilization via a fractional step scheme [9], which is similar to approaches for pressure-velocity coupling in unstructured collocated finite-volume codes. The set of equations is integrated in time using a third-order Runge–Kutta explicit method combined with an eigenvalue-based time-step estimator [10]. As for the LES, the Vreman [11] sub-grid scale (SGS) model is used.

Results

Solutions are obtained in a computational domain of dimensions $15C \times 16C \times 0.2C$ with the leading edge of the airfoil placed at $(0, 0, 0)$. The boundary conditions at the inflow consist of a uniform velocity profile (u,v,w) = ($U_{\mathrm{ref}} \cos\alpha$, $U_{\mathrm{ref}} \sin\alpha$, 0). As for the outflow boundary, a pressure-based condition is imposed. No-slip conditions on the airfoil surface are prescribed. Periodic boundary conditions are used in the spanwise direction. For the actuated cases, an inlet velocity $(u, v, w)_{\mathrm{act}} = (\sin\alpha, \cos\alpha, 0)A_p \sin(2\pi f t)U_{\mathrm{ref}} \sin(2\pi \tau z)$ is imposed. Here, A_p is the maximum amplitude of the jet so as $U_{\max} = A_p U_{\mathrm{ref}}$, f is the actuator frequency so as $F^+ = f U_{\mathrm{ref}}/x_{\mathrm{TE}}$, x_{TE} being the distance from the actuator to the trailing edge and τ is the spanwise period of the signal, in this work $\tau = 0.5L_z$. In the present simulations, a momentum coefficients $c_\mu = h(\rho U^2_{\max})/(C\rho U^2_{\mathrm{ref}}) = 3 \times 10^{-3}$ (with h the actuator width) and a non-dimensional frequency of $F^+ = 1$ are considered. The actuator outlet is located at $1\%C$. The computational meshes used for both the baseline and actuated cases are unstructured grids of about 30 million of degrees of freedom. For obtaining these meshes, an unstructured two-dimensional mesh is extruded in N_{elements} in the spanwise direction. The number of elements in the spanwise direction in all computations has been 64. In all cases, the meshes have been constructed so as in the near wall region the non-dimensional wall distance $y^+ \approx 1$.

Baseline Simulations

In order to validate the current numerical set-up, the un-actuated case has been compared to available solutions in the literature. In Fig. 41.1, a detailed comparison of the pressure coefficient and skin friction at $\alpha = 4°$ and $11°$ is presented. As can be seen from the figure, in all cases results are in good agreement with those of the literature.

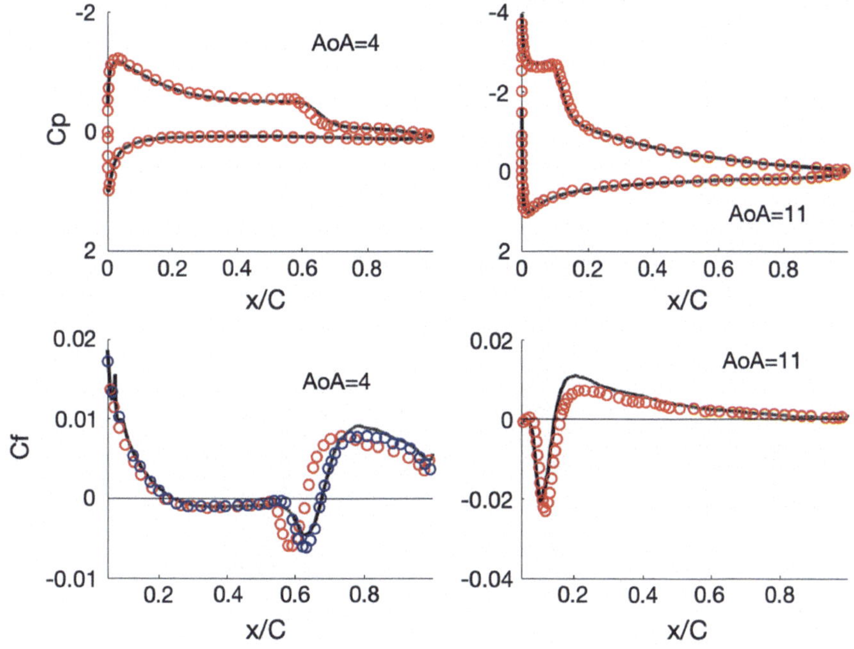

Fig. 41.1 Baseline case. Comparison with the literature. (solid line) present results. (blue circle) LES [12], (red circles) ILES [13]

Actuated Cases

The actuation at AoA = 4, 11, 14° has been considered. In general, at AoA lower than the point of maximum lift (i.e., AoA = 11°), the actuation eliminates the laminar separation bubble and triggers the transition to turbulence just after the location of the actuator. However, it offers no advantages from the airfoil efficiency point of view (see Fig. 41.2). For the stalled airfoil, i.e., AoA = 14°, the airfoil efficiency considerably increases. In Fig. 41.2, the comparison between actuated and baseline cases is given, together with results from the literature for the uncontrolled case.

In Fig. 41.3, instantaneous vortical structures for AoA = 14° are presented for both the baseline and actuated cases. As can be seen from the figure, in the baseline case flow massively separates from almost the leading edge of the airfoil and fails to reattach, producing a large recirculation bubble and momentum deficit in the suction side of the airfoil. As a consequence of the large detachment of the flow, the airfoil is stalled and its efficiency drops. When the actuator is activated, the jet transfer momentum to the flow and produces three-dimensional instabilities in the shear layer which trigger the transition to turbulence. The added momentum forces the shear layer to reattach to the airfoil surface (see also Fig. 41.4) and there is a reduction in the separated zone of the airfoil. Actually, as can be seen from the figure, there is a small separation bubble close to the leading edge of the airfoil and then the flow reattaches completely to the airfoil. Yet, the actuation does not eliminate completely the flow separation. At about $x/C = 0.65$ the flow detaches from the airfoil surface producing a recirculation zone in the last part of it. However, this zone is much smaller than the larger region produced in the uncontrolled case, where the recirculation zone covers completely the suction side of the airfoil. As a consequence, the actuation produces a reduction of about 67% in the drag and a 19% increase in the lift coefficient respect to the uncontrolled case. All in all, this represents that the efficiency of the airfoil is almost doubled (efficiency increases a 98%).

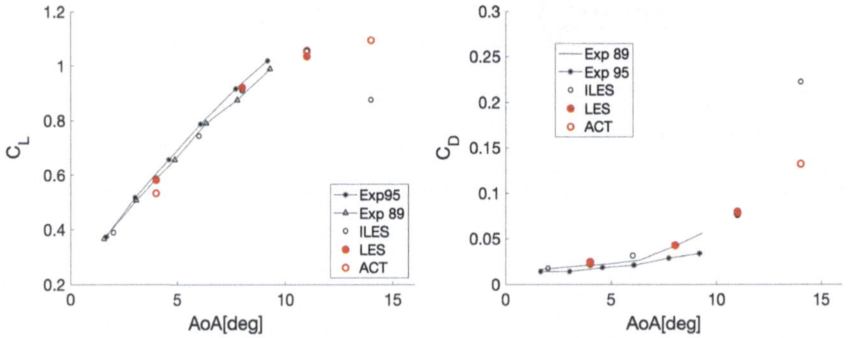

Fig. 41.2 Lift and drag coefficients for the controlled cases. Comparison with the baseline cases and with the literature. EXP95 [14], EXP89 [15], ILES [13]

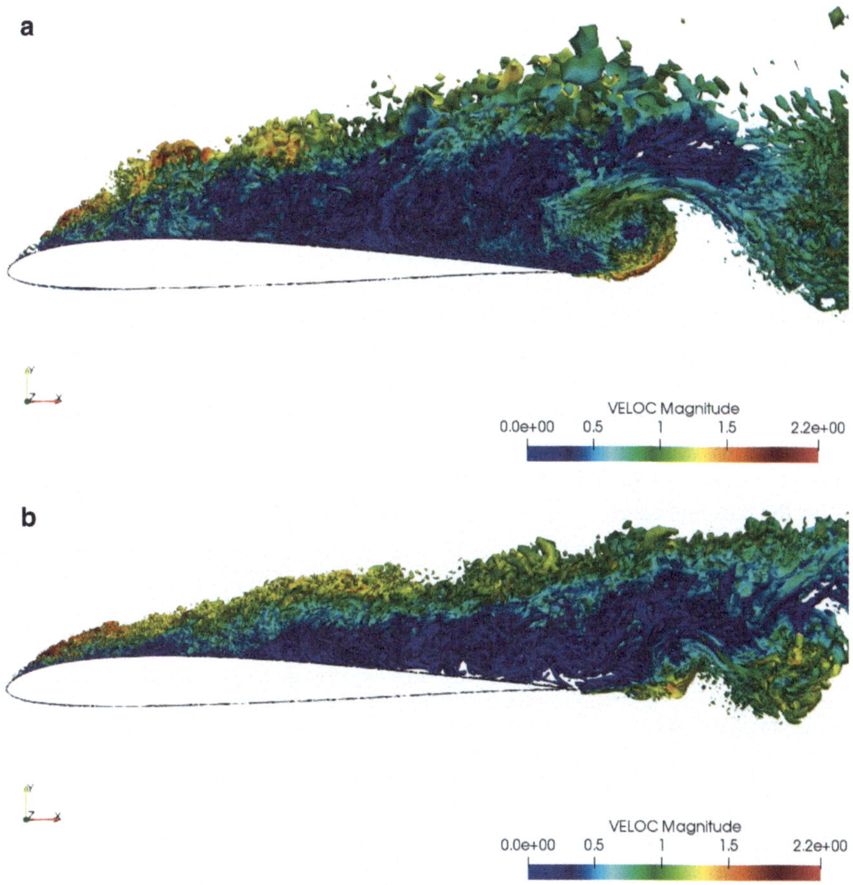

Fig. 41.3 Vortical structures represented by Q-isosurfaces at $\alpha = 14°$ **a** baseline case, **b** actuated case

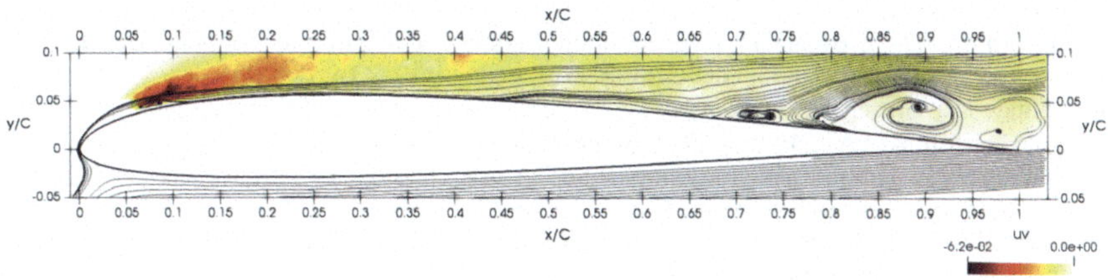

Fig. 41.4 Controlled case at AoA = 14°. Streamlines colored with Reynolds shear stresses

Conclusions

Wall-resolved large-eddy simulations (LESs) of the flow separation over an airfoil at the moderate Reynolds number of Re = 60,000 are performed. At this Reynolds number, the flow separates laminarly from the airfoil forming a laminar separation bubble up to the angle of attack where the lift is maximum; for larger AoA the flow fails to reattach producing a massive flow separation. The effect of the actuation by using sythetic jets is investigated at different AoA. It is shown that while the actuation successfully eliminates the laminar bubble at AoA = 4°, it does not produce an increase in the airfoil efficiency. However, for the stalled airfoil, the actuation not only eliminates the massive flow separation, but also the flow is attached to the airfoil surface in a significant part of the airfoil chord. As a consequence, this kind of actuation almost doubles airfoil efficiency with a reduction of the drag coefficient by a 67%. This kind of technique seems to be promising at delaying flow separation and its associated losses when the angle of attack increases beyond the maximum lift for the baseline case.

Acknowledgements This work has been partially financially supported by the Ministerio de Economía y Competitividad, Secretaría de Estado de Investigación, Desarrollo e Innovación, Spain (Ref. TRA2017-88508-R) and by European Union's Horizon 2020 research and innovation programme (INFRAEDI-02-2018, EXCELLERAT-The European Centre Of Excellence For Engineering Applications H2020.). We also acknowledge Red Española de Surpercomputación (RES) for awarding us access to the MareNostrum IV machine based in Barcelona, Spain (Ref. FI-2018-2-0015 and FI-2018-3-0021).

References

1. Glezer, A.: Some aspects of aerodynamic flow control using synthetic-jet actuation. Philos. Trans. R. Soc. Math. Phys. Eng. Sci. **369**(1940), 1476–1494 (2011)
2. Cattafesta, L.N., Sheplak, M.: Actuators for active flow control. Ann. Rev. Fluid Mech. **43**(1), 247–272 (2011)
3. Mccormick, D.C.: Boundary layer separation control with directed synthetic jets. In 38th Aerospace Sciences Meeting and Exhibit, , vol. 2000–0519. AIAA (2000)
4. Amitay, M., Glezer, A.: Role of actuation frequency in controlled flow reattachment over a stalled airfoil. AIAA J. **40**(2), 209–216 (2002)
5. Gilarranz, J.L., Traub, L.W., Rediniotis, O.K.: A new class of synthetic jet actuators—part I: design, fabrication and bench top characterization. J. Fluids Eng. **127**(2), 367 (2005)
6. You, P., Moin, D.: Active control of flow separation over an airfoil using synthetic jets. J. Fluids Struct. **24**(8), 1349–1357 (2008)
7. Lehmkuhl, O., Houzeaux, G., Owen, H., Chrysokentis, G., Rodriguez, I.: A low-dissipation finite element scheme for scale resolving simulations of turbulent flows. J. Comput. Phys. **390**, 51–65 (2019)
8. Charnyi, S., Heister, T., Olshanskii, M.A., Rebholz, L.G.: On conservation laws of Navier-Stokes galerkin discretizations. J. Comput. Phys. **337**, 289–308 (2017)
9. Codina, R.: Pressure stability in fractional step finite element methods for incompressible flows. J. Comput. Phys. **130**(1), 112–140 (2001)
10. Trias, F.X., Lehmkuhl, O.: A self-adaptive strategy for the time integration of Navier-Stokes equations. Numer. Heat Transfer Part B **60**(2), 116–134 (2011)

11. Vreman, A.W.: An eddy-viscosity subgrid-scale model for turbulent shear flow: algebraic theory and applications. Phys. Fluids **16**(10), 3670–3681 (2004)
12. Schmidt, S., Breuer, M.: Hybrid LES-URANS methodology for the prediction of non-equilibrium wall-bounded internal and external flows. Comput. Fluids **96**, 226–252 (2014)
13. Galbraith, M., Visbal, M.: Implicit large Eddy simulation of low reynolds number flow past the SD7003 airfoil. In: 46th AIAA aerospace sciences meeting and exhibit, pp. 1–17 (2008)
14. Selig, M., Guglielmo, J.J., Broeren, A.P., Giguere, P.: Summary of low-speed airfoil data summary of low-speed airfoil data. Tech. rep., University of Illinois (1995)
15. Selig, M.S., Donovan, J.F., Fraser, D.B.: Airfoils at low speeds. Tech. rep., University of Illinois (1989)

Chapter 42
The Effect of the Sweep Angle to the Turbulent Flow Past an Infinite Wing

C. A. Suardi, A. Pinelli and M. Omidyeganeh

Introduction

Nowadays the majority of civil aircraft employs swept-back wings. This configuration, proposed in the early 30s of last century, has been technologically motivated by the otherwise enhanced drag experienced in transonic cruise condition. Since its introduction, several studies have focused on the assessment of the aerodynamic behaviour of this wing configuration for a flow condition resembling realistic, high Reynolds (Re) number cases of aeronautical interest [1, 2]. In these investigations, the boundary layer interesting the infinite wing undergoes an early transition inhibiting the appearance of any major separated regions for a moderate incidence and can thus considered a developing turbulent boundary layer (TBL) subject to a varying adverse pressure gradient (APG). In the present work, we also consider an infinite wing model at a modest angle of attack. However, because of the prohibiting computational cost of detailed simulations at a high Re number, we prefer to mimic these realistic conditions by superimposing free-stream turbulence to the approaching clean stream. Indeed, this choice induces an early by-pass transition that prevents the eventual separation of the TBL, thus leading to a physical situation similar to the reference ones cited above. The high Re number regime, leading to an almost totally turbulent, attached boundary layer, is a quite interesting case for at least two reasons. The first one was put forwards long ago and concerns the possibility of predicting the aerodynamic performances of a swept wing (lift and drag) in terms of the associated straight one, using a simple trigonometric function of the introduced sweep angle. This property, that has been verified experimentally and numerically by several authors, mainly holds for attached flows and is commonly termed as *Simple Sweep Theory*. Although this theory has been the working horse of swept wing design for decades, its exact applicability limits and its extention to other aerodynamic proper-

C. A. Suardi (✉) · A. Pinelli · M. Omidyeganeh
City, University of London, London, UK
e-mail: carlo.suardi@city.ac.uk

M. García-Villalba et al. (eds.), *Direct and Large Eddy Simulation XII*,
ERCOFTAC Series 27,
https://doi.org/10.1007/978-3-030-42822-8_42

ties are still object of recent investigations. The second aspect that makes the TBL on a swept wing an appealing research topic is related with the comprehension of the structure of the turbulent, wall-bounded flow simultaneously exposed to an APG and to an imposed crossflow. An understanding on how these concomitant effects manipulate the structure of the TBL is the central topic of the present contribution. The region of primary interest is the one close to the trailing edge of the wing, where the pressure gradient is a strong adverse one and where the impact of the cross-flow generated by the sweep angle is more evident. In particular, is accounted the effect of the sweep adoption on the appearence of *reversed flow nuclei*, similar to those reported into the literature for the high Reynolds TBL subject to zero pressure gradient (ZPG) [3] and for the moderate Re under an APG [7]. In general, our simulations on a straight wing confirm the appearance of localised and unsteady local separated cores even for the case of an apparently fully attached boundary layer, due to the intense pressure gradient in the latter stage of the wing suction side for the aerofoil and incidence considered. When the 30° sweep-back angle is considered, we observe a mitigation of these reversed flow nuclei, alongside with a not trivial bending of the principal flow structures participating in the wall turbulence cycle. Due to the idea put forwards by several authors suggesting an eventual link between the mentioned flow structures and the appearance of trailing edge flow separation on a wing, it is believed of great importance to understand how the sweep angle would affect this link. This in order to provide essential guidelines for the development of flow separation devices for the widely adopted swept wings.

Problem Formulation

The flow past an infinite swept wing is dealt with the 3D incompressible, large-eddy simulation equations (LES). The region close to the wing surface is fully resolved while the subgrid-scale closure is achieved via the ILSA model proposed by Piomelli et al. [4]. The LES equations are space discretised via a co-located finite volume formulation on a structured mesh. The solver overall accuracy is second order in both space and time and the MPI library is used in the framework of a domain decomposition approach to exploit parallel, memory distributed computer architectures. The solver that incorporates all the aforementioned features (called *SUSA*) has been extensively validated in the past [5]. The computational domain around each 2D aerofoil (NACA 4412) $x - y$ cross section is sketched in Fig. 42.1 alongside with the wall resolution in inner units in Fig. 42.2 (being the friction velocity, u_τ, and the kinematic viscosity, ν). On the $x - y$ plane, the domain is meshed using a body fitted C-grid which 3D extension is achieved extruding the 2D mesh in z using a uniform spacing. A zero velocity boundary condition is enforced at the solid walls, while on the outer boundary (i.e. the surface obtained when extruding the outer 2D boundary in z), we set an inlet/outlet condition that depends on the local, instantaneous direction of the computed flow (Dirichlet condition obtained from an irrotational solution if flow is incoming, non-reflective condition if outgoing) for the x and y velocity

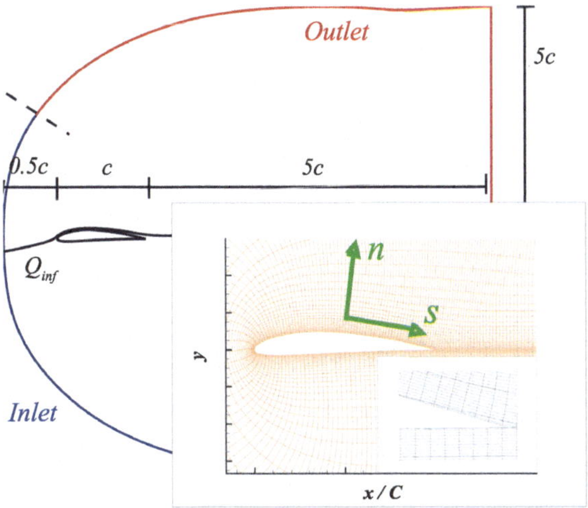

Fig. 42.1 Domain geometry for the computational study. z extension is 0.4 c (c being the chord size). The domain is assumed to be periodic in z

components. The swept/unswept wing configuration is simulated by setting a constant value for the z velocity component (parallel to the leading edge) on the outer boundary equal to $U_\infty \tan(\beta)$, with the sweep angle β, equal to 30°/0° (U_∞ being the free-stream velocity component perpendicular to the aerofoil LE). For both swept and unswept cases, the chord Reynolds number is $\text{Re}_c = U_\infty\, c/\nu = 50 \times 10^3$, and two loading conditions have been considered setting the angle of attack to $\alpha = 5°/10°$. The boundary layer transition is triggered by superimposing to the incoming flow a turbulence field (10% intensity with integral length scale $0.045C$) obtained through a twin, independent DNS of grid generated turbulence (Reynolds number based on grid spacing $\text{Re}_M = 2000$). In Fig. 42.3a is presented the turbulent kinetic energy spectrum of the perturbation introduced, as well as that of the resultant boundary layer in the buffer layer, ($y^+ \approx 25$), at a specific different suction side wing locations, $x/C \approx 0.75$ for the case of un-swept wing at 5° incidence. Inside the in-box, it can be found an illustrative sketch of a portion of the grid turbulence introduced into the domain. The effect of the perturbation introduced is that of triggering the boundary layer transition to turbulence via a by-pass mechanism from the early stage of the wing. The energised boundary layer does not present a major separation from the wall for the considered incidences, condition otherwise faced in the absent of boundary layer perturbation [6]. Instead, it remains attached, by statistical mean, until the trailing edge. The coefficient of friction, shown in Fig. 42.3b for the case of un-swept wing and 5° incidence, clearly support this statement.

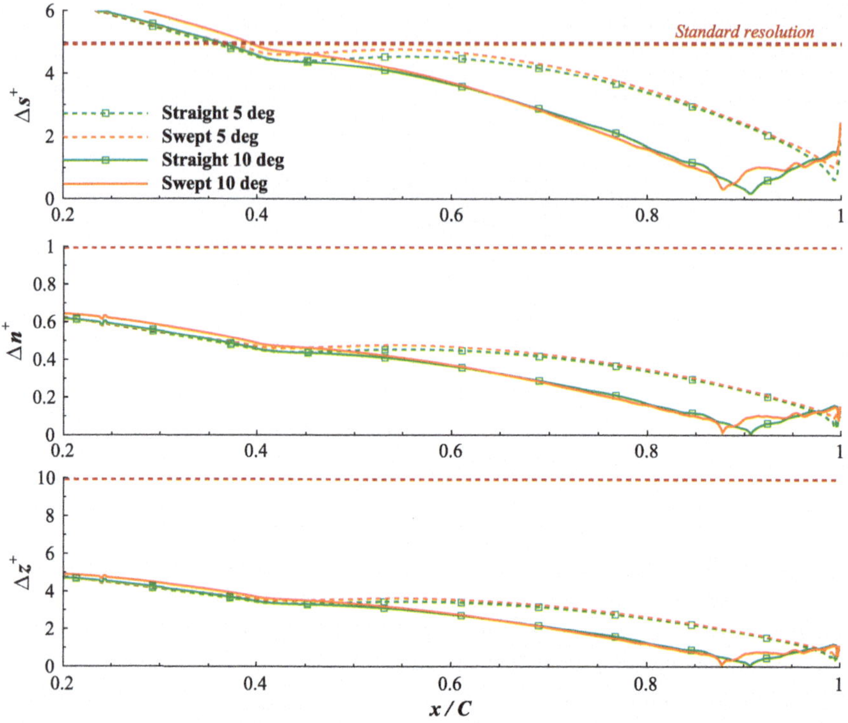

Fig. 42.2 Wall resolution in plus units for all the cases considered in this investigation compared with the standard for a turbulent channel flow

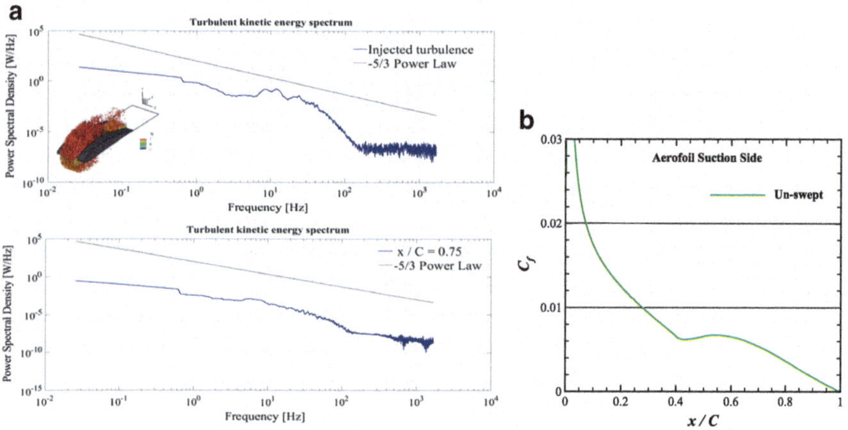

Fig. 42.3 **a** Turbulent kinetic energy spectrum for the introduced perturbetion (Top) and into the buffer layer for late location on the wing suction side (Bottom). An illustration of the introduced grid turbulence and its effect on the boundary layer is presented into the the squared box. **b** Suction side wall friction coefficient along the chord

Results

A comparative study for the effect of a 30° sweep-back adoption on the flow field is made for the two different loading condition considered. Figure 42.4a presents the mean wall coefficient of pressure, whereas Fig. 42.4b presents some chord-wise velocity profile for two consequent chord-wise locations, $x/C = 0.4, 0.95$. The velocity profiles have been scaled with the inner units to make a comparison with the law of the wall in the case of a turbulent channel flow. A good match for

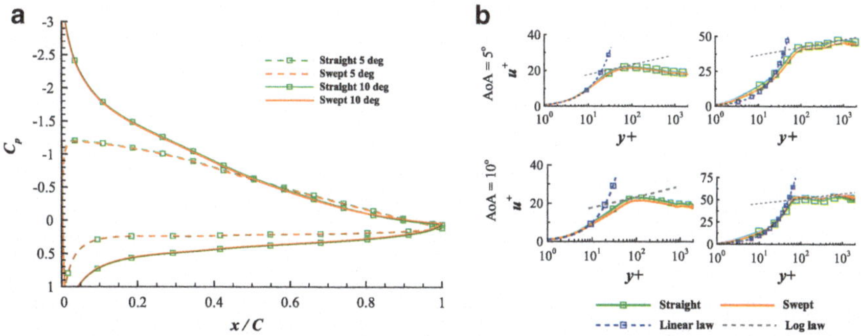

Fig. 42.4 **a** Distribution of the wall coefficient of pressure for all the cases investigated. **b** Profiles of the chord-wise velocity in plus units for the 5° (Top) and 10° (Bottom) incidences, at the chord-wise location $x/C = 0.4$ (Left) and $x/C = 0.95$ (Right), respectively

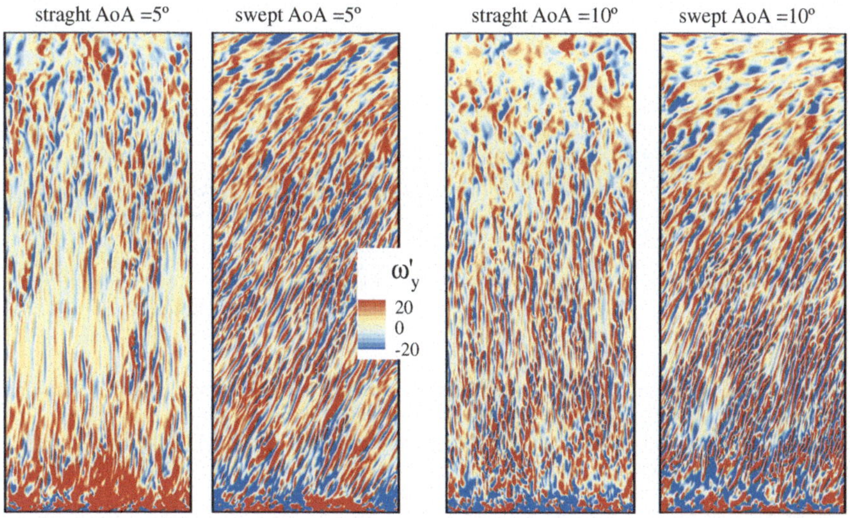

Fig. 42.5 Instantaneous iso-contours of the wall-normal component of the vorticity perturbation for the (Left) 5° and (Right) 10° case. The black line identifies the trailing edge. Suction side view. The flow is from bottom to top

both the incidences is found regardless of the sweep angle adopted, supporting what has been already postulated as the simple sweep theory for the low order statistics of the turbulent attached flow past an infinite wing [1, 2].

Even though some flow statistics has been found mildly affected by the adoption of the sweep angle, the flow dynamics of the turbulent boundary layer results deeply modified. It is reported a varying distortion of the wall turbulence streaks moving towards the trailing edge, especially on the suction side. In Fig. 42.5 can be found the contours of the instantaneous wall-normal vorticity fluctuation on a surface above the wing suction side within the buffer layer for both the loading conditions, highlighting the streaks. The effect of the intensifying adverse pressure gradient seems to have an increasingly stronger impact as the streaks are thicker and present a more meandering pattern. Furthermore, a freshly new turbulent content is detected via the Reynolds stresses due to the introduced crosswind, for both the incidences, Fig. 42.6.

The sweep angle is found to mitigate the reversed flow nuclei appearing inside the wing suction side boundary layer, and thus the portion of the wing interested by reversed flow. Figure 42.7 presents a map of the instantaneous skin friction fluctuation on the suction side varying the sweep adopted, alongside with a quantitative account on the same side of the total probability to detect a reversed flow, with respect to the mean flow direction, moving along the chord, for the 5° (Top) and 10° (Bottom) incidence case. What can be qualitatively be observed by the friction map is clearly supported by the mitigated reversed flow region detected on the swept wing, regardless of the incidence.

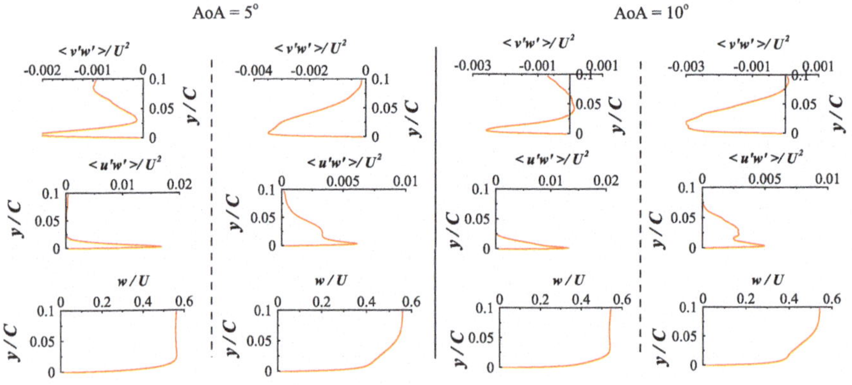

Fig. 42.6 Time and spanwise averaged profiles of quantities introduced purely by the crosswind for the (Left) 5° and (Right) 10° incidence case. Per each incidence are shown the profiles of the crosswind (Bottom), and two Reynold stresses (Centre and Top), extracted on the suction side at two consequent chord-wise locations, $x/C \approx 0.4$ (Left) and at $x/C \approx 0.95$ (Right)

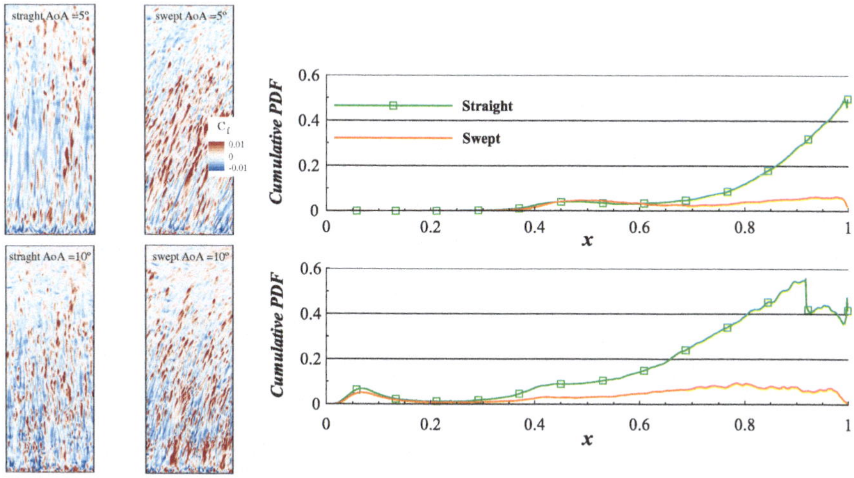

Fig. 42.7 (Left) Instantaneous iso-contours of the wall friction fluctuation, flow from bottom to top, and (Right) total probability to detect a reversed flow along the chord for the suction side wall. The (Top) set refer to the 5° incidence case, the (Bottom) one for the 10° case

Acknowledgements The authors would like to thank City, University of London and Airbus for the funding provided to carry out the current investigation and the EPSRC for the computational time made available on the UK supercomputing facility *ARCHER* via the UK Turbulence Consortium (EP/R029326/1).

References

1. Altman, J.M., Hayter, N.L.F.: A comparison of the turbulent boundary-layer growth on an unswept and swept wing. NACA Technical Note **2500**, (1951)
2. Boltz, F.W., Kenyon, G.C., Allen, C.Q.: Effects of sweep angle on the boundary-layer stability characteristics of an untapered wing at low speeds. NASA D-338 (1960)
3. Brücker, C.: Evidence of rare backflow and skin-friction critical points in near-wall turbulence using micropillar imaging. Phys Fluids. **27**, 031705 (2015)
4. Piomelli, U., Rouhi, A., Geurts, B.J.: A grid-independent length scale for large-eddy simulations. J. Fluid Mech. **766**, 499–527 (2015)
5. Rosti, M.: Direct Numerical Simulation of an Aerofoil at High Angle of Attack and its Control. University of London, London (2016)
6. Suardi, C.A., Pinelli, A., Omidyeganeh, M.: Investigation of the Sweep Independence Principle for Transitional Regime of the Flow Past an Aerofoil. EPSRC UK Turbulence Consortium Meeting, London (2018)
7. Vinuesa, R., Örlü, R., Schlatter, P.: Characterisation of backflow events over a wing section. Jour. of Turb. **18**(2), 170–185 (2017)

Chapter 43
LES Study of the Three-Dimensional Behaviour of Unswept Wing Sections at Buffet Conditions

M. Zauner and N. D. Sandham

Introduction

Recent high-fidelity simulations of airfoils at transonic buffet conditions have shown great potential to study the detailed mechanism of this low-frequency unsteadiness, which limits the flight envelope of modern aircraft [1, 9]. However, due to the high resolution of direct numerical simulations (DNS) that is required even at moderate Reynolds numbers [9], both the run time and domain width were limited. The aim of the current contribution is to explore methods that are capable of reproducing the low-frequency lift oscillations with a well-resolved approach, but cutting down the computational cost by a factor of 16 relative to DNS. Implicit large-eddy simulations have the potential to capture the main flow phenomena without solving additional equations. In order to maintain the numerical stability, however, filtering may be required. Due to the sensitivity of the buffet phenomenon to filtering, a targeted filter technique is developed, which uses spectral-error indicators [3] to identify regions that are prone to grid-to-grid point oscillations and hence vary the filter activity within the domain.

Methodology

All present simulations are carried out using the well-validated high-order fully parallelised multi-block finite difference in-house code SBLI. Details of the solver can be found in [2, 4, 5]. The basic set-up outlined in this subsection is analogous to the

M. Zauner (✉) · N. D. Sandham
University of Southampton, Southampton SO17 1BJ, UK
e-mail: m.zauner@soton.ac.uk

N. D. Sandham
e-mail: n.sandham@soton.ac.uk

M. García-Villalba et al. (eds.), *Direct and Large Eddy Simulation XII*,
ERCOFTAC Series 27,
https://doi.org/10.1007/978-3-030-42822-8_43

direct numerical simulations in [9] and provides the underlying framework for all simulations discussed in this contribution. The dimensionless three-dimensional Navier–Stokes equations are solved applying a fourth-order finite difference scheme and a low-storage Runge–Kutta scheme for the spatial and temporal discretisation, respectively. Zonal characteristic boundary conditions are applied at the outlet, whereas integral characteristic boundary conditions are applied at the remaining outer boundaries. Due to the appearance of shock waves, a total variation diminishing (TVD) scheme is applied in the entire flow field, except for a small radius around the leading edge, due to numerical reasons [9]. To avoid the formation of spurious structures in the freestream, a sixth-order filter, similar to the one used by [6] is incorporated in the farfield, but disabled within a distance of $\Delta y \approx 0.2$ to the airfoil surface, so that well-resolved turbulent scales are not affected. For the filtering, we solve the tridiagonal system

$$\alpha_f \breve{q}_{(i-1)} + \breve{q}_i + \alpha_f \breve{q}_{(i+1)} = \sum_{n=0}^{N} \frac{a_n}{2}(q_{(i+n)} + q_{(i-n)}) \text{ with } N = 3, \tag{43.1}$$

where q and $\breve{q}$ denote unfiltered and filtered components of the solution matrix, respectively. The adjustable parameter is set to $\alpha_f = 0.45$ and the coefficients a_0, a_1, a_2 and a_3 are derived in terms of α_f with Taylor-and Fourier-series analyses [6]. An additional coefficient a_{lim} is applied to better control the dissipation according to $q_{new} = q - a_{lim} \cdot (q - \breve{q})$. For direct numerical simulations, a_{lim} is set to $a_{lim} = 0.1$ in the freestream region, where the filter is applied, but it can also be used as a parameter to ensure numerical stability of simulations with decreased resolutions. More details on the numerical set-up are provided in [7] and [9].

The present contribution considers two implicit large-eddy simulation approaches. The first method (here referred to as ILES) uses the 6th-order filter to tune the implicit subgrid-scale (SGS) model, with the filter coefficient set to $a_{lim} = 0.5$ for the entire domain to maintain numerical stability. For the second approach, namely, the spectral-error-based implicit large-eddy simulation (SE-ILES), the filter is only applied in regions that are prone to grid-to-grid point oscillations, without affecting structures in well-resolved regions. Every $N_E = 10$ time steps, spectral-error indicators are calculated for each processor block (PB) according to [3] using the vorticity magnitude $|\omega|$. Instead of computing the full spectra for each 3D PB, the Fourier amplitudes A_2, A_4 and A_8 (proportional to the square root of the spectral energy) of selected modes ($N_e/2$, $N_e/4$, $N_e/8$) for stencils of $N_e = 16$ grid points are approximated by

$$A_2 = \frac{2^{-2r}}{N_e} \sum_{j=0}^{N_e-1} (-1)^j |\omega_j|, \quad A_4 = \frac{2^{1-r}}{N_e} \sum_{j=0}^{N_e-1} (-i)^j |\omega_j|,$$
$$A_8 = \frac{2}{N_e} \sum_{j=0}^{N_e-1} \exp(-\frac{\pi}{4} i)^j |\omega_j|. \tag{43.2}$$

The parameter r denotes the minimum acceptable decay rate of amplitudes in the high-frequency part of the spectrum and set to $r = -1$ (Kolmogorov's law would be $r = -5/3$). The average of Fourier amplitudes ($\overline{A}$) of a set of similarly oriented stencils, which are homogeneously distributed within the PB, is calculated and the mean error indicator $\overline{I_f}$ is obtained by

$$\overline{I_f} = \ln(1 + \lfloor\frac{\overline{A_2}}{\overline{A_4} + \epsilon}\rfloor + \lfloor\frac{\overline{A_4}}{\overline{A_8} + \epsilon}\rfloor + \lfloor\frac{\overline{A_2}}{\overline{A_8} + \epsilon}\rfloor), \quad (43.3)$$

$$\text{with } \overline{A_2} = \sum_{n=1}^{N_{stencil}} A_{2,n}\,, \; \overline{A_4} = \sum_{n=1}^{N_{stencil}} A_{4,n}\,, \; \text{and } \overline{A_8} = \sum_{n=1}^{N_{stencil}} A_{8,n}, \quad (43.4)$$

where $\lfloor ... \rfloor$ is a 'floor' operation and $\epsilon = 0.03$ used to avoid division by zero and tune the error indicator sensitivity [3]. In case $\overline{I_f}$ exceeds a value of $\varepsilon_1 = 0.6$, the error in this PB is considered as high. The threshold is based on a grid study of the Taylor–Green vortex test case [3]. In addition to the instantaneous error, an average error is computed according to

$$\varepsilon_{av} = (1 - C_{av})\varepsilon_{av} + C_{av}1 \qquad \text{if } \overline{I_f} > \varepsilon_1, \quad (43.5)$$

$$\varepsilon_{av} = (1 - C_{av})\varepsilon_{av} + C_{av}\frac{N_{high}}{N_{check}} \qquad \text{else}, \quad (43.6)$$

which indicates the probability of $\overline{I_f}$ exceeding the threshold ε_1 in this PB using $C_{av} = 0.1$. The number of checks, where the spectral-error exceeded the threshold ε_1 is denoted by N_{high}, while N_{check} denotes the total number of checks. Every $N_F = 30$ time steps, the filter is applied in a PB, if either $\overline{I_f} > \varepsilon_1$ or $\varepsilon_{av} > \varepsilon_2$ (for the present study $\varepsilon_2 = 0.2$). Blending is used to transition from $a_{lim} = 0.4$ (in PBs with an high spectral error) to $a_{lim} = 0.0$ in adjacent PBs with low error estimates. This is repeated for three stencil sets within each PB to calculate error indicators corresponding to the curvilinear coordinates and to apply the filter in each direction independently. More details on spectral-error indicators and the chosen settings can be found in [3] and [10], respectively.

Results

The unsteady flow around Dassault Aviation's V2C profile at an angle of attack $\alpha = 4°$, a moderate Reynolds number of Re $= 500{,}000$ and a Mach number of $M = 0.7$ is discussed in detail in [9] and [8]. Figure 43.1a shows strong pressure gradients as black contours and the sonic line in red. Besides strong acoustic radiation at the rear of the airfoil, upstream-propagating shock waves are also observed. The lift coefficient as a function of time in Fig. 43.1b clearly shows a low-frequency oscillation associated with transonic buffet at a distinct frequency at $St = 0.12$.

Upstream-propagating shock waves, however, are generated at higher frequencies in the range of $St = 0.3 - 0.7$. We will focus the present study on the suction side, where shock waves appear. Figure 43.2a shows the grid spacing on the airfoil surface in the circumferential direction as a function of x, where the blue line corresponds to the DNS grid A, and the red and black curves to ILES grids B and C. Grid B is significantly coarser than the DNS grid in all sections of the airfoil, while grid C is only coarsened in the transitional and turbulent rear part of the airfoil ($x > 0.5$). As shown in Fig. 43.2b, high wall-normal resolution is maintained near the airfoil surface, while the spacing between grid points in the farfield is significantly stretched in the wall-normal (j) direction. The spanwise domain extent (5% of the chord length) of the DNS and both ILES grids is resolved by 150 and 50 grid points, respectively. The blue line in Fig. 43.2c shows the lift coefficient C_L as a function of time for the DNS reference case. The red and orange lines correspond to ILES using grids B and C, respectively, both underestimating the amplitude of the low-frequency oscillation associated with transonic buffet. The full signal of the ILES using grid B is shown in the inset of Fig. 43.2c. Particularly the ILES using grid B shows significantly different flow characteristics compared to the SE-ILES approach using Grid C, which agrees well with the DNS. This is also shown in Fig. 43.2d, plotting contours of the streamwise wall-density gradient as a function of x and time. Comparing the ILES with the SE-ILES, the unsteadiness in the transition region is weaker in the former case. Considering the ILES results, the forepart of the airfoil shows only traces of upstream-propagating pressure waves, whereas the right-hand plot in Fig. 43.2d, corresponding to the SE-ILES approach, also shows low-frequency events associated with transonic buffet at $x \approx 0.25$. Under-resolution in the circumferential direction of the mainly laminar flow in that region attenuates these low-frequency events and affects the lift oscillation significantly, as shown in Fig. 43.2c comparing the red and orange lines. Furthermore, the filtering of the farfield limits the amplitude of the low-frequency lift oscillations, comparing the orange and black lines in Fig. 43.2c, where the latter corresponds to the SE-ILES applying filters only in turbulent regions, as shown in Figs. 43.2a and b, indicating high filter activity by red circles in the circumferential (i) and wall-normal (j) direction, respectively. High filter activity is also observed around the leading edge in the wall-normal direction, where the boundary layer is very thin.

The application of explicit SGS models was also tested using grid B but leads to significantly underestimated lift oscillations. Figure 43.3c shows the lift coefficient as a function of time for the DNS (blue line), ILES (red line) and LES with WALE (green) and mixed-timescale (orange) SGS models. While the computational costs increase using SGS models, low-frequency oscillations in the lift seem to decay even faster compared to the ILES method. The application of the mixed-timescale model (MTS) also led to numerical instabilities. Therefore, the application of SGS models has not been further considered in the scope of this study.

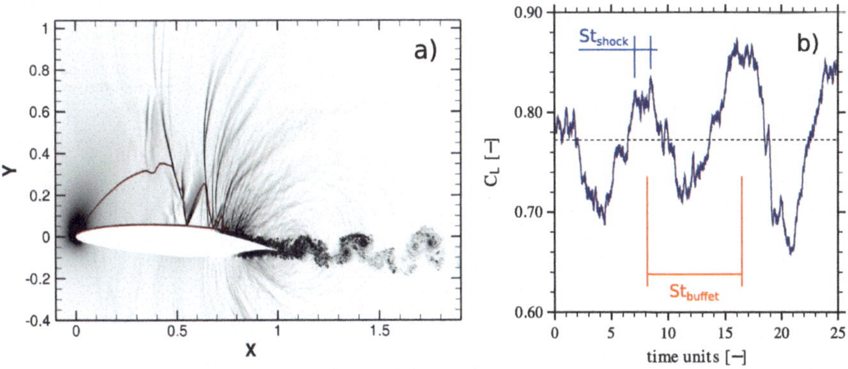

Fig. 43.1 **a** Streamwise pressure gradient and **b** lift coefficient of DNS

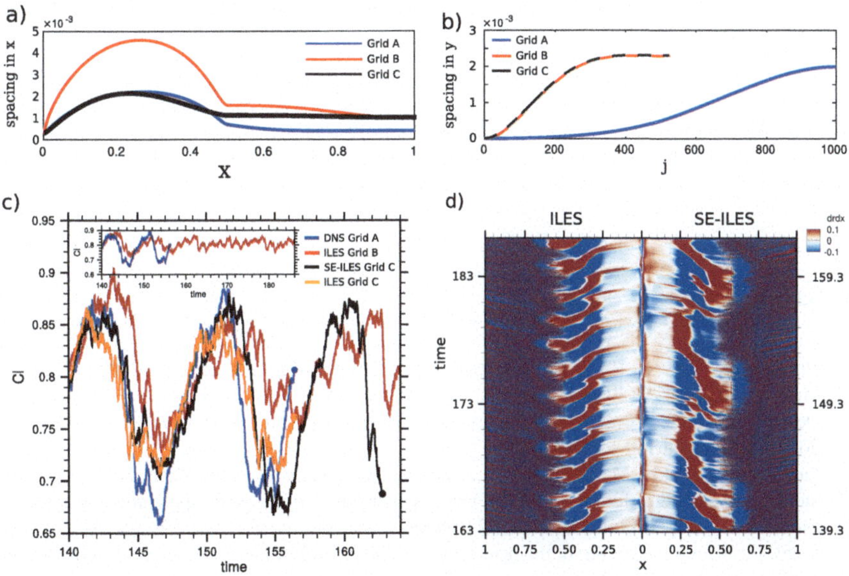

Fig. 43.2 Grid spacing in the **a** circumferential and **b** wall-normal direction, **c** Lift coefficient and **d** space/time diagram

Conclusions

Numerical resolution, including laminar regions, has been shown to significantly affect the simulation of shock motion and acoustic waves, which play an important role in the buffet mechanism. Insufficient resolution mainly impacts the low-lift phases, while the C_L peaks are hardly affected. In the present work, a novel spectral-error-based ILES has been shown to be capable of maintaining numerical stability, applying a filter only where necessary. The DNS result could be satisfactorily reproduced by the SE-ILES method, reducing the overall costs by a factor of 16.

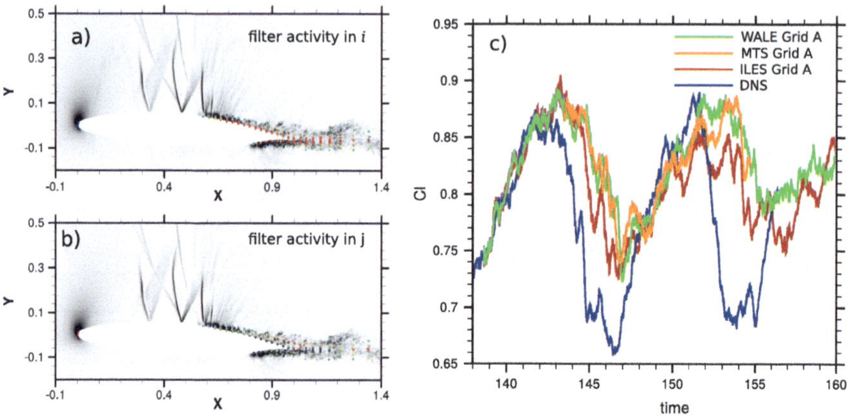

Fig. 43.3 Filter activity in the **a** circumferential and **b** wall-normal direction. **c** Lift coefficient considering also SGS models

Acknowledgements The authors want to acknowledge the Partnership for Advanced Computing in Europe (PRACE) for provided computational Resources. MZ was supported by EPSRC (grant EP/R037167/1).

References

1. Dandois, J., Mary, I., Brion, V.: Large-eddy simulation of laminar transonic buffet. J. Fluid Mech. **850**, 156–178 (2018)
2. De Tullio, N., Sandham, N.D.: Direct numerical simulations and modal analysis of subsonic flow over swept airfoil sections (2019)
3. Jacobs, C.T., Zauner, M., De Tullio, N., Jammy, S.P., Lusher, D.J., Sandham, N.D.: An error indicator for finite difference methods using spectral techniques with application to aerofoil simulation. Comput. Fluids **168**, 67–72 (2018)
4. Jones, L.: Numerical studies of the flow around an airfoil at low Reynolds number. Ph.D. thesis (2008)
5. Sansica, A.: Stability and unsteadiness of transitional shock-wave/boundary-layer interactions in supersonic flows, 179–184 pp. Ph.D. thesis (2015). University of Southampton
6. Visbal, M.R., Gaitonde, D.V.: On the use of higher-order finite-difference schemes on curvilinear and deforming meshes. J. Comput. Phys. 181(1), 155–185 (2002)
7. Zauner, M.: Direct numerical simulation and stability analysis of transonic flow around airfoils at moderate Reynolds numbers. Ph.D. thesis (2019). University of Southampton
8. Zauner, M., De Tullio, N., Sandham, N.: Unsteady behaviour in direct numerical solutions of transonic flow around an airfoil. In: AIAA Aviation Forum. American Institute of Aeronautics and Astronautics, Reston, Virginia, 1–14 Jun 2018
9. Zauner, M., De Tullio, N., Sandham, N.D.: Direct numerical simulations of transonic flow around an airfoil at moderate Reynolds numbers. AIAA J. **57**(2), 597–607 (2019)
10. Zauner, M., Jacobs, C.T., Sandham, N.D.: Grid refinement using spectral error indicators with application to airfoil DNS. In: ECCM-ECFD Conference Proceedings, Glasgow 2018

Chapter 44
Large-Eddy Simulation of Boundary Layer Transition in a Compressor Cascade

J. Fang, S. Rolfo, X. Yu, C. Moulinec, X. Xu and D. R Emerson

Introduction

The spike flow is often observed around the lead-edge of a blade in an axial compressor cascade, which would lead to the sudden expansion and pressure diffusion. A strong spike flow might create a local separation bubble near the leading-edge and even trigger the boundary layer transition immediately [1]. Therefore, the spike flow can be regarded as a source or an amplifier of fluctuations to the boundary layer transition [2]. It was once believed that a thick leading-edge would reduce the sensitivity of the blade to the angle of attack (AoA) [3]. However, the experiments in 1950s have found that a sharp lead-edge could accept a larger AoA and have smaller loss [4]. Further research has revealed that the reason is due to the spike flow at the lead-edge [1]. In 2010, Goodhand and Miller [5] from the Whittle Laboratory have pointed out that the strength of the spike should be one of the principles for the design of a blade, and they defined a diffusion factor for the leading-edge, D_{spike}, to guide the blade design. They proposed that when $D_{spike} < 0.1$, the geometry of the leading-edge has little effect on the performance of the blade. However, the recent research pointed out that even $D_{spike} < 0.1$, the development of spike could be affected by the shape of the leading-edge, and the aerodynamic performance of the blade could be therefore influenced [6]. Therefore, the connection between the geometry of the leading-edge and the boundary layer transition on the surface of the blade is still not well clarified. The objectives of this work are to investigate how well LES can capture the transition

J. Fang (✉) · S. Rolfo · C. Moulinec · D. R. Emerson
Computational Engineering Group, Scientific Computing Department,
STFC Daresbury Laboratory, Warrington WA4 4AD, UK
e-mail: jian.fang@stfc.ac.uk

X. Yu · X. Xu
National Key Laboratory of Science and Technology on Aero-Engine Aero-Thermodynamics,
School of Energy and Power Engineering, Beihang University, Beijing 100191, China
e-mail: yuxj@buaa.edu.cn

M. García-Villalba et al. (eds.), *Direct and Large Eddy Simulation XII*,
ERCOFTAC Series 27,
https://doi.org/10.1007/978-3-030-42822-8_44

in compressor cascades. The LES data are compared with the experimental measurement, and the influences of mesh resolution, incoming turbulence, and angle of attack are studied. The data are then analysed to study the transitional characteristics of the boundary layer on the surfaces of the blade.

Methodology

A linear Controlled Diffusion Aerofoil (CDA) compressor blade cascade is simulated in the work. The leading and trailing edges of the blade are designed as circle-shapes, as shown in Fig. 44.1. The parameters of the blade are given in Table 44.1, and the spanwise length is 10% of the chord, which is narrower than some DNS and LES. The narrow spanwise domain used here is to save the computational cost, and the effect of the spanwise length will be investigated in the following research. The cascade operates in the air with an inflow velocity of 44.24 ms^{-1}, and the Reynolds number based on the chord is $Re_c = 525{,}000$. Five cases have been studied to check the influences of mesh, incoming turbulence and angle of attack. The parameters are summarised in Table 44.1. Two sets of meshes have been studied, the coarse mesh has 10.85 million hexahedral cells, whereas the fine mesh contains 81.81 million cells with a refinement in the x-y plane, as shown in Fig. 44.1. The refinement is to enhance the resolutions in both the near-wall region and the streamwise direction to tackle possible small flow structures during the boundary layer transition. The number of cells in the spanwise direction for both mesh is 128. Inflow and outflow

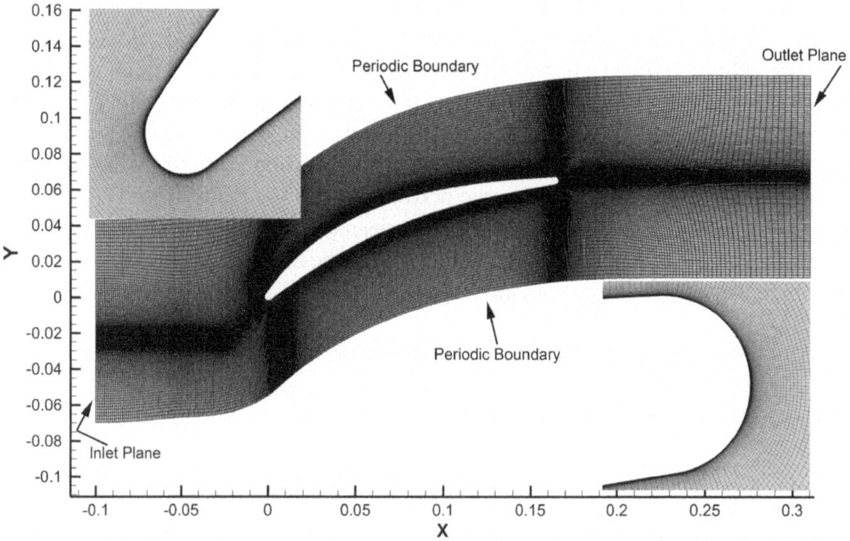

Fig. 44.1 Geometry and mesh of the cascade

Table 44.1 The parameters of the CDA cascade and computational parameters

Chord	Pitch width	Inlet metal angle	Camber angle
0.18 m	0.1125 m	47°	41°
Case	Turbulence intensity (%)	Angle of attack	Mesh size
Case1	0.0	0°	10.85M
Case2	2.8	0°	10.85M
Case3	0.0	0°	81.81M
Case4	2.8	−5°	10.85M
Case5	2.8	+8°	10.85M

boundary conditions are applied in the inlet and outlet planes, and periodic conditions are applied in the y- and z-directions. Both cases, with and without incoming turbulence, have been studied. The inflow turbulence is generated with a synthetic eddy method [7], and introduced to the computational domain via the inlet plane. The turbulence intensity (Ti) is 2.8%, and the integral length scale is 1% of the chord's length. The simulations were performed using the open-source finite volume software *Code_Saturne*, which is developed and maintained by EDF R&D. The code is second-order accurate in space and time and the velocity-pressure coupling is ensured through a prediction/correction method based on a SIMPLEC algorithm. The filtered, incompressible Navier–Stokes equations are resolved and the sub-grid terms are modelled using the dynamic Smagorinky model [8].

Results and Disscusion

Instantaneous Flow Field

The instantaneous turbulent coherent structures visualised using the Q-criterion in Fig. 44.2 shows that a bypass transition occurs at around mid-chord on the suction side of the blade. Comparing Case1 and Case2 in Fig. 44.2, it can be seen that the incoming turbulence has little influence on the transition location and does not trigger any transition in the boundary layer. With the refined mesh in Case3, we can see the transition mode predicted is different from the coarse mesh case. Some 2-D large-scale spanwise vortices are captured in Case3, indicating the local separation, roll-up of vortices and breaking down to turbulence. Comparing Case4 with a negative angle of attack (AoA), and Case5 with a positive AoA, we can see that the negative AoA can stabilise the boundary layer and the incoming turbulence can involve to very large-scale streamwise vortices without breaking down. As expected, the boundary layer in the positive AoA case is less stable and the boundary layer is filled with small-scale structures and the transition is located upstream with respect to Case1. The

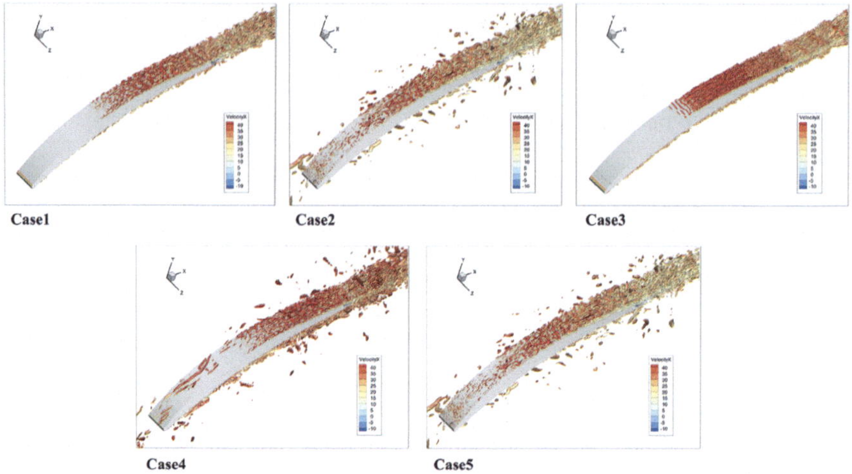

Fig. 44.2 Instantaneous turbulent coherent structures visualised using the Q-criterion and coloured with x-velocity

instantaneous wall-tangential velocity in the near-wall region are shown in Fig. 44.3, from which we can see the streaky velocity structures on the latter half part of the blade, indicating the wall turbulence generated after the boundary layer transition. It can be noted that Case1 and Case2 present different transition processes, indicating current results are not mesh converged. For Case3, a clear separation region can be seen and the transition is clearly induced by the instability of the transition bubble. No separation region can be identified in Case1. Comparing Case1 and Case2, we can see the influence of incoming turbulence. It seems the incoming turbulence delays the appearance of the velocity streaks. Regarding the influence of AoA, the negative AoA delays the transition and positive AoA promotes the transition, as expected. It's worth noting that there is a small separation bubble together with an immediate boundary layer transition downstream in Case5. This verified that the strong spike flow near the leading-edge causes the local separation bubble and trigger a fast boundary layer transition.

Mean Flow Statistics

The mean pressure coefficients on the blade surface are presented in Fig. 44.4, together with the data from the experiment. The boundary layer transition can be identified as a pressure jump on the suction side, but only Case3 with the fine mesh is able to capture it. As the analysis above has shown that only Case3 resolves the instantaneous flow separation on the suction side, therefore, it can be inferred that the boundary layer transition in the experiment should also be triggered by the flow separation. However, the mean streamlines (not presented here) shows that there is

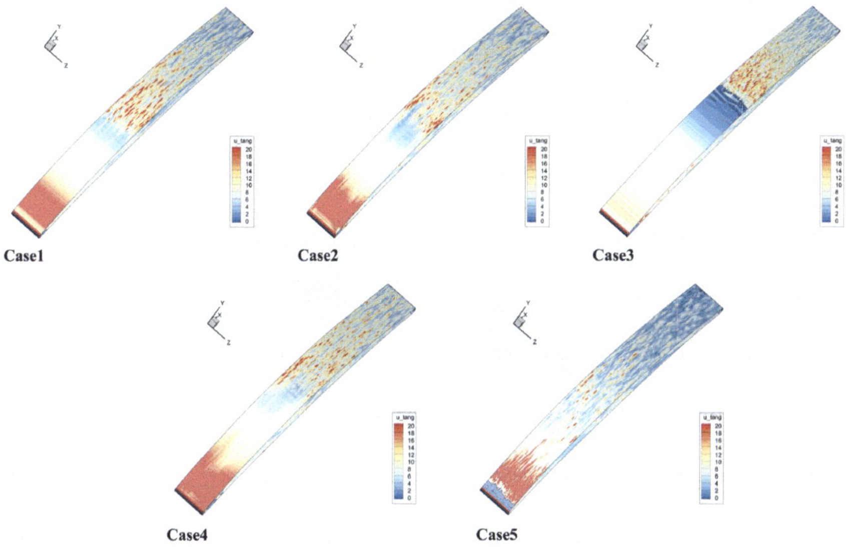

Fig. 44.3 Instantaneous wall-tangential velocity at the slice of 0.05 mm away from wall. The black thick lines are the 0-velocity contour lines

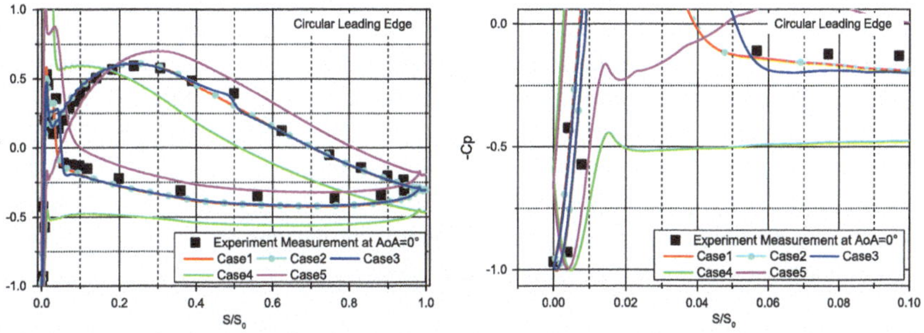

Fig. 44.4 Mean pressure coefficients (left), and a close-up look at the leading-edge (right)

no mean flow separation, which indicates the boundary layer is in a critical state. The pressure coefficients of Case1 and Case2 are on top of each other, meaning the influence of the incoming turbulence is negligible. With the deviation of the AoA from 0°, the widening of the spike flow strength on the suction side can be observed in Fig. 44.4 (right). A local peak of C_P can be identified in Case4 and Case5, creating an adverse pressure gradient locally. By check the mean streamlines around the leading-edge in Fig. 44.5, it can be confirmed that a local separation bubble appears on the pressure side for negative and 0° AoA, and on the suction side for the positive AoA. Also, Case4 has a much larger separation bubble on the pressure side than the 0° cases. This explains why the transition on the suction side is delayed at negative AoA and promoted at positive AoA.

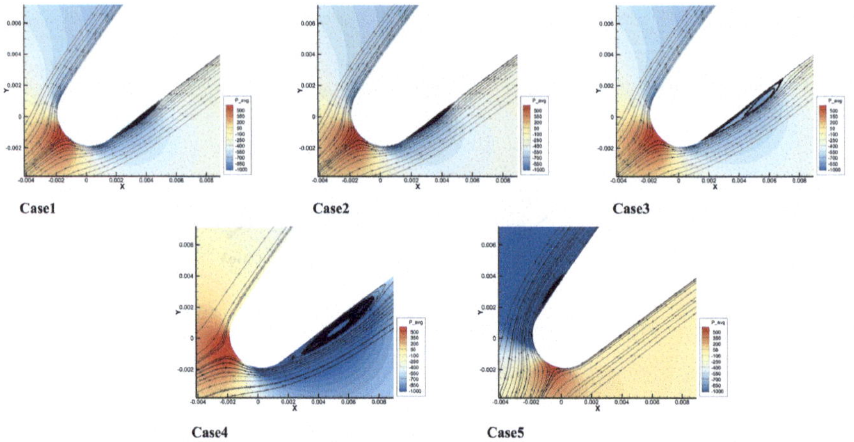

Fig. 44.5 Mean streamlines near the leading-edge

Summary and Concluding Remarks

LES studies of flow in a compressor cascade with a CDA blade are conducted, focusing on the influence of the leading-edge spike flow to the boundary layer transition on the suction side of the blade. The LES indicates that the boundary layer transition is mainly caused by the instability of the separated shear layer, although the mean flow is still attached on the suction surface. At 0° and negative AoA, the leading-edge spike creates a local separation bubble on the pressure side, however, the local bubble appears on the suction side when AoA is positive, which triggers the boundary layer transition immediately on the suction side. More simulations are needed to show the independence of grid, domain size and SGS model.

Acknowledgements The project is supported by the UK Turbulence Consortium (EP/L000261/1 and EP/R029326/1) and the UK Consortium of Turbulent Reacting Flows (EP/K024574/1 and EP/R029369/1) under EPSRC, and the National Natural Science Foundation of China (51420105008, 51790513 and 51476004). The simulations were conducted on the ARCHER supercomputer.

References

1. Walraevens, R.E., Cumpsty, N.A.: Leading edge separation bubbles on turbomachine blades. J. Turbomach. **117**, 115–125 (1995)
2. Wheeler, A.P.S., Sofia, A., Miller, R.J.: The effect of leading-edge geometry on wake interactions in compressors. J. Turbomach. **117**, (2009)
3. Cumpsty, N.A.: Compressor Aerodynamics. Longman, United Kingdom (1989)
4. Andrews, S.J.: Tests related to the effect of profile shape and camber-line on compressor cascade performance. Aeronaut. Res. Counc.**2743**, (1949)

5. Goodhand, M.N., Miller, R.J.: Compressor leading edge spikes: a new performance criterion. J. Turbomach. **133**, 041013 (2011)
6. Liu, B.J., Yuan, C.-X., Yu, X.-J.: Effects of leading-edge geometry on aerodynamic performance in controlled diffusion airfoil. J. Propuls. Technol. **7**, 890–897 (2011)
7. Jarrin, N., Benhamadouche, S., Laurence, D., Prosser, R.: A synthetic-eddy-method for generating inflow conditions for large-eddy simulations. Int. J. Heat Fluid Fl. **27**, 585–593 (2006)
8. Germano, M., Piomelli, U., Moin, P., Cabot, W.H.: A dynamic subgrid-scale eddy viscosity model. Phys. Fluids A-Fluid **3**, 1760–1765 (1991)

Chapter 45
Critical Analysis of the Numerical Setup for the Large-Eddy Simulation of the Low-Pressure Turbine Profile T106C

Christian Morsbach and Michael Bergmann

Introduction

The increasing availability of computational resources has led to numerous publications on large eddy and even sporadically on direct numerical simulations of low-pressure turbine (LPT) flows with Reynolds numbers around 10^5, e.g. [4, 11]. In need for an appropriate validation case, the community has focussed on the T106 profile series for which experimental data are publicly available [6]. The T106A and T106C variants at $\mathrm{Re} = 60,000$ and $\mathrm{Re} = 80,000$, respectively, at low inflow turbulence have been selected as an advanced test case in the *High-Order CFD Methods* workshop series. While the experiment was arranged as a linear cascade of six prismatic blades with an aspect ratio of 2.4, the computational geometry is simplified assuming pitchwise and spanwise periodicity and laminar inflow conditions. The spanwise domain size amounts to 10% of the chord length. Besides the influence of 3D effects, uncertainties exist concerning the exact inflow and stagger angles and have been addressed in several studies, cf. [2, 4]. Even if the nominal boundary conditions of the workshops are used, different numerical methods and solvers show a significant variation in the blade pressure distribution and wake loss profile [3, 5, 7].

In order to fairly assess advanced numerical methods such as high-order discretisation or synthetic turbulence inflow conditions, a good understanding of the numerical setup and its influence on the quantities of interest is required. In an LES, most of these are based on statistical moments. For practical reasons of available computing resources, these moments are estimated using a finite number of samples and are, hence, subject to statistical error. In order to be able to draw any meaningful conclusions when comparing two simulation results, this error has to be considered. Following this rationale, we performed a series of large-eddy simulations of

C. Morsbach (✉) · M. Bergmann
Numerical Methods, Institute of Propulsion Technology,
German Aerospace Center (DLR), Cologne, Germany
e-mail: christian.morsbach@dlr.de

M. García-Villalba et al. (eds.), *Direct and Large Eddy Simulation XII*,
ERCOFTAC Series 27,
https://doi.org/10.1007/978-3-030-42822-8_45

the T106C LPT profile at isentropic exit conditions of $\mathrm{Re} = 80,000$ and $M = 0.65$ using DLR's solver for turbomachinery flows TRACE and carefully analysed the results.

Numerical Method

TRACE solves the filtered compressible Navier–Stokes equations using a second-order accurate, density-based finite volume scheme applying MUSCL reconstruction with $\kappa = 1/3$ [13]. A fraction of 10^{-3} of Roe's numerical flux [10] is added to a central flux to avoid odd-even decoupling. Time integration is performed using a third-order accurate explicit Runge–Kutta method. The subgrid stresses are computed by the WALE model [8].

The following methods have been applied as boundary conditions:

Unsteady 1D Characteristics	Time and surface average boundary state-driven towards prescribed boundary values (stagnation pressure and temperature, flow angles at inflow, static pressure at outflow) by means of incoming characteristics. 1D non-reflecting [12].
Riemann	Time and space local state forced to meet prescribed boundary values (as above) based on Riemann invariants. Prone to reflections [1].
Dirichlet	Directly prescribe time and space local state computed with prescribed velocity vector and static temperature while static pressure is taken from inner cell. Reflecting.

If not otherwise stated, the 1D non-reflecting boundary condition is used for both inflow and outflow.

The primary data obtained from LES are time series of the filtered quantities at the solution points. In the following averaging process over a finite length sample, any filtered quantity f can be split into a time mean $\overline{f}$ and a fluctuating part f'. The error on the mean is given by

$$\sigma_{\overline{f}} = \sqrt{\frac{\sigma_f^2}{n}}, \quad n = \frac{T_{\mathrm{average}}}{2T_{\mathrm{int}}}, \quad T_{\mathrm{int}} = \int_0^\infty \frac{\overline{f'(t)f'(t+\tau)}}{\sigma_f^2} d\tau \tag{45.1}$$

with the variance σ_f^2 and the number of independent samples n. The latter can be estimated by relating the integral time scale T_{int} to the length of the averaging interval T_{average} [9]. A measure for the integral time scale can be obtained from the integral over the auto-correlation function. For practical reasons, the integral is truncated at the first zero crossing of the integrand.

Errors on the derived quantities of engineering interest are obtained using error propagation. The isentropic Mach number and total pressure loss coefficient are given by

$$\overline{M_{\text{is}}} = \sqrt{\frac{2}{\gamma-1}\left(\left(\frac{p_t^{\text{ref}}}{\overline{p}}\right)^{\frac{\gamma-1}{\gamma}} - 1\right)}, \quad \overline{\zeta} = 1 - \frac{\overline{p_t}}{p_t^{\text{ref}}} \tag{45.2}$$

with the heat capacity ratio γ, reference stagnation pressure p_t^{ref}, stagnation pressure p_t and static pressure p treated as independent variables.

Results of the grid study are not reported due to space constraints. The final grid was resolved with 40 points per 10% chord in spanwise direction, giving a total of $10.723 \cdot 10^6$ cells with non-dimensional cell sizes of $\Delta x^+_{\text{max}} = 40.5$, $\Delta x^+_{\text{avg}} = 9.0$, $\Delta z^+_{\text{max}} = 24.6$, $\Delta z^+_{\text{avg}} = 8.6$ and $y^+ < 0.91$ along the blade surface.

Results and Discussion

The flow field was initialised using a RANS solution from which the simulation shows a transient phase of 2D vortex shedding. Eventually, the vortices break down into turbulent structures. All times are expressed in terms of the convective time $T_{\text{through}} = c/u_{\text{exit}}$ with a chord length $c = 0.093$ m and an approximate exit velocity $u_{\text{exit}} \approx 200$ m/s. A quantity to assess the passing of the initial transient is the blade force F_y averaged over 10 throughflows normalised by the average over the last 90 throughflows as shown in Fig. 45.1 (*left*). It can be said that the initial transient has washed out after about 10 throughflows.

Figure 45.1 (*right*) shows the statistical convergence of the mean total pressure loss coefficient ζ and its error σ_ζ with increasing averaging time t_{av}. Each line corresponds to a different pitchwise position traversing the pressure side of the wake at $x = x_{\text{TE}} + 0.7c_{\text{ax}}$ with the axial position of the trailing edge x_{TE} and the axial chord length c_{ax}. Even after 90 throughflows, some drift can be observed especially in the centre of the wake. It remains, however, within the statistical error. Due to limited computational resources, the following studies were conducted with averages over 90 throughflows.

The choice of inflow boundary condition formulation revealed a significant influence on the blade pressure distribution. It has to be mentioned that RANS simulations are basically insensitive to this choice for this case, hence, the effect must be due to unsteady reflections of upstream travelling waves. Figure 45.2 (*left*) shows the blade pressure distribution in terms of isentropic Mach number. The locations at which probe data were available to estimate the error are marked with crosses. Only within this range is the error indicated by shaded regions. The three methods produce significantly different results downstream of the suction side separation point. The results obtained with the *Dirichlet* boundary condition are characterised by a very thin closed separation bubble while. Changing to the *Riemann* formulation with the inlet 0.1 m away from the leading edge thickens the still closed separation bubble.

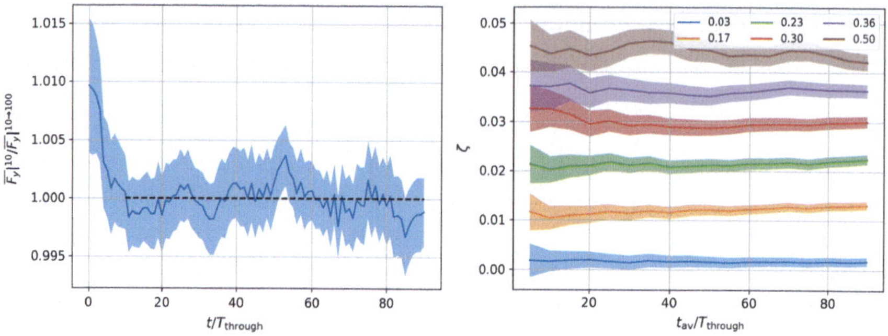

Fig. 45.1 Statistical convergence of pitchwise blade force averaged over 10 throughflows (*left*) and total pressure loss coefficient at $x = x_{TE} + 0.7c_{ax}$ at different relative pitchwise positions (*right*)

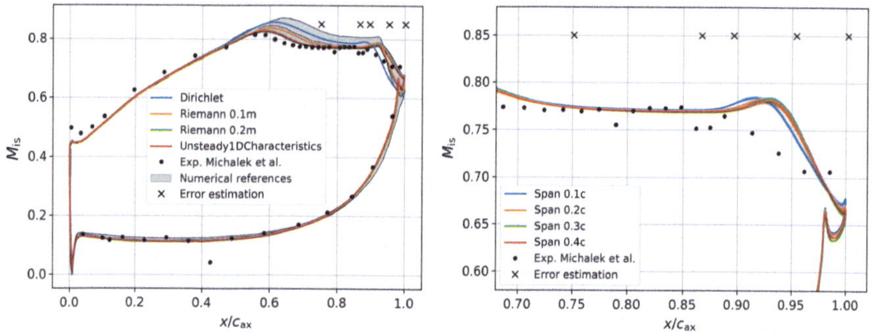

Fig. 45.2 Blade pressure distribution in terms of isentropic Mach number for different inflow boundary conditions with numerical references [3, 5, 7] (*left*) and spanwise domain sizes (*right*)

Moving the inlet upstream to 0.2 m to weaken the influence of reflections or employing the 1D non-reflecting formulation at 0.1 m leads to an open separation bubble and the best agreement with experiments. The great spread in the results suggests that most of the observed variations in the literature could be due to different formulations of boundary conditions.

At a first glance at Fig. 45.2 (*right*), the influence of the spanwise domain size is rather small yet statistically significant. The different pressure distribution can be attributed to a slight change in shape of the separation bubble, especially a small secondary recirculation region within the bubble at $x/c_{ax} \approx 0.9$. The differences become more obvious downstream in the wake. Figure 45.3 is based on data from a 1D-probe near its centre at $x = x_{TE} + 0.4c_{ax}$. The normalised two-point correlation along the span is shown for the axial u' and spanwise w' component of the fluctuating velocity (*left*). For 10% span the latter reaches zero within half the domain size. With increasing domain size, the zero-crossing gradually moves to below a quarter. It has to be mentioned though that the two-point correlation does not show a zero value for large Δz. For the axial velocity component, zero is not even reached for greater domains and a very high correlation value above 0.6 remains for 10% span. Looking

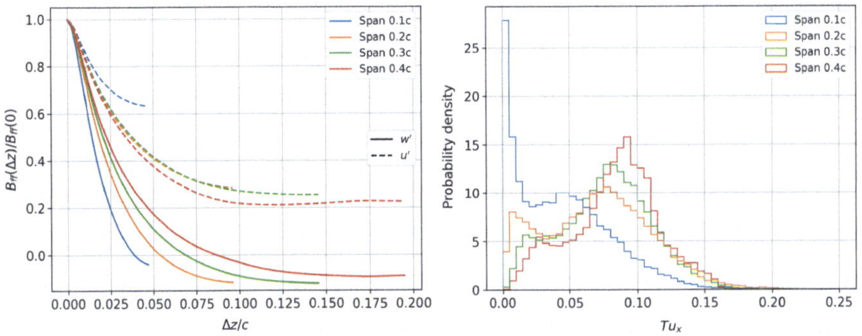

Fig. 45.3 Two-point correlation of axial and spanwise velocity (*left*) and histogram of instantaneous turbulence intensity (*right*) at $x = x_{\mathrm{TE}} + 0.4c_{\mathrm{ax}}$ in the centre of the wake

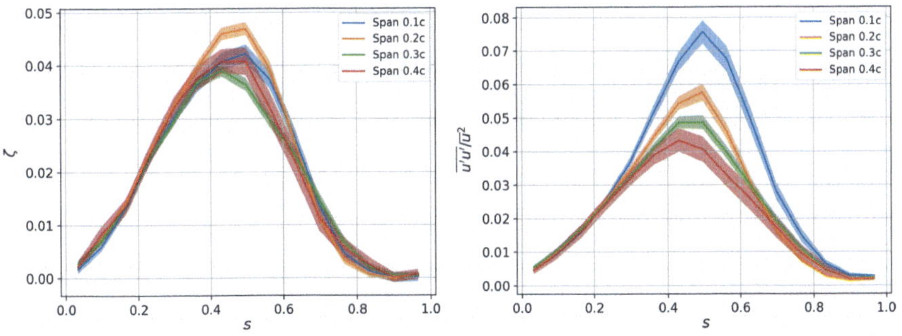

Fig. 45.4 Total pressure loss coefficient (*left*) and axial Reynolds stress component (*right*) at $x = x_{\mathrm{TE}} + 0.7c_{\mathrm{ax}}$

at the raw time traces, this can be attributed to significant periods of spanwise fully correlated flow. To quantify this, an instantaneous turbulence intensity

$$Tu_x(t) = \frac{\sqrt{\langle u'^2(x,t)\rangle}}{\langle u(x,t)\rangle} \tag{45.3}$$

can be defined using a spanwise average denoted by $\langle . \rangle$. The probability density distribution of this quantity is shown in Fig. 45.3 (*right*). For every span, we observe a distribution with two peaks—one at low and one at high turbulence intensity. The anomalous peak at 0 can be seen as a hint for too small a spanwise domain. Increasing the span significantly changes the distribution towards a state where no fully correlated phases across the span exist. Therefore, the assumption of spanwise periodicity in combination with too small domain sizes can lead to wrong conclusions.

The total pressure loss wake profile in Fig. 45.4 (*left*) shows an oscillatory dependence on the domain size while the axial Reynolds stress component (*right*) decreases with increasing domain size. Both quantities show a statistically significant influence of the domain size. For this case, we conclude that the domain should cover at least $0.3c$.

Conclusion

The numerical setup of an LES of the T106C LPT profile has been assessed considering the statistical error on mean values due to finite averaging times. Both inflow boundary conditions and the spanwise domain size show a significant influence on the predicted blade pressure distribution and wake profiles. Most of the variation of numerical results for the pressure distribution can be attributed to the choice of inflow boundary condition. For the considered Reynolds number of 80,000, typically assumed spanwise domain sizes of $0.1c$ are insufficient to capture the decay of 2D structures into turbulence appropriately.

References

1. Carlson, J.R.: Inflow/outflow boundary conditions with application to FUN3D. Technical Report NASA TM-2011-217181, NASA Langley Research Center, Hampton, VA (2011)
2. Carton de Wiart, C., Hillewaert, K., Lorriaux, E., Verheylewegen, G.: Development of a discontinuous Galerkin solver for high quality wall-resolved/modelled DNS and LES of practical turbomachinery flows on fully unstructured meshes. In: ASME Turbo Expo. Montreal, Quebec, Canada (2015)
3. Garai, A., Diosady, L.T., Murman, S.M., Madavan, N.K.: DNS of low-pressure turbine cascade flows with elevated inflow turbulence using a discontinuous-Galerkin spectral-element method. In: ASME Turbo Expo. Seoul, Korea (2016)
4. Hillewaert, K., Carton de Wiart, C., Verheylewegen, G., Arts, T.: Assessment of a high-order discontinuous Galerkin method for the direct numerical simulation of transition at low-reynolds number in the T106C high-lift low pressure turbine cascade. In: ASME Turbo Expo. Düsseldorf, Germany (2014)
5. Hillewaert, K., Hartmann, R., Leicht, T., Couaillier, V., Wang, Z.J., Cagnone, J.S.: Summary and conclusions of the 4th international workshop on high order CFD methods. In: ECCOMAS. Crete, Greece (2016). https://how4.cenaero.be/content/ms910-eccomas-cfd-2016
6. Michálek, J., Monaldi, M., Arts, T.: Aerodynamic performance of a very high lift low pressure turbine airfoil (T106C) at low Reynolds and high Mach number with effect of free stream turbulence intensity. J. Turbomach. **134**(6), 061009 (2012)
7. Mitra, P., Mathew, J.: Large eddy simulation of turbine aerodynamics by explicit filtering. In: GPPS Montreal18 Proceedings. Montreal, Canada (2018)
8. Nicoud, F., Ducros, F.: Subgrid-scale stress modelling based on the square of the velocity gradient tensor. Flow Turbul. Combus. **62**(3), 183–200 (1999)
9. Ries, F., Nishad, K., Dressler, L., Janicka, J., Sadiki, A.: Evaluating large eddy simulation results based on error analysis. Theor. Comp. Fluid Dyn. **32**(6), 733–752 (2018)
10. Roe, P.L.: Approximate Riemann solvers, parameter vectors, and difference schemes. J. Comput. Phys. **43**(2), 357–372 (1981)
11. Sandberg, R.D., Michelassi, V., Pichler, R., Chen, L., Johnstone, R.: Compressible direct numerical simulation of low-pressure turbines–part I: methodology. J. Turbomach. **137**(5), 051011 (2015)
12. Schlüß, D., Frey, C., Ashcroft, G.: Consistent non-reflecting boundary conditions for both steady and unsteady flow simulations in turbomachinery applications. In: ECCOMAS. Crete, Greece (2016)
13. van Leer, B.: Towards the ultimate conservative difference scheme. V. A second-order sequel to Godunov's method. J. Comput. Phys. **32**(1), 101–136 (1979)

Chapter 46
The Coupling of a Synthetic Turbulence Generator with Turbomachinery Boundary Conditions

S. Leyh and C. Morsbach

Introduction

The power of modern computers and massive parallelization on clusters makes it possible to do scale-resolving simulations at *Reynolds* numbers relevant for turbomachinery. Since the computational domain can only represent a finite part of the physical world, it is essential for numerical simulation codes to provide high quality boundary conditions (BC) in the sense of non-reflective properties and the possibility of unsteady inflow.

If the inlet of the considered configuration is located in a turbulent region, the solver has to feature an appropriate boundary treatment. Especially in turbomachinery, the flow has a highly unsteady and turbulent character. Hence, the inlet boundary has to deal with both wakes of previous stages and turbulence. Since precursor simulations are very expensive and impractical for industrial configurations, there are numerous approaches to creating artificial (synthetic) turbulence. Popular examples are the digital filter method [4], the synthetic eddy method [3], and spectral methods based on the superposition of *Fourier* modes [5].

The latter method has been implemented into DLR's in-house turbomachinery code TRACE [8] and tested in combination with a source term formulation and a BC formulation [7]. Since TRACE provides a number of different boundary conditions, the present paper investigates the interaction of selected BCs with the turbulence generator.

S. Leyh (✉) · C. Morsbach
German Aerospace Center (DLR), Linder Höhe, Cologne, Germany
e-mail: sascha.leyh@dlr.de

C. Morsbach
e-mail: christian.morsbach@dlr.de

M. García-Villalba et al. (eds.), *Direct and Large Eddy Simulation XII*,
ERCOFTAC Series 27,
https://doi.org/10.1007/978-3-030-42822-8_46

Synthetic Turbulence Generator

The synthetic turbulence generator (STG) used in TRACE is described by Shur et al. [15] and was implemented by Morsbach and Franke [8] who adapted it for periodic domains. The turbulence dissipation rate ε in the original formulation is unknown and can be written as $\varepsilon = \beta k \omega$ [17, Eq. A4b]. With the turbulent length scale $L_T = \sqrt{k}/\omega$ [17, Eq. 24] the dissipation rate can be replaced by

$$\varepsilon = \frac{\beta k^{\frac{3}{2}}}{L_T} \tag{46.1}$$

For the present work, the length scale L_T and the *Reynolds* stress tensor are prescribed. To calculate the *Reynolds* stresses of isotropic turbulence from a given turbulence intensity Tu, the initialized velocity as a result of the prescribed *Mach* number Ma, total temperature T_0, and the isentropic relation, as well as $a = \sqrt{\gamma R_s T}$ is used and assumed constant during the simulation.

Boundary Conditions

In the finite-volume solver of TRACE, the BCs determine a set of primitive variables according to the respective theory, located at the face of the boundary cell. The used ghost cells are filled by an extrapolation polynomial of order zero or one. Furthermore, the inner cells are updated first and subsequently the BCs are evaluated with the new cell values.

The considered ***Dirichlet*** BC prescribes the velocity components and the static temperature. The velocity consists of a constant given mean part with superposed perturbations of the STG. The static pressure of the first inner cell and the given static temperature are used to calculate the density (ideal gas). The ghost cells are filled by linear extrapolation using the state of the first inner cell and the face state.

The BC of Sandhu and Sandham [12] is called **integrated characteristic** in this work and is based on the characteristic concept described by Thompson [16]. The basic equation is given as [12, Eq. 3.1]

$$\mathbf{q} = \mathbf{q}_0 + \int \left(\frac{\partial \hat{F}}{\partial x}\right)_{\lambda_1} dt, \quad \left(\frac{\partial \hat{F}}{\partial x}\right)_{\lambda_1} = \begin{pmatrix} \frac{1}{2a^2} \\ \frac{u}{2a^2} - \frac{1}{2a} \\ \frac{v}{2a^2} \\ \frac{w}{2a^2} \\ \frac{1}{2}\,|\mathbf{v}|^2 \frac{1}{2a^2} - \frac{u}{2a} + \frac{1}{2}\frac{1}{\gamma - 1} \end{pmatrix} \mathscr{L}_1 \tag{46.2}$$

According to the authors, $\mathbf{q}_0$ in Eq. (46.2) "represents the forced basic flow behavior at the inflow boundary" [12, p. 6] and the integral is a "correction made to the inflow boundary [so that] the characteristic velocity λ_1 leaves the computational domain smoothly". $\frac{\partial \hat{F}}{\partial x}$ is the *Euler* part x-component of the conservation equations and is determined by a characteristic analysis of the 1D *Euler* equations. The outgoing wave amplitude $\mathscr{L}_1$ can be found in Poinsot and Lele [10, Eq. 19]. The integral in Eq. (46.2) is approximated with the same time scheme as used in the interior domain. Therefore the integrand is calculated with inner cell values before the time step update of the interior cells. Unlike the other BCs in this paper, the resulting values are directly written into the ghost cells. Treating them as face values and filling the ghost cells by extrapolating leads to unstable behavior.

The ***Riemann*** BC is described, e.g., by Carlson [1, p. 7] and is implemented with changed sequence. The total pressure and temperature, as well as the flow angle are given at the inflow. The outgoing *Riemann* invariant $j_- = u - \frac{2a}{\gamma-1}$ [1, Eq. 12] is calculated with values of the first inner cell and is extrapolated to the boundary face. Instead of assuming adiabatic flow, the total enthalpy h_t is calculated from the given total temperature. The difference of the invariants can be written as

$$j_+ - j_- = \frac{4}{\gamma - 1} a \; , \tag{46.3}$$

and since the sonic speed a is a function of total enthalpy and the outgoing invariant, the incoming invariant $j_+ = u + \frac{2a}{\gamma-1}$ can be written as

$$j_+ = \left(1 - \frac{4}{\gamma - 1}\right) j_- \pm \frac{4}{\gamma + 1}\sqrt{(\gamma + 1)\, h_t - \frac{\gamma - 1}{2} j_-^2} \; . \tag{46.4}$$

"The physically consistent result is the larger of the two roots" [1, p. 8]. With known values of both invariants the *Mach* number Ma and the sonic speed a are calculated. The velocity components are determined from the given inflow direction. The static pressure is calculated with the given total pressure and the known *Mach* number using the isentropic relation. This static pressure and the known sonic speed are used to calculate the density $\rho = \frac{\gamma p}{a^2}$. Since the turbulent perturbations are added to the calculated face state, the effective total pressure is larger than the prescribed value. Therefore, the total pressure has to be corrected for the turbulent kinetic energy.

The BC called **unsteady 1D characteristics** is described by Giles [2] and was implemented in TRACE by Schlüß et al. [13]. In this BC the eigenvector matrices L_{1D} [13, Eq. 20] and its inverse R_{1D} [13, Eq. 22] are used to transform primitive into characteristic variables and vice versa. The flow at the boundary is decomposed into three parts. The first part is a mean part designed to match the specified mean flow values using relaxation and a residual vector [13, Eq. 29]. A second part contains the instantaneous difference of the area-time averaged and area-averaged values. After transforming this primitive vector of perturbations into characteristic variables, the incoming characteristics are set to zero to obtain a non-reflective behavior. Another part consists of the difference of area-averaged and current values of the first inner cell and the incoming characteristic values are set to zero. Following the transformation of all parts into primitive variables, the STG output is added and the ghost cells are filled by linear extrapolation. As for the ***Riemann*** BC, the total pressure has to be corrected.

Test Cases

To evaluate the basic properties of the coupled BCs, one of the two test cases is **homogeneous isotropic turbulence** in a cuboid domain. For given total pressure of $p_0 = 323555.53$ Pa, total temperature of $T_0 = 300.9$ K and *Mach* number $Ma = 0.1$, the flow field is initialized with the static values of pressure and temperature. The velocity vector has only one non-zero component in x-direction with $u = Ma\sqrt{\gamma RT}$. The simulated time is prescribed by multiples of the throughflow time as ratio of domain size and mean velocity. To avoid averaging over the initial transient phase, recording statistics is started after 30 throughflows. The used subgrid-scale model is the Smagorinsky [14] model and the grid has $128 \times 32 \times 32$ points.

The second case is the **low-pressure-turbine profile T106C** designed by *MTU Aero Engines*. The simulation restarts from a RANS solution and uses a third-order explicit *Runge-Kutta* scheme and the WALE subgrid-scale model [6]. The throughflow time is determined with a stream trace including the point of smallest distance of two adjacent blade surfaces and evaluating the needed time to pass the line part between the x-coordinates of the leading- and trailing edge. Using 0.005% of this time as time step, 20 throughflows are used to record statistics after 10 throughflows of transient phase. The span is equivalent to 40% of the chord length.

Results

The turbulence intensity Tu and length scale L_T, normalized with their respective prescribed values are presented in Fig. 46.1a. The *Dirichlet* and integrated character-

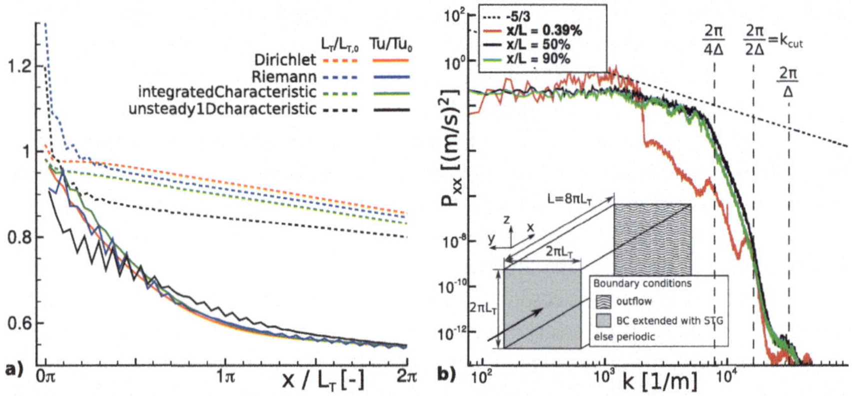

Fig. 46.1 **a** Turbulent length scale and turbulence intensity for all considered BCs, **b** turbulence spectra for three streamwise positions

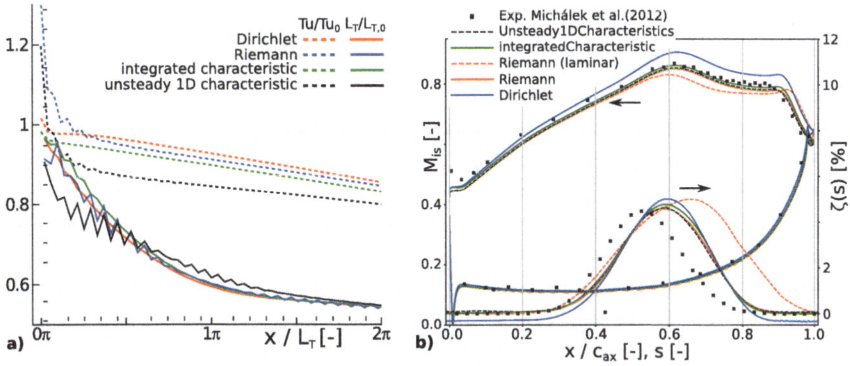

Fig. 46.2 T106C, $Tu = 0.032$, $L_T = 12$ mm, **a** probe position and turbulence intensity for a laminar and turbulent simulation with the ***Riemann*** BC, **b** isentropic *Mach* number (left axis) and total pressure loss (right axis) for various BCs

istic BC reach a value near one ($x/L_T = 0$). A certain development length is apparent for the ***Riemann*** BC until the turbulence intensity distribution is located between the *Dirichlet* and integrated characteristic line. The differences obtained with the unsteady 1D characteristics BC can be explained by a too small normal component of the *Reynolds* stress in x-direction (not displayed). In general, the decay rates are similar. Hence, in Fig. 46.1b only spectra for the *Dirichlet* BC are shown. In the first cell ($x/L = 0.39\%$) the spectrum has an undeveloped shape, but further downstream, the spectrum has a distinct section parallel to the dashed $-5/3$ line. Near the outlet, the spectrum shows oscillations in the $-5/3$ region, since the turbulence does not pass the outlet boundary cleanly. Furthermore, three wavenumbers associated with the edge length Δ of the cubic cells are marked. It can be seen, that wavelengths smaller than four cell sizes are strongly damped. These qualitative results are obtained with all tested BCs.

To meet the given experimental inflow values of the T106C profile, the decay law by Roach [11] is used, as done by Michálek et al. [9]. In this way, the value at the inlet can be reconstructed. In Fig. 46.2a, the simulation results show that the turbulence intensity at a line shifted by half of the pitch reaches a value near the target one (green circle). The turbulence intensity of nearly 1% for laminar inflow shows the effect of upstream traveling acoustic waves, produced by the wake flow. These waves are the reason why the isentropic *Mach* number in Fig. 46.2b of the reflective *Dirichlet* simulation differs from all other BCs. Furthermore, with exception of the *Dirichlet* BC, the general level of the total pressure loss matches the exp. data provided by Michálek et al. [9], although the location of the wake does not coincide.

Conclusions

The STG has been coupled with four BCs. The results show a good agreement with the prescribed values of turbulent intensity and turbulent length scale. The simulation of homogeneous isotropic turbulence in a cuboid domain shows a physical spectrum downstream a certain development length. The analysis of a turbine profile isentropic *Mach* number distribution proves the importance of turbulent inflow conditions, since otherwise no meaningful results are obtained. Particularly for the pressure loss distribution, large discrepancies are obtained if the inflow is approximated as laminar. A next step could be the injection of non-isotropic turbulence and the specification of non-homogeneous input distributions.

References

1. Carlson, J.-R.: Inflow/outflow boundary conditions with application to FUN3D. Technical Report. NASA TM-2011-217181, NASA Langley Research Center (2011)
2. Giles, M.: UNSFLO: a numerical method for the calculation of unsteady flow in turbobmachinery. Technical Report. Gas Turbine Lab Report GTL 205, MIT Dept. of Aero. and Astro. (1991)
3. Jarrin, N., Benhamadouche, S., Laurence, D., Prosser, R.: A synthetic-eddy-method for generating inflow conditions for large-eddy simulations. Int. J. Heat Fluid Flow **27**(4), 585–593 (2006)
4. Klein, M., Sadiki, A., Janicka, J.: A digital filter based generation of inflow data for spatially developing direct numerical or large eddy simulations. J. Comp. Phys. **186**(2), 652–665 (2003)
5. Lee, S., Lele, S.K., Moin, P.: Simulation of spatially evolving turbulence and the applicability of Taylor's hypothesis in compressible flow. Phys. Fluids A Fluid Dyn. **4**(7), 1521–1530 (1992)
6. Nicoud, F., Ducros, F.: Subgrid-scale stress modelling based on the square of the velocity gradient tensor. Flow Turbul. Combus. **62**(3), 183–200 (1999)
7. Matha, M., Morsbach, C., Bergmann, M.: A comparison of methods for introducing synthetic turbulence. In: 7th European Conference on Computational Fluid Dynamics (2018)
8. Morsbach, C., Franke, M.: Analysis of a synthetic turbulence generation method for periodic configurations. In: Proceedings ERCOFTAC Direct and Large Eddy Simulations, vol. XI, pp. 61–67 (2017)
9. Michálek, J., Monaldi, M., Arts, T.: Aerodynamic performance of a very high lift low pressure turbine airfoil (T106C) at low Reynolds and high Mach number with effect of free stream turbulence intensity. J. Turbomach. **134** (2012)
10. Poinsot, T., Lele, S.K.: Boundary conditions for direct simulations of compressible viscous flows. J. Comput. Phys. **101**, 104–129 (1992)
11. Roach, P.E.: The generation of nearly isotropic turbulence by means of grids. Int. J. Heat Fluid Flow **7**(2), 117–125 (1986)
12. Sandhu, H., Sandham, N.: Boundary conditions for spatially growing compressible shear layers. Technical Report. QMW-EP-1100, QM & Westfield College (1994)
13. Schlüß, D., Frey, C., Ashcroft, G.: Consistent non-reflecting boundary conditions for both steady and unsteady flow simulations in turbomachinery applications. In: ECCOMAS Congress 2016 (2016)
14. Smagorinsky, J.: General circulation experiments with the primitive equations. Mon. Weather Rev. **3**(91), 99–164 (1963)

15. Shur, M.L., Spalart, P.R., Strelets, M.K., Travin, A.K.: Synthetic turbulence generators for RANS-LES interfaces in zonal simulations of aerodynamic and aeroacoustic problems. Flow Turbul. Combus. **93**(1), 63–92 (2014)
16. Thompson, K.W.: Time dependent boundary conditions for hyperbolic systems. J. Comp. Phys. **68**(1), 1–24 (1987)
17. Wilcox, D.C.: Reassessment of the scale-determining equation for advanced turbulence models. AIAA J. **26**(11), 1299–1310 (1988)

Chapter 47
Implementation of a Wall-Distance-Free Composite RANS-ILES Model in a High-Order Discontinuous Galerkin Solver

F. Bassi, A. Colombo, A. Ghidoni and G. Noventa

Introduction

In recent years, the interest for composite RANS-LES models, wall-modelled LES approaches and their application to the high-fidelity simulation of turbulent flows has been growing rapidly. This interest is motivated by the need to overcome the well known limitations of RANS models in simulating massively separated flows and to mitigate the cost of a full LES, which nowadays can be still not feasible for many industrial applications.

In this work the Extra-Large Eddy Simulation (X-LES) composite RANS-LES model [1] has been revisited, coupling the same RANS model with the implicit LES, i.e. a LES where the numerical dissipation behaves like a spectral cut-off filter and replace the Subgrid-Scale (SGS) model [2].

The proposed approach, called iX-LES, has the same switch of the original X-LES, but the zones that are predicted to be in LES mode are actually modelled by the Navier-Stokes equations (without any explicit SGS model), and hence modelled in a ILES fashion. In particular the model switches between URANS and ILES according to the local values of the turbulent scales and of the filter width parameter Δ.

The aim of this contribution is to present the implementation of the iX-LES model in a high-order Discontinuous Galerkin (DG) solver and to assess its predictive capability computing the turbulent flow around a square cylinder at Reynolds number Re = 22,000.

F. Bassi · A. Colombo
Dipartimento di Ingegneria, Università degli Studi di Bergamo, Bergamo, Italy

A. Ghidoni (✉) · G. Noventa
Dipartimento di Ingegneria Meccanica e Industriale,
Università degli Studi di Brescia, Brescia, Italy
e-mail: antonio.ghidoni@unibs.it

M. García-Villalba et al. (eds.), *Direct and Large Eddy Simulation XII*,
ERCOFTAC Series 27,
https://doi.org/10.1007/978-3-030-42822-8_47

The iX-LES Model

The set of governing equations for the iX-LES of a compressible flow can be written as

$$\frac{\partial \rho}{\partial t} + \frac{\partial}{\partial x_j}(\rho u_j) = 0, \tag{47.1}$$

$$\frac{\partial}{\partial t}(\rho u_i) + \frac{\partial}{\partial x_j}(\rho u_j u_i) = -\frac{\partial p}{\partial x_i} + \frac{\partial \widehat{\tau}_{ji}}{\partial x_j}, \tag{47.2}$$

$$\frac{\partial}{\partial t}(\rho E) + \frac{\partial}{\partial x_j}(\rho u_j H) = \frac{\partial}{\partial x_j}\left(u_i \widehat{\tau}_{ij} - \widehat{q}_j\right) - P_k + D_k, \tag{47.3}$$

$$\frac{\partial}{\partial t}(\rho k) + \frac{\partial}{\partial x_j}(\rho u_j k) = \frac{\partial}{\partial x_j}\left[\left(\mu + \sigma^* \hat{\mu}_t\right)\frac{\partial k}{\partial x_j}\right] + P_k - D_k, \tag{47.4}$$

$$\begin{aligned}\frac{\partial}{\partial t}(\rho \widetilde{\omega}) + \frac{\partial}{\partial x_j}(\rho u_j \widetilde{\omega}) &= \frac{\partial}{\partial x_j}\left[\left(\mu + \sigma \hat{\mu}_t\right)\frac{\partial \widetilde{\omega}}{\partial x_j}\right] + \left(\mu + \sigma \hat{\mu}_t\right)\frac{\partial \widetilde{\omega}}{\partial x_k}\frac{\partial \widetilde{\omega}}{\partial x_k} \\ &+ P_\omega - D_\omega + C_D,\end{aligned} \tag{47.5}$$

where ρ is the fluid density, E and H the total energy and enthalpy per unit mass, respectively, and $\widetilde{\omega}$ the logarithm of the specific dissipation rate ω. The pressure p, the mean strain-rate tensor S_{ij}, the turbulent and total stress tensor, $\hat{\tau}_{ij}$ and τ_{ij}, the heat flux vector $\hat{q}_j$ and the eddy viscosity coefficient $\hat{\mu}_t$ are given by

$$p = (\gamma - 1)\rho\left(E - \frac{1}{2}u_k u_k\right), \tag{47.6}$$

$$S_{ij} = \frac{1}{2}\left(\frac{\partial u_i}{\partial x_j} + \frac{\partial u_j}{\partial x_i}\right), \tag{47.7}$$

$$\tau_{ij} = 2\hat{\mu}_t\left(S_{ij} - \frac{1}{3}\frac{\partial u_k}{\partial x_k}\delta_{ij}\right) - \frac{2}{3}\rho\hat{k}\delta_{ij}, \qquad \widehat{\tau}_{ij} = 2\mu\left(S_{ij} - \frac{1}{3}\frac{\partial u_k}{\partial x_k}\delta_{ij}\right) + \tau_{ij}, \tag{47.8}$$

$$\widehat{q}_j = -\left(\frac{\mu}{\mathrm{Pr}} + \frac{\hat{\mu}_t}{\mathrm{Pr}_t}\right)\frac{\partial h}{\partial x_j}, \tag{47.9}$$

$$\hat{\mu}_t = \alpha^* \frac{\rho \hat{k}}{\hat{\omega}}, \tag{47.10}$$

where γ is the ratio of gas specific heats and Pr and Pr_t are the molecular and turbulent Prandtl numbers, respectively. The production, destruction and cross diffusion terms are given by

$$P_k = \tau_{ij}\frac{\partial u_i}{\partial x_j}, \qquad P_\omega = \alpha\left[\alpha^*\frac{\rho}{e^{\widetilde{\omega}_r}}\left(S_{ij} - \frac{1}{3}\frac{\partial u_k}{\partial x_k}\delta_{ij}\right) - \frac{2}{3}\rho\delta_{ij}\right]\frac{\partial u_i}{\partial x_j}, \tag{47.11}$$

$$D_k = \beta^*\rho\hat{k}\hat{\omega}, \qquad D_\omega = \beta\rho e^{\widetilde{\omega}_r}, \tag{47.12}$$

$$C_D = \sigma_d\frac{\rho}{e^{\widetilde{\omega}_r}}\max\left(\frac{\partial k}{\partial x_k}\frac{\partial \widetilde{\omega}}{\partial x_k}, 0\right), \tag{47.13}$$

where $\alpha, \alpha^*, \beta, \beta^*$, and σ_d are the coefficients of the Wilcox k-ω turbulence model [3]. To discriminate between URANS and ILES zones, iX-LES shares the same automatic switch of X-LES, based on the 'composite' specific dissipation rate $\hat{\omega}$ [4], and a limited turbulence kinetic energy value, $\hat{k}$:

$$\hat{\omega} = \max\left(e^{\widetilde{\omega}_r}, \omega_l\right), \qquad \hat{k} = \theta\,\bar{k}, \tag{47.14}$$

where

$$\omega_l = \frac{\sqrt{\bar{k}}}{C_1\Delta}, \qquad \bar{k} = \max(k, 0), \qquad \theta = \frac{1}{2} - \text{sign}\left(\frac{1}{2}, \frac{\omega_l}{e^{\widetilde{\omega}_r}} - 1\right). \tag{47.15}$$

In Eqs. (47.14)–(47.15) the values Δ and C_1 are the filter width parameter and a coefficient of the model ($C_1 = 0.06$ in this work). The function $\text{sign}(A, B)$ returns $|A|$ if $B \geq 0$ else $-|A|$. In ILES mode only the convection and laminar diffusion terms are active in the k-equation, being $\theta = 0$ and, as a consequence, $\hat{k} = 0$. The variable $\widetilde{\omega}_r$ in the source terms and in the $\hat{\omega}$ definition indicates that $\widetilde{\omega}$ must fulfil a 'realizability' condition $\widetilde{\omega}_r = \max(\widetilde{\omega}, \widetilde{\omega}_{r0})$, which sets a lower bound on $\widetilde{\omega}$. The 'realizability' constraint ensures that in URANS mode positive normal turbulent stresses are predicted and the Schwarz inequality for shear stresses is satisfied. The limit value $\widetilde{\omega}_{r0}$ is a function of the mean velocity gradients only, see [5, 6] for details.

Space and Time Discretization

In compact form the set of equations for the compressible iX-LES can be written as

$$\mathbf{P}(\mathbf{w})\frac{\partial \mathbf{w}}{\partial t} + \nabla\cdot\mathbf{F}_c(\mathbf{w}) + \nabla\cdot\mathbf{F}_v(\mathbf{w}, \nabla\mathbf{w}) + \mathbf{s}(\mathbf{w}, \nabla\mathbf{w}) = \mathbf{0}, \tag{47.16}$$

where $m = 7$ is the number of variables, $d = 3$ the number of dimensions, $\mathbf{w} \in \mathbb{R}^m$ the unknown solution vector, $\mathbf{F}_c, \mathbf{F}_v \in \mathbb{R}^m \otimes \mathbb{R}^d$ the convective and viscous flux functions and $\mathbf{s} \in \mathbb{R}^m$ the vector of the source terms of the model. Matrix $\mathbf{P}(\mathbf{w}) \in \mathbb{R}^m \otimes \mathbb{R}^m$ takes into account of the change of variables from the conservative set $\mathbf{w}_c = [\rho, \rho E, \rho u_i, \rho k, \rho\widetilde{\omega}]^T$ to the primitive set $\mathbf{w} = [\tilde{p}, \tilde{T}, u_i, k, \widetilde{\omega}]^T$. The logarithms of pressure $\tilde{p}$ and temperature $\tilde{T}$ are used in place of p and T to ensure the positivity of the thermodynamic variables at the discrete level [7, 8].

All the governing equations are discretized in space, at the same order, according to the DG method [4]. The weak form of the equations is obtained by multiplying Eq. (47.16) by an arbitrary test function and integrating by parts. The solution and the test function are then replaced with a finite element approximation and a discrete test function. The convective numerical flux is computed with an exact Riemann solver [9], while the BR2 [10] method is adopted to discretize the viscous flux.

Assembling together the elemental contributions of the spatial discretization leads to a system of nonlinear ordinary differential equations that can be written as

$$\frac{\mathrm{d}\mathbf{W}}{\mathrm{d}t} + \widetilde{\mathbf{R}}(\mathbf{W}) = \mathbf{0}, \quad \text{with} \quad \widetilde{\mathbf{R}}(\mathbf{W}) = \mathbf{M_P}^{-1}(\mathbf{W})\,\mathbf{R}(\mathbf{W}), \tag{47.17}$$

where $\mathbf{M_P}$ is the global block diagonal mass matrix and $\mathbf{R}(\mathbf{W})$ is the vector of residuals.

Multi-stage, linearly implicit, Rosenbrock-type, Runge–Kutta schemes [7] are adopted for the accurate time integration. Rosenbrock schemes require the solution of a linear system at each stage, here solved by a preconditioned GMRES algorithm [11], while the Jacobian matrix needs to be assembled only once per time step

$$\mathbf{W}^{n+1} = \mathbf{W}^{n} + \sum_{j=1}^{s} m_j \mathbf{Y}_j, \tag{47.18}$$

$$\left(\frac{\mathbf{I}}{\gamma \Delta t} + \widetilde{\mathbf{J}}\right)^{n} \mathbf{Y}_i = -\widetilde{\mathbf{R}}\left(\mathbf{W}^{n} + \sum_{j=1}^{i-1} a_{ij}\mathbf{Y}_j\right) + \sum_{j=1}^{i-1} \frac{c_{ij}}{\Delta t}\mathbf{Y}_j, \quad i = 1, \ldots, s, \tag{47.19}$$

where

$$\mathbf{J} = \frac{\partial \mathbf{R}}{\partial \mathbf{W}}, \qquad \widetilde{\mathbf{R}} = \mathbf{M_P}^{-1}\mathbf{R}, \qquad \widetilde{\mathbf{J}} = \frac{\partial \widetilde{\mathbf{R}}}{\partial \mathbf{W}} = \mathbf{M_P}^{-1}\left(\mathbf{J} - \frac{\partial \mathbf{M_P}}{\partial \mathbf{W}}\widetilde{\mathbf{R}}\right),$$

and m_i, a_{ij}, c_{ij} are the coefficients of the scheme.

Numerical Results

The proposed approach has been assessed computing the turbulent flow over a square cylinder at $\alpha = 0°$ and Reynolds number $\mathrm{Re} = 22{,}000$ and $M = 0.1$ for a $\mathbb{P}^2$ solution approximation. The three-stages order three ROS3P scheme [12] has been used to advance in time the solution. According to numerical experiments, the filter width value was set equal to $\Delta = 10^{-2}$.

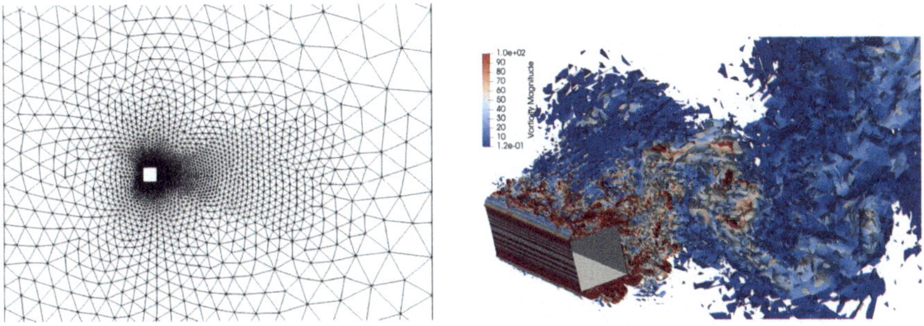

Fig. 47.1 Computational mesh (left) and Q-criterium coloured by vorticity magnitude contours

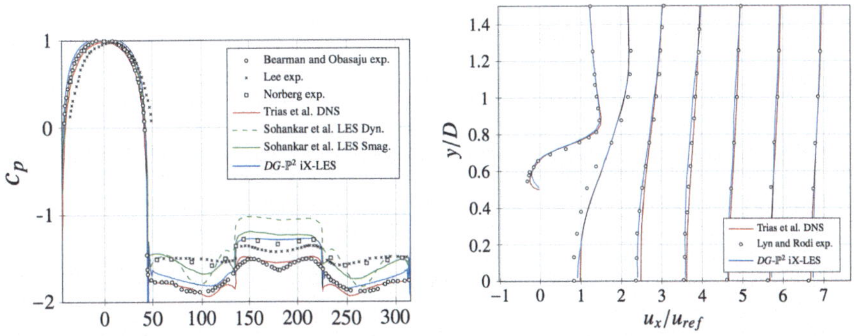

Fig. 47.2 Span-wise and time averaged pressure coefficient (left) and stream-wise velocity along the wake (right), $\mathbb{P}^2$ solution approximation

The mesh has been generated with a 2D high-order version of a fully automated hybrid mesh generator based on the advancing-Delaunay strategy [13] and extruded in the span-wise direction. The mesh was extruded in the span-wise direction using 12 elements and a span-wise periodicity of $4D$ was assumed, where D is the edge of the square. The mesh consists of 83,016 hybrid elements (hexahedra in the boundary layer and prism outside) with linear edges.

Figure 47.1 shows the mesh refinement in the wake of the square (left), and the Q-criterium coloured by the vorticity magnitude contours (right).

Figure 47.2 shows a comparison of the computed results with experimental and DNS data in terms of pressure coefficient and stream-wise velocity distribution in the wake. The span-wise and time-averaged pressure coefficient distribution on the square boundary is in good agreement with DNS results of Trias [14] and experimental data of Bearman [15], which are the most recent results available in literature. Some discrepancies are evident on the square rear face, probably due to a lack of resolution in the wake zone. The stream-wise velocity distribution along the wake is in good agreement with Trias results at all stations.

Conclusions

The implementation of the iX-LES hybrid approach in a high-order DG method has been presented. iX-LES has been chosen, being appealing for the high-fidelity simulation of turbulent flows characterised by high Reynolds numbers. iX-LES proved to be robust and able to correctly deal with separated flows, as shown comparing the computed results with experimental and DNS data. Future work will be addressed to further investigate the filter width influence on results accuracy.

Acknowledgements We acknowledge the CINECA award, under the ISCRA initiative (grant number HP10BUDJC7), for the availability of high performance computing resources and support.

References

1. Kok, J., Dol, H., Oskam, B., van der Ven, H.: Extra-large eddy simulation of massively separated flows. In: 42nd AIAA Aerospace Sciences Meeting and Exhibit, pp. 181–184. American Institute of Aeronautics and Astronautics (2004). https://doi.org/10.2514/6.2004-264
2. Bassi, F., Botti, L., Colombo, A., Crivellini, A., Ghidoni, A., Massa, F.: On the development of an implicit high-order discontinuous Galerkin method for DNS and implicit LES of turbulent flows. Eur. J. Mech. B Fluids **55**, 367–379 (2016). https://doi.org/10.1016/j.euromechflu.2015.08.010
3. Wilcox, D.C.: Formulation of the $k-\omega$ turbulence model revisited. AIAA J. **11**(46), 2823–2838 (2008). https://doi.org/10.2514/1.36541
4. Bassi, F., Botti, L., Colombo, A., Crivellini, A. Ghidoni, A. Nigro, A. Rebay, S.: Time integration in the discontinuous Galerkin code MIGALE—unsteady problems. In: Kroll, N., Hirsch, C., Bassi, F., Johnston, C., Hillewaert, K. (eds.) IDIHOM: Industrialization of High-Order Methods—A Top-Down Approach, vol. 128 of Notes on Numerical Fluid Mechanics and Multidisciplinary Design. Springer International Publishing. Springer, pp. 205–230 (2015). https://doi.org/10.1007/978-3-319-12886-311
5. Bassi, F., Crivellini, A., Rebay, S., Savini, M.: Discontinuous Galerkin solution of the Reynolds-averaged Navier-Stokes and k-ω turbulence model equations. Comput. Fluids **34**, 507–540 (2005)
6. Bassi, F., Botti, L., Colombo, A., Ghidoni, A., Massa, F.C., Noventa, G.: On the development of an implicit high-order discontinuous Galerkin solver for a hybrid RANS-LES model. In: Salvetti, M.V., Armenio, V., Fröhlich, J., Geurts, B.J., Kuerten, H. (eds.) Direct and Large-Eddy Simulation XI. Springer International Publishing, pp. 75–82 (2019). https://doi.org/10.1007/978-3-319-12886-311
7. Bassi, F., Botti, L., Colombo, A., Ghidoni, A., Massa, F.C.: Linearly implicit Rosenbrock-type Runge-Kutta schemes applied to the discontinuous Galerkin solution of compressible and incompressible unsteady flows. Comput. Fluids **118**, 305–320 (2015)
8. Bassi, F., Botti, L., Colombo, A., Crivellini, A., Franchina, N., Ghidoni, A.: Assessment of a high-order accurate discontinuous Galerkin method for turbomachinery flows. Int. J. Comput. Fluid Dyn. **4**(30), 307–328 (2016)
9. Gottlieb, J.J., Groth, C.P.T.: Assessment of Riemann solvers for unsteady one-dimensional inviscid flows of perfect gases. J. Comput. Phys. **78**(2), 437–458 (1988)
10. Bassi, F., Rebay, S., Mariotti, G., Pedinotti, S., Savini, M.: A high-order accurate discontinuous finite element method for inviscid and viscous turbomachinery flows. In: Proceedings of the 2nd European Conference on Turbomachinery Fluid Dynamics and Thermodynamics, pp. 99–108 (1997)

11. PETSc Web Page. http://www.mcs.anl.gov/petsc
12. Lang, J., Verwer, J.: ROS3P—an accurate third-order Rosenbrock solver designed for parabolic problems. BIT **4**(41), 731–738 (2001)
13. Ghidoni, A., Pelizzari, E., Rebay, S., Selmin, V.: 3D anisotropic unstructured grid generation. Int. J. Numer. Methods Fluids **51**, 1097–1115 (2006)
14. Trias, F.X., Gorobets, A., Oliva, A.: Turbulent flow around a square cylinder at Reynolds number 22,000: a DNS study. Comput. Fluids **123**, 87–98 (2015)
15. Bearman, P.W., Obasaju, E.D.: An experimental study of pressure fluctuations on fixed and oscillating square-section cylinders. J. Fluid Mech. **119**, 297–321 (1982)

Chapter 48
Assessment of Split Form Nodal Discontinuous Galerkin Schemes for the LES of a Low Pressure Turbine Profile

M. Bergmann, C. Morsbach and G. Ashcroft

Introduction

Due to the favorable dispersion and dissipation characteristics combined with local data and algorithm structures on unstructured grids, the discontinuous Galerkin (DG) method is a promising candidate for scale-resolving simulations of turbomachinery flows, c.f. [5]. Especially the dissipation characteristic of the upwind DG operator is an interesting aspect for large eddy simulation (LES) as it affects only the largest wave numbers and leaves lower and medium wave numbers uninfluenced. This inherent property mimics the behavior of subgrid-scale models and, thus, a common approach within the DG community is to perform implicit large eddy simulation (iLES) [9].

Unfortunately, without relying on the extra dissipation provided by subgrid-scale models, marginally resolved flows in combination with high-order DG approximations with their lower inherent numerical dissipation lead to instabilities. These instabilities are caused by aliasing errors due to insufficient precision of the numerical integration of the non-linear advective flux terms. A stabilization mechanism for the nodal discontinuous Galerkin spectral element method (DGSEM) with collocated Legendre–Gauss–Lobatto (LGL) nodes was proposed by Gassner et al. [4]. This approach relies on the reformulation of the non-linear advective flux into split forms. Several different formulations are possible and each choice results in a novel variant of the DG scheme with unique properties.

M. Bergmann (✉) · C. Morsbach · G. Ashcroft
Institute of Propulsion Technology, German Aerospace Center (DLR),Cologne, Germany
e-mail: michael.bergmann@dlr.de

C. Morsbach
e-mail: chrstian.morsbach@dlr.de

G. Ashcroft
e-mail: graham.ashcroft@dlr.de

M. García-Villalba et al. (eds.), *Direct and Large Eddy Simulation XII*,
ERCOFTAC Series 27,
https://doi.org/10.1007/978-3-030-42822-8_48

The split form DG schemes have been recently integrated into DLR's solver for turbomachinery flows TRACE and showed promising results for academic test cases, such as the iLES of a nearly incompressible turbulent channel flow [2]. In this paper, we increase the level of challenge and investigate the robustness and accuracy of split form DG schemes for a marginally resolved LES of the T106C low pressure turbine (LPT) profile at Re = 80,000 with an isentropic exit Mach number of $M_{\text{is},2} = 0.65$ featuring curved boundaries, density variations, and separation-induced transition. Following that, we examine the influence of the accuracy order on the spectral resolution and time-averaged results.

Numerical Method

The split form DG scheme for the Navier–Stokes equations is based on the strong formulation of DGSEM, which uses a tensor-product nodal Ansatz for the solution vector combined with collocated LGL nodes as interpolation and quadrature points [1]. The resulting discrete DG operator satisfies the diagonal-norm summation-by-parts (SBP) property, which is a key feature in order to construct conservative split form schemes. By rewriting the advective volume terms into a sub-cell finite volume formulation with the use of consistent and symmetric sub-cell numerical flux functions, one can design novel DG schemes with unique properties, c.f. [3]. Gassner et al. presented a dictionary to translate split forms, which are already well known in finite difference methods, into numerical sub-cell volume fluxes, e.g. the splitting of Ducros, Kennedy and Gruber, Morinishi and Pirozzoli [4]. It was also shown that the latter three preserve the kinetic energy in the sense that only pressure work is responsible for changes in the kinetic energy. Roe's approximate Riemann solver is used as a numerical flux function at the interfaces. In addition to the above split forms, it is also possible to construct an entropy-stable scheme by using the numerical volume flux of Ismail and Roe and the respective stabilization term on faces [3].

The viscous terms of the Navier–Stokes equations are discretized by the Bassi–Rebay 1 method. Furthermore, an explicit third-order accurate Runge–Kutta time-integration scheme is applied. A constant time step size based on numerical experiments is used. Every crashed simulation was repeated with a decreased time step size.

LES of the T106C Low Pressure Turbine Profile

We use the T106C profile as a realistic test case to investigate the robustness and accuracy of the split form DG schemes. The highly loaded LPT cascade has been studied extensively in experiments and numerical simulations, c.f. [5] and is an advanced test case in the current *High-Order CFD Methods* workshop series (HOW). A numerical setup based on the experimental operation point with (nearly) laminar inflow conditions in [7] was chosen. Fourth-order polynomials have been used to

Table 48.1 Simulation parameters for the presented results. The effective mesh spacing is equal to $hu_\tau/\nu(\mathrm{N}+1)$, where h is the cell size on the wall, ν the kinematic viscosity, and N the polynomial approximation order

Mesh	Polynomial order N	Number of elements	Number of DOF in M	Average effective mesh spacing on the suction side of the blade		
				Δx^+	$\Delta y^+_{\mathrm{min}}$	Δz^+
1	1	203,400	1.62	22.3	0.90	17.9
2	3	24,480	1.57	20.5	0.80	16.0
3	5	7,068	1.51	19.8	0.71	17.4
4	7	3,051	1.56	18.9	0.96	16.6

discretize the blade geometry. 20% of the chord c is set as a spanwise domain, which is discretized with 40 degrees of freedom (DOF). Furthermore, we apply periodicity in spanwise and pitchwise direction. Consistent, 1D non-reflecting boundary conditions based on characteristics of a running average are prescribed at the inflow and outflow of the domain to avoid spurious reflections from the boundaries [10]. The mesh parameters for all considered polynomial orders are given in Table 48.1. The starting point for each simulation is a flow field initialized with a RANS solution. After an initial transient phase of 10 throughflow periods, the flow field will be averaged temporally over 90 throughflow periods and, afterward, spatially in the spanwise direction to obtain the statistically converged solution. One throughflow period is defined as $c/U_0 = 0.093\,\mathrm{m}/200\,\mathrm{m/s} = 4.65\cdot 10^{-4}\,\mathrm{s}$.

Analysis of the Split Forms

The analysis of the split forms is performed on the mesh 4 of Table 48.1 containing 3051 hexahedra with an eighth-order accurate scheme. We consider this as a coarse LES grid as typical grids for scale-resolving simulations of the T106C found in the literature contain around 10^7 DOF [5]. When you start a simulation without any stabilization mechanism, the simulation crashes immediately. Despite the geometric nonlinearities and the underresolution, both kinetic energy preserving (KEP) split forms with cubic splitting and the entropy-stable scheme of Ismail and Roe are stable. The split form of Morinishi is, despite its KEP, the least stable scheme as the simulation crashes in the initial transient phase. The split form of Ducros is stable for over 20 throughflow periods but crashes afterward without any sign of instability. This was also observed for the compressible Taylor–Green–Vortex [4]. The reasons for that are the density variations in combination with the quadratic splitting of the momentum equations [6]. If a cubic splitting is applied for the momentum equations, i.e. split the density and the velocities separately, but the splitting of the continuity and energy equation is kept, the simulation finishes successfully.

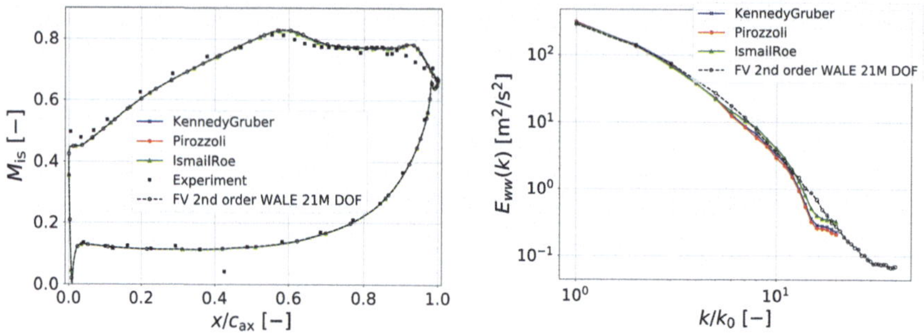

Fig. 48.1 Results of the iLES with an eighth-order accurate DGSEM and different split forms on mesh 4. Time-averaged isentropic Mach number M_{is} distribution (left) and spanwise one-dimensional energy spectra of the spanwise velocity E_{ww} obtained in the wake (right)

Although a coarse mesh is used, typical flow structures are present for all stable schemes, e.g. the incoming separated laminar shear-layer and the Kelvin–Helmholtz type roll-up with a separation-induced transition process. The averaged isentropic Mach number around the blade is shown in Fig. 48.1 (left). Similar results can be obtained for all stable schemes, which are overall in good accordance with the experimental data. The largest differences to the experiment are present in the leading edge region of the suction side and in the region of the laminar separation between $x/c_{ax} = [0.58, 1.0]$. Similar discrepancies can be found in other numerical studies of the T106C flow and can, therefore, not be linked to the use of split form DG, c.f. [5]. To further substantiate this, the result of a reference LES on a fine mesh, based on studies in [8], is shown in Fig. 48.1 (left), which matches perfectly with the high-order split form DG results on a coarser grid. In Fig. 48.1 (right), the one-dimensional energy spectra of the spanwise velocity in spanwise direction are shown. The spectra have been obtained from an FFT of the two-point correlation probed $0.3c_{ax}$ downstream of the trailing edge inside the wake. After a normalized wave number of $k/k_0 = 11$, a significant drop-off of the energy can be noted for all stable schemes compared to the numerical reference, which can be related to the dissipation properties of the high-order DG scheme. The spectra of the split forms Kennedy and Gruber and Pirozzoli are identical. Similar findings are also presented in [2] for the iLES of the channel flow. As the same surface dissipation term is used, it is likely that the dissipation characteristics for underresolved simulations are more affected by the numerical surface flux than the numerical volume flux. The energy level of the entropy-stable form is slightly higher for wave numbers over $k/k_0 \geq 7$, which is supposedly due to the different surface flux formulation.

Influence of the Accuracy Order

With the robust implementation for very high-order approximations on a coarse mesh, we want to review the benefit of using a high-order DG scheme to lower order approximations on a given number of DOF. This is still an open question for the DG-iLES approach of nontrivial flows. Therefore, we consider the four presented simulation setups of Table 48.1 with four different approximation orders. The Kennedy and Gruber splitting is used as a stabilization mechanism in all simulations. The distribution of the isentropic Mach number for the simulations N3, N5, and N7 are identical to the numerical reference solution (not shown here). However, the lowest order approximation N1 shows significant discrepancies in the region of the separation. Therefore, the power spectral density of the static pressure fluctuations in the separation bubble is shown in Fig. 48.2 (left). In comparison to the other simulations, the spectrum of N1 drops off sharply at a relatively low frequency of $f = 5 \cdot 10^3$ Hz, where highly energetic structures are still present. This indicates that the mesh resolution is too coarse for the resolution capability of DGSEM with linear approximations. With increasing polynomial order, the drop-off in the spectrum moves toward higher frequencies. A notable bump in the spectrum before the drop-off point can be observed for the highest approximation order N7, which is not present for lower orders. This was also observed in [11] for the inviscid Taylor–Green–Vortex and has been related to the sharper dissipation in wave number space of very high-order DG schemes for underresolved simulations. Overall, the N5 solution shows the best match to the reference spectrum on a fine mesh. In order to investigate the transition behavior of different orders, the maximum turbulent kinetic energy of blade-normal cuts along the suction side is presented in Fig. 48.2 (right). The level of the turbulent kinetic energy is significantly higher in the incoming shear-layer for the N1 simulation, which is linked to an earlier separation of the flow field. Thus, the transition characteristics are completely different compared to the remaining results. However, the results of the high-order approximations follow similar trends. The level of the turbulent kinetic energy slowly increases in the shear-layer with nearly identical values

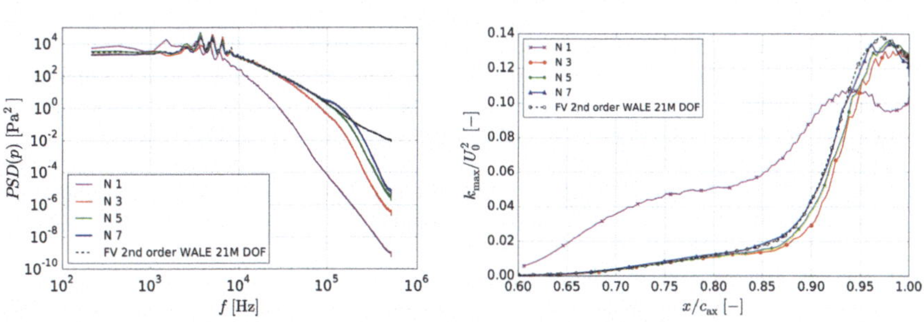

Fig. 48.2 Results of the iLES DGSEM with different approximation orders for a given number of DOF. Power spectral density of the static pressure $PSD\,(p)$ in the separation (left) and Maximum turbulent kinetic energy of blade-normal cuts along the suction side (right)

up to $x/c_{ax} = 0.80$ for N3, N5, and N7. The transition point, which is defined through a significant rise of the turbulent kinetic energy, is located further upstream for an increasing polynomial order. This can be related to the lower dissipation level of high-order DG schemes. The highest approximation order is the closest to the given reference distribution.

Conclusion

In this work, we investigated the robustness and accuracy of split form DG schemes for the iLES of the low pressure turbine cascade T106C. The split forms of Kennedy and Gruber, Pirozzoli and the entropy-stable scheme of Ismail and Roe are able to handle the underresolved turbulence, density variations, and geometric nonlinearities and found to be stable for this test case. In contrast, the split form of Ducros and Morinishi crash during the simulation. Although a very coarse mesh resolution has been used, we showed that the stable split form DG schemes match the results of the reference LES on a fine mesh in terms of the isentropic Mach number and boundary layer representation. Moreover, a study of different accuracy orders highlights the quality improvements when using a high-order DG scheme compared to a low-order scheme on the same number of DOF, i.e. the same mesh resolution. In future research, we plan to extend our framework to sliding interfaces in order to investigate rotor–stator interactions with high-order split form DGSEM-iLES scheme and focus on an in-depth comparison with an ordinary LES approach using a second-order FV scheme and explicit subgrid-scale model for realistic test cases.

References

1. Bergmann, M., Drapkina, S., Ashcroft, G., Frey, C.: A comparison of various nodal discontinuous Galerkin methods for the 3D Euler equations. In: ECCOMAS Congress 2016—Proceedings of the 7th European Congress on Computational Methods in Applied Sciences and Engineering, vol. 4, pp. 7956–7966 (2016). https://doi.org/10.7712/100016.2389.6203
2. Bergmann, M., Gölden, R., Morsbach, C.: Numerical investigation of split form nodal discontinuous galerkin schemes for the implicit les of a turbulence channel flow. In: Proceedings of the 7th European Conference on Computational Fluid Dynamics (ECFD), Glasgow, UK, pp. 1982–1993 (2018)
3. Carpenter, M.H., Fisher, T.C., Nielsen, E.J., Frankel, S.H.: Entropy stable spectral collocation schemes for the Navier-Stokes equations: discontinuous interfaces. SIAM J. Sci. Comput. **36**(5), B835–B867 (2014). https://doi.org/10.1137/130932193
4. Gassner, G.J., Winters, A.R., Kopriva, D.A.: Split form nodal discontinuous Galerkin schemes with summation-by-parts property for the compressible euler equations. J. Comput. Phys. **327**, 39–66 (2016). https://doi.org/10.1016/j.jcp.2016.09.013
5. Hillewaert, K., Carton de Wiart, C., Verheylewegen, G., Arts, T.: Assessment of a high-order discontinuous Galerkin method for the direct numerical simulation of transition at low-Reynolds number in the T106C high-lift low pressure turbine cascade. In: ASME Turbo Expo 2014:

Turbine Technical Conference and Exposition, vol. 2B: Turbomachinery (2014). https://doi.org/10.1115/GT2014-26739
6. Kennedy, C.A., Gruber, A.: Reduced aliasing formulations of the convective terms within the Navier-Stokes equations for a compressible fluid. J. Comput. Phys. **227**, 1676–1700 (2008). https://doi.org/10.1016/j.jcp.2007.09.020
7. Michálek, J., Monaldi, M., Arts, T.: Aerodynamic performance of a very high lift low pressure turbine airfoil (T106C) at low Reynolds and high mach number with effect of free stream turbulence intensity. J. Turbomach. **134**(6), 061,009–061,009–10 (2012). https://doi.org/10.1115/1.4006291
8. Morsbach, C., Bergmann, M.: Critical analysis of the numerical setup for the large eddy simulation of the low pressure turbine profile T106C. In: Submitted to Proceedings of Direct and Large Eddy Simulation, 12 Sept, Madrid, Spain (2019)
9. Moura, R., Mengaldo, G., Peiró, J., Sherwin, S.: On the eddy-resolving capability of high-order discontinuous Galerkin approaches to implicit LES/under-resolved DNS of Euler turbulence. J. Comput. Phys. **330**(C), 615–623 (2017). https://doi.org/10.1016/j.jcp.2016.10.056
10. Schlüß, D., Frey, C., Ashcroft, G.: Consistent non-reflecting boundary conditions for both steady and unsteady flow simulations in turbomachinery applications. In: ECCOMAS Congress 2016–Proceedings of the 7th European Congress on Computational Methods in Applied Sciences and Engineering, vol. 4, pp. 7403–7422 (2016). https://doi.org/10.7712/100016.2342.5411
11. Winters, A.R., Moura, R.C., Mengaldo, G., Gassner, G.J., Walch, S., Peiro, J., Sherwin, S.J.: A comparative study on polynomial dealiasing and split form discontinuous Galerkin schemes for under-resolved turbulence computations. J. Comput. Phys. **372**, 1–21 (2018). https://doi.org/10.1016/j.jcp.2018.06.016

Chapter 49
Effective Viscosities of Smoothed Dissipative Particle Dynamics in 3D

M. Borreguero and N. A. Adams

Introduction

We aim to extend simpler stochastic models for turbulence such as generalized Langevin model [1, 2]. In this work, a particular focus is placed on the Smoothed Dissipative Particle Dynamic (SDPD) model. It is the combination of the meshless Lagrangian discretization of the Navier–Stokes equations (smoothed particle hydrodynamics, SPH) and additional thermal fluctuations. SDPD satisfies the fluctuation-dissipation theorem.

We explore SDPD beyond typical parameter ranges to gain further understanding of the influence of thermal fluctuations on the statistics of fluid flow at mesoscales by analyzing the effect of the random force on the effective viscosity. Smoothed particle hydrodynamics (SPH) is the deterministic version of SDPD and is known to exhibit numerical viscosities [3]. These numerical viscosities may be interpreted as an effective viscosity with three contributions which may act as implicit turbulence model, see [4]. Many interesting phenomena for highly compressible solvent have been recently observed in two-dimensional SDPD simulations. Among other encouraging results from the 2D simulations, the acceleration probability density function (PDF) at high viscosity exhibits interesting features. We have observed that for SDPD the acceleration PDF deviates from the Gaussian by developing stretched tails as the Reynolds number decreases. In [5], it was reported that the Lagrangian acceleration PDF in fully developed turbulence follows

$$P(a) = C\exp(-a^2/((1+|a\beta/\sigma|^{\gamma})\sigma^2)). \tag{49.1}$$

M. Borreguero (✉) · N. A. Adams
Chair of Aerodynamics and Fluid Mechanics, Technical University of Munich, Munich, Germany
e-mail: m.borreguero@tum.de

M. García-Villalba et al. (eds.), *Direct and Large Eddy Simulation XII*, ERCOFTAC Series 27,
https://doi.org/10.1007/978-3-030-42822-8_49

The parameters obtained by fitting two-dimensional data with Eq. 49.1 are in good agreement with the experimental data.

The present paper presents three-dimensional results for homogeneous shear flow produced by the Lees–Edwards boundary condition. We discuss results of the fluctuating field structure and analyze its agreement with established results for homogeneous sheared turbulence. We also report on the observed onset of the kinetic regime, interpreted as "phase-transition" phenomenon.

Model Description

The equation of motion for an SDPD particle i is given by the sum of three interparticle forces

$$m\dot{\mathbf{v}}_i = \sum_j \mathbf{F}^C_{ij} + \sum_j \mathbf{F}^D_{ij} + \sum_j \mathbf{F}^R_{ij}, \tag{49.2}$$

a conservative, $\mathbf{F}^C$, a dissipative, $\mathbf{F}^D$, and a random force, $\mathbf{F}^R$ where summation j extends over neighboring particles within a cutoff radius $0 \leq r_{ij} < r_c, r_{ij} = |\mathbf{r}_i - \mathbf{r}_j|$ denotes the normed distance between particle i and j. The conservative and the dissipative force are directly derived from the Navier–Stokes equation using SPH interpolation and are thus deterministic

$$m\dot{\mathbf{v}}_i = \sum_{j \neq i} \underbrace{-m^2\left(\frac{p_i}{\rho_i^2} + \frac{p_j}{\rho_j^2}\right)\omega'_{ij}\hat{\mathbf{e}}_{ij}}_{=\mathbf{F}^C_{ij}} + \sum_{j \neq i} \underbrace{\frac{1}{3}\eta_0 m^2 \frac{\omega'_{ij}}{\rho_i \rho_j r_{ij}} \left[4(\hat{\mathbf{e}}_{ij} \cdot \mathbf{v}_{ij})\hat{\mathbf{e}}_{ij} + 5\mathbf{v}_{ij}\right]}_{=\mathbf{F}^D_{ij}} + \sum_{j \neq i} \mathbf{F}^R_{ij},$$

where p is the pressure and the vectors $\hat{\mathbf{e}}_{ij}$ and $\mathbf{v}_{ij}$ are the unit direction vector and the velocity difference between particle i and j, respectively. The random force is a stochastic-force model for thermal fluctuations

$$\mathbf{F}^R_{ij} = B_{ij}(1/dt)\mathbf{dW}_{ij} \cdot \hat{\mathbf{e}}_{ij}. \tag{49.3}$$

The matrix $\mathbf{dW}_{ij}$ is a traceless symmetric matrix $\mathbf{dW}_{ij} = \frac{1}{2}\left(\mathbf{d}\mathcal{W}_{ij} + \mathbf{d}\mathcal{W}^T_{ij}\right) - Tr[\mathbf{d}\mathcal{W}_{ij}]\frac{I}{d}$, and $\mathbf{d}\mathcal{W}_{ij}$ is an increment of a Wiener process. Note that $\mathbf{dW}$ includes a factor $dt^{1/2}$. dt is the time step for the velocity-Verlet integration scheme. The matrix $\mathbf{B_{ij}}$ is computed such that it satisfies the fluctuation-dissipation balance

$$\mathbf{B}_{ij} = \left[-4dk_B T\eta_0 \left(\frac{1}{\rho_i^2} + \frac{1}{\rho_j^2}\right)\frac{\omega'_{ij}}{r_{ij}}\right]^{1/2}, \tag{49.4}$$

where ω'_{ij} is the derivative of the kernel function $\omega_{ij} = \omega(r_{ij})$ and d is the spatial dimensionality of the problem. Following previous work and [3] we use the Lucy kernel. Since the mean square value of the time integral of the random force have to be proportional to time, see [6], we have that $\mathbf{F}_{ij}^R \propto dt^{-1/2}$. Note that the temperature as well as the input viscosity η_0 set the level of the fluctuations. In order to close the system of equations, a weakly compressible equation of state (EOS) is used

$$p(\rho) = p_0 \left[\left(\frac{\rho}{\rho_0} \right)^{\gamma} - 1 \right], \tag{49.5}$$

where γ is a model parameter.

Microscopic Analysis

We employ a concept from nonequilibrium molecular dynamics in order to evaluate effective viscosities of SDPD model for homogeneous shear flow. Similarly to the two-dimensional case [3], the pressure tensor resulting from a flow of momentum induced by a perturbation is measured. The use of Lees–Edwards boundary condition generates a constant mean shear $\dot{\gamma}$. For the analysis of effective viscosities, we follow the Irving–Kirkwood procedure [7] as continued in [3]. In the case of SDPD, the local average of the momentum flux gives the instantaneous microscopic tensor $\mathbf{P}$

$$\mathbf{P} = \frac{1}{V} \left(m \sum_i \mathbf{c}_i \mathbf{c}_i + \sum_{i,j>i} \mathbf{r}_{ij} \mathbf{F}_{ij}^C + \sum_{i,j>i} \mathbf{r}_{ij} \mathbf{F}_{ij}^D + \sum_{i,j>i} \mathbf{r}_{ij} \mathbf{F}_{ij}^R \right),$$

where $\mathbf{c_i} = (v_i^x - \dot{\gamma} y_i, v_i^y)$ is the relative velocity, and V is the sample volume. The total shear viscosity is obtained by averaging over time and over the third dimension z the off diagonal component of the pressure tensor and upon dividing by the shear rate $\dot{\gamma}$

$$\eta = -\frac{\langle P_{yx} \rangle}{\dot{\gamma}} = \eta_{\mathrm{kin}} + \eta_{\mathrm{pot}} + \eta_{\mathrm{diss}} + \eta_{\mathrm{rand}}, \tag{49.6}$$

which is decomposed into four contributions. The subscripts kin, pot, diss, and rand stand for the kinetic, potential, dissipative, and random contribution, respectively.

Analysis of Simulation Data

We consider the three-dimensional case of a homogeneous mean shear flow in x-direction produced by means of Lees–Edwards boundary conditions, Fig. 49.1a. At the other boundaries of the squared box of size L, periodic conditions are imposed, Fig. 49.1b. This configuration produces a homogeneous shear flow which is one of

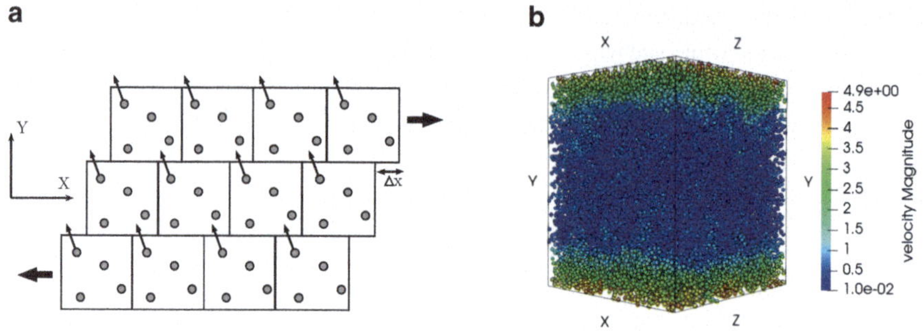

Fig. 49.1 **a** Lees–Edwards boundary condition for a shear flow. **b** Screenshot of the numerical domain after 10 time steps for the case $k_B T = 1.0$, Re = $2.5 \cdot 10^2$

the canonical configurations for investigation of anisotropic turbulence. The Mach number is defined as $Ma = (\dot{\gamma} L)/c_s$ where c_s is the speed of sound. We emphasize that with the weakly compressible equation of state Eq. (49.5), the Mach number characterizes fluid compressibility and does not classify the considered flow situation in a macroscopic sense. We investigate a highly compressible case with $Ma = 3.0$ as in previous 2D-simulations unconventional behavior was observed. Unless specified differently, simulations presented in this paper have been performed with parameters presented in Table 49.1. The time-step size dt has to satisfy

$$\Delta t \leq \min[\Delta t_{\text{visc}}, \Delta t_{\text{CFL}}] \tag{49.7}$$

where $\Delta t_{\text{visc}} = \frac{h^2}{8\eta_0 \rho_0}$ defines the viscous time scale, and $\Delta t_{\text{CFL}} = \frac{h}{4(c_s + v_{\text{ref}})}$ is the Courant–Friedrichs–Lewy (CFL) condition. The time-step size is kept constant for all realizations.

We ensure that the system reaches a stationary state before we extract statistics. In Fig. 49.2, different contributions to the total SDPD viscosity are normalized by the input viscosity η_0 and reported as function of the Reynolds number Re = $\frac{L^3 \dot{\gamma}}{\eta_0}$. As expected, the results are very similar to the two-dimensional case: The dissipative viscosity dovetails with the input value and thus this contribution does not produce error in the computation of the physical fluid viscosity. The potential contribution is small over almost the entire range. The random contribution increases with the increasing Reynolds number though it does not contribute significantly to the total fluid viscosity. We observe that with increasing Reynolds number, the kinetic contribution becomes dominant. We characterize the cross-over to the kinetic regime by a critical Reynolds number $\text{Re}_c = 5 \cdot 10^3$. The transition to the kinetic regime coincides with a sudden increase in the mean density which suggests the interpretation of a "phase transition". An important observation is the increase of the kinetic contribution as the level of thermal fluctuation varies. This means that the additional Gaussian noise indirectly influences the kinetic contribution. This may be explained by two possible ways. First, it may be a consequence of the "stiffness" of the equation of

Table 49.1 Input parameters of the SDPD solvent

Domain size ($V_{LE} = L_x \times L_z \times (L_y + 2\Delta L_y)$)	$1 \times 1 \times 1.5$
Sample size ($V = L_x \times L_z \times L_y$)	$1 \times 1 \times 1$
Total number of particles (N_p)	32^3
Mean interparticle distance (Δx)	$1/\sqrt[3]{N_p}$
Density (ρ_0)	1
Mass (m)	$\rho_0 \cdot \Delta x^3$
Pressure (p_0)	1
EOS parameter (γ)	7
Shear rage ($\dot{\gamma}$)	5
Speed of sound (c_s)	1.6
Compressibility (κ)	9
Time step (Δt)	$2 \cdot 10^{-4}$
Simulation length in time step (N_t)	$2.5 \cdot 10^6$
Cutoff radius (h)	$3\Delta x$
Thermal fluctuation ($k_B T$)	1.0

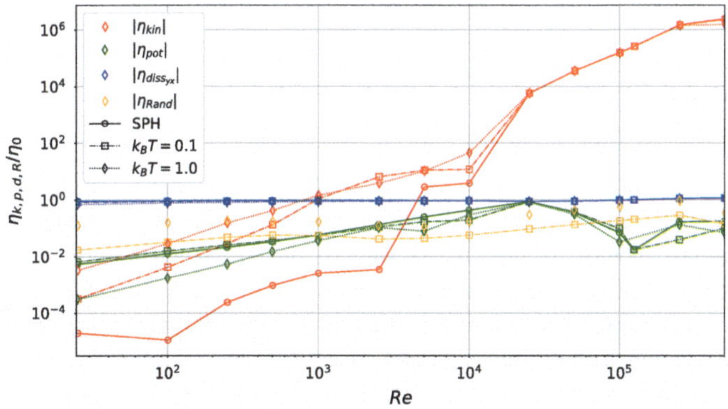

Fig. 49.2 Comparison of the implicit viscosity contributions for different levels of thermal fluctuation

state characterized by the exponent γ. Second, the non-linearity of the kinetic part (see Eq. (49.3)) may be the origin of the observed behavior. Moreover since in SDPD the mesoscopic scale is determined by the level of thermal fluctuations, as the latter increases the physical scales under consideration decrease. We have been able to rule out the first possibility by varying the stiffness of the EOS. The overall behavior was not affected by a decrease of γ in Eq. (49.5).

In Fig. 49.3, stretched exponential tails are observed in the probability density function (PDF) accelerations at Re = $2.5 \cdot 10^5$ for both SPH and SDPD. At a rather low Reynolds number, Re < Re_c, while for SPH, the acceleration PDF is Gaussian,

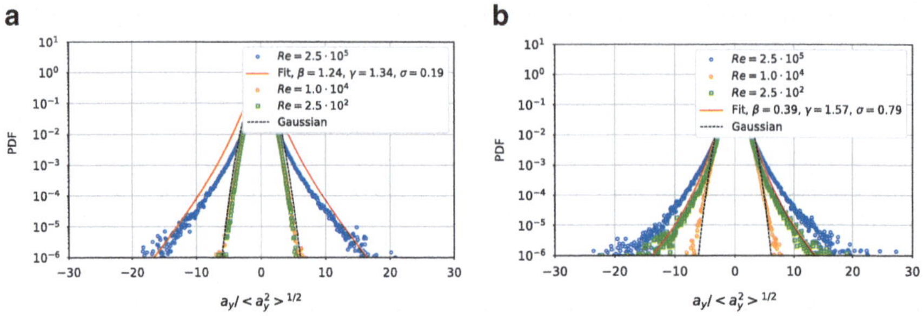

Fig. 49.3 **a** Probability density function of the acceleration for SPH. The realization for Re = $2.5 \cdot 10^5$ is fit with function (49.1). **b** Probability density function of the acceleration. The realization for Re = $2.5 \cdot 10^2$ is fit with function

for SDPD, Fig. 49.3b, the probability of finding rare high accelerations increases with decreasing Reynolds number resulting in non-Gaussian PDF. For SDPD, the realization Re = $2.5 \cdot 10^2$ has been fitted with Eq. (49.1), the parameters are shown in the legend. We have ensured that this behavior is not a numerical artifact by further decreasing the time-step size dt, we did not note changes in the statistics.

Conclusion

A three-dimensional microscopic analysis of SDPD allowed us to further investigate on the differences between SPH and SDPD. The statistics of the 3D realizations show similar qualitative behavior than in two dimension. The stochastic force contributes indirectly to a variation of the kinetic viscosity due to its non-linear nature. Relying on these promising results, we will further investigate the structure functions and also consider a benchmark microchannel case.

Acknowledgements We would like to acknowledge the support by the Deutsche Forschungsgemeinschaft within the Priority Programme Turbulent Superstructures under Grant no. AD 186/30.

References

1. Pope, S.B.: Transport equation for the joint probability density function of velocity and scalars in turbulent flow. Phys. Fluids **24**, 588 (1981)
2. Haworth, D.C., Pope, S.B.: A generalized Langevin model for turbulent flows. Phys. Fluids **29**, 387 (1986)
3. Ellero, M., Español, P., Adams, N.A.: Implicit atomistic viscosities in smoothed particle hydrodynamics. **82** (2010)
4. Pope, S.: Turbulent Flows. Cambridge University Press (2000)

5. La Porta, A., Voth, G., Crawford, A., Alexander, J., Bodenschatz, E.: Fluid particle accelerations in fully developed turbulence. Nature **409**, 1017–1019 (2001)
6. Groot, R.D., Warren, P.B.: Dissipative particle dynamics: bridging the gap between atomistic and mesoscopic simulation. J. Chem. Phys. **107**, 4423–4435 (1997)
7. Irving, J. H., Kirkwood, J. G.: The statistical mechanical theory of transport processes. IV. The equations of hydrodynamics. J. Chem. Phys. **18**, 817 (1950)

Chapter 50
Multiscale and Directional Approach to Rotating Shear Flow

Y. Zhu, C. Cambon and F. S. Godeferd

Introduction

Turbulence and stability in rotating shear flows is essential in many contexts ranging from engineering—e.g. turbomachinery or hydroelectric power generation—to geophysics and astrophysics. A statistical approach remains relevant, even assuming statistical homogeneity restricted to fluctuations [8], if a very large range of Reynolds numbers is investigated, until the highest ones, in contrast with DNS, and if strong anisotropy is accurately accounted for, in addition to a multi-scale approach, ignored in RANS. In order to improve calculations in both our recent three-dimensional spectral models [5, 11] and in classical pseudo-spectral DNS adapted to shear-driven flows by [7] (Engineering) and by [3] (Astrophysics), we propose a Finite-Differences Scheme (FDS) for the convection operator.

Finite-Differences Scheme for Advection in Fourier Space

The generic structure of our typical equations in three-dimensional (3D) Fourier space is

$$\dot{\Phi}(\boldsymbol{k},t) + \nu k^2 \Phi(\boldsymbol{k},t) + \mathscr{L}(\Phi)(\boldsymbol{k},t) = \mathscr{T}(\Phi)(\boldsymbol{k},t)\ , \tag{50.1}$$

Y. Zhu (✉) · C. Cambon · F. S. Godeferd
LMFA, Écully, France
e-mail: ying.zhu@doctorant.ec-lyon.fr

C. Cambon
e-mail: claude.cambon@ec-lyon.fr

F. S. Godeferd
e-mail: fabien.godeferd@ec-lyon.fr

M. García-Villalba et al. (eds.), *Direct and Large Eddy Simulation XII*,
ERCOFTAC Series 27,
https://doi.org/10.1007/978-3-030-42822-8_50

in which Φ is a state-vector related to the fluctuating flow, that will be specified in following sections, together with linear $\mathscr{L}$ and nonlinear $\mathscr{T}$ operators, $\boldsymbol{k}$ is the 3D wave-vector, ν is the kinematic viscosity, or any equivalent diffusion parameter. The crucial operator which translates the advection by a simplified *extensional* mean flow U_i with uniform mean velocity gradient $A_{ij} = \frac{\partial U_i}{\partial x_j}$ is

$$\dot{\Phi}(\boldsymbol{k}, t) = \left(\frac{\partial}{\partial t} - A_{mn} k_m \frac{\partial}{\partial k_n} \right) \Phi(\boldsymbol{k}, t) \,. \tag{50.2}$$

From classical linear solutions in Rapid Distortion Theory to nonlinear pseudo-spectral DNS, this operator is solved following characteristic lines, that are equivalent to mean trajectories in physical space: this is also equivalent to considering the wave-vector as time-dependent, according to $\dot{\Phi}(\boldsymbol{k}(t), t) = \frac{\partial \Phi}{\partial t} + \frac{\partial \Phi}{\partial k_n} \frac{\mathrm{d}k_n}{\mathrm{d}t}$, with $\dot{k}_n = -A_{mn} k_m$ an Eikonal equation. This method is well adapted to DNS when initialized with a particular so-called 'turbulent' velocity field [7], with a choice of random phases and orientations, that can 'jump' from a direction of $\boldsymbol{k}$ to another one, even very close. On the other hand, related strong distortion of meshes from initial time t_0 to final t needs interpolation and remeshing, as illustrated in Fig. 50.1. A FDS scheme in 3D Fourier space is preferred here for solving the $\frac{\partial}{\partial k_n}$-derivative in Eq. (50.2), for three reasons: (i) There is no initial discontinuity (jump) with respect to the orientation of $\boldsymbol{k}$ for smooth quantities, if Φ represents smooth statistical quantities, or quasi-deterministic realizations; (ii) the FDS scheme is much more accurate and versatile, with respect to any chosen discretization of the fixed $\boldsymbol{k}$-grid; (iii) spherical averaging and thereby extraction of angular harmonics of any degree, are much easier with a fixed $\boldsymbol{k}$-grid consistent with polar-spherical coordinates.

For this purpose, given a system of polar-spherical coordinates (k, θ_k, ϕ_k) for $\boldsymbol{k}$, the $\boldsymbol{k}$-derivative is simply expressed as the sum of three terms, the radial one along k, the meridional one along θ_k and the azimuthal (or zonal) one along ϕ_k. One recovers the local, radial/meridional/zonal, frame, called Craya-Herring in the turbulence community. This frame of reference was particularly useful for expressing the 3D velocity component $\widehat{\boldsymbol{u}}(\boldsymbol{k}, t)$ in Fourier space, where the divergence-free constraint amounts to $\widehat{\boldsymbol{u}} \cdot \boldsymbol{k} = 0$ [8].

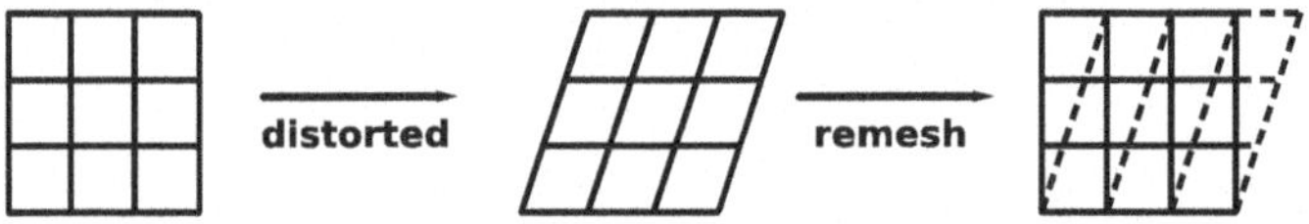

Fig. 50.1 Moving grid for uniform shear flow

Spectral Statistical Model. Brief Description and Typical First Results

In our elaborate model, the state-vector Φ in Eq. (50.1) represents the spectral tensor $\widehat{R}_{ij}(\boldsymbol{k},t)$ of two-point second-order velocity correlations. This tensor is the 3D Fourier transform of the two-point velocity correlation tensor $R_{ij}(\boldsymbol{r},t) = \langle u_i(\boldsymbol{x},t)u_j(\boldsymbol{x}+\boldsymbol{r},t)\rangle$; it includes the same level of information in homogenized turbulence, but can be expressed with a minimum set of statistical descriptors, as a set ($\mathscr{E}$, Z), given the Hermitian symmetry, incompressibility and homogeneity, for arbitrary isotropy. $\mathscr{E}$ is the 3D energy spectrum, which includes the directional anisotropy, and Z holds for the polarization anisotropy. The equations for ($\mathscr{E}$, Z) can be obtained as:

$$
\begin{aligned}
&(\dot{k\mathscr{E}}) + 2\nu k^3\mathscr{E} + \Re\left(kZ(\boldsymbol{k},t)S_{ij}N_i(\boldsymbol{\alpha})N_j(\boldsymbol{\alpha})\right) = kT^{(\mathscr{E})}(\boldsymbol{k},t)\\
&(\dot{kZ}) + 2\nu k^3 Z + k\mathscr{E}(\boldsymbol{k},t)S_{ij}N_i(-\boldsymbol{\alpha})N_j(-\boldsymbol{\alpha}) - \mathrm{i}kZ(\boldsymbol{k},t)\left((\boldsymbol{W}+4\boldsymbol{\Omega})\cdot\boldsymbol{\alpha} - 2\Omega_E\right)\\
&= kT^{(Z)}(\boldsymbol{k},t),
\end{aligned}
\tag{50.3}
$$

where $\boldsymbol{\alpha} = \boldsymbol{k}/k$, S_{ij} and $\boldsymbol{W}$ represent the symmetric and antisymmetric part of mean-velocity gradient respectively with $A_{ij} = S_{ij} + 1/2\epsilon_{imj}W_m$, and $\boldsymbol{\Omega}$ is the system angular velocity vector (readers can find all the details of Eq. (50.3) and its numerical procedure in [11]). The linear part of Eq. (50.3), if considered alone, reflects the exact RDT. This linear part is solved without assumption, using the FDS technique, a fixed $\boldsymbol{k}$-grid with a large number of spheres and angles over them, and a high-order scheme. The non-linear source terms, corresponding to the right-hand-side of Eq. (50.3), are closed by an anisotropic EDQNM (Eddy-Damped Quasi-Normal-Markovian, from [6] in the isotropic case.) This model called 'hybrid' couples the closure using a truncated expansion (degree 2) in terms of angular harmonics by [5], called 'ZCG', once the angular dependence is restored in nonlinear transfer terms, with a spectral Rotta's model [10] restricted to clipping angular harmonics of higher degree. Our first application is the case of plane shear with constant rate S ($A_{ij} = S\delta_{i1}\delta_{i3}$), rotating around its spanwise direction with angular speed Ω, that is a useful model from engineering (classical rotating channel) to astrophysics (e.g. toy-model for accretion discs in astrophysics, according to the Shearing Sheet Approximation, e.g. Balbus and Hawley, Ref. in [8])

Promising results are found for various cases of the ratio R of system vorticity 2Ω to shear vorticity S (or $-S$ depending on the choice of axes), from destabilizing, $-1 \leq R \leq 0$, to stabilizing, $R \leq -1$ or $R \geq 0$, cases. The development of turbulent kinetic energy is given for the case without system rotation on Fig. 50.2, in which various models, MCS [5], Weinstock [10], 'ZCG' and 'hybrid' [11, 12] are compared to DNS [9].

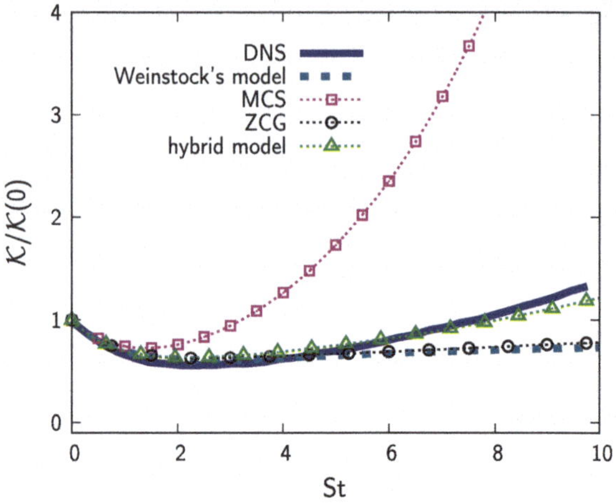

Fig. 50.2 Time evolution of turbulent kinetic energy

Perspectives Towards Pseudo-spectral DNS for Shear-Driven Flows

The new FDS scheme will be used in pseudo-spectral DNS, as an alternative to solving the advection operator with coordinates comoving with the mean flow [3, 7]. New DNS will be performed on a large variety of turbulent flows, in the presence of various mean flows, with hyperbolic, linear and elliptic streamlines. Typical initial data for the fluctuating flow will be used, with continuous angular distribution, possibly after a pre-computation from a 'random' initialization. Cross-validation between DNS results and results of our present models [11] will be extended to the highest Reynolds number allowed by our numerical facilities.

Further Applications to Modelling, with Multiscale and/or Directional Aspects

Improvement of Single-Point Closures, Possibly Structure-Based

Our spectral modelling approach will lead to improve single-point closure models, from classical RSM to more elaborate structure-based models by Kassinos, Reynolds and coworkers (see [2] and Ref. in [8].) Among several attempts to improve the 'rapid' part of the pressure-strain rate tensor in RSM, in line with homogenized RDT, the recent study by Mishra and Girimaji [4] offers new insights about turbulent flows in the presence of mean flows much more general than the pure plane shear, with both hyperbolic and elliptic streamlines: our modelling approach could give useful assessments and improvements, since it is exact, regarding the linear response of turbulence with such various instabilities, and can also give information on their non-linear saturation.

On the one hand, it was shown in [4] that transport of additional structure-based tensors, is not optimal from the viewpoint of global efficiency and especially realisability. A very simple model of 'rapid' pressure-strain rate tensor was emphasized in [4], in which a single parameter $A_5(\psi)$ was directly linked to the mean flow ellipticity. Going back to [1, 5], it is shown that this parameter generalizes the tuned constant (e.g. C_2 in Launder, Reece and Rodi (1975), Ref. in [8]) and has a spectral counterpart denoted $a(k, t)$ in [1]. The latter parameter controls the partition between directional anisotropy and polarisation anisotropy, so that it offers new prospects for advanced modelling.

On the other hand, the approach by [2] (KRR) foreshadows useful aspects, which are underlined by our spectral modelling approach. Directional anisotropy is reflected by the *dimensionality tensor*, and all other new tensors (with respect to the Reynolds Stress one) can be obtained from our $(\mathscr{E}, Z)$ decomposition, which provides their 3D and 1D spectra as well, if needed. The KRR *gradient* vector in the Particle Representation Model is very similar to the direction of the wave vector in RDT: it could be generalized as the gradient of a phase purely advected by the mean flow. The main difficulty for reproducing RDT in a single-point model is now clarified: this is due to the vortical part of the mean flow, or mean vorticity $\boldsymbol{W}$, which possibly combines with the system vorticity $2\boldsymbol{\Omega}$. Because KRR introduced a *stropholysis tensor* for breaking mirror symmetry (not to be confused with helicity), that is relevant only in the presence of rotational mean flow, we propose to coin *dynamical stropholysis term* the term $\imath Z\,(\boldsymbol{W} + 4\boldsymbol{\Omega}) \cdot \boldsymbol{\alpha}$ which concentrates the whole rotational contribution in Eq. (50.3) for polarization anisotropy. In the case of purely rotating flow, a typical damping of the Coriolis effect in single-point modelling is called *rotational randomization* in KRR: this corresponds to the *phase-mixing by dispersive waves* in our fully spectral approach. Our dynamical stropholysis term involves twice the dispersion law $2\boldsymbol{\Omega} \cdot \boldsymbol{\alpha}$ of inertial waves in the case of dominant system rotation, but it can exist even in the absence of waves, e.g. in the destabilizing and neutral cases of the rotating shear, and even for the pure shear flow.

Towards a More General Use of Angular Harmonics, from Spectral Space to Physical One

In our present model, the linear part of the equations for the second-order spectral tensor for velocity, or in a more compact form $(\mathscr{E}, Z, \mathscr{H})$-equations, is performed in 3D Fourier space, without truncation in terms of angular harmonics. Only the contribution of general transfer terms, mediated by nonlinearity, is expanded at the degree two, in agreement with the MCS model, and includes a clipping approximation for damping the modes of higher degree, in the hybrid model. A general and systematic a posteriori decomposition of the set $(\mathscr{E}, Z)$ resulting from both linear and fully nonlinear evolution is in progress. Classical *scalar* spherical harmonics can be used for $\mathscr{E}$, mainly in polar-spherical coordinates for the degree larger than 2, in order to use the properties of orthogonality of generalized Legendre polynomials. The same decomposition applies to $\mid Z \mid$, and to the helicity spectrum, if re-introduced, but not to Z itself, so that the original proposals in [12] ought to be checked and validated. This reflects the fact that polarization could involve vectorial spherical harmonics. Finally, such a systematic use of angular harmonics is interesting since it offers the possibility to extract information for correlations in physical space, such as vectorial structure functions, using Wiener-Kinchin transforms, mode by mode, and not the full 3D inverse Fourier transform.

Contribution to a General Strategy for Modelling Multiscale Inhomogeneous Flows

For future application, we hope to integrate at least a simplified part of our fully nonlinear, multi-scale and directional, present approach, now restricted to an academic homogeneous model entirely in Fourier space, to a more general, and more practical, inhomogeneous flow. Some inhomogeneous RDT is recovered by the *resolvent analysis* (e.g. keynote lectures from Beverley McKeon in the last ETMM12 colloquium), which expresses the linear interaction mean-to-fluctuating. In turn, the feedback from the generalized Reynolds stresses is accounted for, as the fluctuating-to-mean interaction, to be incorporated in a weakly inhomogeneous version of our model. The most crucial improvement is expected in the last fluctuating-to-fluctuating interaction, which is either ignored or roughly mimicked by an effective (turbulent) diffusivity. In this context, a contribution to the new COST (last ERCOFTAC Autumn Festival in Warsaw, October 18, 2018) is of interest. Last but not least, a relevant multiscale model is addressed by the PANS (Partially Averaged Navier-Stokes) approach by Sharath Girimaji and coworkers, and important connections can be exploited with our present model/strategy.

References

1. Cambon, C., Jeandel, D., Mathieu, M.: Spectral modelling of homogeneous non-isotropic turbulence. J. Fluid Mech. **104**, 247–262 (1981)
2. Kassinos, S.C., Reynolds, W.C., Rogers, M.M.: One-point turbulence structure tensors. J. Fluid Mech. **428**, 213–248 (2001)
3. Lesur, G., Longaretti, P.Y: On the relevance of subcritical hydrodynamic turbulence to accretion disk transport. Astron. Astrophys. **444**, 25–44 (2005)
4. Mishra, A.A., Girimaji, S.S.: Toward approximating non-local dynamics in single-point pressure-strain correlation closures. J. Fluid Mech. **811**, 168–188 (2017)

5. Mons, V., Cambon, C., Sagaut, P.: A spectral model for homogeneous shear-driven anisotropic turbulence in terms of spherically averaged descriptors. J. Fluid Mech. **788**, 147–182 (2016)
6. Orszag, S.A.: Analytical theories of turbulence. J. Fluid Mech. **41**, 363–386 (1969)
7. Rogallo, R.S.: Numerical experiments in homogeneous turbulence. NASA technical memorandum 81315, National Aeronautics and Space Administration (1981)
8. Sagaut, P., Cambon, C.: Homogeneous Turbulence Dynamics, Monograph, 2nd edn. Springer (2018)
9. Salhi, A., Jacobitz, F.G., Schneider, K., Cambon, C.: Nonlinear dynamics and anisotropic structure of rotating sheared turbulence. Phys. Rev. E **89**(1), 013020 (2014)
10. Weinstock, J.: Analytical theory of homogeneous mean shear turbulence. J. Fluid Mech. **727**, 256–281 (2013)
11. Zhu, Y., Cambon, C., Godeferd, F., Salhi, A.J.: Non linear spectral model for rotating sheared turbulence. Fluid Mech. **866**, 5–32 (2019)
12. Zhu, Y.: Modelling and calculation for shear-driven rotating turbulence, with multiscale and directional approach. Ph.D. thesis, Ecole Centrale de Lyon, France (2019)

Chapter 51
Under-Resolved DNS of Non-trivial Turbulent Boundary Layers via Spectral/*hp* CG Schemes

R. C. Moura, J. Peiró and S. J. Sherwin

Introduction

This study assesses the suitability of spectral/*hp* continuous Galerkin (CG) schemes [1] for model-free under-resolved simulations of a non-trivial turbulent boundary layer flow. We consider a model problem proposed by Spalart in [2] that features a rotating free-stream velocity and admits an asymptotic solution with significant crossflow effects. Note this test case is substantially more complex than typical turbulent boundary layer canonical problems owing to its unsteadiness and enhanced small-scale anisotropy. Reported LES-based solutions to this problem are known to require sophisticated modelling and relatively fine grids to achieve meaningful results, with traditional models exhibiting poor performance. The model-free CG-based approach advocated, on the other hand, yields surprisingly good results with considerably less degrees of freedom for higher order discretisations. Usefully accurate results for the mean flow quantities could even be obtained with half as many degrees of freedom per direction (in comparison to reference LES solutions). Usage of high-order spectral element methods (CG in particular) is therefore strongly motivated for wall-bounded turbulence simulations via under-resolved DNS (uDNS), sometimes called implicit LES (iLES), approaches.

R. C. Moura (✉)
Instituto Tecnológico de Aeronáutica, São José dos Campos, Brazil
e-mail: moura@ita.br

J. Peiró · S. J. Sherwin
Imperial College London, London, UK
e-mail: j.peiro@imperial.ac.uk

S. J. Sherwin
e-mail: s.sherwin@imperial.ac.uk

M. García-Villalba et al. (eds.), *Direct and Large Eddy Simulation XII*,
ERCOFTAC Series 27,
https://doi.org/10.1007/978-3-030-42822-8_51

The Rotating Boundary Layer Problem

The considered test case is defined as follows. A flat plate rests (at $y = 0$) while the wind blows parallel to it at infinity ($y = y_\infty$), having fixed magnitude V_o, but rotating at constant angular frequency f, so that, at $y = y_\infty$,

$$u(x, z, t) = V_o \cos\phi, \quad w(x, z, t) = V_o \sin\phi, \quad \phi = ft. \tag{51.1}$$

Laterally, the domain extends to infinity, whereby periodic boundary conditions can be used. A rotating pressure gradient is also necessary for an asymptotic solution to exist, namely,

$$\frac{\partial p}{\partial x} = f\rho_o V_o \sin\phi, \quad \frac{\partial p}{\partial z} = -f\rho_o V_o \cos\phi, \tag{51.2}$$

where ρ_o is the free-stream density (incompressible flow is assumed as in [2]). This pressure gradient is always normal to the free-stream velocity and should be applied everywhere in the domain, thus generating crossflow effects close to the wall.

The flow is homogeneous in planes parallel to the wall and statistically stationary for a rotating reference frame aligned to the free-stream velocity. The rotating, asymptotic solution is sketched in Fig. 51.1 with the axis notation adopted. Different reference frames are required. Frame $\tilde{x} \times \tilde{z}$ rotates with the free-stream velocity, with $\tilde{x}$ aligned to V_o. Frame $\bar{x} \times \bar{z}$ rotates with the (averaged) wall shear stress vector, with $\bar{x}$ aligned to it. Finally, $\hat{x} \times \hat{z}$ rotates with the (averaged) Reynolds stress vector at the wall, with $\hat{x}$ aligned to it. Angles α and ϕ^* denote the misalignment between pairs $\bar{x}$–$\tilde{x}$ and $\hat{x}$–$\tilde{x}$, respectively.

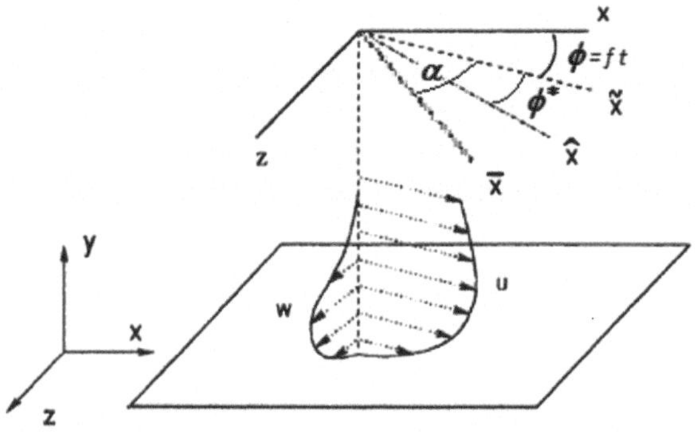

Fig. 51.1 Sketch of the model problem (adapted from [3])

The velocity components for the different reference frames are defined in the usual way, e.g.

$$\widetilde{u} = u \cos\phi + w \sin\phi, \quad \widetilde{w} = w \cos\phi - u \sin\phi, \tag{51.3}$$

which are the velocity components in the frame rotating with V_o. The laminar solution is known analytically from [2], with the laminar boundary layer thickness given by $\delta_\ell = (2\nu/f)^{1/2}$, where ν is the fluid's kinematic viscosity. A laminar Reynolds number can thus be defined as $R_\ell = V_o\,\delta_\ell/\nu = 2\,V_o/(f\,\delta_\ell)$. We note that $R_\ell = 500$ is about the lowest Reynolds number for a turbulent solution to exist [2].

For turbulent solutions, the viscous length scale δ_* is used to normalise near-wall quantities (e.g. $y^+ = y/\delta_*$), namely $\delta_* = \nu/V_*$, with V_* being the friction velocity,

$$V_* = \left[\nu\,\partial_y\sqrt{u^2+w^2}\right]_{wall}^{1/2}, \tag{51.4}$$

in which the statistical averaging $\langle\sqrt{u^2+w^2}\,\rangle$ is implied.

The relevant length for the large scales, δ_t, also taken to be the turbulent boundary layer thickness [2], is $\delta_t = V_*/f$ and defines a turbulent Reynolds number R_t, namely $R_t = V_*\,\delta_t/\nu = (R_\ell\,V_*/V_o)^2\,/2$, which is analogous to the friction Reynolds number used in classical turbulent boundary layer flows. According to [2], angle ϕ^* can be estimated from α as $\phi^* \approx \alpha - 52/R_\tau$, radian values being assumed. In Table 51.1, a summary of relevant quantities is given, based on computations conducted in [2–4]. Here, only the highest Reynolds number case shown in Table 51.1 is considered.

Numerical Setup and Discussion of Results

Now we present results of CG-based uDNS for the model problem in question. We compare our results to classical LES based on centred second-order finite-difference (FD) schemes with either the dynamic Smagorinsky (DS) or the more sophisticated mixed Lagrangian (ML) model from [4]. DNS results are also available from [2].

Our simulations, after being initialised from the analytical laminar solution with added random noise, are evolved for ten rotation periods and then probed for statistics, as in [3, 4]. We also followed these studies by adopting a cubic domain of side

Table 51.1 Summary of Reynolds numbers and length scales

R_ℓ	R_t	δ_t/δ_ℓ	V_*/V_o	δ_*/δ_ℓ	α (°)	ϕ^* (°)
500	466	15.3	0.0610	0.0328	26.24	19.88
620	653	18.1	0.0583	0.0277	23.20	18.62
767	914	21.4	0.0561	0.0232	20.91	17.70

$1.6\delta_t$, cf. Table 51.1. Their grids had about 64^3 points which, in the wall-parallel plane, resulted in $\Delta x^+ = \Delta z^+ \approx 23$. In the wall-normal direction, they employed ten equispaced grid points within $y^+ \leq 10$ and a constant geometric stretching for $y^+ > 10$.

In our simulations, although similar grids could have been used, comparable results were already obtained in preliminary runs on grids twice as coarse per direction. Our results were obtained with the spectral/*hp* CG-based incompressible flow solver of *Nektar++* [5]. Although spectral vanishing viscosity (SVV) is sometimes used in CG-based uDNS as a stabilisation/regularisation technique [6], SVV was not used here. Still, full polynomial dealising was employed via over-integration [7].

First, a low-order run ($p = 1$, linear elements) was carried out on 32^3 elements, which is akin to a second-order finite element computation on 32^3 DOFs, since CG has p independent polynomial DOFs per element (and per direction). Here, 8 elements were placed within $y^+ \leq 10$, with the remaining layers following a constant

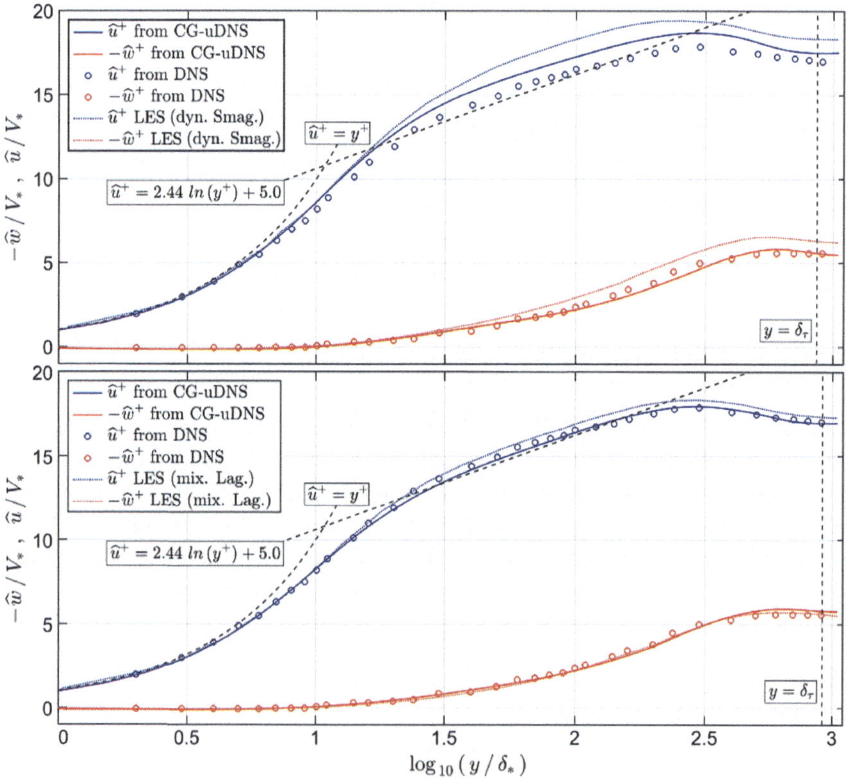

Fig. 51.2 Rotated velocity components $\widehat{u}^+$ (blue) and $-\widehat{w}^+$ (red) versus wall-normal distance y^+. DNS results at $R_t = 767$ (circles) are compared against low-order (top) and high-order (bottom) CG-based uDNS (solid line) on 32^3 DOFs. 2nd order FD-based LES results obtained on 64^3 DOFs with dynamic Smagorinsky (top) and a mixed Lagragian model (bottom) are shown as dotted lines

geometric stretching. The mean velocity profiles obtained are shown in Fig. 51.2 (top) and are better than those yielded by the DS model on 64^3 DOFs. Comparison of further statistics (not shown) confirmed this pattern.

Second, a high-order run ($p = 4$, fifth-order discretisation) was conducted on 8^3 elements, corresponding to the same 32^3 DOFs. In this case, only 2 elements were placed within $y^+ \leq 10$, with the remaining layers following the usual stretching. The mean velocity profiles are shown in Fig. 51.2 (bottom) and are comparable to those yielded by the sophisticated ML model on 64^3 DOFs. Comparison of further statistics once more confirmed this pattern, as shown in Fig. 51.3.

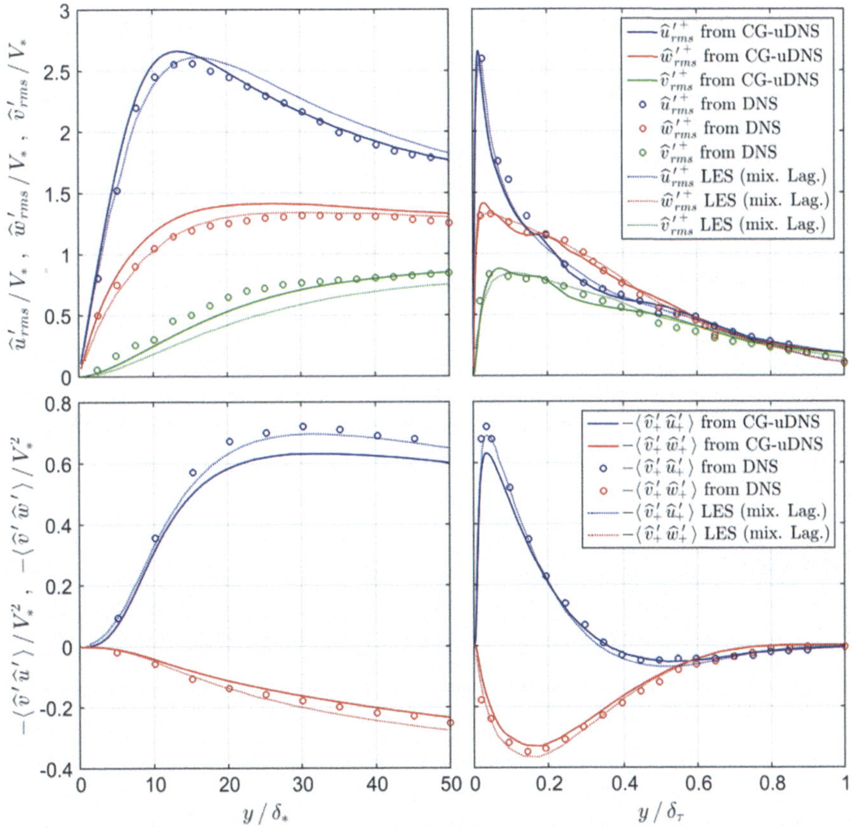

Fig. 51.3 CG-uDNS results of high order case (solid lines) shown against DNS (circles) and fine-grid LES (dashed lines) obtained with the mixed Lagragian model. Turbulent intensities (top) and Reynolds stresses (bottom) indicate similar solution quality between CG-uDNS and fine-grid LES

Concluding Remarks and Future Work

These results strongly support the suitability of high-order CG-based uDNS/iLES approaches for flows involving complex turbulence features. The reason for this performance is likely related to the fact that spectral/*hp* methods have a superior resolution power per DOF at higher orders, which effectively makes a broader range of scales to follow the governing equations precisely.

While different theoretical techniques can be applied to assess the suitability of spectral/*hp* methods for iLES/uDNS [8, 9], the most insightful technique so far is arguably the so-called eigensolution analysis (ESA), where numerical dispersion and diffusion characteristics are quantified in wavenumber/frequency space. Although mostly feasible for simple model problems, such as one-dimensional linear advection, ESA's estimates for spectral/*hp* methods have been shown to hold more generally for nonlinear problems and even turbulence simulations [10–14]. These estimates quantify how the numerical error (dissipation in particular) increases at large wavenumbers/frequencies, affecting especially the smallest, poorly-resolved scales. Although most ESA studies have focused on discontinuous spectral/*hp* methods, the continuous Galerkin scheme has also been assessed [6, 15], and the application of ESA to CG in connection with the rotating boundary layer case considered in the present work is planned for the near future.

Acknowledgements R. C. Moura would like to acknowledge funding from Coordenação de Aperfeiçoamento de Pessoal de Nível Superior (CAPES), Finance Code 001, under the Brazilian Science without Borders scheme. S. J. Sherwin acknowledges support from EPSRC Platform grant EP/R029423/1. The simulations in this work were performed using the Imperial College Research Computing Service (https://doi.org/10.14469/hpc/2232).

References

1. Karniadakis, G.E., Sherwin, S.J.: Spectral/*hp* element methods for computational fluid dynamics. Oxford University Press, 2nd edn. Springer (2005)
2. Spalart, P.R.: Theoretical and numerical study of a three-dimensional turbulent boundary layer. J. Fluid Mech. **205**, 319–340 (1989)
3. Abbà, A., Cercignani, C., et al.: A 3D turbulent boundary layer test for LES models. In: Proceedings of the 1st International Conference on Computational Fluid Dynamics—ICCFD1, pp. 485–490 (2001)
4. Wu, X., Squires, K.D.: Large eddy simulation of an equilibrium three-dimensional turbulent boundary layer. AIAA J. **35**(1), 67–74 (1997)
5. Cantwell, C.D., Moxey, D., Comerford, A., Bolis, A., Rocco, G., et al.: Nektar++: an open-source spectral/*hp* element framework. Comput. Phys. Commun. **192**, 205–219 (2015)
6. Moura, R.C., Sherwin, S.J., Peiró, J.: Eigensolution analysis of spectral/*hp* continuous Galerkin approximations to advection-diffusion problems: insights into spectral vanishing viscosity. J. Comput. Phys. **307**, 401–422 (2016)
7. Kirby, R.M., Karniadakis, G.E.: De-aliasing on non-uniform grids: algorithms and applications. J. Comput. Phys. **191**, 249–264 (2003)

8. Moura, R.C., Sherwin, S.J., Peiró, J.: Modified equation analysis for the discontinuous Galerkin formulation. In: Spectral and High Order Methods for PDEs—ICOSAHOM 2014. Springer (2015)
9. Moura, R.C., Mengaldo, G., Peiró, J., Sherwin, S.J.: An LES setting for DG-based implicit LES with insights on dissipation and robustness. In: Spectral and High Order Methods for PDEs—ICOSAHOM 2016. Springer (2017)
10. Moura, R.C., Sherwin, S.J., Peiró, J.: Linear dispersion-diffusion analysis and its application to under-resolved turbulence simulations using discontinuous Galerkin spectral/*hp* methods. J. Comput. Phys. **298**, 695–710 (2015)
11. Moura, R.C., Mengaldo, G., Peiró, J., Sherwin, S.J.: On the eddy-resolving capability of high-order discontinuous Galerkin approaches to implicit LES/under-resolved DNS of Euler turbulence. J. Comput. Phys. **330**, 615–623 (2017)
12. Mengaldo, G., Moura, R.C., Giralda, B., Peiró, J., Sherwin, S.J.: Spatial eigensolution analysis of discontinuous Galerkin schemes with practical insights for under-resolved computations and implicit LES. Comput. Fluids **169**, 349–364 (2018)
13. Mengaldo, G., De Grazia, D., Moura, R.C., Sherwin, S.J.: Spatial eigensolution analysis of energy-stable flux reconstruction schemes and influence of the numerical flux on accuracy and robustness. J. Comput. Phys. **358**, 1–20 (2018)
14. Fernandez, P., Moura, R.C., Mengaldo, G., Peraire, J.: Non-modal analysis of spectral element methods: towards accurate and robust large-eddy simulations. Comput. Methods Appl. Mech. **346**, 43–62 (2019)
15. Moura, R.C., Aman, M., Peiró, J., Sherwin, S.J.: Spatial eigenanalysis of spectral/*hp* continuous Galerkin schemes and their stabilisation via DG-mimicking spectral vanishing viscosity for high Reynolds number flows. J. Comput. Phys. **406** (2020). https://doi.org/10.1016/j.jcp.2019.109112

Chapter 52
Mesh Optimization Using Dual-Weighted Error Estimators: Application to the Periodic Hill

N. Offermans, A. Peplinski and P. Schlatter

Introduction

Simulations have matured to be an important tool in the design and analysis of many modern industrial devices, in particular in the process, vehicle and aeronautical industries. Most of the considered flows are turbulent, for which it is impossible to perform straightforward grid-convergence studies without considering long-term statistics. Therefore, when performing simulations in computational fluid dynamics (CFD), the use of adaptive mesh refinement (AMR) is an effective strategy to increase the reliability of the solution at a reduced computational cost. Such tools allow for error control, reduced simulation time, easier mesh generation, better mesh quality and better resolution of the a priori unknown physics. In the present contribution, the design of an optimal mesh, using AMR, is investigated in Nek5000 [3], a highly scalable code based on the spectral element method (SEM) [9] and aimed at the direct numerical simulation (DNS) of the incompressible Navier–Stokes equations (further discussed in section "Numerical Method").

The development of AMR capabilities in a CFD code requires the implementation of two main features: tools for mesh refinement and error estimators, which drive the refinement process. In this study, we discuss both aspects. First, h-refinement is used for mesh adaptation. Then, we consider both traditional local error indicators and newly developed adjoint error estimators, based on the computation of a dual

N. Offermans (✉) · A. Peplinski · P. Schlatter
Linné FLOW Center, KTH Mechanics, Royal Institute of Technology, Stockholm, Sweden
e-mail: nof@mech.kth.se

A. Peplinski
e-mail: adam@mech.kth.se

P. Schlatter
e-mail: pschlatt@mech.kth.se

M. García-Villalba et al. (eds.), *Direct and Large Eddy Simulation XII*, ERCOFTAC Series 27,
https://doi.org/10.1007/978-3-030-42822-8_52

problem, as detailed in section "Spectral Error Indicators". Our objective is to design a final, optimal mesh in the case of 3D turbulent simulations. The test case considered is that of a periodic, constricted channel, also called periodic hill, at four different Reynolds numbers, see section "Adjoint Error Estimators". In particular, we study the effect of both error estimators on the refinement pattern and on the resolution of relevant flow quantities.

Numerical Method

In Nek5000, the numerical solution to the Navier–Stokes equations is computed by the SEM, where the numerical domain is split into a number of non-overlapping elements. On each element, the solution is expanded in terms of high-order polynomials up to order N on the Gauss–Lobatto–Legendre (GLL) quadrature points. The code offers low dissipation and high accuracy for the computation of turbulent flows. By default, the code only allows for conforming meshes, made of hexahedral (3D) or quadrilateral (2D) elements, and constant polynomial order, meaning that mesh adaptation is not supported by default.

Therefore, we modify the code to enable h-refinement, where selected elements are refined by an oct-tree structure, thus locally increasing the number of gridpoints. The first change is the introduction of interpolation operators at the interface between coarse and fine elements [5], where the solution from the coarse face of the interface is interpolation onto the fine faces. Moreover, external libraries are employed for the mesh management in parallel: `p4est` is used to keep track of the grid hierarchy and connectivity [2], and ParMETIS is used for the grid partitioning [4]. Other changes include the modification of the preconditioner for the pressure equation, which combines an overlapping Schwarz decomposition, resulting in small local problems, and a global coarse-grid solver [10]. It has been verified that those changes preserve the accuracy and stability of the method, the efficiency of the pressure solver and the overall good scalability of the code.

Spectral Error Indicators

The spectral error indicators are a local measure of the error, based on the spectral properties of the solution [6]. Considering a 1D case for illustration, a solution $u(x)$ can be expanded as a sum of Legendre polynomials as $u(x) = \sum_{k=0}^{\infty} \hat{u}_k L_k(x)$, where L_k is the Legendre polynomial at order k, $\hat{u}_k = \frac{1}{\gamma_k} \int_{-1}^{1} u(x)L_k(x)\mathrm{d}x$ is the spectral coefficient at order k and $\gamma_k = ||L_k||_{L^2}^2$.

Assuming an expansion of the solution up to order N only, as well as an exponential decay for the spectral coefficients of the form $\hat{u}_k \sim \hat{u}(k) = c \exp(-\sigma k)$, and accounting for an additional contribution to the error due to quadrature, an estimate of the total error is given by

$$\varepsilon_{\text{ind}} = \left(\int_N^\infty \frac{\hat{u}(k)^2}{\frac{2k+1}{2}} \mathrm{d}k + \frac{\hat{u}_N^2}{\frac{2N+1}{2}} \right)^{\frac{1}{2}}.$$

This expression is readily expanded to 2D and 3D configurations and the error is averaged in time on each spectral element. These estimators provide an indication of the local error and are cheap to compute, amounting to less than 1% of the total computational cost. However, they might induce over-refinement in unnecessary regions, e.g. in the wake far away from a bluff body, where the resolution is poor but does not affect the computation of relevant quantities.

Adjoint Error Estimators

Contrary to the previous method, the adjoint error estimators are goal-oriented and based on the dual-weighted residual method [1]. These estimators provide a measure of the error committed on a functional of interest, which takes the form of a surface (e.g. drag or lift) or volume (e.g. the solution in a specific region) integral, via the resolution of a global, dual problem, linearized around the forward, or direct, solution. The choice of this functional also determines the forcing terms and the boundary conditions for the adjoint system of equations. The expression for the adjoint error estimator on each spectral element is given by

$$\varepsilon_{\text{adj}} = \boldsymbol{\rho}_1 \cdot \boldsymbol{\omega}_1 + \boldsymbol{\rho}_2 \cdot \boldsymbol{\omega}_2 + \rho_3\, \omega_3,$$

where the terms $\boldsymbol{\rho}_1$ and ρ_3 are measures of the strong residuals on the momentum and continuity equations, respectively. The term $\boldsymbol{\rho}_2$ is a measure of the jump in the stresses at the interface between elements, which we skip at the moment. The terms $\boldsymbol{\omega}_1$, $\boldsymbol{\omega}_2$ and ω_3 are adjoint weights, i.e. a measure of the interpolation error committed on the dual solution [8]. As with the spectral error indicators, the adjoint error estimators are averaged in time and provide an error map of the sensitivity of the functional of interest with respect to mesh resolution. The main drawback of the method is the costly computation of the backward time integration, which requires the recomputation of the forward solution. Assuming that both forward and backward time integrations bear the same computational cost, the adjoint error estimators are approximately three times more expensive to compute than the spectral error indicators. In practice, let us note that the computation of the backward time integration is performed by an appropriate checkpointing scheme [11]. The adjoint error estimators are designed to prevent the unnecessary refinement that might arise from the spectral error indicators, when the primary focus of the simulation is the computation of an integral quantity of interest.

Application and Results

Test Case

The test case used to compare both error estimators is that of the periodic hill [7]. The configuration is 3D and four Reynolds numbers based on the height of the hill and the mean velocity at the crest of the hill are considered: Re $= 700, 1400, 2800$ and 5600. The flow is turbulent at all regimes and is characterized by a recirculation bubble, which separates on the curved surface of the hill and reattaches on the flat bottom wall further downstream. A 2D representation of the configuration (the case is uniform in the spanwise direction) is shown in Fig. 52.1, where the dimensions considered for the case are $L_x = 9h$, $L_y = 3.035h$ and $w_h = 1.929h$. The spanwise dimension is $L_z = 4.5h$ and the polynomial order for the simulation is fixed at $N = 8$.

A forcing term maintains the flow at bulk velocity U_b, while the functional of interest is chosen to be the drag on the flat part of the bottom wall. This translates into an adjoint Dirichlet condition u_{adj} illustrated in red in the figure. In both cases, the error estimators are averaged over a relatively short time interval, which is still sufficiently long to capture relevant flow physics. Then, mesh refinement is performed and statistical data is gathered on the refined mesh over a long time interval.

Error Maps

The error maps when using the adjoint error estimators are shown in Fig. 52.2a for the initial mesh and in Fig. 52.2b for the mesh after two rounds of refinement. As an

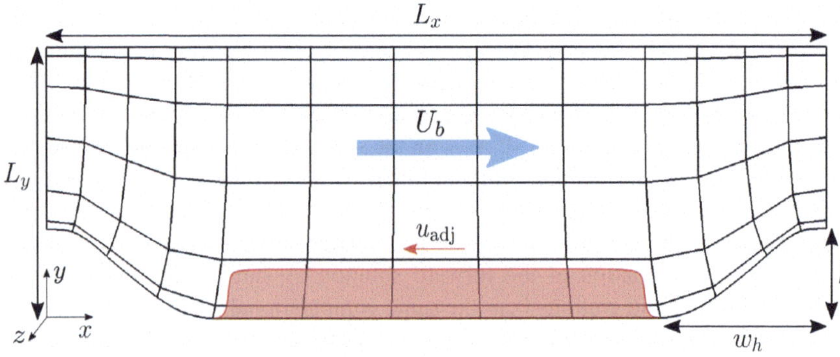

Fig. 52.1 Configuration of the flow around a periodic hill (2D profile shown). The original spectral grid is shown, along with the adjoint Dirichlet boundary condition u_{adj}; the velocity is applied on the wall and the red curve only shows the smoothing at the transition. In the spanwise direction, the mesh is made of 6 uniform element layers

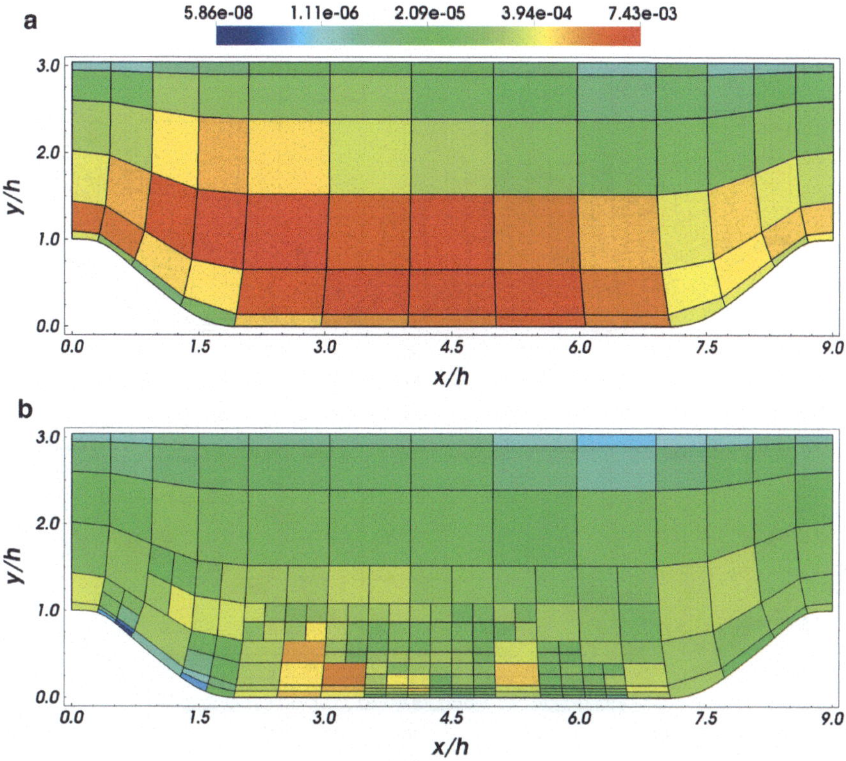

Fig. 52.2 **a** Re = 1400—adjoint error estimators on the initial mesh **b** Re = 1400—adjoint error estimators after two rounds of refinement

effect of mesh refinement, the error in the domain is more uniform and significantly reduced after two rounds of refinement compared to the original mesh. Moreover, refinement is focused toward the bottom flat wall, in the region of the recirculation bubble and as well in the vicinity of the separation point. As opposed to the spectral error indicators (not shown), the refinement does not propagate on the uphill section of the hill and in the upper half of the channel.

Vorticity

We look at instantaneous vorticity fields at Re = 5600 to compare the effect of the two error estimators. As can be observed in Fig. 52.3a, the spectral error indicators tend instead to resolve the mesh uniformly, such that the turbulent features are well-captured everywhere in the domain. In Fig. 52.3b, it is seen that the adjoint error estimators tend to refine the area of the separation point and the region of interest,

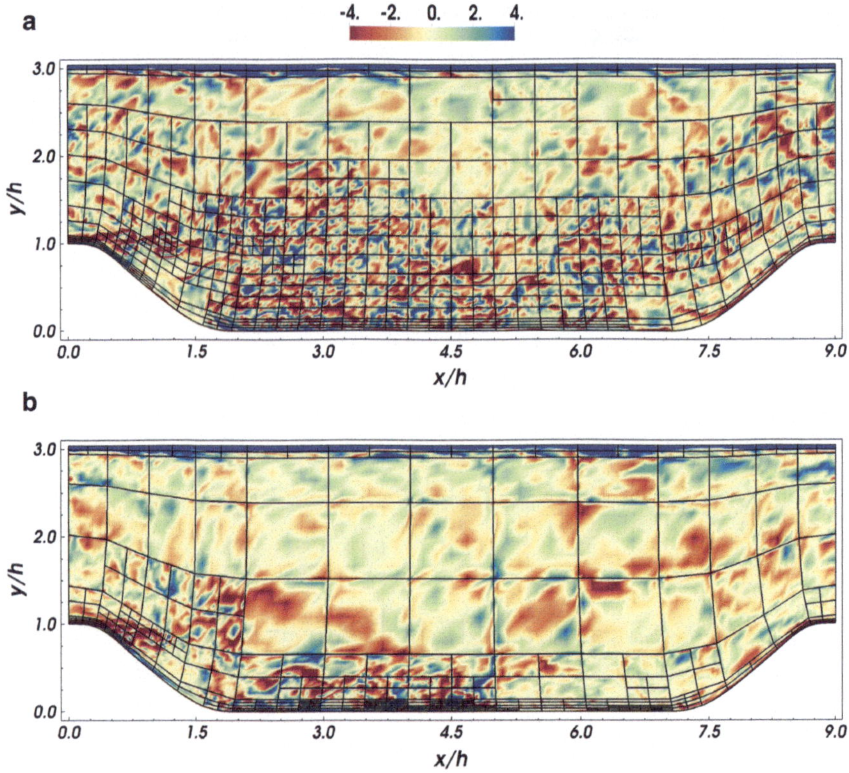

Fig. 52.3 **a** Re = 5600—spectral error indicators—vorticity in the spanwise direction ω_z **b** Re = 5600—adjoint error estimators—vorticity in the spanwise direction ω_z

while leaving the bulk of the flow in the upper half of the channel highly under-resolved.

Conclusions and Outlook

We have performed 3D turbulent flow simulations using AMR in Nek5000, a code based on the spectral element method. Two types of error estimators were compared: local spectral error indicators and goal-oriented adjoint error estimators. The former method propagates refinement evenly in the domain while the latter focuses refinement for the optimal computation of a quantity of interest, thus avoiding unnecessary over-resolution in less relevant regions.

Future work will focus on higher Reynolds numbers and fully 3D geometries, as well as on the validation of the adjoint error estimators as a reliable estimate of the error on the functional.

References

1. Bangerth, W., Rannacher, R.: Adaptive Finite Element Methods for Differential Equations. Birkhäuser, Basel (2002)
2. Burstedde, C., Wilcox, L.C., Ghattas, O.: p4est: scalable algorithms for parallel adaptive mesh refinement on forests of octrees. SIAM J. Sci. Comput. **33**(3), 1103–1133 (2011)
3. Fischer, P.F., Lottes, J.W., Kerkemeier, S.G.: http://nek5000.mcs.anl.gov (2008)
4. Karypis, G., Kumar, V.: MeTis: Unstructured Graph Partitioning and Sparse Matrix Ordering System, Version 4.0. http://www.cs.umn.edu/~metis (2009)
5. Kruse, G.W.: Parallel nonconforming spectral element solution of the incompressible Navier–Stokes equations in three dimensions. Ph.D. thesis, Brown University (1997)
6. Mavriplis, C.: A posteriori error estimators for adaptive spectral element techniques. In: Wesseling, P. (ed.) Notes on Numerical Fluid Mechanics, pp. 333–342 (1990)
7. Mellen, C., Fröhlich, J., Rodi, W.: Large eddy simulation of the flow over periodic hills. In: Deville, M., Owens, R. (eds.) Proceedings of the IMACS World Congress (2000)
8. Offermans, N.: Aspects of adaptive mesh refinement in the spectral element method. Ph.D. thesis, KTH Royal Institute of Technology (2019)
9. Patera, A.T.: A spectral element method for fluid dynamics: laminar flow in a channel expansion. J. Comput. Phys. **54**(3), 468–488 (1984)
10. Peplinski, A., Offermans, N., Fischer, P.F., Schlatter, P.: Non-conforming elements in Nek5000: pressure preconditioning and parallel performance. In: Proceedings of the International Conference On Spectral And High Order Methods 2018. To appear
11. Schanen, M., Marin, O., Zhang, H., Anitescu, M.: Asynchronous two-level checkpointing scheme for large-scale adjoints in the spectral-element solver Nek5000. Procedia Comput. Sci. **80**(Supplement C), 1147–1158 (2016)

Chapter 53
Assessment of LES Using Sliding Interfaces

G. Sáez-Mischlich, G. Grondin, J. Bodart and M. C. Jacob

Introduction

Large Eddy Simulations (LES) of complex turbomachinery applications require specific numerical and computational techniques to handle both rotating and static geometries. For finite volume solvers, several different algorithms have been proposed such as chimaera grids and sliding mesh interface methods. The latter method is studied here in combination with an ALE method to treat the moving regions with the Navier–Stokes equations. Due to the relative movement of the grids along the sliding interfaces, the sliding mesh computational method needs to manage non-conformal flux computations and time-dependent processor connectivity, leading to unavoidable modifications of the performance and accuracy of the existing numerical methods. In order to perform LES of rotating geometries, it is important to ensure that turbulent structures are neither distorted nor dissipated by the presence of both the sliding mesh interfaces and the moving grid. Several studies [1–3] have described various techniques to compute the flow around complex rotating and moving geometries, but the numerical dissipation and performance of these methods have not been fully characterised for turbulent flows.

This study aims at characterising the behaviour of a newly developed sliding mesh technique in the unstructured and massively parallel compressible LES solver CharLES^X [4, 5]. The accuracy of the method is measured for a set of canonical turbulent flows. In the analysis, several aspects of the overall implementation are reviewed, in particular robustness and accuracy regarding the objective of computing high fidelity turbulent flows.

G. Sáez-Mischlich (✉) · G. Grondin · J. Bodart · M. C. Jacob
ISAE-SUPAERO, Toulouse, France
e-mail: gonzalo.saez-mischlich@isae-supaero.fr

M. García-Villalba et al. (eds.), *Direct and Large Eddy Simulation XII*,
ERCOFTAC Series 27,
https://doi.org/10.1007/978-3-030-42822-8_53

Method

The compressible Navier–Stokes equations are solved by coupling a set of instances of the original CharLES^X code assigned to each static and moving regions of the mesh. The coupling of these regions along their interfaces is performed using a sliding mesh method inspired by the works of [1] and [2].

Moreover, due to relative mesh movement of the different grid zones across the sliding interfaces, there exists non-conformal connectivity between the faces located at each side of the sliding mesh interfaces. Polygonal intersections of faces need to be computed at each time step to obtain the interpolation and gradient operators and also to ensure the flux conservation among the sliding interfaces. These operators are built by using virtual opposite cells for each face of the sliding interfaces. The value of the conservative variables at these cells is computed by introducing a weighting of the intersecting states. To ensure mass, momentum and energy flux conservation the method proposed by [2] has been implemented. In this algorithm, the flux is computed using left and right states at each side of the intersection polygons for each face of the sliding interface. This method ensures flux conservation on fully covered sliding interfaces. Furthermore, due to the unstructured nature of the mesh, the computation of these intersections requires the search of the opposite neighbour-intersecting faces using a set of non-ordered elements. So-called "brute force" search algorithms are not considered due to their high computational cost. Therefore, to minimise the computational cost of this algorithm, the search of the nearest neighbours is performed using the k-d binary tree search method developed by [6].

Grid movement is taken into account using the Arbitrary Lagrangian Eulerian (ALE) method proposed by [7], in which extra terms dependent on the grid velocity and the conservative variables are added to the discretised fluxes of the Navier–Stokes equations expressed in the absolute reference frame.

As there exist moving regions of the grid, the parallel connectivity between the cores owning sliding interface faces needs to be recomputed at each time step. To implement this procedure, a 2D bounding box (using a surface coordinate system associated with the sliding interface) is assigned to each processor. This bounding box is defined as the rectangular limits of the interface faces which belong to the given core. The intersections of the bounding boxes of the processor at each side of the sliding interface define the connectivity of the processors at each time step. This algorithm is built to minimise data communication between processors. We refer to [1] for more details on this implementation.

Results

To evaluate the numerical dissipation of the sliding interfaces in simulations of inviscid and turbulent flows, two different test cases are proposed: an inviscid isentropic vortex case (section "Inviscid Isentropic Vortex") and a turbulent pipe channel flow case (section "Pipe Channel Flow").

Inviscid Isentropic Vortex

The inviscid isentropic vortex case is a case involving the Euler equations, for which there exists a known analytical solution [8]. The mean flow is initialised with density $\rho = 1$, pressure $p = 1$, x-velocity $u = 0$ and y-velocity $v = 0$. The computational domain is $[0, L] \times [0, L]$ where the length L of the domain is set to 10 times the radius R_v of the vortex. The domains is discretised using a uniform Cartesian mesh with number of cells $N_y = N_x = N$ and cell size $\Delta y = \Delta x = h$. An isentropic vortex is added to the mean flow, which corresponds to perturbations for each component of the velocity and the temperature T given by

$$(\delta u, \delta v) = \frac{\sigma}{2\pi R_v} e^{\frac{1}{2}(1-(r/R_v)^2)}(-\bar{y}, \bar{x}) \tag{53.1}$$

$$\delta T = -\frac{(\gamma - 1)\sigma^2}{8\gamma\pi^2 R_g} e^{1-(r/R_v)^2} \tag{53.2}$$

where $(\bar{x}, \bar{y}) = (x - x_c, y - y_c)$, (x_c, y_c) is the location of the center of the vortex, $r = \sqrt{\bar{x}^2 + \bar{y}^2}$ is the distance to the center of the vortex, σ is the intensity of the vortex with value $\sigma = 5$ and R_g is the gas constant.

To evaluate the error introduced by the ALE methods and the non-conformal discretisation in sliding interfaces, four different cases are proposed:

1. Conformal case or reference case;
2. Non-conformal case in which a small non-conformity is introduced by vertically displacing a block of the domain located at $[0, L/2] \times [0, L]$. The vertical displacement (along the y-direction) is set as $\Delta y/L = \alpha/N$ with N being the number of cells in each direction and α a fixed parameter with value $\alpha = 0.5$;
3. Moving mesh case in which the whole mesh is displaced at a given grid y-velocity with an associated Mach number M_g;
4. Sliding mesh case which is a combination of the previous two cases. In this configuration, grid y-velocity is imposed in a block of the domain located at $[0, L/2] \times [0, L]$.

The errors are analysed after approximately 10 characteristic times of the vortex defined as $t_c = \frac{R_v}{\sigma}$. The L_2-norm error of the density of the solution for the different cases are shown in Fig. 53.1a, b for different mesh sizes. In both figures it can be

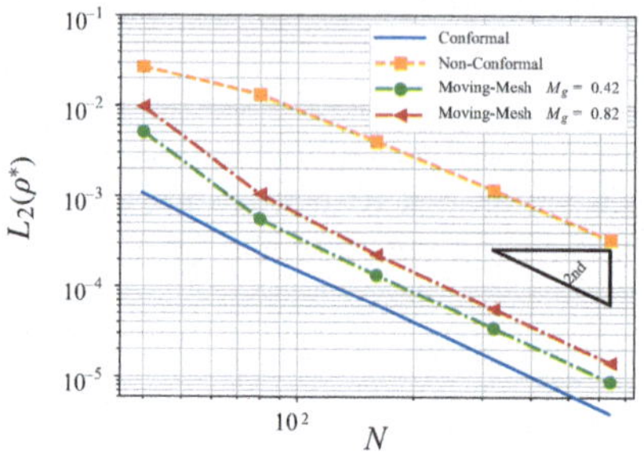

(a) Error comparison of the conformal, non-conformal and moving mesh cases

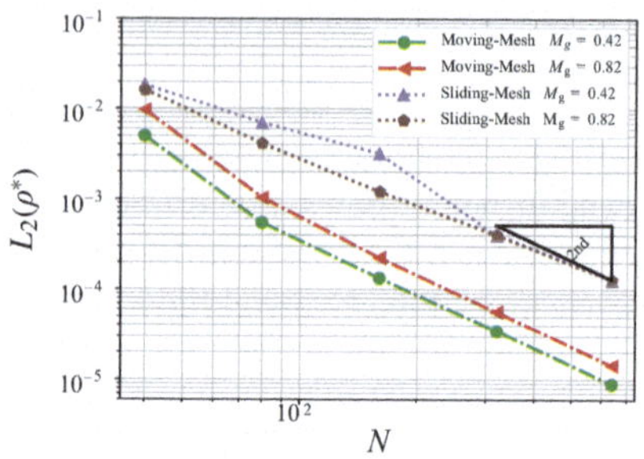

(b) Error comparison of the moving mesh and sliding mesh cases

Fig. 53.1 L_2-errors on density for the conformal, non-conformal, moving and sliding mesh cases

observed that the L_2-norm error shows a second-order relation with grid spacing for all the cases. The analysis with the L_∞-norm yields similar results. Regarding the error levels we can observe that both the addition of non-conformal mesh and grid velocity (ALE formulation) substantially increase the L_2-norm of the error. Furthermore, the error increases when increasing the Mach number of the grid velocity. This can be explained due to the higher magnitude value of the ALE terms that need to be taken into account in the flux computation. Moreover, the sliding mesh shows an error evolution similar to the one found in the non-conformal case for low cell count values. However, for higher cell count values the error behaviour resembles the one found for both the non-conformal and moving mesh cases.

Pipe Channel Flow

DNS simulations of a turbulent pipe flow test case at $\text{Re}_\tau = 360$, $\text{Re}_b = 5300$ and $M_b \sim 0.1$ have been performed using the configuration shown in Fig. 53.2 and have been compared to the incompressible reference solution of [9]. Reynolds numbers are computed based on the diameter D of the pipe. The length L of the pipe along the z axis is set equal to $5D$. In this test case, the domain is divided into two zones: the first zone extends from $z = 0$ to $z = 1.25D$ and from $z = 3.75D$ to $z = 5D$, while the second zone is located between $z = 1.25D$ and $z = 3.75D$. In the second zone, mesh rotation (with rotation speed Ω) is imposed such that $\Omega D/U_b = 1$ where U_b is the bulk velocity. The wall boundary condition of the second zone ensures that the wall absolute velocity is zero even if the grid moves. To perform this simulation it is necessary to impose two rotating sliding mesh interfaces which separate the two mesh zones. This test case allows to analyse the effect of the sliding mesh discretisation methods on the main turbulence quantities of the turbulent pipe flow. Moreover, in order to compare the numerical dissipation obtained with the sliding mesh interfaces

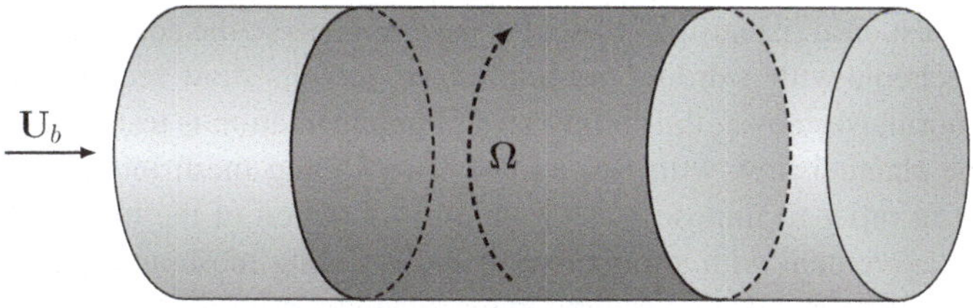

Fig. 53.2 Pipe flow test case configuration using two sliding mesh interfaces

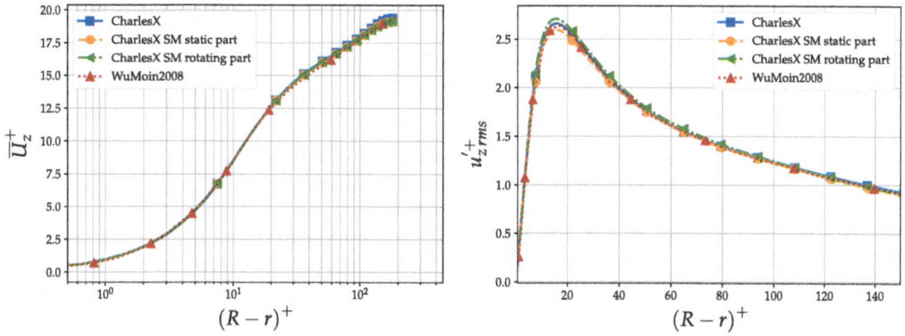

(a) Time-averaged axial velocity (b) Time-averaged axial turbulent intensity

Fig. 53.3 Time-averaged axial velocity profile and axial turbulent intensity, comparison with the reference solution of [9]

against that obtained without, this test case is also simulated without sliding mesh interfaces using the original CharLES^X code.

Figure 53.3 shows the time-averaged velocity profile and the axial turbulent intensity comparing the simulations with and without sliding mesh interfaces and the results from [9]. Velocity profiles are in excellent agreement with the results from the reference simulation. Peak intensities as well as the location of the axial turbulence intensity maximum are well reproduced by the simulations using sliding meshes. Further studies are required to analyse the characteristics of the turbulent structures obtained with and without the use of sliding mesh interfaces.

Conclusions

A sliding mesh method has been implemented in a compressible LES FVM code which allows to simulate test cases in which there exist static and moving mesh zones such as rotor–stator interaction cases. This includes the development of non-conformal numerical schemes, ALE methods to take into account mesh movement in the Navier–Stokes equations and parallel communication algorithms to assess the connectivity between the different processors in the static and moving regions.

An inviscid test case (isentropic vortex) is used to analyse the accuracy of the non-conformal scheme with a sliding mesh interface. Results show second-order errors for the L2 norm. Moreover, the sliding mesh implementation is tested in a turbulent case: a pipe channel flow with $Re_\tau = 360$. Two sliding mesh interfaces are used in this case in order to impose mesh rotation in a region of the mesh. The results are in good agreement with simulations without sliding mesh and reference DNS data. These validation test cases allow to evaluate the influence of the sliding mesh interfaces in different variables of turbulent flows simulations.

References

1. Ganine, V., Amirante, D., Hills, N.: Enhancing performance and scalability of data transfer across sliding grid interfaces for time-accurate unsteady simulations of multistage turbomachinery flows. Comput. Fluids **115**, 140–153 (2015)
2. Rinaldi, E., Colonna, P., Pecnik, R.: Flux-conserving treatment of non-conformal interfaces for finite-volume discretization of conservation laws. Comput. Fluids (2015)
3. Gaofeng, W., Papadogiannis, D., Florent, D., Nicolas, G., Gicquel, L.Y.: Towards massively parallel large eddy simulation of turbine stages, vol. 55249, p. V06CT42A021 (2013)
4. Bermejo-Moreno, I., Campo, L., Larsson, J., Bodart, J., Helmer, D., Eaton, J.K.: Confinement effects in shock wave/turbulent boundary layer interactions through wall-modelled large-eddy simulations. J. Fluid Mech. **758**, 5–62 (2014)
5. Grébert, A., Bodart, J., Jamme, S., Joly, L.: Simulations of shock wave/turbulent boundary layer interaction with upstream micro vortex generators. Int. J. Heat Fluid Flow **72**, 73–85 (2018)
6. Kennel, M.B.: KDTREE 2: fortran 95 and C++ software to efficiently search for near neighbors in a multi-dimensional Euclidean space. ArXiv Physics e-prints, p 8 (2004)
7. Luo, H., Baum, J.D., Löhner, R.: On the computation of multi-material flows using ALE formulation. J. Comput. Phys. **194**, 304–328 (2004)
8. Zhang, R., Zhang, M., Shu, C.W.: On the order of accuracy and numerical performance of two classes of finite volume WENO schemes. Commun. Comput. Phys. (2011)
9. Wu, X., Moin, P.: A direct numerical simulation study on the mean velocity characteristics in turbulent pipe flow. J. Fluid Mech. **608**, 81–112 (2008)

Chapter 54
A Priori Assessment of Subgrid-Scale Models and Numerical Error in Forced Convective Flow at High Prandtl Numbers

L. Sufrà and H. Steiner

Introduction

The accurate modelling of the subgrid-scale (sgs) fluxes of momentum and heat near heated/cooled walls still represents a challenging task in Large-Eddy Simulations (LES). This is specially the case at molecular Prandtl numbers (Pr) well beyond unity, where the relevant length scales of the smallest turbulent thermal structures locally differ significantly from their dynamical counterparts. This disparity strongly conflicts with the Reynolds analogy between the transport of momentum and heat, which is inherently assumed by most sgs models for the turbulent heat flux. The present study carries out an extensive a priori analysis on data from fully resolved DNS (Direct Numerical Simulation) fields of fully developped heated turbulent pipe flow at high molecular Prandtl numbers $\mathrm{Pr} = 10/20$, considering three popular modelling candidates for subgrid-scale closure. Aside from assessing the models' capabilities to describe quantitatively the unresolved turbulent fluxes, a special focus is also put on the role of the numerical error, which arises from the discretization of the filtered advective fluxes on a coarse LES grid. The present analysis extends here previous studies on subgrid-scale momentum transport in isothermal mixing layer and channel flow carried out by [1] and [6], respectively, to the subgrid-scale transport of heat at high Prandtl numbers. The statistical dependence between the individual contributions (resolved, subgrid-scale, numerical discretization error) constituting the filtered advective flux terms in the LES formulation is investigated in terms of corresponding cross-çorrelations, as well.

L. Sufrà (✉) · H. Steiner
Institute of Fluid Mechanics and Heat Transfer, Graz University of Technology, Graz, Austria
e-mail: lorenzo.sufra@tugraz.at

H. Steiner
e-mail: helfried.steiner@tugraz.at

M. García-Villalba et al. (eds.), *Direct and Large Eddy Simulation XII*, ERCOFTAC Series 27,
https://doi.org/10.1007/978-3-030-42822-8_54

Filtered Governing Equations

For the considered incompressible pipe flow with constant material properties, the spatially filtered conservation equations of mass, momentum and energy solved in LES read

$$\overline{\nabla} \cdot \overline{\mathbf{U}} = 0, \tag{54.1}$$

$$\frac{\partial \overline{\mathbf{U}}}{\partial t} + \overline{\nabla} \cdot (\overline{\mathbf{U}}\,\overline{\mathbf{U}}) = -\overline{\nabla}\,\overline{P} + \frac{1}{\mathrm{Re}_\tau}\overline{\nabla}^2\overline{\mathbf{U}} - \overline{\nabla} \cdot \underline{\tau}_{\mathrm{sgs}}, \tag{54.2}$$

$$\frac{\partial \overline{\theta}}{\partial t} + \overline{\nabla} \cdot (\overline{\mathbf{U}}\,\overline{\theta}) = \frac{1}{\mathrm{Re}_\tau \mathrm{Pr}}\overline{\nabla}^2\overline{\theta} - \overline{\nabla} \cdot \mathbf{q}_{\mathrm{sgs}}. \tag{54.3}$$

The pipe diameter D, the wall friction velocity $w_\tau = \sqrt{\tau_w/\rho}$ and temperature $T_\tau = q_w/\rho c_p w_\tau$ have been used as length, velocity and temperature scales, respectively, for non-dimensionalization. $\overline{\mathbf{U}} = (\bar{u}, \bar{v}, \bar{w})^T$ and $\overline{\theta}$ represent the non-dimensional filtered velocity vector and temperature difference from the wall value, respectively. The overlined operator $\overline{\nabla}$ denotes the gradient vector computed on the LES grid. $\underline{\tau}_{\mathrm{sgs}}$ and $\mathbf{q}_{\mathrm{sgs}}$ represent the unresolved subgrid-scale (sgs) stress tensor and heat flux vector, respectively, arising from the spatial filtering of the non-linear advection terms. They are accordingly defined as

$$\underline{\tau}_{\mathrm{sgs}} = \overline{\mathbf{U}\,\mathbf{U}} - \overline{\mathbf{U}}\,\overline{\mathbf{U}}, \tag{54.4}$$

$$\mathbf{q}_{\mathrm{sgs}} = \overline{\mathbf{U}\,\theta} - \overline{\mathbf{U}}\,\overline{\theta}. \tag{54.5}$$

For closing the set of LES equations the deviatoric part of sgs stress tensor and the sgs heat flux vector are modelled with the Boussinesq Ansatz as

$$\underline{\tau}^d_{\mathrm{sgs}} = -\nu_{\mathrm{sgs}}(\overline{\nabla}\,\overline{\mathbf{U}} + \overline{\nabla}\,\overline{\mathbf{U}}^T), \tag{54.6}$$

$$\mathbf{q}_{\mathrm{sgs}} = -\frac{\nu_{\mathrm{sgs}}}{Pr_{\mathrm{SGS}}}\overline{\nabla}\bar{\theta}, \tag{54.7}$$

respectively. The subgrid-scale eddy viscosity ν_{sgs} in (54.6) is provided by the adopted sgs model, and the subgrid-scale Prandtl number in (54.7) is assumed as a constant parameter.

A Priori LES

Fully resolved instantaneous velocity **U** and temperature θ fields obtained from DNS of heated turbulent pipe flow have been used, considering four cases at wall friction

Reynolds number $\mathrm{Re}_{\tau} = 360/500$ and $Pr = 10/20$. The DNS results are computed in cylindrical coordinates on a mesh of $256 \times 512 \times 1024$ grid points in the radial r, azimuthal φ and axial directions z, respectively. The total axial length of the computational domain is five pipe diameters. A fourth-order accurate Finite Volume scheme and a second-order explicit Adams–Bashforth scheme have been used for spatial and temporal discretization, respectively. The wall of the pipe is uniformly heated with a constant averaged heat flux $\langle q_w \rangle = \text{const.}$, assuming no-slip wall boundary conditions for velocity, $\mathbf{U}_w = 0$, and Dirichlet-type thermal boundary conditions $\theta_w = 0$. Periodic boundary conditions are applied in the axial and the azimuthal directions. The DNS data have been filtered with a spatial box filter, using a filter width $\bar{\Delta} = 8\Delta$, which is eight times the cell size used in the DNS. This implies a spatial resolution for the a priori LES obtained on a mesh with $32 \times 64 \times 128$ grid points in the radial r, azimuthal φ and axial direction z, respectively. The present analysis splits the spatially filtered advective divergence terms appearing in the LES transport equations of momentum and heat into three contributions, being the resolved component, the unresolved subgrid-scale component and the numerical error due to the discretization on the coarse LES grid. Accordingly, the filtered divergence of the advective momentum and heat fluxes are decomposed into

$$\nabla \cdot \overline{\mathbf{U}\mathbf{U}} = \overline{\nabla} \cdot \overline{\mathbf{U}}\,\overline{\mathbf{U}} + \overline{\nabla} \cdot \underline{\tau}_{\text{sgs}} + \beta_{\mathbf{U}} \,, \tag{54.8}$$

$$\nabla \cdot \overline{\mathbf{U}\theta} = \overline{\nabla} \cdot \overline{\mathbf{U}}\,\overline{\theta} + \overline{\nabla} \cdot \mathbf{q}_{\text{sgs}} + \beta_{\theta} \,, \tag{54.9}$$

respectively. The vector $\beta_{\mathbf{U}}$ and scalar β_{θ} representing the numerical error are estimated as proposed by [6], assuming the DNS solution on the fine DNS grid as representative for the exact solution, such that

$$\beta_{\mathbf{U}} = \nabla \cdot \left(\overline{\mathbf{U}}\,\overline{\mathbf{U}}\right) - \overline{\nabla} \cdot \left(\overline{\mathbf{U}}\,\overline{\mathbf{U}}\right), \tag{54.10}$$

$$\beta_{\theta} = \nabla \cdot \left(\overline{\mathbf{U}}\,\overline{\theta}\right) - \overline{\nabla} \cdot \left(\overline{\mathbf{U}}\,\overline{\theta}\right), \tag{54.11}$$

where the operator ∇ refers to the fourth-order accurate discretization used on the fine DNS-grid, while $\overline{\nabla}$ refers to the second-order accurate discretization used on the coarse LES-grid. The subgrid-scale contribution to be delivered by the sgs model is consistently computed on the coarse grid, applying alternatively the assessed candidates, namely the Standard Smagorinsky model with Van Driest-type wall dampening (SSM) [4], the Wall-Adapting Local Eddy-viscosity model (WALE) [3] and the Coherent Structure Model (CSM) [2].Neither explicit test-filtering procedures for determining model parameters is required, nor any subsequent statistical averaging of these for the sake of numerical stability. CSM and WALE additionally provide the benefit of not requiring any artificial wall damping functions to enforce the decrease of the eddy viscosity to zero near the wall. These features make the models attractive candidates in complex technical engineering applications far apart from simplified generic test flow configurations. In Eq. (54.7), the thermal eddy diffusivity was always related to the eddy viscosity ν_{sgs} through a subgrid-scale Prandtl number, assumed as constant $Pr_{SGS} = 0.5$.

Results

Figure 54.1 shows the radial variations of the statistically averaged budgets for the advective heat flux terms for the four considered cases, as obtained from the decomposition (54.9). The angular brackets indicate statistical averaging into the homogeneous spatial directions z and φ. A pronounced negative peak always occurs near the upper limit of the viscous sublayer. The peak increases in magnitude for increasing Reynolds and Prandtl numbers, and it is located closer to the wall for the higher Prandtl number.

The numerical error evidently adds on average very little to the resolved contribution. The remaining considerable gap to the blue (DNS-based) target line is filled and partly overcompensated only by the Standard Smagorinsky subgrid-scale model (SSM), while the WALE model delivers a comparatively much smaller contribution and the CSM model produces even less. As follows from Eq. (54.7), the contribution from the subgrid-scale model is certainly also affected by the choice of a constant subgrid-scale Prandtl number $\mathrm{Pr}_{sgs} = 0.5$. For the considered isothermal boundary condition and molecular Prandtl numbers, Pr_{sgs} would rather increase towards the wall to values beyond unity, which would effectively reduce sgs contribution. Since the variation of Pr_{sgs} affects only a narrow region very next to the wall, its effect on the advective heat flux budgets is still minor.

In contrast to the small average contribution observed in Fig. 54.1, the instantaneous fluctuating representations of the numerical discretization error are signficantly high. This is clearly seen from Fig. 54.2, showing exemplarily the axial component of the error contribution to the momentum transport β_w and the error contribution to the heat transport β_θ. The rms values evidently exceed markedly the statistical means with increasing tendency for higher Reynolds and Prandtl numbers.

The mutual statistical dependence of the individual contributions is further investigated in terms of their normalized cross-correlations, as exemplarily shown for the advective heat flux at $Re_\tau = 360$ and $\mathrm{Pr} = 10$ in Fig. 54.3. The correlation coefficient between the resolved contribution and the numerical error becomes strongly negative towards the wall. This indicates the dissipative nature of the numerical error, which effectively dampens the resolved turbulent fluctuations in this region. On the other hand, the resolved contribution appears as positively correlated with the contribution from subgrid-scale model in the diffusive sub- and buffer layer $y^+ < 10$, implying that the applied subgrid-scale model tends to enhance the resolved turbulent fluctuations in this region. This positive correlation is obviously most pronounced for the SSM, which also explains the over-predicted subgrid-scale contribution of SSM near the wall, as seen in the budgets in Fig. 54.1, right column.

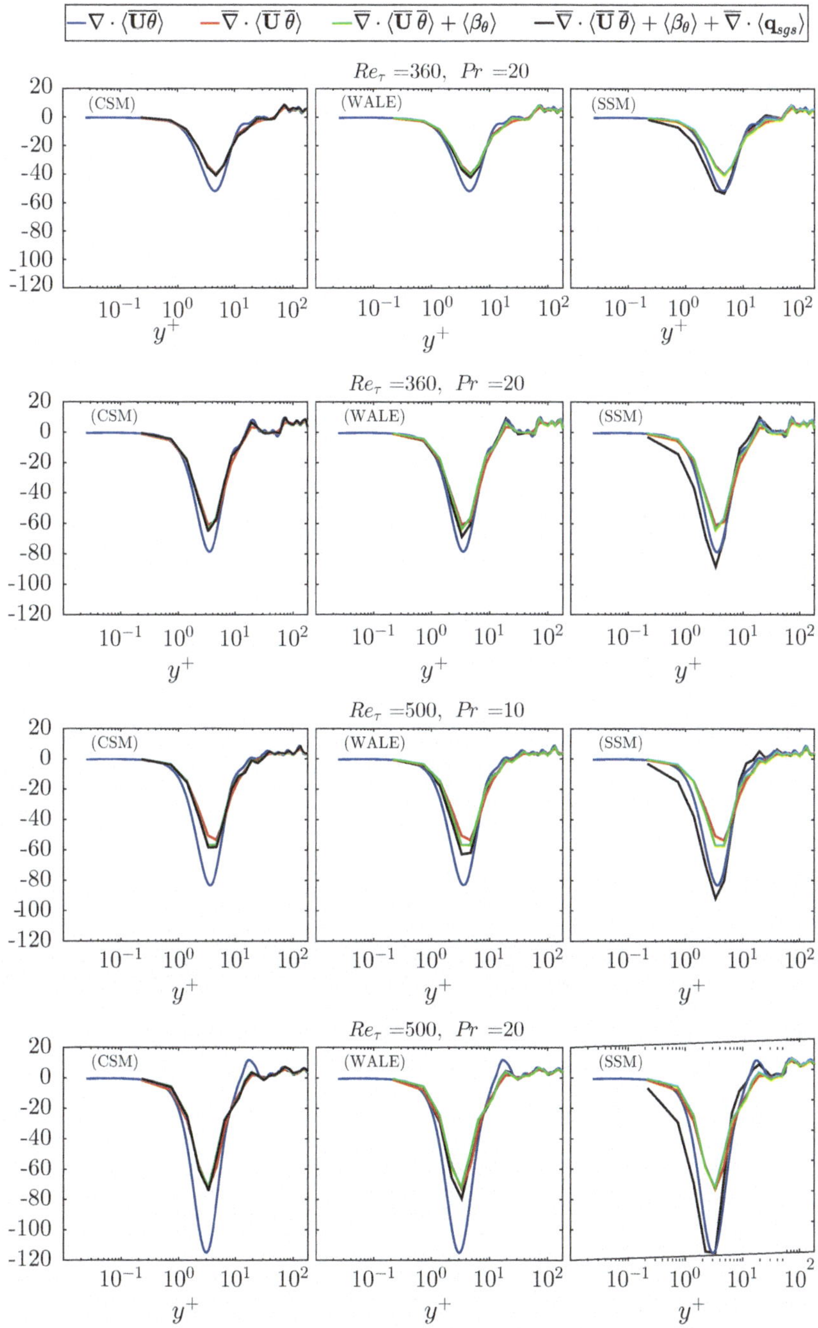

Fig. 54.1 Budgets of the averaged advective divergence terms for the heat transport comparing the CSM, WALE and SSM subgrid-scale model contributions for different $\mathrm{Re}_\tau = 360/500$, $\mathrm{Pr} = 10/20$

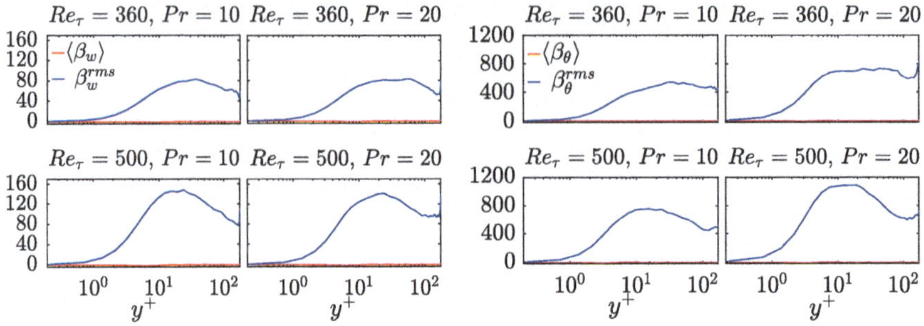

Fig. 54.2 Averaged and RMS values of the numerical error contributions β_w and β_θ for different Re_τ and Pr numbers

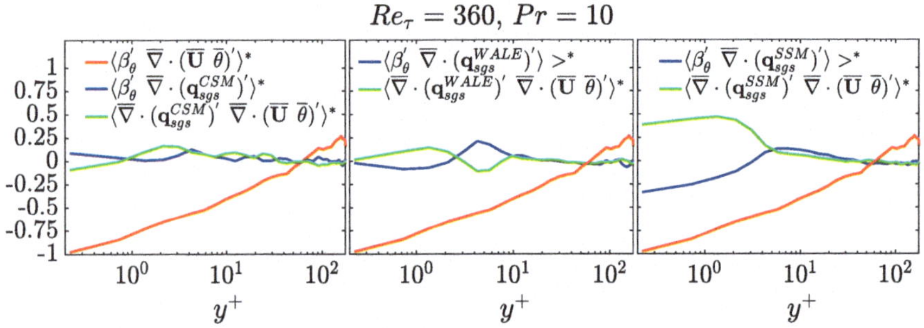

Fig. 54.3 Cross-correlation coefficient $\langle\ \rangle^*$ between numerical error and resolved contribution (red line), numerical error and sgs-model contribution (blue line), resolved and sgs-model contribution (green line) for $\mathrm{Re}_\tau = 360$ and $\mathrm{Pr} = 10$

References

1. Brandt, T.: A priori tests on numerical errors in large eddy simulation using finite differences and explicit filtering. J. Numer. Meth. Fluids **51**, 635–657 (2006). https://doi.org/10.1002/fld.1144
2. Kobayashi, H.: The subgrid-scale models based on coherent structures for rotating homogeneous turbulence and turbulent channel flow. Flow Turb. Comb. **17**, 45–104 (2005). https://doi.org/10.1063/1.1874212
3. Nicoud, F., Ducros, F.: Subgrid-scale stress modelling based on the square of the velocity gradient tensor. Flow Turb. Comb. **62**, 183–200 (2006). https://doi.org/10.1023/A:1009995426001
4. Smagorinsky, J.: General circulation experiments with the primitive equations. I. The basic experiment. Mon. Weather Rev. **91**, 99–164 (1963). https://doi.org/10.1175/1520-0493
5. Van Driest, E.R.: On turbulent flow near a wall. Mon. Weather Rev. **23**, 1007–1011 (1956). https://doi.org/10.2514/8.3713
6. Vreman, B., Geurts, B., Kuerten, H.: A priori tests of large eddy simulation of the compressible mixing layer. J. Eng. Math. **29**, 299–327 (1995). https://doi.org/10.1007/BF00042759

Chapter 55
On a Proper Tensor-Diffusivity Model for Large-Eddy Simulations of Buoyancy-Driven Flows

F. X. Trias, F. Dabbagh, A. Gorobets and A. Oliva

Introduction

In this work, we plan to shed light on the following research question: *can we find a nonlinear subgrid-scale (SGS) heat flux model with good physical and numerical properties, such that we can obtain satisfactory predictions for buoyancy-driven turbulent flows?* This is motivated by our findings showing that the classical (linear) eddy-diffusivity assumption fails to provide a reasonable approximation for the SGS heat flux. This was shown in our work [1] where SGS features have been studied a priori for a Rayleigh–Bénard convection (RBC). We also concluded that nonlinear (or tensorial) models can give good approximations of the actual SGS heat flux. Briefly, the large-eddy simulation (LES) equations arise from applying a spatial commutative filter, with filter length δ, to the incompressible Navier–Stokes and thermal energy equations,

F. X. Trias (✉) · F. Dabbagh · A. Oliva
Heat and Mass Transfer Technological Center, Technical University of Catalonia,
C/Colom 11, 08222 Terrassa (Barcelona), Spain
e-mail: xavi@cttc.upc.edu

F. Dabbagh
e-mail: firas.dabbagh@jku.at; firas@cttc.upc.edu

A. Oliva
e-mail: oliva@cttc.upc.edu

F. Dabbagh
Christian Doppler Laboratory for Multi-Scale Modeling of Multiphase Processes,
Johannes Kepler University, Altenbergerstraße 69, 4040 Linz, Austria

A. Gorobets
Keldysh Institute of Applied Mathematics, 4A, Miusskaya Sq.,
Moscow 125047, Russia
e-mail: cherepock@gmail.com

M. García-Villalba et al. (eds.), *Direct and Large Eddy Simulation XII*,
ERCOFTAC Series 27,
https://doi.org/10.1007/978-3-030-42822-8_55

$$\partial_t \overline{\boldsymbol{u}} + (\overline{\boldsymbol{u}} \cdot \nabla)\overline{\boldsymbol{u}} = (\mathrm{Pr}/\mathrm{Ra})^{1/2}\nabla^2\overline{\boldsymbol{u}} - \nabla\overline{p} + \overline{\boldsymbol{f}} - \nabla \cdot \tau, \tag{55.1}$$

$$\partial_t \overline{T} + (\overline{\boldsymbol{u}} \cdot \nabla)\overline{T} = (\mathrm{Ra}\,\mathrm{Pr})^{-1/2}\nabla^2\overline{T} - \nabla \cdot \boldsymbol{q}, \tag{55.2}$$

where $\overline{\boldsymbol{u}}$, $\overline{T}$, and $\overline{p}$ are respectively the filtered velocity, temperature, and pressure, and the incompressibility constraint reads $\nabla \cdot \overline{\boldsymbol{u}} = 0$. The SGS stress tensor, $\tau = \overline{\boldsymbol{u} \otimes \boldsymbol{u}} - \overline{\boldsymbol{u}} \otimes \overline{\boldsymbol{u}}$, and the SGS heat flux vector, $\boldsymbol{q} = \overline{\boldsymbol{u}T} - \overline{\boldsymbol{u}}\overline{T}$, represent the effect of the unresolved scales, and they need to be modeled in order to close the system. The most popular approach is the eddy-viscosity assumption, where the SGS stress tensor is assumed to be aligned with the local rate-of-strain tensor, $\mathsf{S} = 1/2(\nabla\overline{\boldsymbol{u}} + \nabla\overline{\boldsymbol{u}}^t)$, i.e. $\tau \approx -2\nu_e \mathsf{S}(\overline{\boldsymbol{u}})$. By analogy, the SGS heat flux, $\boldsymbol{q}$, is usually approximated using the gradient-diffusion hypothesis (linear modeling), given by

$$\boldsymbol{q} \approx -\kappa_t \nabla\overline{T} \qquad (\equiv \boldsymbol{q}^{\mathrm{eddy}}). \tag{55.3}$$

Then, the Reynolds analogy assumption is applied to evaluate the eddy-diffusivity, κ_t, via a constant turbulent Prandtl number, Pr_t, i.e. $\kappa_t = \nu_e/\mathrm{Pr}_t$. These assumptions have been shown to be erroneous to provide accurate predictions of the SGS heat flux [1]. Namely, a priori analysis showed that the eddy-diffusivity assumption, $\boldsymbol{q}^{\mathrm{eddy}}$ (Eq. 55.3), is completely misaligned with the actual subgrid heat flux, $\boldsymbol{q}$ (see Fig. 55.1, left). In contrast, the tensor diffusivity (nonlinear) Leonard model [2], which is obtained by taking the leading term of the Taylor series expansion of $\boldsymbol{q}$,

$$\boldsymbol{q} \approx \frac{\delta^2}{12}\mathsf{G}\nabla\overline{T} \qquad (\equiv \boldsymbol{q}^{\mathrm{nl}}), \tag{55.4}$$

provides a much more accurate a priori representation of $\boldsymbol{q}$ (see Fig. 55.1, left). Here, $\mathsf{G} \equiv \nabla\overline{\boldsymbol{u}}$ represents the gradient of the resolved velocity field. It can be argued that the rotational geometries are prevalent in the bulk region over the strain slots, i.e. $|\Omega| > |\mathsf{S}|$ (see Refs. [1, 3]). Then, the dominant antisymmetric tensor, $\Omega = 1/2(\mathsf{G} - \mathsf{G}^T)$, rotates the thermal gradient vector, $\nabla\overline{T}$, to be almost perpendicular to $\boldsymbol{q}^{\mathrm{nl}}$ (see Eq. 55.4). Therefore, the eddy-diffusivity paradigm is only supported in the not-so-frequent strain-dominated areas.

Nonlinear SGS Heat Flux Models for Large-Eddy Simulation

Since the eddy-diffusivity, $\boldsymbol{q}^{\mathrm{eddy}}$, cannot provide an accurate representation of the SGS heat flux, we turn our attention to nonlinear models. As mentioned above, the Leonard model [2] given in Eq. (55.4) can provide a very accurate a priori representation of the SGS heat flux (see Fig. 55.1, left). However, the local dissipation (in the L2-norm sense) is proportional to $\nabla T \cdot \mathsf{G}\nabla T = \nabla T \cdot \mathsf{S}\nabla T + \nabla T \cdot \Omega\nabla T = \nabla T \cdot \mathsf{S}\nabla T$. Since the velocity field is divergence-free, $\lambda_1^{\mathsf{S}} + \lambda_2^{\mathsf{S}} + \lambda_3^{\mathsf{S}} = \nabla \cdot \boldsymbol{u} = 0$, the eigenvalues of S can be ordered $\lambda_1^{\mathsf{S}} \geqslant \lambda_2^{\mathsf{S}} \geqslant \lambda_3^{\mathsf{S}}$ with $\lambda_1^{\mathsf{S}} \geqslant 0$ (extensive eigendi-

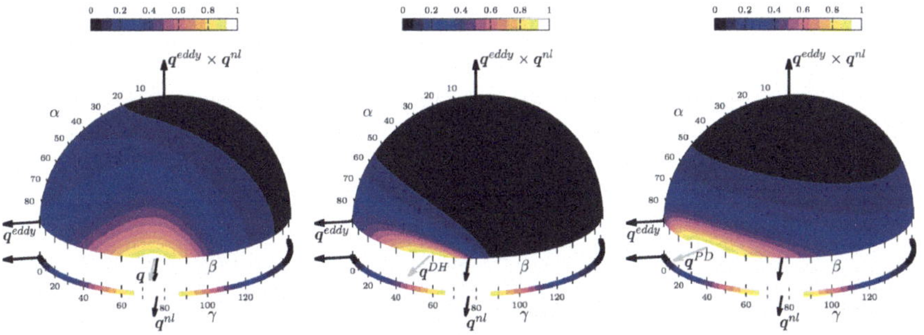

Fig. 55.1 Joint probability distribution functions (PDF) of the angles (α, β) plotted on a half unit sphere to show the orientation in the space of the mixed model. From left to right, alignment trends of the actual SGS heat flux, $\boldsymbol{q}$, the Daly and Harlow [4] model (Eq. 55.6) and the Peng and Davidson [5] model (Eq. 55.5). For simplicity, the JPDF and the PDF magnitudes are normalized by its maximal. For details the reader is referred to [1]

rection) and $\lambda_3^{\mathsf{S}} \leqslant 0$ (compressive eigendirection), and λ_2^{S} is either positive or negative. Hence, the local dissipation introduced by the model can take negative values; therefore, the Leonard model cannot be used as a standalone SGS heat flux model, since it can produce a finite-time blow-up. A similar problem is encountered with the nonlinear tensorial model $\boldsymbol{q}^{PD}$ proposed by Peng and Davidson [5],

$$\boldsymbol{q} \approx C_t \delta^2 \mathsf{S} \nabla T \qquad (\equiv \boldsymbol{q}^{\mathrm{PD}}), \tag{55.5}$$

$$\boldsymbol{q} \approx -\mathscr{T}_{\mathrm{SGS}} \tau \nabla T = -\frac{1}{|\mathsf{S}|} \frac{\delta^2}{12} \mathsf{G}\mathsf{G}^T \nabla T \qquad (\equiv \boldsymbol{q}^{\mathrm{DH}}), \tag{55.6}$$

whereas the nonlinear model $\boldsymbol{q}^{\mathrm{DH}}$ proposed by Daly and Harlow [4] relies on the positive semi-definite tensor $\mathsf{G}\mathsf{G}^T$. Here, $\mathscr{T}_{\mathrm{SGS}} = 1/|\mathsf{S}|$ is the SGS timescale. Notice that the model proposed by Peng and Davidson, $\boldsymbol{q}^{\mathrm{PD}}$, can be viewed in the same framework if the SGS stress tensor is estimated by an eddy-viscosity model, i.e. $\tau \approx -2\nu_e \mathsf{S}$ and $\mathscr{T}_{\mathrm{SGS}} \propto \delta^2/\nu_e$. These two models have shown a much better a priori alignment with the actual SGS heat flux, especially the DH model (see Fig. 55.1, middle). Moreover, the DH is numerically stable since the tensor $\mathsf{G}\mathsf{G}^T$ is positive semi-definite. Hence, it seems appropriate to build models based on this tensor. However, the DH model does not have the proper near-wall behavior, i.e. $\boldsymbol{q} \propto \langle v'T' \rangle = \mathscr{O}(y^3)$ where y is the distance to the wall. An analysis of the DH model leads to $\mathsf{G}\mathsf{G}^T \nabla \overline{T} \propto \mathscr{O}(y^1)$. Therefore, the near-wall cubic behavior is recovered if $\mathscr{T}_{\mathrm{SGS}} \propto \mathscr{O}(y^2)$. This is not the case of the timescale used in the Daly and Harlow [4] model, i.e. $\mathscr{T}_{\mathrm{SGS}} = 1/|\mathsf{S}| = \mathscr{O}(y^0)$.

At this point it is interesting to observe that new timescales, $\mathscr{T}_{\mathrm{SGS}}$, can be derived by imposing restrictions on the differential operators they are based on. For instance, let us consider models that are based on the invariants of the tensor $\mathsf{G}\mathsf{G}^T$

$$\boldsymbol{q} \approx -C_M \left(P^p_{\mathsf{G}\mathsf{G}^T} Q^q_{\mathsf{G}\mathsf{G}^T} R^r_{\mathsf{G}\mathsf{G}^T} \right) \frac{\delta^2}{12} \mathsf{G}\mathsf{G}^T \nabla T \qquad (\equiv \boldsymbol{q}^{S2}) \tag{55.7}$$

where P_{GG^T}, Q_{GG^T}, and R_{GG^T} are the first, second, and third invariant of the GG^T tensor. This tensor is proportional to the gradient model [6] given by the leading term of the Taylor series expansion of the subgrid stress tensor $\tau(\overline{\boldsymbol{u}}) = (\delta^2/12)\mathsf{GG}^T + \mathcal{O}(\delta^4)$. Then, the exponents p, q, and r in Eq. (55.7), must satisfy the following equations

$$-6r - 4q - 2p = 1; \qquad 6r + 2q = s, \tag{55.8}$$

to guarantee that the differential operator has units of time, i.e. $[P^p_{\mathsf{GG}^T} Q^q_{\mathsf{GG}^T} R^r_{\mathsf{GG}^T}] = [T^1]$ and a slope s for the asymptotic near-wall behavior, i.e. $\mathcal{O}(y^s)$. Solutions for $q(p, s) = -(1+s)/2 - p$ and $r(p, s) = (2s+1)/6 + p/3$ are displayed in Fig. 55.2. If we restrict ourselves to solutions with the proper near-wall scaling, i.e. $s = 2$ (blue lines in Fig. 55.2), a family of p-dependent models follows. Restricting ourselves to solutions involving only two invariants of GG^T three models follow

$$\boldsymbol{q}^{S2PQ} = -C_{s2pq} P^{-5/2}_{\mathsf{GG}^T} Q_{\mathsf{GG}^T} \frac{\delta^2}{12} \mathsf{GG}^T \nabla\overline{T}, \tag{55.9}$$

$$\boldsymbol{q}^{S2PR} = -C_{s2pr} P^{-3/2}_{\mathsf{GG}^T} R^{1/3}_{\mathsf{GG}^T} \frac{\delta^2}{12} \mathsf{GG}^T \nabla\overline{T}, \tag{55.10}$$

$$\boldsymbol{q}^{S2QR} = -C_{s2qr} Q^{3/2}_{\mathsf{GG}^T} R^{5/6}_{\mathsf{GG}^T} \frac{\delta^2}{12} \mathsf{GG}^T \nabla\overline{T}, \tag{55.11}$$

for $p = -5/2$, $p = -1.5$, and $p = 0$, respectively. These three solutions are represented in Fig. 55.2. Apart from being unconditionally stable, these models display very good a priori alignment trends in the bulk (see Fig. 55.3) similar to the PD model (see Fig. 55.1, middle) but also in the near-wall region. Hence, we consider that they are very good candidates for a posteriori LES simulations of buoyancy-driven flows.

Assessment of Eddy-Viscosity Models at Very Low *Pr* Numbers

At this stage, a posteriori results are necessary to assess the performance of the newly proposed SGS heat flux models. However, apart from the underlying numerics, such results will be strongly influenced by the SGS stress tensor model. Hence, we first aim to answer the following research question: *are eddy-viscosity models for momentum able to provide satisfactory results for turbulent Rayleigh–Bénard convection?* In order to shed light on this, a set of simulations at $\mathrm{Pr} = 0.005$ (liquid sodium) has been carried out at $\mathrm{Ra} = 7.14 \times 10^6$ and $\mathrm{Ra} = 7.14 \times 10^7$. Figure 55.4 displays a snapshot of the temperature and velocity magnitude for the DNS simulation at the highest Ra. This clearly illustrates the separation between the smallest scales of temperature and velocity, i.e. the ratio between the Kolmogorov length scale and the Obukhov–Corrsin length scale is given by $\mathrm{Pr}^{1/2}$ [8]. Therefore, for a $\mathrm{Pr} = 0.005$

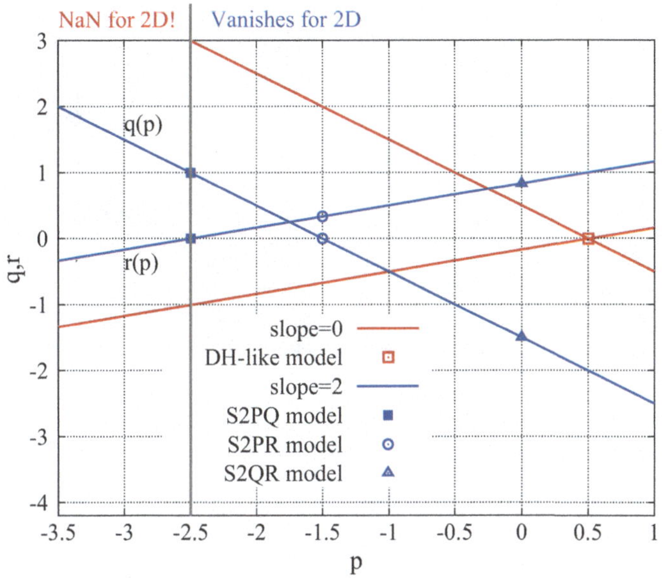

Fig. 55.2 Solutions of the linear system of Eq. (55.8) for $s = 0$ (red lines) and $s = 2$ (blue lines). Each (r, p, q) represents an tensor-diffusivity model with the form of Eq. (55.7)

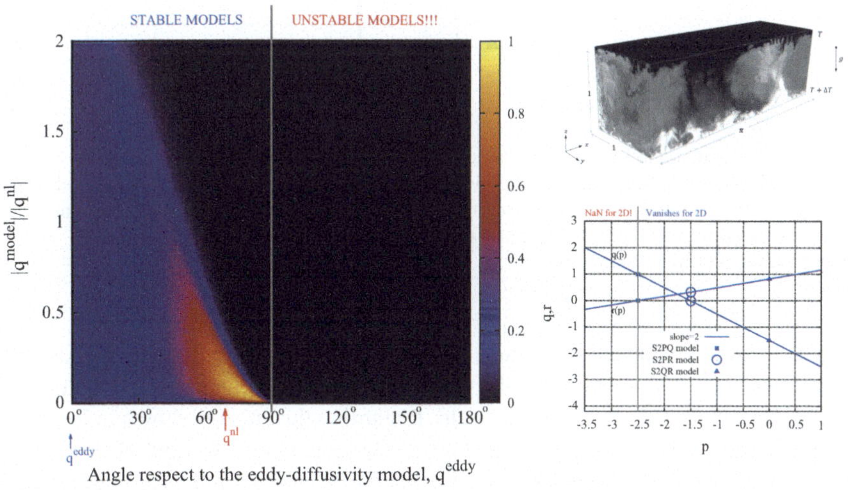

Fig. 55.3 Joint PDF for the S2PR model (Eq. 55.10) in the space $(|\boldsymbol{q}^{PD}|/|\boldsymbol{q}^{nl}|, \beta)$ where the angle β is defined in Fig. 55.1. The analyzed data corresponds to the bulk region of the air-filled Rayleigh–Bénard configuration at Ra $= 10^{10}$ studied in Refs. [1, 3]

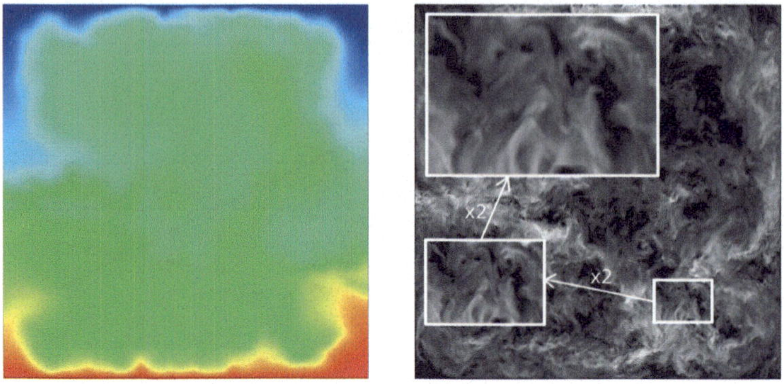

Fig. 55.4 An instantaneous picture of the temperature field (left) and velocity magnitude (right), $|\boldsymbol{u}|$, of the DNS simulation of RBC at $\mathrm{Ra} = 7.14 \times 10^7$ and $\mathrm{Pr} = 0.005$ (liquid sodium) carried out using a mesh of $966 \times 966 \times 2048 \approx 1911M$ grid points

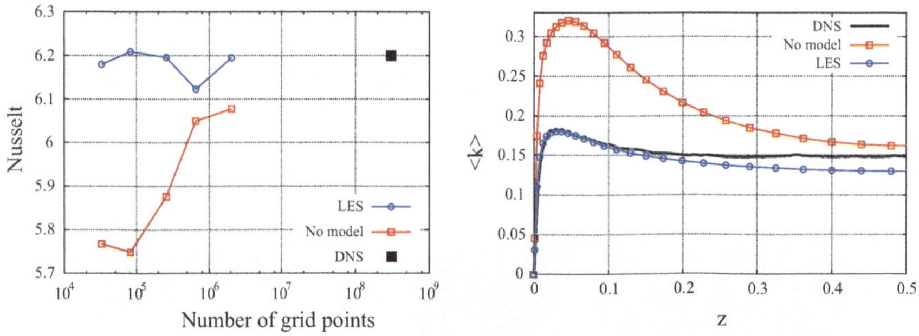

Fig. 55.5 Comparison of LES (and no-model) versus DNS results of RBC at $\mathrm{Ra} = 7.14 \times 10^6$ and $\mathrm{Pr} = 0.005$ (liquid sodium). LES results have been obtained using the eddy-viscosity model S3PQ proposed in Ref. [7]. Left: average Nusselt for different meshes. Right: turbulent kinetic energy at cavity mid-width for a $96 \times 52 \times 52$ mesh compared with the DNS results obtained with a mesh of $488 \times 488 \times 1280 \approx 305M$

(liquid sodium) we have a separation of more than one decade. Hence, it is possible to combine a LES simulation for the velocity field with the numerical resolution of all the relevant scales of the thermal field. Regarding this, results shown in Fig. 55.5 seem to confirm the adequacy of eddy-viscosity models for this kind of flows. Our future research plans include the extension of this analysis to higher Ra-numbers and testing a posteriori the new nonlinear SGS heat flux models for air-filled RBC.

Acknowledgements F. X. T., F. D., and A. O. have been financially supported by the *Ministerio de Economía y Competitividad*, Spain (ENE2017-88697-R). F. X. T. is supported by a *Ramón y Cajal* postdoctoral contract (RYC-2012-11996). F. D. is supported by the Austrian Federal Ministry for Digital and Economic Affairs, the National Foundation for Research, Technology and Development, and the K1MET center for metallurgical research in Austria (www.k1-met.com). Calculations have been performed on the MareNostrum 4 supercomputer at the BSC (PRACE 15th

Call, Ref. 2016163972, "Exploring new frontiers in Rayleigh-Bénard convection"). The authors thankfully acknowledge these institutions.

References

1. Dabbagh, F., Trias, F.X., Gorobets, A., Oliva, A.: A priori study of subgrid-scale features in turbulent Rayleigh-Bénard convection. Phys. Fluids **29**, 105103 (2017)
2. Leonard, A.: Large-eddy simulation of chaotic convection and beyond. AIAA paper, 97-0304 (1997)
3. Dabbagh, F., Trias, F.X., Gorobets, A., Oliva, A.: On the evolution of flow topology in turbulent Rayleigh-Bénard convection. Phys. Fluids **28**, 115105 (2016)
4. Daly, B.J., Harlow, F.H.: Transport equations in turbulence. Phys. Fluids **13**, 2634 (1970)
5. Peng, S., Davidson, L.: On a subgrid-scale heat flux model for large eddy simulation of turbulent thermal flow. Int. J. Heat Mass Transf. **45**, 1393–1405 (2002)
6. Clark, R.A., Ferziger, J.H., Reynolds, W.C.: Evaluation of subgrid-scale models using an accurately simulated turbulent flow. J. Fluid Mech. **91**, 1–16 (1979)
7. Trias, F.X., Folch, D., Gorobets, A., Oliva, A.: Building proper invariants for eddy-viscosity subgrid-scale models. Phys. Fluids **27**(6), 065103 (2015)
8. Sagaut, P.: Large Eddy Simulation for Incompressible Flows: An Introduction, 3rd edn. Springer (2005)

Chapter 56
Implicit Wall-Layer Modelling in Turbulent Pipe Flow

R. Vicente Cruz, E. Lamballais and R. Perrin

Introduction

In the context of large eddy simulations (LES) of wall-bounded turbulence, the explicit calculation of all the energy-containing near-wall eddies is a key point. To accomplish this goal, it is recommended the employment of a mesh resolution between 10 and 100 wall units for the streamwise direction discretization, from 10 to 20 wall units in the spanwise direction, and typically one wall unit in the wall normal direction through a near-wall mesh refinement. When these resolution criteria are too demanding in terms of computational resources, an alternative is to use a discretization that bypasses the inner layer with the aid of a wall-layer model [1, 2].

However, it was shown by [3] that excellent basic turbulent statistics in a pipe flow can be obtained with a mesh resolution coarser than the viscous sublayer without using any explicit wall-layer modelling. In this context, the near-wall regularization is simply performed by means of an original and robust implicit LES (ILES) technique that—without any extra cost or adaptation in the near-wall region—displays a wall modelling feature. Two Reynolds numbers $\mathrm{Re}_\tau = 180{,}550$[1] were considered when comparing ILES results to unfiltered DNS data of [4].

The term implicit LES refers here to the use of regularization as a substitute of subgrid-scale (SGS) modelling [5, 6]. The originality of the method is found

[1] Based on the friction velocity u_τ and the pipe radius R.

R. Vicente Cruz · E. Lamballais (✉)
Incompressible Turbulence and Control Group, Pprime Institute,
CNRS-Univ-Poitiers-ISAE/ENSMA, Poitiers, France
e-mail: eric.lamballais@univ-poitiers.fr

R. Perrin
Department of Mechanical Engineering, Faculty of Engineering at Sriracha,
Kasetsart University Sriracha Campus, Sriracha, Chonburi 20230, Thailand

M. García-Villalba et al. (eds.), *Direct and Large Eddy Simulation XII*,
ERCOFTAC Series 27,
https://doi.org/10.1007/978-3-030-42822-8_56

in the built-in numerical dissipation that does not come from the discretization of the convective term, but from the viscous term in the Navier–Stokes equation. This strategy has also been used successfully by [7, 8] for turbulent plane channel and impinging jet configurations with regular Cartesian meshes, nonetheless, the near-wall behaviour of the implicit SGS modelling was not specifically investigated.

Until now, the aforementioned conclusions about the ILES technique performance were based on comparisons of basic statistics with unfiltered DNS data, which means that even if a good agreement was observed, no rigorous assessment could be made. In the present investigation, this matter is examined at higher Reynolds number ($Re_\tau = 1000$) while determining to what extent it can be generalized in terms of near-wall region bypass. To do so, a quasi-DNS database is presented with an original method to easily produce filtered statistics in order to enable rigorous comparisons with ILES at any mesh resolution. The implicit LES results are, thus, analysed in terms of ability to predict near-wall statistics despite the use of very coarse grid for which even the near-wall turbulent production region is fully bypassed.

Numerical Method and Computational Configuration

As a generic numerical tool to investigate this concept of implicit LES, the sixth-order flow solver Incompact3d is used. Thanks to its features of kinetic energy conservation[2] and flexibility for the application of numerical dissipation, this code enables an easy calibration of the implicit SGS modelling. In particular, the role of its scale selectivity has been shown in [6] through rigorous comparisons with DNS results based on a priori and a posteriori analysis with a particular attention to distant triad interactions between SGS and very large scales [5]. In addition, it was shown by [6] that the present implicit LES technique is equivalent to the use of spectral vanishing viscosity (SVV), ensuring high-order accuracy and flexible application. In previous studies using the method [5, 6], the validation framework was based on the Taylor–Green Vortex problem as a prototype of wall-free turbulent flow.

As in [3], a regular Cartesian mesh is combined with the use of an immersed boundary technique to discretize the pipe geometry while considering periodic boundary conditions in the three spatial directions. A one-dimensional reconstruction method [9] is used inside the solid to ensure the smoothness of the solution everywhere in the computational domain. This feature clearly improves the accuracy of compact finite difference schemes as used in the code Incompact3d.

[2] In the discrete and inviscid sense up to the time advancement error.

Filtered Quasi-DNS Database

Quasi-DNS of pipe flow is performed at $\mathrm{Re}_\tau = 1000$ in order to, afterwards, produce the filtered benchmark database for comparison with ILES results. A mesh of $n_x \times n_y \times n_z = 768 \times 768 \times 1920$ cells was used, corresponding, in wall units to $\Delta x^+ = \Delta y^+ = 2.9$ and $\Delta z^+ = 13$. The term "quasi" is naturally used due to the employed mesh resolution, however, as shown in Fig. 56.1, thanks to the success of the technique, a very good agreement with DNS data of [4] is found.

As in [5], the filter is chosen to take the implicit SGS dissipation into account using a spectral Pao-like closure. By specifying the ratio between DNS and LES mesh resolutions, this Pao-like closure provides the spectral transfer functions that are used to define the targeted LES solution. Thus, this targeted solution is determined rigorously and consistently with the artificial dissipation implicitly introduced by the error differentiation in the present ILES framework [6]. The use of periodic boundary conditions enables the application of the present homogeneous filter defined in the Fourier space. The filter is applied once in every direction to take the anisotropy of the LES mesh into account. Following this procedure, a consistent framework is established for rigorous comparisons between LES and filtered DNS data.

The filter effect over streamwise-z and normal xy-directions separately is shown in Fig. 56.2. Acting as a pre-analysis, this information guided the choice of interesting mesh resolutions based on the levels of energy $k = \frac{1}{2}\langle u_i u_i \rangle$ to be potentially captured when the viscous and even near-wall production regions are bypassed. It is important to emphasize that the application of filter in normal xy-directions is only possible thanks to the solution smoothness provided by the reconstruction. As shown in Fig. 56.3, if no extension of the fluid solution is made into the solid zone before filtering, the no-slip condition is poorly ensured and near-wall region results are not reliable, a problem that is strongly reduced with the reconstruction. Nonetheless, reconstruction's ability to ensure the success of normal direction filtering is limited to not so high values of the filter width, as shown in Fig. 56.2 left by the abnormal near-wall behaviour when data is filtered at a corresponding cell

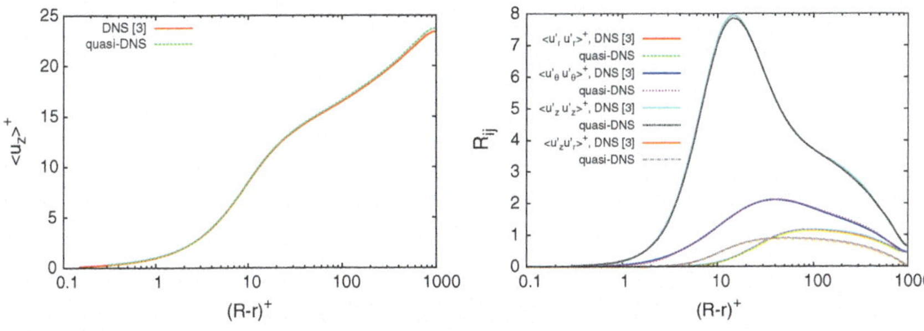

Fig. 56.1 Mean velocity (left) and Reynolds stresses profiles (right) for the quasi-DNS at $\mathrm{Re}_\tau = 1000$ compared to DNS statistics of [4]

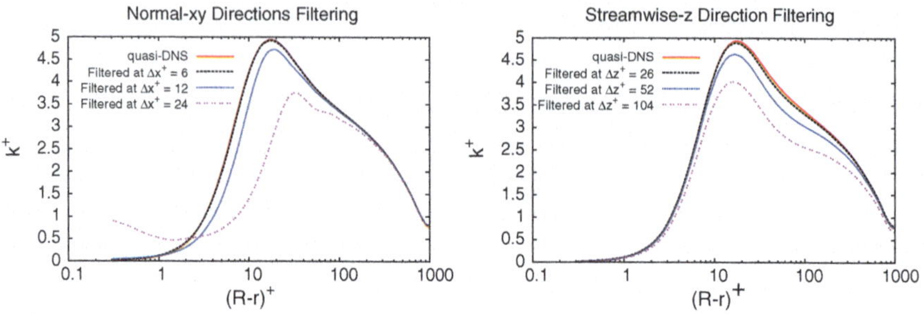

Fig. 56.2 Turbulent kinetic energy from filtered quasi-DNS at different mesh resolutions. The drop of energy is analysed for filtering along (left) streamwise-z direction and (right) normal-xy directions

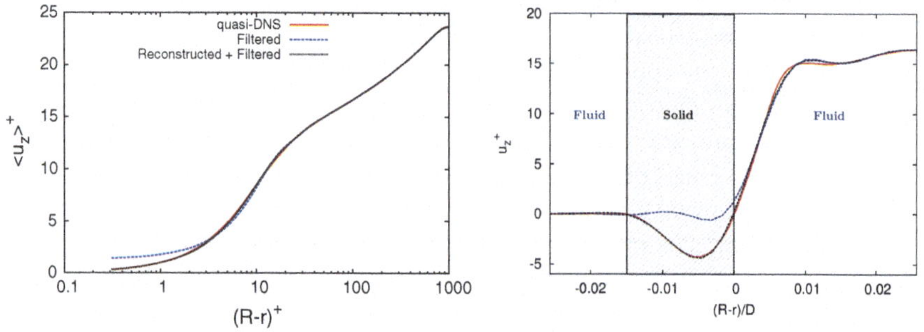

Fig. 56.3 Comparison of mean (left) and instantaneous (right) velocity profiles demonstrates the effect/necessity of reconstruction on the reliability of data filtered in the normal Cartesian directions (x, y). In this illustration, quasi-DNS data have been filtered at a corresponding mesh resolution of $\Delta x^+ = \Delta y^+ = 12$

size $\Delta x^+ = \Delta y^+ = 24$. For $\Delta x^+ = \Delta y^+ = 15$, the unrealistic near-wall rising of the turbulent kinetic energy k can be avoided, but persistent velocity fluctuations are observed close to the wall with for instance $k^+ \approx 0.156$ at $(R - r)^+ \approx 1.03$ to be compared to $k^+ \approx 0.124$ for the reference non-filtered value. Here, we have chosen to limit our analysis to $\Delta x^+ = \Delta y^+ = 12$ for which the filtering procedure produces also $k^+ \approx 0.124$ at this near-wall location.

LES Results

Before proceeding to the results, it is important to highlight that the term "implicit wall-layer modelling" is used to emphasize the ability of the present methodology to regularize the solution in the near-wall region without any explicit wall model. The improvement brought by the technique is the simple result of a targeted choice of

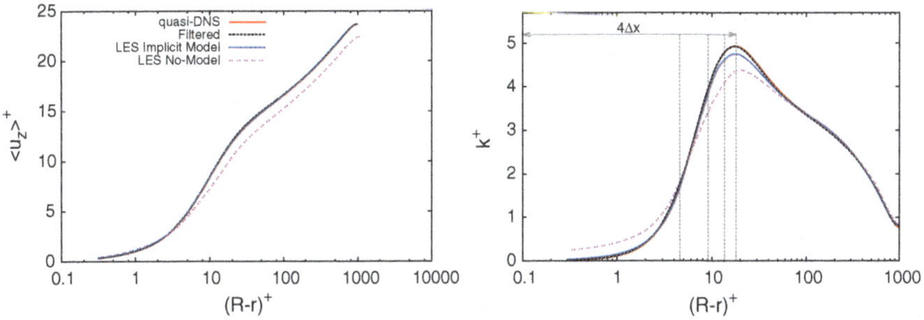

Fig. 56.4 Mean velocity profile (left) and Reynolds stresses profiles (right) for the Marginal Resolution ILES at $\mathrm{Re}_\tau = 1000$. Results are compared to: no-model LES, unfiltered and filtered quasi-DNS statistics

coefficients for the centred compact finite difference scheme used for the discretization of the viscous term in the Navier–Stokes equations [3].

For a Reynolds number of $\mathrm{Re}_\tau = 1000$, two implicit LES have been performed using: (i) Marginal Resolution: $n_x \times n_y \times n_z = 512 \times 512 \times 1280$ cells, corresponding in wall units to $\Delta x^+ = \Delta y^+ \approx 4.5$, $\Delta z^+ \approx 19.5$; and (ii) Low Resolution: $n_x \times n_y \times n_z = 256 \times 256 \times 960$ cells, expressed in wall units: $\Delta x^+ = \Delta y^+ = 12$, $\Delta z^+ = 26$. Mesh resolutions were chosen in order to evaluate the method performance in terms of viscous and near-wall production region bypass, respectively. The implicit wall modelling feature of the present approach is evidenced by comparing ILES to no-model LES results performed without any (implicit or explicit) SGS/wall-layer modelling.

Beginning with the Marginal Resolution comparisons, in Fig. 56.4-right, the deviation between ILES and filtered kinetic energy profiles at the peak points to the existence of numerical/modelling errors. However, the overall precision and agreement of the statistics are the confirmation at high Reynolds number that accurate turbulent statistics can be obtained despite the use of a mesh resolution unable to capture the viscous sublayer. In particular, it can be observed that the peak of k^+ is remarkably well predicted at a level of detail clearly higher than the minimal scale $4\Delta x$ computed with accuracy. When compared to "No-Model" results, numerical dissipation is found definitely necessary so that it can be seen as playing the role, in this context, of an implicit viscous sublayer modelling. To explain this improvement, it must be mentioned that the no-model solution is subjected to small-scale oscillations which make inaccurate the prediction of friction velocity with an overestimation of about $+6\%$. Here, the use of numerical dissipation can remove these spurious oscillations making clearly more accurate the friction velocity prediction with only a deviation of -1% by comparison to DNS.

For the low resolution results, for which the mesh choice was based on the results of Fig. 56.2, a good prediction of the mean velocity is recovered by the ILES with an acceptable -2% underprediction of u_τ (see Fig. 56.5-left). The filtered profile of k^+ in Fig. 56.5-right shows that good level of energy could still be captured despite the

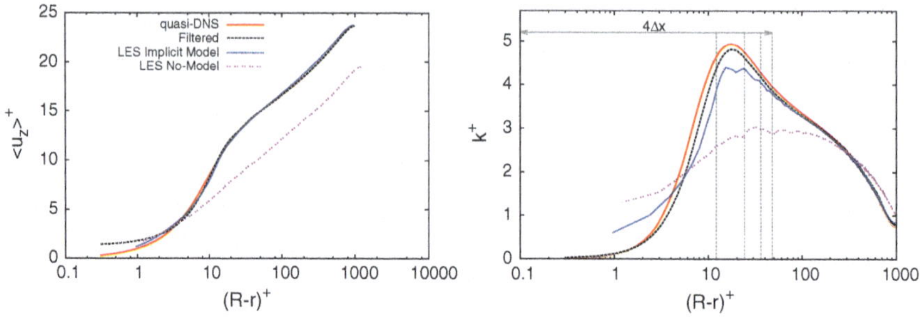

Fig. 56.5 Mean velocity profile (left) and Reynolds stresses profiles (right) for the low resolution ILES at $\text{Re}_\tau = 1000$. Results are compared to: no-model LES, unfiltered and filtered quasi-DNS statistics

astonishing value of 12 wall units used for the normal xy-discretization. Even though the profile of k^+ is somewhat irregular in the region of the peak, its amplitude is only underestimated by -11% with a shape captured at a scale evidently smaller than the minimum scale of accuracy $4\Delta x^+ = 48$. In the outer layer, the agreement with filtered DNS is almost perfect. The same cannot be said about the no-model LES for which both mean velocity and kinetic energy profiles are completely unrealistic when such a coarse mesh resolution is used. Based on these observations, it can be concluded that the present implicit SGS modelling has also a feature of implicit wall-layer model through its ability to restore the quality of the wall friction from $+24\%$ to -2% while enabling realistic prediction of the fluctuating velocity profiles. Note, however, that for the present low resolution, near-wall turbulent statistics are subjected to errors related to the azimuthal location of the nodes, especially those close to the cardinal points. These errors, due to the immersed boundary technique, tend to compensate each other leading to the acceptable k^+ profile presented in Fig. 56.5-right. A development is under progress to remove these immersed boundary artefacts making a room for further improvement of the present implicit wall-layer modelling.

Acknowledgements This work was Granted access to the HPC resources of TGCC/CINES/IDRIS under the allocation A0012A07624/A0032A07624/A0052A07624 made by GENCI.

References

1. Sagaut, P.: Large Eddy Simulation of Incompressible Flow: An Introduction, 2nd edn. Springer (2005)
2. Piomelli, U., Balaras, E.: Wall-layer models for large-eddy simulations. Ann. Rev. Fluid Mech. **34**, 349–374 (2002)
3. Dairay, T., Lamballais, E., Benhamadouche, S.: Mesh node distribution in terms of wall distance for large-eddy simulation of wall-bounded flows. Flow Turbul. Combust. **100**(3), 617–626 (2018)

4. El Khoury, G.K., Schlatter, P., Noorani, A., Fischer, P.F., Brethouwer, G., Johansson, A.V.: Direct numerical simulation of turbulent pipe flow at moderately high Reynolds numbers. Flow Turbul. Combust. **91**(3), 475–495 (2013)
5. Lamballais, E., Dairay, T., Laizet, S., Vassilicos, C.: Implicit/explicit spectral viscosity and large-scale SGS effects. In: Proceedings of the DLES-11, (Pisa, Italy) (2017)
6. Dairay, T., Lamballais, E., Laizet, S., Vassilicos, C.: Numerical dissipation vs. subgrid-scale modelling for large eddy simulation. J. Comput. Phys. **337**, 252–274 (2017)
7. Lamballais, E., Fortuné, V., Laizet, S.: Straightforward high-order numerical dissipation via the viscous term for direct and large eddy simulation. J. Comput. Phys. **230**, 3270–3275 (2011)
8. Dairay, T., Fortuné, V., Lamballais, E., Brizzi, L.: LES of a turbulent jet impinging on a heated wall using high-order numerical schemes. Int. J. Heat Fluid Flow **50**, 177–187 (2014)
9. Gautier, R., Laizet, S., Lamballais, E.: A DNS study of jet control with microjets using an immersed boundary method. Int. J. Comput. Fluid Dyn. **28**(6–10), 393–410 (2014)

Chapter 57
Improved Near-Wall Flow Prediction Combining Immersed Boundary Method and Data Assimilation

M. Meldi

Introduction

The accurate prediction of numerous bulk flow features of unstationary flows such as aerodynamic forces is driven by the precise representation of localized near-wall dynamics. Thus, the improvement in the prediction of near-wall flow features is a central target for the development of new investigative tools in Computational Fluid Dynamics (CFD). This aspect is particularly relevant for the flow prediction around complex geometries. In this case, classical body-fitted approaches may have to deal with high deformation of the mesh elements, usually providing poor numerical prediction. Additionally, the simulation of moving bodies may require prohibitively expensive mesh updates. In the last decades, the Immersed Boundary Method (IBM) [1–5] has emerged as one of the most popular strategies to handle these two problematic aspects. Among the numerous proposals reported in the literature, discrete methods [6, 7] can rely on Lagrangian markers to describe the discretized surface of the body. The IBM forcing is obtained through communication between the Eulerian flow system and the markers via an interpolation/spreading procedure.

The main difficulty with IBM is the accurate representation of near-wall flow features, which is a governing aspect in most engineering cases. The wall resolution required increases with the Reynolds number leading to a faster rise of the resources demanded when compared with body-fitted methods. One possible solution to reduce such computational costs for both body-fitted tools and IBM is the use of wall functions, which provide estimate statistics of the near-wall flows with reduced mesh

M. Meldi (✉)
Institut Pprime, Department of Fluid Flow, Heat Transfer and Combustion,
CNRS-ENSMA-Université de Poitiers, UPR 3346, Futuroscope Chasseneuil Cedex,
86962, Chasseneuil-du-Poitou, France
e-mail: marcello.meldi@ensma.fr

CNRS, Centrale Marseille, Aix-Marseille Univ., M2P2, Marseille, France

M. García-Villalba et al. (eds.), *Direct and Large Eddy Simulation XII*,
ERCOFTAC Series 27,
https://doi.org/10.1007/978-3-030-42822-8_57

requirements. However, wall functions are usually not efficient in capturing essential features such as boundary layer separation.

The present work aims for the advancement of the IBM method recently developed by the Author via Data Assimilation (DA). DA includes a wide spectrum of tools, referred to as estimators, which derive an optimized state integrating a model and available observation which are affected by uncertainties. Among these tools, the estimator recently proposed by the Author [8, 9] successfully provided an augmented state estimation integrating CFD and experimental data for the analysis of turbulent flows for engineering applications. Thus, the objective is to improve the accuracy of the IBM method (model) integrating high-fidelity data (observation) via DA. More precisely, the Eulerian information interpolated on the Lagrangian markers is not produced by the CFD model only. The DA state estimation obtained via combination of the CFD solver and the observation is instead employed. Thus, a full DA cycle is performed before the IBM step. This technique is expected to improve the estimation of the volume force on the Lagrangian markers, providing a more accurate prediction in the near-wall region. In future works, this strategy will provide information to infer the optimized behavior of free parameters in the IBM model.

Test Case: Flow Around a Circular Cylinder, $\mathrm{Re}_D = 3900$

The study is performed via the analysis of the flow around a circular cylinder for $\mathrm{Re}_D = U_\infty D/\nu = 3900$ [10–12]. This flow configuration has been identified for multiple reasons. The flow exhibits turbulent features in the wake but the boundary layer is laminar. This aspect allows to exclude the representation of wall turbulence effects in this initial analysis. Second, this configuration does not exhibit a boundary layer separation driven by the geometry, allowing to assess the capabilities of the IBM-DA approach to correctly reconstruct essential physical features.

Three classical simulations have been initially performed using the open-source code *OpenFOAM*. For every simulation, centered second-order schemes have been used for the discretization in space. Time advancement has been performed using a second-order backward scheme. The three simulations are now described:

1. A body-fitted direct numerical simulation, which will be referred in the following as DNS. The calculation has been performed in cylindrical physical domain of radius $25D$ and width (spanwise direction) $2\pi D$. The physical domain is discretized in 40×10^6 mesh elements and the near-wall resolution is on average $\Delta_r = 1.4 \times 10^{-3} D$, $\Delta_\theta \approx 10^{-2} D$, and $\Delta_z = 2.5 \times 10^{-2} D$. The width of the laminar boundary layer just before separation includes the first seven mesh elements.
2. A body-fitted large-eddy simulation using the WALE model [13], referred to as LES. The physical domain is the same employed for the DNS calculation, which is discretized in 10×10^6 mesh elements. The near-wall resolution is in this case $\Delta_r = 1.4 \times 10^{-3} D$, $\Delta_\theta \approx 2 \times 10^{-2} D$, and $\Delta_z = 1.3 \times 10^{-1} D$.

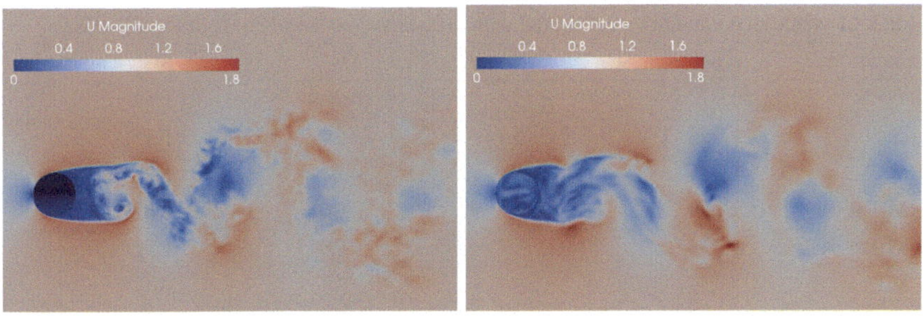

(a) (b)

Fig. 57.1 Isocontours of the instantaneous velocity magnitude. Results for **a** the body-fitted DNS simulation and **b** the IBM simulation are shown, respectively

3. A classical IBM simulation (without subgrid scale model), referred to in the following as IBM. In this case, the physical domain is a rectangular box of size $[-8D, 24D] \times [-8D, 8D] \times [0, 2\pi D]$ in the streamwise, normal, and spanwise direction. The discretization is performed via hexahedral elements for a total of 2.4×10^6 mesh elements. The near-wall resolution is here rather coarse: $\Delta_x = \Delta_y = 2 \times 10^{-2} D$, $\Delta_z = 1.3 \times 10^{-1} D$. The reasons why are going to be described in the following.

All the three simulations capture the physical behavior of the flow, as qualitatively shown for the simulations DNS and IBM in Fig. 57.1. However, the analysis of the predicted bulk flow quantities reported in Table 57.1 highlights significant differences. The results from the DNS are in agreement with high-precision results reported in the open literature [10–12]. For this reason, it will be employed as reference data for the assessment of the other simulations. The results for the simulation LES show an accurate prediction of the drag coefficient, due to the sufficient near-wall resolution and the subgrid model effect disappearing in the laminar boundary layer. However, the effect of the subgrid scale model is arguably responsible for the lack of agreement of the bulk flow quantities associated with the behavior of the flow in the wake region, namely, the standard deviation of the lift coefficient C'_L, the Strouhal number St, and the length of the recirculation bubble. Almost opposite conclusions are observed for the simulation IBM. Wake bulk flow quantities are here well predicted, despite the mesh resolution of LES and IBM in the wake is similar. This suggests that the addition of an LES model could in this case reduce the accuracy of the prediction for these quantities. On the other hand, the lack of near-wall resolution is responsible for a significant error in the prediction of the C_D.

Table 57.1 Bulk flow quantities calculated with the numerical simulations of the database

Simulation	C_D	C'_L	St	Rec. length
DNS	0.977	0.118	0.209	1.53
LES	1.093	0.25	0.2	1.15
IBM	1.225	0.101	0.2	1.7
IBM-DA-W	1.109	0.107	0.206	1.62
IBM-DA-K	1.233	0.117	0.207	1.52

Combining IBM and DA: Operative Strategy

Previous discussion highlighted how the coarse near-wall resolution of the IBM simulation was responsible for the lack of accuracy in the prediction of C_D, when compared with the DNS results. This coarse resolution was chosen in order to appreciate a significant difference between DNS and IBM. Thus, the IBM-DA strategy is going to be used to improve the performance of IBM integrating the high precision DNS data. Formally, this means that the Kalman Filter strategy relies on:

- A continuous low-precision model, which is the IBM over the coarse mesh
- High confidence local observation, which is provided in the form of instantaneous velocity DNS data sampled over a limited number of sensors.

The integration of the IBM and DA tools is consistently performed for the PISO solver of *OpenFOAM*, for which the two modules were separately developed [7, 8]. In practice, the Kalman filter provides a precise state estimation which is used to calibrate the IBM volume forcing term introduced in the discretized equations. The performance of this strategy is tested estimating its sensitivity to one of the most challenging aspects in DA studies, namely, the positions of the sensors. Two different distributions have been chosen for a cloud of 4000 sensors over which DNS data have been sampled. The first configuration, which is shown in Fig. 57.2a, consists of positioning the sensors in the circular crown for $0.5 < r/D < 0.55$, $z \in [0, 2\pi D]$. In this case, the sensors are positioned inside or at the surface of the boundary layer. This simulation will be referred to as IBM-DA-W. In the second case, shown in Fig. 57.2b, sensors are positioned in a box of size $[D,\ 5D] \times [-D,\ D] \times [0,\ 2\pi D]$, i.e., the region including the first shedding cycle. This case will be referred to as IBM-DA-K. The simulations are initialized interpolating the DNS field for $t = 0$ on the coarse mesh, so that the initial error covariance matrix is considered to be zero. For both simulations, the Kalman filter operates each 25 time steps. Considering that $\Delta t = 0.004L/U = 0.004t_A$, this implies that 10 Data Assimilation cycles are performed for a full advection time t_A. This also implies that approximately 50 assimilations are performed over a shedding cycle.

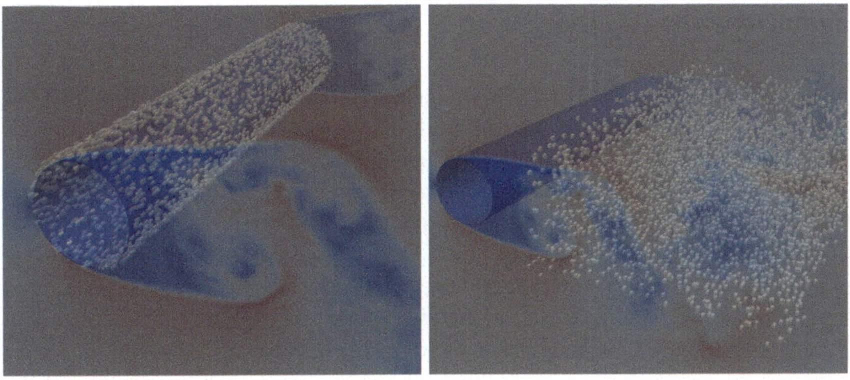

(a) (b)

Fig. 57.2 Positioning of the sensors providing information to the case (**a**) IBM-DA-W and (**b**) IBM-DA-K, respectively

Results

The performance of the IBM-DA simulations is assessed via observation of the computed bulk flow quantities, which are reported in Table 57.1. Results for the simulation IBM-DA-W are remarkably good for all the physical quantities, were an improvement of around 50% is observed. This result is a good starting point to further perform improvement of the IBM method via parametric inference. In particular, the pre-multiplying coefficient linearly driving the value of the volume forcing will be targeted to perform a model optimization, which will be able to provide accurate near-wall representation even without available observation. On the other hand, the results for the simulation IBM-DA-K are less satisfying. An excellent agreement is obtained for C_L', St, and the recirculation length, which are quantities associated with the behavior of the wake. This is consistent with the assimilation of high-precision data in the wake region. However, the drag coefficient C_D is mostly unchanged when compared to the classical IBM simulation. This implies that the accurate information obtained integrating the DNS data is not efficiently propagated upstream. The reason for this observation can be explained by looking at Fig. 57.1. The IBM is already able to provide a qualitative representation of the physical phenomena at play in the boundary layer, so that the reconstructed information in the wake complies with the standard model prediction. For this reason, the effect of the Kalman filter reconstruction upstream is negligible. These results stress how important it is to perform an accurate choice of the positions of the sensors in order to provide improvement of the results. However, they also show that the worst possible scenario obtained by the Kalman estimator is equivalent to pure application of the IBM model.

Conclusions

The accuracy of a discrete IBM method has been improved via application of Data Assimilation strategies in the present work. More precisely, a sequential reduced-order estimator has been used to provide an augmented state estimation, combining a coarse IBM simulation with local high-accuracy DNS samples. This prediction is used to obtain a higher accuracy calculation of the IBM forcing.

The IBM-DA strategy has been applied to the analysis of the flow around a circular cylinder, $\mathrm{Re} = 3900$. Two simulations have been performed, in which the position of the sensors has been changed. The methodology shows that a significant improvement of the predicted bulk flow quantities can be obtained. However, near-wall improvements are difficult to be obtained when sensors are positioned downstream, if the model is already able to capture the main physical mechanisms at play.

Future works will be devoted to further analysis of the physical solution obtained by the IBM-DA method, such as the analysis of the boundary layer separation and the transition to turbulence.

Acknowledgements Numerical simulations have been performed using computational resources of the project EDARI A0052A07590. The *Association française de mécanique* is also acknowledged for sponsoring the participation to the conference DLES12 to present this work.

References

1. Peskin, C.S.: Flow patterns around heart valves: a numerical method. J. Comput. Phys. **10**(2), 252–271 (1972)
2. Mittal, R., Iaccarino, G.: Immersed boundary methods. Ann. Rev. Fluid Mech. **37**, 239–261 (2005)
3. Uhlmann, M.: An immersed boundary method with direct forcing for the simulation of particulate flows. J. Comput. Phys. **209**(2), 448–476 (2005)
4. Taira, K., Colonius, T.: The immersed boundary method: a projection approach. J. Comput. Phys. **225**(10), 2118–2137 (2007)
5. Riahi, H., Meldi, M., Favier, J., Serre, E., Goncalves, E.: A pressure-corrected immersed boundary method for the numerical simulation of compressible flows. J. Comput. Phys. **374**(1), 361–383 (2018)
6. Pinelli, A., Naqavi, I., Piomelli, U., Favier, J.: Immersed-boundary methods for general finite-difference and finite-volume Navier-Stokes solvers. J. Comput. Phys. **229**(24), 9073–9091 (2010)
7. Constant, E., Favier, J., Meldi, M., Meliga, P., Serre, E.: An immersed boundary method in openfoam: verification and validation. Comput. Fluids **157**(3), 55–72 (2017)
8. Meldi, M., Poux, A.: A reduced order model based on Kalman filtering for sequential data assimilation of turbulent flows. J. Comput. Phys. **347**, 207–234 (2017)
9. Meldi, M.: Augmented prediction of turbulent flows via sequential estimators: sensitivity of state estimation to density of time sampling for available observation. Flow Turbul. Combust. **101**, 389–412 (2018)
10. Breuer, M.: Large eddy simulation of the subcritical flow past a circular cylinder: numerical and modeling aspects. Int. J. Numer. Methods Fluids **28**, 1281–1302 (1998)
11. Kravchenko, A., Moin, P.: Numerical studies of flow over a circular cylinder at $re_d = 3900$. Phys. Fluids **12**, 403 (2000)

12. Parnaudeau, P., Carlier, J., Heitz, D., Lamballais, E.: Experimental and numerical studies of the flow over a circular cylinder at Reynolds number 3900. Phys. Fluids **20**(8), 085101 (2008)
13. Nicoud, F., Ducros, F.: Subgrid-scale stress modelling based on the square of the velocitygradient tensor. Flow Turbul. Combust. **3**(62), 183–200 (1999)

Chapter 58
Hybrid RANS-LES Methods with Continuous Mode Variation

S. Heinz, R. Mokhtarpoor and M. K. Stoellinger

Introduction

The most convenient way to overcome the cost issue of large eddy simulation (LES) for high Reynolds number (Re) turbulent wall flows and the reliability problem of Reynolds-averaged Navier–Stokes (RANS) equations is the development of hybrid RANS-LES methods [2, 3]. The essential inherent feature of hybrid RANS-LES is the appearance of modeled motions (via the RANS model and modeled LES contribution) and resolved motions (fluctuating modes produced by the model equations if the grid is sufficiently fine). The basic problem of existing hybrid RANS-LES is the lack of control of modeled and resolved motions, which has serious consequences. A rudimentary expectation is that such models are functioning well for flows at arbitrarily high Re and arbitrary (appropriate) grids. This requires a smooth model response to Re and grid variations, which needs a balance between resolved and modeled motions. On top of that, the lack of hybrid mode control was shown to be the reason for an incorrect reflection of wall physics (the logarithmic law of the wall cannot be obtained) and other relevant physics (like flow separation [9]).

A mathematical exact solution to this problem, which can be applied in the framework of several popular turbulence models, was recently presented in Ref. [4]. The goal of this paper is to describe the potential of these novel simulation tech-

S. Heinz (✉)
Department of Mathematics, University of Wyoming, 1000 E. Univ. Ave., Laramie, WY 82071, USA
e-mail: heinz@uwyo.edu

R. Mokhtarpoor · M. K. Stoellinger
Department of Mechanical Engineering, University of Wyoming, 1000 E. Univ. Ave., Laramie, WY 82071, USA
e-mail: rmokhtar@uwyo.edu

M. K. Stoellinger
e-mail: mstoell@uwyo.edu

M. García-Villalba et al. (eds.), *Direct and Large Eddy Simulation XII*, ERCOFTAC Series 27,
https://doi.org/10.1007/978-3-030-42822-8_58

niques in applications to high Re flows. The paper is organized in the following way. Section "Modeling Approach" describes the theoretical approach, and section "Periodic Hill Flow Simulations" deals with the description of turbulent flow simulations. Flow simulation results are presented in section "Simulation Results", and conclusions are derived in section "Summary".

Modeling Approach

The incompressible continuity and momentum equations read $\partial \widetilde{U}_i/\partial x_i = 0$ and

$$\frac{\widetilde{D}\widetilde{U}_i}{\widetilde{D}t} = -\frac{1}{\rho}\frac{\partial \widetilde{p}}{\partial x_i} + \frac{\partial(2\nu \widetilde{S}_{ij})}{\partial x_j} - \frac{\partial \tau_{ij}}{\partial x_j}. \tag{58.1}$$

Here, the tilde refers to space-averaged variables, and $\widetilde{D}/\widetilde{D}t = \partial/\partial t + \widetilde{U}_j \partial/\partial x_j$ denotes the filtered Lagrangian time derivative. $\widetilde{U}_i$ denotes components of the velocity vector, $\widetilde{p}$ is the pressure, ρ is the constant fluid density, ν is the constant kinematic viscosity, and $\widetilde{S}_{ij} = (\partial \widetilde{U}_i/\partial x_j + \partial \widetilde{U}_j/\partial x_i)/2$ is the rate-of-strain tensor. The sum convention is used throughout this paper. The subgrid-scale (SGS) stress tensor τ_{ij} appears as an unknown on the right-hand side of Eq. (58.1). The stress is modeled by a usual eddy viscosity model, $\tau_{ij} = 2k\delta_{ij}/3 - 2\nu_t \widetilde{S}_{ij}$. Here, δ_{ij} is the Kronecker symbol, k is the SGS kinetic energy, and $\nu_t = C_\mu k/\omega$ is the SGS turbulent viscosity, which involves the turbulence frequency ω and the model parameter C_μ [5, 6].

The modeling approach can be applied in several two-equation setups [4]. It is used here in the framework of $k - \omega$ models [1],

$$\frac{Dk}{Dt} = P - \varepsilon + D_k, \qquad \frac{D\omega}{Dt} = C_{\omega_1}\omega^2\Big(\frac{P}{\varepsilon} - \beta\Big) + D_\omega + D_{\omega c}. \tag{58.2}$$

Here, $P = \nu_t S^2$ is the turbulence production involving the characteristic strain rate $S = (2\widetilde{S}_{mn}\widetilde{S}_{nm})^{1/2}$, and $\varepsilon = k\omega$ is the dissipation rate. We have $\beta = C_{\omega_2}/(C_\mu C_{\omega_1})$, where C_{ω_1} and C_{ω_2} are model parameters. The turbulent transport terms, which involve the parameter σ_ω, are given by $D_k = \partial[(\nu + \nu_t)\,\partial k/\partial x_j]/\partial x_j$ and $D_\omega = \partial[(\nu + \nu_t/\sigma_\omega)\,\partial\omega/\partial x_j]/\partial x_j$. The cross-diffusion term $D_{\omega c}$ reads $D_{\omega c} = C_\omega k^{-1}(\nu + \nu_t)[\partial k/\partial x_j][\partial\omega/\partial x_j]$. Here, we apply $(C_{\omega 1}, C_{\omega 2}, C_\omega, \sigma_\omega) = (0.49, 0.072, 1.1, 1.8)$. The integration of the ω equation through the viscous sublayer can cause numerical errors. To avoid this problem, $\omega = 2\nu/d^2$ was used at the first cells above the wall, where d refers to the distance from the wall to the cell center of the first cell.

There are several ways to hybridize Eq. (58.2) [4]. We consider here three variants,

$$\text{CES-KOS:} \quad \frac{Dk}{Dt} = P - \varepsilon + D_k, \quad \frac{D\omega}{Dt} = C_{\omega_1}\omega^2\Big(\frac{P}{\varepsilon} - \beta^*\Big) + D_\omega + D_{\omega c}. \tag{58.3}$$

$$\text{CES-KOK:} \quad \frac{Dk}{Dt} = P - \psi_\beta\varepsilon + D_k, \quad \frac{D\omega}{Dt} = C_{\omega_1}\omega^2\Big(\frac{P}{\varepsilon} - \beta\Big) + D_\omega + D_{\omega c}. \tag{58.4}$$

$$\text{CES-KOKU: } \frac{Dk}{Dt} = \psi_{\beta u}^{-1}(P - \psi_{\beta u}^{2}\varepsilon + D_k), \frac{D\omega}{Dt} = C_{\omega 1}\omega^2\left(\frac{P}{\varepsilon} - \beta\right) + D_\omega + D_{\omega c}. \tag{58.5}$$

In particular, the hybridization is accomplished via a modification of the dissipation term in the ω equation, k equation, and all terms involving the turbulent time scale, respectively. The corresponding functions are given by $\beta^* = 1 + R(\beta - 1)$, $\psi_\beta = 1 + \beta - \beta^*$, and $\psi_{\beta u}^2 = \psi_\beta$. The calculation of the model parameter R will be addressed in the next paragraph. The writing of $\psi_{\beta u}^{-1}$ in front of the diffusion term is symbolic: this function is applied inside the outer derivative. Regarding the model notation, CES refers to continuous eddy simulation, and CES-KO refers to CES performed with the $k - \omega$ model. In particular, -KOS (-KOK, -KOKU) refer to CES-KO performed by modifying the scale equation (the k equation, the turbulent time scales in the k equation according to unified RANS-LES). These equations were applied in conjunction with the elliptic blending approach applied in our previous work to account for wall damping [5, 6] (in RANS-mode for all the flow field).

Let us consider next the calculation of the parameter R in Eqs. (58.3), (58.4), (58.5). In partially averaged Navier–Stokes (PANS) and partially integrated transport modeling (PITM) methods, R is assumed to be equal to the kinetic energy ratio $k_+ = k/k_{tot}$ (k and k_{tot} refer to modeled and total $k_{tot} = k + k_{res}$ energy contributions, k_{res} being the resolved energy) [2]. k is defined via the turbulence model, and the resolved contribution is calculated by $k_{res} = (\langle\widetilde{U}_i\widetilde{U}_i\rangle - \langle\widetilde{U}_i\rangle\langle\widetilde{U}_i\rangle)/2$, where the bracket refers to averaging in time. In the approach presented here, R was theoretically calculated as $R = L_+^2$, where $L_+ = L/L_{tot}$ is the turbulence length scale resolution ratio involving modeled (L) and total contributions (L_{tot}) [4]. The modeled contribution is calculated as usual by $L = k^{3/2}/\varepsilon$. The total length scale is calculated correspondingly by $L_{tot} = k_{tot}^{3/2}/\varepsilon_{tot}$. Corresponding to $k_{tot} = k + k_{res}$, ε_{tot} is the sum of modeled and resolved contributions, $\varepsilon_{tot} = \varepsilon + \varepsilon_{res}$. Here, the resolved contribution is calculated by $\varepsilon_{res} = \nu(\langle\partial\widetilde{U}_i/\partial x_j \partial\widetilde{U}_i/\partial x_j\rangle - \langle\partial\widetilde{U}_i/\partial x_j\rangle\langle\partial\widetilde{U}_i/\partial x_j\rangle)$. The significant differences between the concept presented here and corresponding PANS and PITM approaches are analyzed in Ref. [4].

The mechanism of equations applied in this way is such that the degree of flow resolution is measured via L_+. The model receives information about the flow resolution via $R = L_+^2$. The modeled viscosity is adjusted to the degree of flow resolution such that the total motion is unchanged. This enables simulations where grid coarsenings and increased Re imply a smooth variation of modeled and resolved modes. It is worth noting that L_+, which is the actual ratio taken here from the simulation, can be related to the filter width [2, 4]. No attempt of doing this is made here to keep the approach as simple as possible.

Periodic Hill Flow Simulations

All models that are discussed in this study have been implemented in the OpenFOAM CFD Toolbox [7]. Details regarding the numerical solution of equations can be found elsewhere [5, 6]. We considered 2-D periodic hill flow including separation.

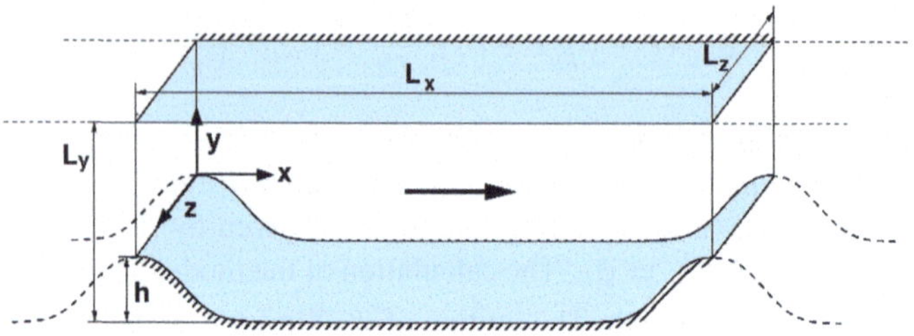

Fig. 58.1 The geometry of two-dimensional periodic hill flows

Figure 58.1 shows the streamwise-periodic hill flow configuration. The experimental study of this flow at the Reynolds number Re = 37000 (based on the hill height and bulk velocity above the hill crest) has been carried out by Rapp and Manhart [8]. The size of the computational domain is $L_x = 9h$, $L_y = 3.035h$, and $L_z = 4.5h$ in the streamwise x, wall normal y, and spanwise z directions, respectively, where h is the height of the hill. The hill crest is located at $(x/h, y/h) = (0, 1)$. At the bottom and top, the channel is constrained by solid walls. No-slip and impermeability boundary conditions are used at these walls. Periodic boundary conditions are employed in streamwise and spanwise directions.

Throughout of paper, reference quantities for length and velocity are h and U_b. All data presented are made dimensionless with these quantities. After 20 flow-through times, mean quantities were averaged over 140 flow-through times and averaged in the spanwise direction. The time step was chosen to result in maximum CFL number of 0.5 and an averaged CFL number of about 0.1. The filter width is defined by the volume filter width $\Delta_{vol} = (\Delta_x \Delta_y \Delta_z)^{1/3}$ Standard simulations are carried out on a grid of $N_x \times N_y \times N_z = 128 \times 80 \times 48$ using $500K$ cells for Re $= 37K$. We also used two coarser (120 and $250K$) grids, and we also performed simulations at two higher Reynolds numbers Re $= (100, 500K)$.

Simulation Results

A comparison of the three hybrid models considered here for the Re = 37 and $500K$ grid case is shown in Fig. 58.2. It may be seen that the performance of all three models regarding the flow simulation is almost equal and excellent. The design of these models supports the idea that the models should behave equivalently regarding their resolution ability. The L_+ plot shows that this is almost but not exactly the case: we see minor model differences. We see that these models, basically, work in LES mode over all the flow field with the exception of near-wall regions, which are treated in RANS mode. The L_+ plot compared to the k_+ plot shows that k_+ under-estimates the flow resolution. It may be seen that the assumption $\varepsilon_+ = 1$ often used in PANS and PITM methods becomes inappropriate in the relevant near-wall regions.

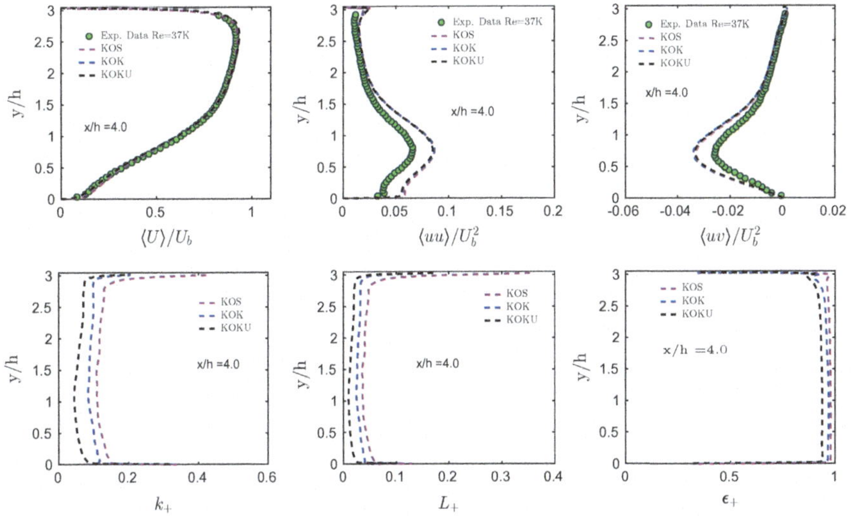

Fig. 58.2 Periodic hill flow simulations on the $500K$ grid at $\text{Re} = 37K$ for $x/h = 4$, comparison of CES-KOS, CES-KOK, CES-KOKU models: mean velocity, streamwise normal, Reynolds shear stress, and the modeled-to-total ratios of k, L, ε

The effect of Reynolds number and grid variations on the flow resolution and model performance is shown in Fig. 58.3 by only applying the CES-KOKU model. The effect of increasing Reynolds numbers is shown in the first row. The experimental data for $\text{Re} = 37K$ serve as benchmark data (experimental data are unavailable for higher Re). The relevant observation is that the amount of modeled motion smoothly increases with Re, as desired. Grid variation effects are shown in the second row for $\text{Re} = 37K$ in comparison to the experimental data. Similar to the Re effect, we see that a grid coarsening implies a smooth increase of modeled motion, as desired. The model performance is hardly affected by the increased amount of modeled motion. A drastic effect of grid variations considered on the flow resolution can be seen for the $\text{Re} = 500K$ case (last row) which shows again a consistent and stable model response. It is remarkable that the model performance is basically unchanged in response to these mode redistributions (the experimental data for $\text{Re} = 37K$ serve as a reference, they cannot be used for model validation).

Summary

A rudimentary expectation is that hybrid models are properly functioning for flows at arbitrarily high Re using arbitrary (appropriate) grids. The requirement for that is a smooth model response to Re and grid variations, which needs a balance between resolved and modeled motions. Existing hybrid method cannot deal with this chal-

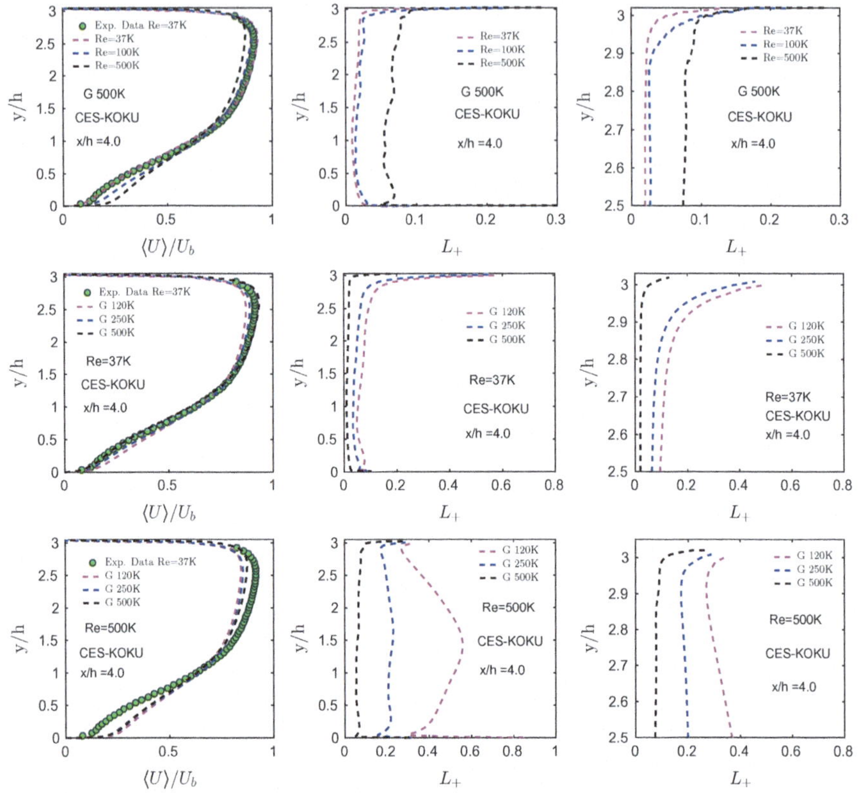

Fig. 58.3 CES-KOKU simulations, $\langle U\rangle/U_b$ and L_+ at $x/h = 4$. The first row shows for the $500K$ grid Re $= (37, 100, 500K)$ variations. The second and third rows show for Re $= 37K$ and Re $= 500K$, respectively, results for $(120, 250, 500K)$ grids

lenge. Based on a mathematical solution to this problem presented recently [4], we show that implied hybrid RANS-LES can deal with the required mode balance. Evidence for this view is provided via periodic hill flow simulations at high Re including flow separation. It is shown that the hybridization mechanism works almost equivalently for different turbulence model structures. The model performance is hardly affected by redistributions of resolved and modeled motions.

Acknowledgements The authors would like to acknowledge support through NASA's NRA research opportunities in aeronautics program (Grant No. NNX12AJ71A) and support from the National Science Foundation (DMS-CDS & E-MSS, Grant No. 1622488). We are very thankful for computational resources provided by the Advanced Research Computing Center and the Wyoming-NCAR Alliance at the University of Wyoming. S. Heinz gratefully acknowledges support of the Hanse-Wissenschaftskolleg (HWK).

References

1. Bredberg, J., Peng, S.H., Davidson, L.: An improved k- ω turbulence model applied to recirculating flows. Int. J. Heat Fluid Flow **23**(6), 731–743 (2002)
2. Chaouat, B.: The state of the art of hybrid RANS/LES modeling for the simulation of turbulent flows. Flow Turbul. Combust. **99**(2), 279–327 (2017)
3. Fröhlich, J., Terzi, D.V.: Hybrid LES/RANS methods for the simulation of turbulent flows. Prog. Aerosp. Sci. **44**(5), 349–377 (2008)
4. Heinz, S.: The large eddy simulation capability of Reynolds-averaged Navier–Stokes equations: analytical results. Phys. Fluids **31**(2), 021702/1–021702/6 (2019)
5. Mokhtarpoor, R., Heinz, S., Stoellinger, M.: Dynamic unified RANS-LES simulations of high Reynolds number separated flows. Phys. Fluids **28**(9), 095101/1–095101/36 (2016)
6. Mokhtarpoor, R., Heinz, S.: Dynamic large eddy simulation: stability via realizability. Phys. Fluids **29**(10), 105104/1–105104/22 (2017)
7. OpenFOAM: The open source CFD toolbox. http://www.openfoam.com (2015). Accessed 08 May 2015
8. Rapp, C., Manhart, M.: Flow over periodic hills—an experimental study. Exp. Fluids **51**, 247–269 (2011)
9. Spalart, P.R.: Detached-eddy simulation. Annu. Rev. Fluid Mech. **41**, 181–202 (2009)

Chapter 59
Average-Based Adaptive Grid Refinement in Hybrid LES

S. Mozaffari, M. Visonneau and J. Wackers

Introduction

Hybrid LES models were proposed as a solution to the high computational costs of Large Eddy Simulation (LES) for complex geometries at high Reynolds numbers. To take advantage of their capabilities, meshes with sufficient resolution around the turbulent regions of simulation are necessary. Adaptive Grid Refinement (AGR) creates these meshes by starting the solution on a coarse grid and refining this grid locally to accurately resolve areas of interest. This is an ideal way to efficiently solve flow problems that have strong local structures whose position is not known a priori, without excessive computational effort.

In hybrid LES flows which are highly unsteady, following all the structures of the turbulent flow with AGR (Fig. 59.1) means that the mesh is modified every few time steps. Unfortunately, this introduces extra truncation errors due to the interpolation which is performed during the grid refinement in order to transfer the solution between grids. To avoid these rapid changes in the adapted mesh, the refinement criterion which determines which part of the mesh should be refined, could be based on the averaged quantities instead of the instantaneous solutions. Thus, the refinement is restricted to capturing the large energy-containing eddies, without following all the turbulent structures. Toosi and Larsson [3] also tested an averaged error estimator for grid adaptation in LES simulation.

S. Mozaffari (✉) · M. Visonneau · J. Wackers
LHEEA Centrale Nantes/CNRS, 44300 Nantes, France
e-mail: sajad.mozaffari@ec-nantes.fr

M. Visonneau
e-mail: michel.visonneau@ec-nantes.fr

J. Wackers
e-mail: jeroen.wacker@ec-nantes.fr

M. García-Villalba et al. (eds.), *Direct and Large Eddy Simulation XII*,
ERCOFTAC Series 27,
https://doi.org/10.1007/978-3-030-42822-8_59

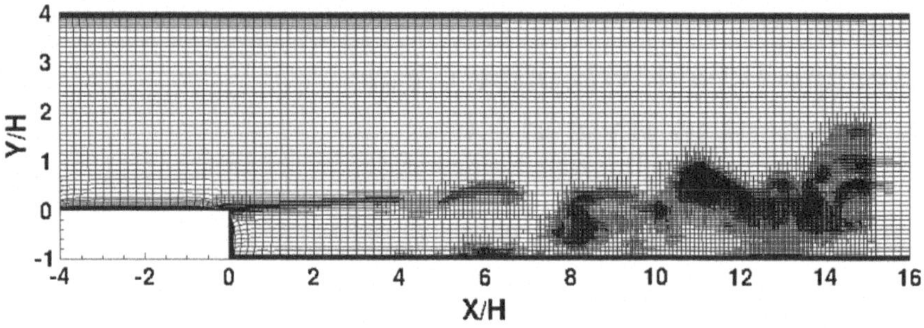

Fig. 59.1 Refined mesh based on instantaneous solutions (AGR-100, see Table 59.1)

In this paper, AGR with averaged criteria is tested for the backward-facing step flow of Vogel and Eaton [4]. Improved Delayed Detached-Eddy Simulation (IDDES) [2] is used in the ISIS-CFD solver, developed at Centrale Nantes/CNRS.

Adaptive Grid Refinement with Averaged Quantities

The AGR algorithm [5] begins with the computational domain covered with a coarse grid. As the calculation progresses, the mesh is regularly adapted based on the refinement criterion: a tensor field computed from the second derivatives of the solution, whose values determine which parts of the mesh should be refined or derefined. The global mesh size is imposed by a threshold parameter T_r which is proportional to the desired cell sizes.

To limit the mesh evolution, a criterion is desired which evolves slowly in time. Therefore it is based on a time-averaging process, either by averaging the instantaneous flow field or the computed criterion. In the first case, a time-averaged solution is computed, the refinement criterion is defined based on this averaged solution, and the mesh is adapted accordingly. In the second case, the criterion is computed every time step from the instantaneous solutions and the refinement is based on the average of the computed criteria.

Backward-Facing Step Simulation

The Backward-Facing Step (BFS) studied here is based on [1, 2, 4]. Behind the step, the flow has three main regions (Fig. 59.2): the shear layer region with the recirculation, the reattachment zone, and the relaxation region where the turbulence spreads through the entire channel. The computational domain based on the step height (H), runs from $x = -4H$ to $x = 20H$ in streamwise direction, with size $4H$ spanwise direction and channel expansion ratio is $5/4$.

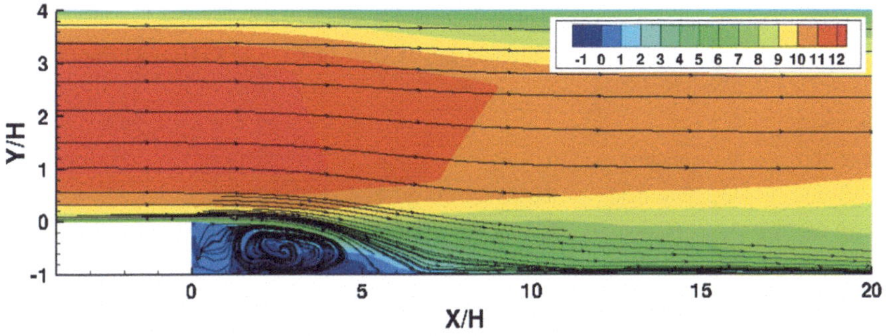

Fig. 59.2 Average axial flow and streamlines for the BFS (Fine grid, see Table 59.1)

At the inlet, velocity and turbulence from a precursor RANS calculation are imposed. In ISIS-CFD, periodic boundary condition is not available with AGR, so mirror boundary conditions are specified in spanwise direction. A no-slip condition is specified on the solid walls with $y^+ = 0.5$ and on the outlet boundary a frozen pressure is applied. The Reynolds number based on the step height, $H = 0.038\,\text{m}$, is equal to 28,000 and the non-dimensional time step is 0.018, for 10,000 time steps in total.

Results and Discussion

Several averaging strategies (Table 59.1) are tested and compared in terms of recirculation length, wall shear stress (Fig. 59.3), and turbulent structures (Fig. 59.7). Reference computations are performed on a fine fixed mesh without AGR, and by AGR without averaging (AGR-100), with refinement every two time steps, unlike the average-based computations that have refinement every 25 time steps.

While the fine mesh works well, AGR-100 creates perturbations in the skin friction on the lower wall. Although AGR follows the turbulent structures and refines the mesh where unsteadiness is high (Fig. 59.7), the frequent mesh changes perturb the solution. This confirms the interest of average-based refinement.

Average of the Flow Fields (CA)

For CA@-25 the criterion is based on the averaged flow. The final mesh in Fig. 59.4 shows refinement around the free-shear layer, but the refined mesh is not regular. This is because a part of the average solution was interpolated from the original mesh to the refined mesh. Due to the linear interpolation used, this contribution to

Table 59.1 Description of the computations, with computed recirculation length

Computation	Average method	Threshold T_r	Cells ($\times 10^6$)	L_r (X/H)	Error (%)
Exp.	–	–	–	6.66	–
Fine mesh	–	–	9.3	7.3	9.6
AGR-100	No averaging	0.01	6.6	8.35	25
CA@-25	Avg. flow field	0.0025	2.4	7.14	7.2
AC@-25	Avg. criterion	0.0025	1.3	7.9	18.7
AC@-20	Avg. criterion	0.0020	2.5	7.0	5.1
AC@-15	Avg. criterion	0.0015	9.9	6.9	3.6
AC@-25B†	Avg. criterion	0.0025	2.0	6.5	−2.5

† A refinement box is applied in the recirculation zone

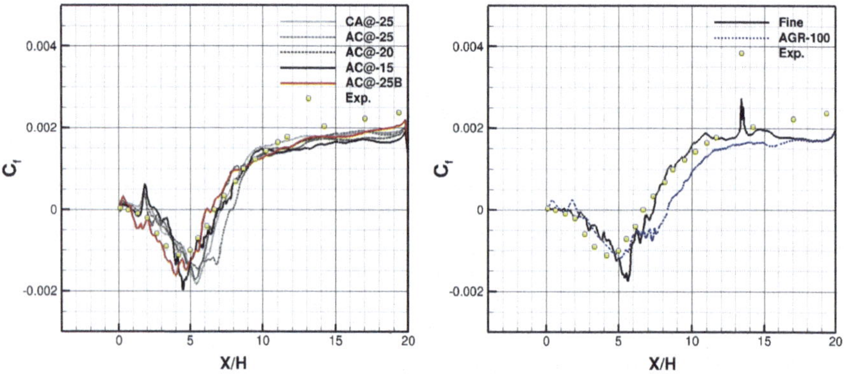

Fig. 59.3 Mean friction coefficients along the lower wall, for average-based (left) and reference (right) computations—CA: average of the flow fields and AC: average of the criterion

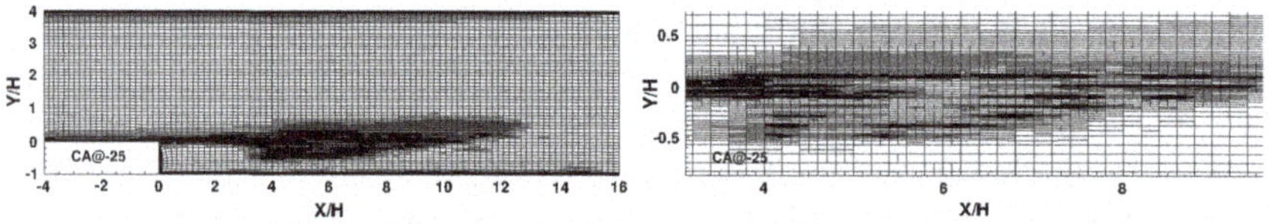

Fig. 59.4 Final refined mesh for the time-averaged flow field (CA@-25), with mesh detail

the average solution is not smooth enough to allow the computation of a second-derivative refinement criterion.

For other parts of the domain, the mesh remains coarse especially behind the step (recirculation region) where the velocity gradients are low. Figure 59.3 shows an overestimation for the mean friction coefficient distribution before the reattachment point, while the distance to the step tip (reattachment length) for this case is more than 7% larger compared to the experiment. Thus, the AGR based on the averaged flow fields, with some degree of error, can predict the behavior of the flow, but the irregular refinement makes the mesh invalid.

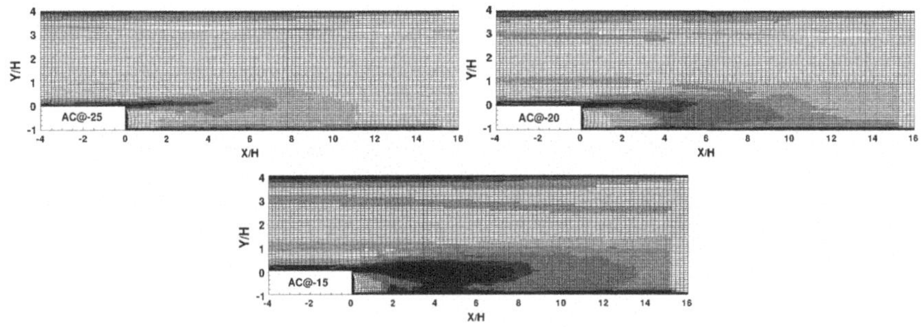

Fig. 59.5 The final refined meshes for the computations with time-averaged criterion

Average of the Criterion (AC)

For AC@-25, the averaging is applied over the computed criteria. For the same threshold value as the previous case, the final mesh is coarser (Table 59.1), so the difference between the computed mean friction coefficient and the experimental data increases (Fig. 59.3). However, the mesh is of better quality (Fig. 59.5).

Threshold Effect. Since the AC@-25 mesh is too coarse to capture the flow details, the threshold which globally indicates the requested amount of mesh refinement, is lowered to $T_r = 0.0020$ and 0.0015. As the threshold is decreased, larger regions of the domain are refined (Fig. 59.5) and the final mesh converges to a larger number of cells. This increase in the refinement occurs mostly around the shear layer and the reattachment region. The AGR also captures the edges of the attached boundary layers.

Using lower thresholds improves the solutions, particularly the wall shear stress in the reattachment region (Fig. 59.3). Table 59.1 shows that for the lowest threshold ($T_r = 0.0015$), the difference between computed reattachment length and the experiment decreased to 3.6%.

Influence of the Initial Mesh. In none of the previous cases, the mesh is refined in the recirculation zone. In the LES simulation of a BFS by Toosi and Larsson [3], adaptive refinement did not refine the grids in the recirculation zone as well. However, secondary vortices established in the recirculation zone can perturb the shear layer and might have an effect on the downstream flow. Therefore, the computation AC@-25B is performed using an initial mesh with a refinement box behind the step (Fig. 59.6). The other settings are as for AC@-25.

By using the refinement box, more turbulent structures are captured and a larger region is refined compared to AC@-25. The wall shear stress in the recirculation region is improved and the reattachment length error reduced to less than 3%. Figure 59.7 shows that, compared with other AGR cases, more turbulent structures in the recirculation zone are captured and the free-shear layer is represented better. These improvements could be the result of the small vortices in the recirculation zone which are captured, thanks to the refinement box.

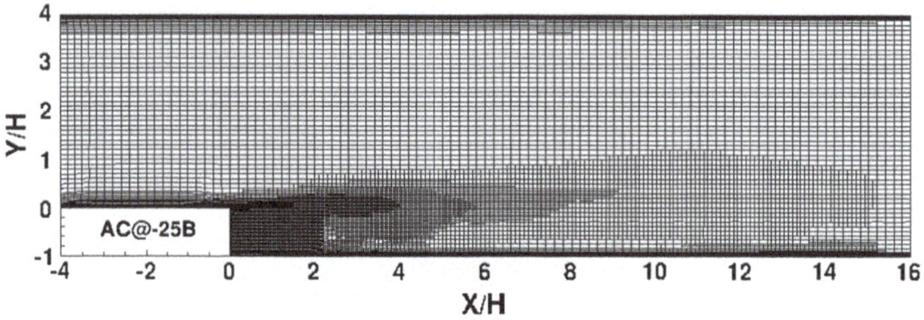

Fig. 59.6 The effect of the refinement box $x \in [0, 2]$, $y \in [-1, 0]$ in the recirculation zone

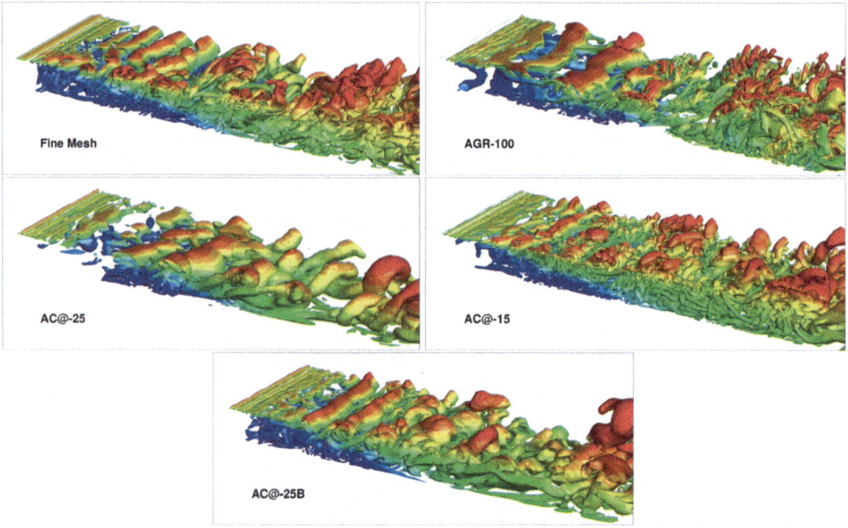

Fig. 59.7 Iso-surface second invariant $Q = 2500$; colored by averaged velocity X

Conclusion

Adaptive grid refinement was coupled with a hybrid LES turbulence model to simulate the flow over a backward-facing step. Using AGR with a refinement criterion based on the averaged solution is ineffective, since this solution is not smooth enough to allow the evaluation of its second derivative. However, using the averaged result of the computed criterion at each time step gives an advantage over AGR with frequent refinements without averaging, in terms of the accuracy of the solutions and also the computational cost. By reducing the threshold sufficiently, an accurate solution in the reattachment and relaxation regions is obtained. However, the refinement box in the recirculation zone shows that turbulent structures which are too small to create adaptive refinement with any criterion tested, may have an influence on the solution. How to detect and adaptively refine such zones, is an open question.

References

1. Gritskevich, M.S., Garbaruk, A.V., Schütze, J., Menter, F.R.: Development of DDES and IDDES formulations for the k-ω shear stress transport model. Flow Turbul. Combust. **88**(3), 431–449 (2012)
2. Shur, M.L., Spalart, P.R., Strelets M.K.H., Travin, A.K.: A hybrid RANS-LES approach with delayed-DES and wall-modelled LES capabilities. Int. J. Heat Fluid Flow **29**(6), 1638–1649 (2008)
3. Toosi, S., Larsson, J.: Anisotropic grid-adaptation in large eddy simulations. Comput. Fluids **156**, 146–161 (2017)
4. Vogel, J.C., Eaton, J.K.: Combined heat transfer and fluid dynamic measurements downstream of a backward-facing step. J. Heat Transf. **107**(4), 922–929 (1985)
5. Wackers, J., Deng, G.B., Guilmineau, E., Leroyer, A., Queutey, P., Visonneau, M.: Combined refinement criteria for anisotropic grid refinement in free-surface flow simulation. Comput. Fluids **92**, 209–222 (2014)

Chapter 60
Improving DES Capabilities for Predicting Kelvin–Helmholtz Instabilities. Comparison with a Backward-Facing Step DNS

A. Pont-Vílchez, F. X. Trias, A. Duben, A. Revell and A. Oliva

Introduction

The Detached Eddy Simulation (DES) turbulence model, presented by Spalart et al. [1] in the late 90s, was specifically designed to simulate those flow configurations where massive separation was involved. Since then, several authors have focused their efforts on addressing the main DES shortcomings, but there are still unsettled issues. In particular, delays in the transition zone from RANS to LES (Grey Area) severely affect the stimulation of the shear layer instabilities, harming the flow downstream of the step edge. In this regard, two different strategies are generally employed to overcome this issue [2]. The first of these consists of using artificial oscillations in specific areas (zonal approach) in order to trigger turbulence, whereas the other is based on reducing the local subgrid-scale viscosity, ν_{sgs}, given by

$$\nu_{sgs} = (C_m \Delta)^2 D_{sgs}(\bar{u}). \tag{60.1}$$

The second approach is aligned with the initial non-zonal DES philosophy, and retains its advantages; chiefly a more straightforwards implementation and a single numerical mesh. In this context, the subgrid-length scale (SLS), Δ, and the differential operator, $D_{sgs}(\bar{u})$, have been intensively studied in recent years, with particular focus on alleviating problems related to the Grey Area issue. For instance,

A. Pont-Vílchez (✉) · F. X. Trias · A. Oliva
Heat Transfer Technological Center (CTTC),
Universitat Politècnica de Catalunya-BarcelonaTech (UPC), 08222 Terrassa, Spain
e-mail: arnau@cttc.upc.edu

A. Duben
Keldysh Institute of Applied Mathematics, Moscow, Russia

A. Revell
School of MACE, University of Manchester, England, UK

M. García-Villalba et al. (eds.), *Direct and Large Eddy Simulation XII*,
ERCOFTAC Series 27,
https://doi.org/10.1007/978-3-030-42822-8_60

the possibility of providing a kinematic sensitivity to Δ was investigated by different authors. First, an SLS resistant to mesh anisotropies was proposed by Mockett et al. [2], $\tilde{\Delta}_{\omega}$, defending the importance of using the maximum meaningful scale at each LES control volume. Shur et al. [3] proposed another SLS in combination with the Smagorinksy (SMG) model, Δ_{SLA}, wherein $\tilde{\Delta}_{\omega}$ was modified to become inactive in $2D$ flow regions via a blending function, F_{KH}, in what is known as a Shear Layer Adapted (SLA) model. More recently we proposed a new SLS, Δ_{lsq}, inherited from the LES literature [4] which has been satisfactorily demonstrated to address the Grey Area phenomenon in DES [5].

The influence of $D_{sgs}(\bar{u})$ was also studied [6, 7]. In particular, the $\sigma - LES$ [8] was proposed as a good candidate due to its ability for switching off in $2D$ flow regions. The $\sigma - LES$ in combination with $\tilde{\Delta}_{\omega}$ is named as $\sigma - DES$. Other techniques directly inherited from LES could also be applied, such as the $S3PQR$ LES turbulence models presented by Trias et al. [9]. It is interesting to note that in this way the $\Delta_{SLA} = \tilde{\Delta}_{\omega} F_{KH}$ may be interpreted as $\tilde{\Delta}_{\omega}$ with a version of $D_{sgs}(\bar{u})$ sensitive to $2D$ flows, $D_{sgs}^{2D}(\bar{u})$. Namely,

$$\begin{aligned} \nu_{sgs} &= (C_m \Delta_{SLA})^2 D_{sgs}(\bar{u}) \\ &= \left(C_m \tilde{\Delta}_{\omega}\right)^2 (F_{KH}^2 D_{sgs}(\bar{u})) \\ &= \left(C_m \tilde{\Delta}_{\omega}\right)^2 D_{sgs}^{2D}(\bar{u}). \end{aligned} \tag{60.2}$$

This is, indeed, one of the most significant contributions for addressing the Grey Area shortcoming in DES models (mentioned in Shur et al. [3]).

Hence, the present work assesses the capabilities of the standard and new DES strategies for predicting physical instabilities in shear layers. In particular, the Backward-Facing Step (BFS) configuration is chosen as this case is subjected to a massive separation induced by geometry, Fig. 60.1 (suitable for DES models). First, the effect of different SLS (Δ_{SLA}, $\tilde{\Delta}_{\omega}$, Δ_{lsq}) is studied using the same $D_{sgs}(\bar{u})$, SMG. For this purpose, the rms distributions along the stream-wise direction are compared with the DNS data [10], as well as the growth of the Kelvin–Helmholtz instabilities in the stream-wise and normal directions. Once the effect of the SLS is clear, different SLS and $D_{sgs}(\bar{u})$ combinations are tested. These are: Δ_{SLA}, SMG; $\tilde{\Delta}_{\omega}$, $\sigma - LES$; $\tilde{\Delta}_{\omega}$, $S3QR$ and Δ_{lsq}, $S3QR$.

Computational Set-Up

The BFS geometrical domain is $24h \times 2h \times 2h$ in the stream-wise, normal and span-wise directions, respectively. The sudden expansion is located at $4h$ from the inflow. The origin of coordinates is placed at the expansion sharp edge. Regarding the mesh, three different refinement levels at the free shear layer (stream-wise direction,

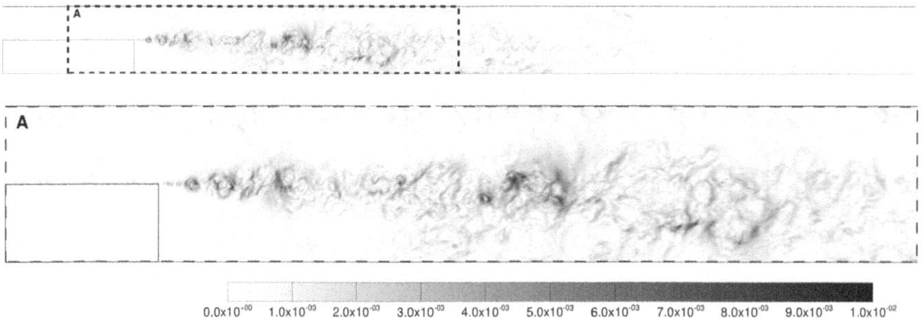

Fig. 60.1 Instantaneous magnitude of the dimensionless pressure gradient in a large part of the BFS domain (top), and a detailed view (*A*) of the sudden expansion (bottom). The colour meaning can be observed in the grayscale. See the film attached in the DNS data base [13]

$x_1 \in [0, h]$) have been considered for evaluating the mesh independence capabilities of the studied strategies. The length of the first node after the step edge in the streamwise direction is 8, 16 and 32 times the inflow y^+. The rest of mesh parameters are kept constant. These are: the Poisson growth ratio equal to 1.1, the number of cells per xy-plane equal to 11,800 and the number of planes in the periodic direction equal to 60.

The DDES turbulence model presented by Spalart et al. [11] has been used in this paper, including the Ψ term specially designed to override the unintended low-Re terms. All simulations carried out in this study have been run using *OpenFOAM-v1706*. The hybrid convection scheme presented by Travin et al. [12] for hybrid *RANS/LES* calculations is used. The max σ value is limited to 0.5. For the temporal discretisation, a 2nd implicit backward scheme is considered with a *Courant* value below 1. The velocity–pressure system is coupled using the well-known *PISO* algorithm. Concerning the boundary conditions, a fully developed turbulent channel flow is set at the inflow at $\mathrm{Re}_\tau = 395$.

Results

Subgrid-Length Scales, Δ

The effect of changing the aspect ratio, $\Delta x_1/\Delta x_2$, of the first cell downstream of the step edge is first discussed using the results shown in Fig. 60.2. For this purpose, three slightly different meshes have been used (see section "Computational Set-Up").

The highest aspect ratio (right) presents some non-physical oscillations at the step edge, close to $x_1/h = 0$. This is not the case for the other two figures (left, centre), where the aspect ratio is considerably smaller. Apart from that, the performance of the SLS in different meshes can also be appreciated. A general good independency of the mesh resolution is observed for Δ_{lsq} and Δ_{SLA}, whereas $\tilde{\Delta}_\omega$ presents a strong

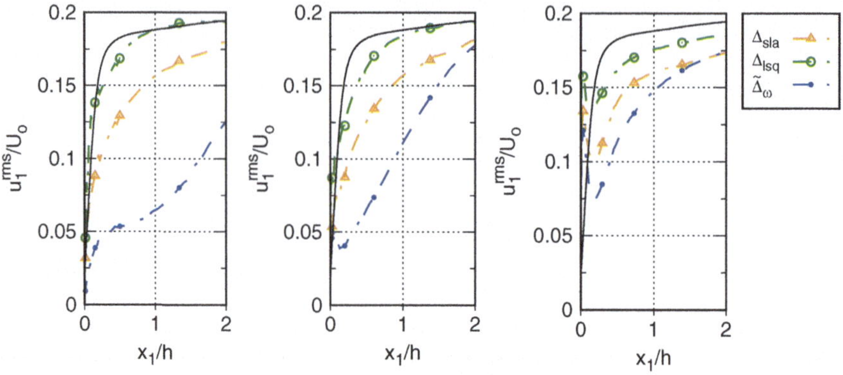

Fig. 60.2 Resolved Reynolds stresses, u_1^{rms}, along the stream-wise direction at $x_2 = 0$, considering similar meshes with different aspect ratios, $\Delta x_1/\Delta x_2$, at the step edge. These are: 8 (left), 16 (middle) and 32 (right). Where U_o refers to the inflow bulk velocity. Reference solid line has been obtained from Pont-Vílchez et al. [10]

mesh dependency. It can be justified, taking into account that $\tilde{\Delta}_\omega$ strongly depends on the stream-wise cell length in 2D flow configurations [4]. This is not the case for Δ_{SLA} and Δ_{lsq}, as the first is deactivated in such areas and the latter depends on both the stream-wise and normal cell dimensions [5]. The growth of instabilities at the shear layer presented by different SLS is also analysed using the same approach described by Pont-Vílchez et al. [10]. The characteristic length of instabilities in the stream-wise direction, $\Delta\delta_1$, is calculated using a set of 2-point correlations of u_2' along the stream-wise direction downstream of the step edge (Fig. 60.3). Unfortunately, this technique cannot be applied for assessing the size of instabilities in the normal direction, $\Delta\delta_2$, as the flow behaves laminarly in some parts along such direction. For this reason, another approach based on mean quantities has been used [10, 14],

$$\Delta\delta_2 = \Delta U_1/(\partial\langle u_1\rangle/\partial x_2)_{\max}. \tag{60.3}$$

Even though the rms profiles present a strong dependence on the *SLS* (Fig. 60.2) along the shear layer, this is not so significant in the $\Delta\delta_1$ distribution (Fig. 60.3, left). In particular, Δ_{lsq}, together with Δ_{SLA}, show the best alignment at $x_1 \in [0, 0.7h]$. However, the strength of correlation with *DNS* data is notably reduced further downstream, leading to a distinct departure of the slope gradient from the reference, which is mainly attributed to the mesh coarsening in this region.

Regarding $\Delta\delta_2$, it seems to be quite sensitive to the SLS (such as rms profiles), presenting strong differences in values and slopes (Fig. 60.3, right). The fact that we are using Eq. 60.3, instead of a 2-point correlation along the normal direction, plays an important role as $(\langle u_1\rangle/\partial x_2)_{\max}$ is highly influenced by the rms. This is clearly indicated in Fig. 60.3 (right) where the diffusion introduced by Δ_{SLA} and $\tilde{\Delta}_\omega$ may be permitted to grow to excessive levels, preventing the KH instabilities from developing along the shear layer.

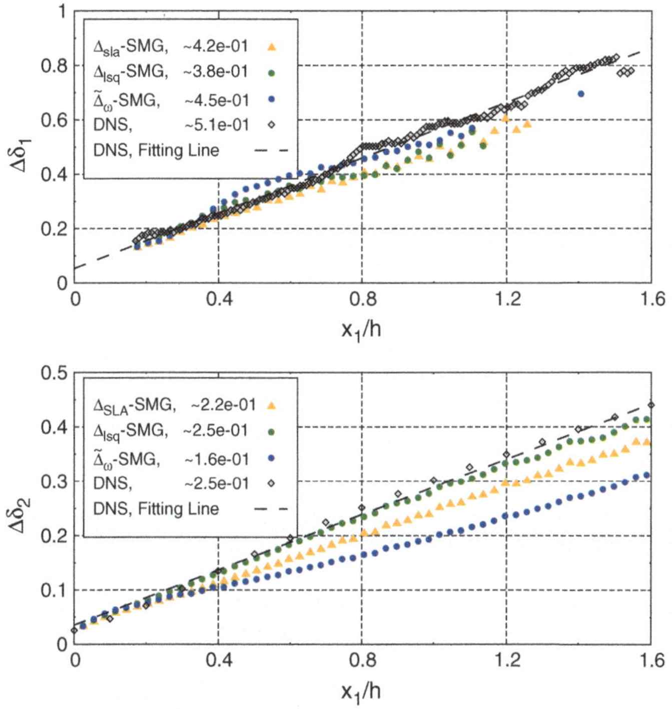

Fig. 60.3 Estimation of the KH rate of growth in the streamwise (left) and normal (right) directions downstream of the step edge using different SLS, $\Delta\delta_1$ and $\Delta\delta_2$, respectively. Reference solid line has been obtained from Pont-Vílchez et al. [10]

Differential Operator, D_{sgs} ($\bar{u}$)

A set of simulations using different operators is presented, including Δ_{SLA}, SMG [3]; $\tilde{\Delta}_{\omega}$, $\sigma - LES$ [6]; $\tilde{\Delta}_{\omega}$, $S3QR$ and Δ_{lsq}, $S3QR$ [5]. In contrast to the previous section, all turbulence models are deactivated in free shear layer 2D flow domains. Their influence can be appreciated in Fig. 60.4, where the rms and the $\Delta\delta_2$ are shown. Regarding $\tilde{\Delta}_{\omega}$, both $S3QR$ and $\sigma - LES$ present a clear improvement in comparison to SMG (Fig. 60.3), providing results similar to Δ_{SLA}, SMG in agreement with the observations made in section "Introduction". Instead, no significant improvements are observed with Δ_{lsq}, $S3QR$ respect to Δ_{lsq}, SMG.

Conclusions

This work shows that the use of the new SLS proposed in Pont-Vílchez et al. [4], Δ_{lsq}, provides substantial benefits in DES simulation of free shear layers respect to the other SLS strategies. The influence of differential operators sensitive to 2D is

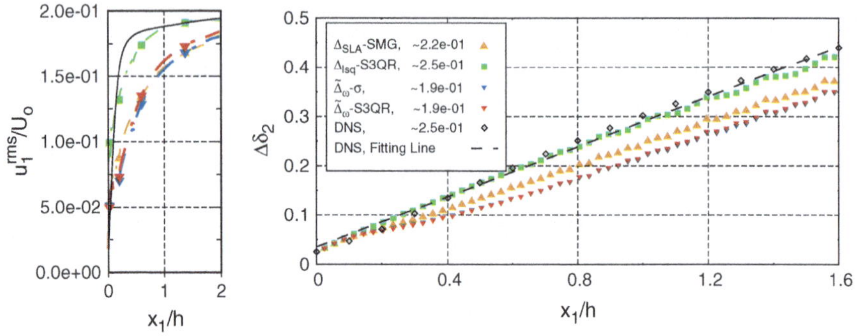

Fig. 60.4 Resolved Reynolds stresses (right), u_1^{rms}, along the stream-wise direction at $x_2 = 0$ ($\Delta x_1/\Delta x_2 = 16$), at the step edge. Estimation of the KH rate of growth in the normal (right) direction downstream of the step edge using different strategies, $\Delta\delta_2$, where U_o refers to the inflow bulk velocity

also noticed, especially in combination with SLS which are sensitive to the stream-wise meshing (such as $\tilde{\Delta}_\omega$). While further work is required to investigate whether these observations hold in other test cases, these initial results indicate that Δ_{lsq} is a promising strategy for naturally mitigating the RANS to LES numerical delay.

References

1. Spalart. P.R., Jou, W.H., Strelets, M., Allmaras, S.: Comments on the Feasibility of LES for Wings, and on a Hybrid RANS/LES Approach (1997)
2. Mockett, C., Fuchs, M., Garbaruk, A., Shur, M., Spalart, P., Strelets, M., Thiele, F., Travin, A.: Progress in Hybrid RANS-LES Modelling, pp. 187–201. Springer International Publishing, Cham (2015)
3. Shur, M.L., Spalart, P.R., Strelets, M.K., Travin, A.K.: Flow, Turbul. Combust. **95**(4) (2015)
4. Trias, F.X., Gorobets, A., Oliva, A.: Phys. Fluids **115109** (2017)
5. Pont-Vílchez, A., Trias, F.X., Revell, A., Oliva, A.: Proceedings of Hybrid RANS-LES Methods 7, 17–19 Sept., Berlin, Germany (2018)
6. Fuchs, M., Sesterhenn, J., Thiele, F., Mockett, C.: 22nd AIAA Computational Fluid Dynamics Conference (2015)
7. Probst, A., Schwamborn, D., Garbaruk, A., Guseva, E., Shur, M., Strelets, M., Travin, A.: Int. J. Heat Fluid Flow (2017)
8. Nicoud, F., Toda, H., Cabrit, O., Bose, S., Lee, J.: Phys. Fluids **23**, 8 (2011)
9. Trias, F.X., Folch, D., Gorobets, A., Oliva, A.: Phys. Fluids **27**, 6 (2015)
10. Pont-Vílchez, A., Trias, F.X., Gorobets, A., Oliva, A.: J. Fluid Mech. **863**, 341363 (2019)
11. Spalart, P.R., Deck, S., Shur, M.L., Squires, K.D., Strelets, M.K., Travin, A.: Theoret. Comput. Fluid Dyn. **20**(3), 181 (2006)
12. Travin, A., Shur, M., Strelets, M., Spalart, P.: Fluid Mechanics and its Applications, vol. 65 (2004)
13. Pont-Vílchez, A., Trias, F.X., Gorobets, A., Oliva, A.: Direct Numerical Simulation dataset. http://www.cttc.upc.edu/downloads/BFS_Ret395_ER2 (2018)
14. Winant, C.D., Browand, F.K.: J. Fluid Mech. **63**(2), 237 (1974)

Chapter 61
FSI Investigations on a Flexible Air-Inflated Thin-Walled Structure: An LES Study with Experimental Validation

G. De Nayer, J. N. Wood, A. Apostolatos and M. Breuer

Introduction

The easy-to-shape and light-weight characteristics of membranous structures form the basis of their success in modern civil engineering. In the case of outside constructions they are exposed to wind loads and consequently undergo fluid–structure interactions (FSI). In order to correctly dimension these flexible buildings and to avoid damages, numerical and experimental investigations are necessary. The current paper describes an LES study of a real-world application relying on a modern numerical simulation methodology and supported by an experimental validation.

Methodology for a Real-World Application

In order to validate the numerical FSI simulation methodology based on the large-eddy simulation (LES) technique developed by Breuer et al. [1], two complementary numerical-experimental FSI validation studies [2–6] were recently successfully carried out. The numerical and experimental data were made available online on the ERCOFTAC database [3, 6]. However, these basic test cases considering flexible sheet-like structures mounted on the backside of a rigid cylinder nominally represent two-dimensional geometries. To extend the application area of the coupled solver to

G. De Nayer · J. N. Wood · M. Breuer (✉)
Professur für Strömungsmechanik, Helmut-Schmidt Universität-Hamburg, Hamburg, Germany
e-mail: breuer@hsu-hh.de

G. De Nayer
e-mail: denayer@hsu-hh.de

A. Apostolatos
Lehrstuhl für Statik, Technische Universität München, Munich, Germany

M. García-Villalba et al. (eds.), *Direct and Large Eddy Simulation XII*,
ERCOFTAC Series 27,
https://doi.org/10.1007/978-3-030-42822-8_61

Fig. 61.1 Thin-walled air-inflated structures of hemispherical shape: Roofing from Architen Landrell in Qatar (left), Gasholders of the company Sattler Ceno TOP-TEX GmbH (right)

real-world three-dimensional cases, a flexible air-inflated membranous structure in form of a hemisphere is chosen, which can be found for example as roofing or gas holder (see Fig. 61.1). A wall-mounted hemisphere is a simple geometry. However, the turbulent airflow around the structure is complex as demonstrated for a rigid hemisphere in [7, 8]. In the present investigation, the approaching flow is a turbulent boundary layer (1/7-th power law) including appropriate fluctuations. Its thickness corresponds to the height of the hemisphere. Three Reynolds numbers based on the diameter of the hemisphere D and the free-stream velocity U_∞ (Re = 50,000, 75,000, and 100,000) are considered. In case of the flexible dome, the inner pressure has to be adjusted for each Re number in order to approximately guarantee a hemispherical shape of the structure (see details in [9]). The dome is made of a highly deformable thin silicone membrane (thickness = 0.165 mm, mass = 6.22 g, E = 7×10^5 Pa). This membrane is exposed to turbulent wind loads. The added-mass effect is negligible. In the simulations, the damping of the material is modeled by the so-called Rayleigh damping and its parameters are determined based on a pure structural test case analyzing the response of the membrane to a local excitation measured in the experiments [9], for further details refer to [10]. To enable a comparison with the experiments, a pre-stress value $n_{\text{membrane}} = 7794.5$ Pa determined by a static structural test case is set for the structure model in the simulations. For this purpose, the flexible hemisphere is exposed to a variable inner pressure evaluating the displacements at the apex, c.f., [9, 10].

Two different FSI simulations both relying on wall-resolved LES for the fluid flow (block-structured grid with about 4.3 million CVs, Smagorinsky model with $C_s = 0.1$) but different formulations for the structure are carried out: For the first (denoted LES-FEM) the thin structure is discretized by 1926 membrane finite elements [10], whereas for the second (denoted LES-IGA) 625 bi-quadratic isogeometric elements are applied [11]. Using the well-established partitioned semi-implicit predictor-corrector coupling scheme especially developed for FSI applications within LES [1], the solver based on either finite elements or on the isogeometric analysis for the structural predictions are coupled to a finite-volume code for the fluid written in an ALE formulation. Both fields (flow and structure) are computed simultaneously exchanging their corresponding data at the FSI interface by the open-source coupling code EMPIRE. In the present case a loose coupling algorithm is sufficient. A CPU-time efficient hybrid grid deformation technique involving inverse distance weighting and transfinite interpolations was developed for block-structured grids [12] to preserve the high grid quality required for LES. To ensure appropriate inflow conditions mimicking the approaching turbulent boundary layer, a digital filter-based inflow generator is used, which was recently extended by a source-term formulation allowing to inject the inflow turbulence in well-resolved flow regions within the numerical domain [13, 14].

For validation purposes, an intensive experimental campaign is carried out in a wind tunnel [7–9]: The flow measurements rely on laser-Doppler anemometry, particle image velocimetry and constant-temperature anemometry. A digital-image correlation method is used to determine the 3D deformations of the membranous surface. Unfortunately, the flow and the structure measurements cannot be done simultaneously leading to non-synchronized experimental data.

Time-Averaged and Instantaneous Results

In order to characterize the different phenomena occurring in this FSI case, time-averaged and instantaneous data are complementary gathered from the simulations and the experiments. In the following, the results are restricted to the most interesting case, i.e., the highest Re number (Re = 100,000), since it involves the largest deformations of the structure.

In a first step the time-averaged deformations of the flexible membrane (see Fig. 61.2) and the time-averaged flow characteristics (see Fig. 61.3) are compared between the LES predictions and the experiments. The dimensionless time-averaging interval is larger than $\Delta T^* = \Delta T \; U_\infty / D = 125$. As visible in Fig. 61.2 a very good agreement with the experiments is reached for LES-FEM and an even better coincidence is found for LES-IGA.

Similar observations can be made concerning the time-averaged first- and second-order moments of the flow. The LES-FEM data coincide well with the measurements. However, a slightly better representation of the recirculation area is observed for LES-IGA: The streamlines and contour levels of the Reynolds stresses fit better to the experimental data. This is possibly due to the better representation of the FSI interface in case of LES-IGA. As demonstrated in [11], the use of finite elements for the structure discretization leads to an artificial roughness of the FSI interface, which alters the flow. Applying IGA the FSI interface is perfectly smooth. Nevertheless, the level of accuracy obtained by LES-FEM is reasonable, and thus both simulations can be used to investigate the appearing instantaneous FSI phenomena.

To understand the FSI phenomena related to the fluid and those related to the structure, power spectra of the displacements at predefined monitoring points are evaluated based on the numerical and experimental data. Accordingly, particular frequency peaks are found corresponding to fluid-related phenomena such as vortex sheddings. Using the advantage of the synchronized numerical results, two types

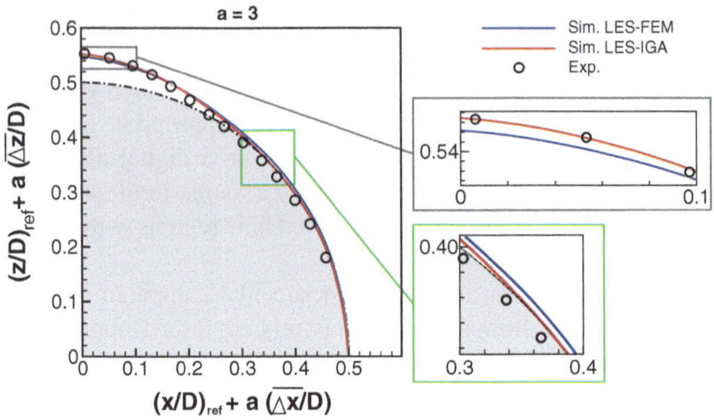

Fig. 61.2 Time-averaged dimensionless deformation of the flexible membrane at Re = 100,000 in the symmetry x–z plane: Simulations LES-FEM [10] and LES-IGA [11] and Experiment [9]. For better visibility the displacements are magnified by a factor of a = 3

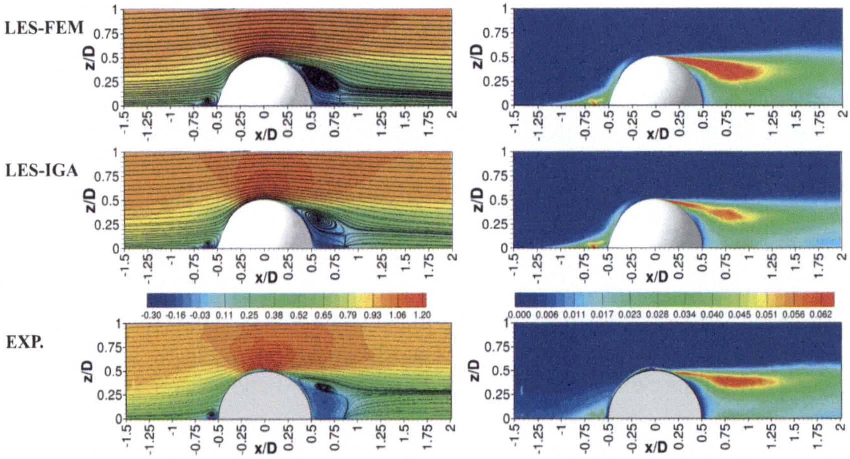

Fig. 61.3 Comparison of the time-averaged dimensionless flow (left: streamlines and streamwise velocity contour, right: streamwise normal Reynolds stress) at Re = 100,000 in the symmetry plane based on the numerical results LES-FEM [10] (top), LES-IGA [11] (middle) and the measurements [9] (bottom)

of vortex sheddings of large vortical structures are identified past the flexible body. A von Kármán vortex shedding occurs at a Strouhal number $\mathrm{St} = f\,D/U_\infty = 0.16$ depicted in the left picture of Fig. 61.4. This shedding type alternates in time with a second shedding type of a large arc-type symmetric vortical structure at a Strouhal number of St = 0.27 depicted in the right picture of Fig. 61.4. Both types imply different deformations of the membrane detailed in [10]. Another interesting flow phenomenon involving medium-sized vortices and generating wave-like deformations on the membrane occurs near the apex of the hemisphere with a higher Strouhal num-

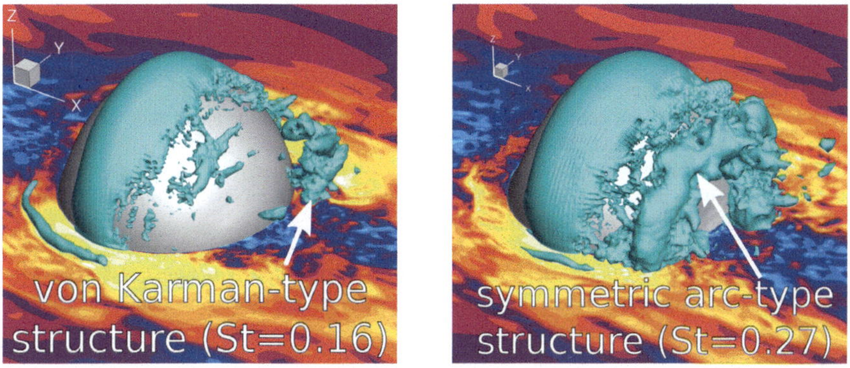

Fig. 61.4 Visualization of the instantaneous 3D vortical structures in the back of the hemisphere in case of the von Kármán (left) and the arc-type symmetric (right) shedding types (velocity magnitude near the bottom wall normalized by the free-stream velocity and iso-surface of the pressure coefficient $c_p = -0.48$) [10]

ber of St = 1.55. This is analyzed to be the result of rapidly merging vortices evolving by the Kelvin–Helmholtz shear layer instability. Furthermore, other frequency peaks in the power spectra of the displacements correspond to the natural eigenfrequencies of the flexible membrane. These eigenfrequencies and the corresponding eigenmodes are obtained by a linear modal analysis of the structure in vacuum and help to understand the complex deformation patterns of the membranous surface.

Conclusions

A thin-walled air-inflated structure of hemispherical shape exposed to turbulent wind loads is numerically and experimentally investigated. The flow prediction based on a wall-resolved LES is coupled to two different setups for the structure. The first, relying on a standard finite element discretization, delivers a reasonable agreement with the measurements. However, the results obtained by the second configuration based on the recent isogeometric analysis even better fit to the experimental data. This can be explained by the smoother representation of the FSI interface in case of IGA.

Besides the validation purposes, the experimental data also serve to better comprehend the FSI case, since the advantages and drawbacks encountered by both simulations and experiments are complementary. For example, during the experiments, it is easy to record a long time-span and therefore to carry out accurate averaging or a frequency analysis, whereas the simulations allow a higher temporal and spatial resolution of a limited time-span [9–11].

Acknowledgements The project was financially supported by the *German Research Foundation* (BR 1847/12-2 and BL 306/26-2).

References

1. Breuer, M., De Nayer, G., Münsch, M., Gallinger, T., Wüchner, R.: Fluid-structure interaction using a partitioned coupled predictor-corrector scheme for the application of large-eddy simulation. J. Fluid Struct. **29**, 107–130 (2012)
2. De Nayer, G., Kalmbach, A., Breuer, M., Sicklinger, S., Wüchner, R.: Flow past a cylinder with a flexible splitter plate: A complementary experimental-numerical investigation and a new FSI test case (FSI-PfS-1a). Comput. Fluids **99**, 18–43 (2014)
3. De Nayer, G., Kalmbach, A., Breuer, M.: Test case on QNET ERCOFTAC database (Underlying Flow Regime 2–13): Fluid-structure interaction in turbulent flow past cylinder/plate configuration I (First swiveling mode). http://www.kbwiki.ercoftac.org/w/index.php/Abstr:UFR_2-13
4. Kalmbach, A., Breuer, M.: Experimental PIV/V3V measurements of vortex-induced fluid-structure interaction in turbulent flow—A new benchmark FSI-PfS-2a. J. Fluid Struct. **42**, 369–387 (2013)
5. De Nayer, G., Breuer, M.: Numerical FSI investigation based on LES: Flow past a cylinder with a flexible splitter plate involving large deformations (FSI-PfS-2a). Int. J. Heat Fluid Flow **50**, 300–315 (2014)
6. Kalmbach, A., De Nayer, G., Breuer, M.: Test case on QNET ERCOFTAC database (Underlying Flow Regime 2-14): Fluid-structure interaction in turbulent flow past cylinder/plate configuration II (Second swiveling mode). http://www.kbwiki.ercoftac.org/w/index.php/Abstr:UFR_2-14
7. Wood, J.N., De Nayer, G., Schmidt, S., Breuer, M.: Experimental investigation and large-eddy simulation of the turbulent flow past a smooth and rigid hemisphere. Flow Turbul. Combust. **97**(1), 79–119 (2016)
8. Wood, J.N., De Nayer, G., Schmidt, S., Breuer, M.: Test case on QNET ERCOFTAC database (Underlying Flow Regime 3-33): Turbulent flow past a wall-mounted hemisphere. http://www.kbwiki.ercoftac.org/w/index.php/Abstr:UFR_3-33
9. Wood, J.N., Breuer, M., De Nayer, G.: Experimental studies on the instantaneous fluid-structure interaction of an air-inflated flexible membrane in turbulent flow. J. Fluid Struct. **80**, 405–440 (2018)
10. De Nayer, G., Apostolatos, A., Wood, J.N., Bletzinger, K., Wüchner, R., Breuer, M.: Numerical studies on the instantaneous fluid-structure interaction of an air-inflated flexible membrane in turbulent flow. J. Fluid Struct. **82**, 577–609 (2018)
11. Apostolatos, A., De Nayer, G., Bletzinger, K., Breuer, M., Wüchner, R.: Systematic evaluation of the interface description for fluid-structure interaction simulations using the isogeometric mortar-based mapping. J. Fluid Struct. **86**, 368–399 (2019)
12. Sen, S., De Nayer, G., Breuer, M.: A fast and robust hybrid method for block-structured mesh deformation with emphasis on FSI-LES applications. Int. J. Numer. Meth. Eng. **111**(3), 273–300 (2017)
13. Schmidt, S., Breuer, M.: Source term based synthetic turbulence inflow generator for eddy-resolving predictions of an airfoil flow including a laminar separation bubble. Comput. Fluids **146**, 1–22 (2017)
14. De Nayer, G., Schmidt, S., Wood, J.N., Breuer, M.: Enhanced injection method for synthetically generated turbulence within the flow domain of eddy-resolving simulations. Comput. Math. Appl. **75**(7), 2338–2355 (2018)

Chapter 62
A New Method for Fully Resolved Simulations of Fracturing in Fluid-Structure Interaction Problems

F. Dalla Barba and F. Picano

Introduction

Fluid-Structure Interaction (FSI) is a complex, nonlinear, and multi-physics matter involving at the same time both fluid and solid mechanics. These are, in general, fully coupled problems where deformable solid media interact with fluid flow through a time-evolving interface. FSI problems are of great interest in a large variety of scientific fields and engineering applications such as multiphase flows, biological flows, and aeroelasticity problems. Despite their scientific and engineering relevance, solutions to FSI problems are still challenging even when approaching the problem from a numerical point of view. A particularly challenging matter deals with the prediction of solid media fracture due to the action of hydrodynamic forces (hydraulic fracture). An example consists in the fracking process, a well development technique used to extract gas and oil from shale rocks. When approaching hydraulic fracture two main issues arise adding further complexity. The former consists of the low-reliability of local theories of solid mechanics in predicting the behavior of materials when the crack formation is taken into account. The latter is caused by the rapid generation of new interfaces, solid–solid contact regions, and the effect of lubrication forces. It is then clear how an accurate prediction and understanding of the physics of hydraulic fracture processes is extremely challenging. In this context, the present work aims to propose a novel numerical approach to FSI problems focusing in particular on hydraulic fracture. The proposed methodology is based on the Navier-Stokes (NS) equations coupled with the Peridynamics theory (PD) through the Immersed Boundary Method (IBM). Peridynamics is a nonlocal continuum theory of solid mechanics

F. Dalla Barba (✉)
CISAS, University of Padova, Padua, Italy
e-mail: federico.dallabarba@phd.unipd.it

F. Picano
CISAS & DII, University of Padova, Padua, Italy
e-mail: francesco.picano@unipd.it

M. García-Villalba et al. (eds.), *Direct and Large Eddy Simulation XII*,
ERCOFTAC Series 27,
https://doi.org/10.1007/978-3-030-42822-8_62

based on integral equations [6]. The advantages introduced by peridynamics consist mainly of its reliable crack detection capability together with automatic tracking of crack propagation. Indeed, describing the problem by means of integral equations avoids issues due to the singularity of derivatives in partial differential equations. The force exchange between solid and fluid is accounted for by means of an IBM approach. The basic principle behind the IBM is that of imposing an additional forcing to the fluid, within a neighborhood of fluid–solid interface, such that the wall boundary conditions (BCs) are satisfied within a certain degree of accuracy. The proposed numerical methodology has been successfully implemented and preliminary quantitative validation benchmarks are provided together with a demonstrative test case.

Numerical Method

The incompressible formulation of the NS equations is used to describe fluid flows

$$\nabla \cdot \mathbf{u} = 0, \tag{62.1}$$

$$\rho_f \left(\frac{\partial \mathbf{u}}{\partial t} + \mathbf{u} \cdot \nabla \mathbf{u} \right) = -\nabla p + \mu_f \nabla^2 \mathbf{u} + \rho_f \mathbf{f}, \tag{62.2}$$

where $\mathbf{u}$ is the fluid velocity, ρ_f the fluid density, p the hydrodynamic pressure, μ_f the dynamic viscosity, and $\mathbf{f}$ the IBM forcing. This latter term is added to the NS equations to indirectly impose the correct no-slip and no-penetration conditions at the solid boundary (see Ref. [1] for more details about the procedure). Solid and fracture mechanics are addressed in the framework of the bond-based peridynamics. The basic idea behind this model is that a solid can be represented as a set of Lagrangian particles interacting among each other via short-range elastic potentials. These pairwise interactions (*bonds*) vanish beyond a threshold distance called *peridynamic horizon*, δ. Let $\mathbf{X}_0$ be the position of each material particle in the unloaded body configuration and $\mathbf{X}$ the position of the particle in the actual one at time t. For each material particle, the so-called *peridynamic family* can be defined as $\mathcal{H}_{\mathbf{X}_0} = \left\{\mathbf{X}_0' \in IR^3 \mid ||\mathbf{X}_0' - \mathbf{X}_0|| \leq \delta\right\}$. In this framework, the peridynamic equation of motion reads:

$$\rho_s \frac{d^2\mathbf{X}(\mathbf{X}_0, t)}{dt^2} = \int\limits_{\mathcal{H}_{\mathbf{X}_0}} \mathbf{F}\left(\mathbf{X}_0', \mathbf{X}_0, t\right) dV_{\mathbf{X}_0'} + \mathbf{B}(\mathbf{X}_0, t), \tag{62.3}$$

where ρ_s is the solid density, $\mathbf{F}\left(\mathbf{X}_0', \mathbf{X}_0, t\right)$ is referred to as *pairwise force density function* while $\mathbf{B}(\mathbf{X}_0, t)$ is the force per unit volume acting by the fluid on the solid particle. For each couple of material particles, the function $\mathbf{F}$ is proportional to the *bond stretch*, s:

$$\mathbf{F}\left(\mathbf{X}_0', \mathbf{X}_0, t\right) = \mathbf{F}(\boldsymbol{\xi}, \boldsymbol{\eta}) = c_0 \; \lambda(s) \; s \; \frac{\boldsymbol{\xi} + \boldsymbol{\eta}}{||\boldsymbol{\xi} + \boldsymbol{\eta}||}, \tag{62.4}$$

with s, $\boldsymbol{\xi}$ and $\boldsymbol{\eta}$ defined as:

$$s(\boldsymbol{\xi}, \boldsymbol{\eta}) = \frac{||\boldsymbol{\xi} + \boldsymbol{\eta}|| - ||\boldsymbol{\xi}||}{||\boldsymbol{\xi}||}, \tag{62.5}$$

$$\boldsymbol{\xi} = \mathbf{X}_0' - \mathbf{X}_0, \tag{62.6}$$

$$\boldsymbol{\eta} = (\mathbf{X}' - \mathbf{X}_0') - (\mathbf{X} - \mathbf{X}_0). \tag{62.7}$$

The parameter c_0 is called *bond micro-modulus* and its value depends on the macroscopic mechanical properties of the material, the Young's modulus, E and Poisson's ratio, ν_s. Crack formation is accounted for via the coefficient λ defined as

$$\lambda(s) = \begin{cases} 1, & s \leq s_0 \quad \forall t \geq 0, \\ 0, & otherwise, \end{cases} \tag{62.8}$$

where the limit bond stretch, s_0, is expressed as a function of the critical fracture energy release rate of the material, G_0. Crack formation occurs when the stretch of a bond overcomes the threshold value, s_0. Once the bond breaks the parameter λ switches from 1 to 0 such that the contribution of the bond to the global internal forces is set to zero. Further details concerning peridynamics and the values of model constants are provided in the references [4, 6]. The temporal integration is performed by an explicit third-order Runge-Kutta algorithm for both the NS and PD equations. The NS equations are discretized by second-order finite difference schemes on a staggered mesh. A multi-direct forcing scheme [1] is adopted to impose BCs on the solid–liquid interface. The force per unit volume, $\mathbf{B}$, is computed via the IBM forcing for each peridynamic particle: $\mathbf{B} = \rho_f \mathbf{f}$. It should be remarked that this approach allows simple and efficient computation of the back reaction of the fluid on the solid but it holds only for high density ratios, $\rho_s/\rho_f \gg 1$, or stationary conditions. The described methodology has been implemented into a massively parallelized MPI code. Peridynamics equations are solved via an in-house solver while the open-source fluid solver CaNS [2] is used for the fluid.

Results and Discussion

Three validation test cases are provided. The first benchmark considers a 2D rigid cylinder in a laminar channel flow. The domain extends for $H \times 6.5H$ in the wall-normal and flow direction, respectively. The computational grid uses 128×832 grid nodes along the same directions. The center of the cylinder lays on the center line of the channel at an axial distance from the inflow section $x_c/H = 1.5$. A Poiseuille inflow BC is prescribed at the inlet, convective outflow at the outlet and no-slip,

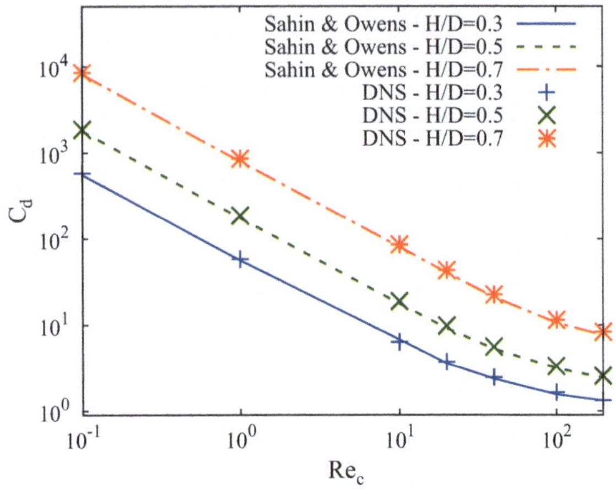

Fig. 62.1 Drag coefficient of a 2D cylinder in a laminar channel versus the Reynolds number, Re_c, for three different blockage ratios D/H. Dots refers to the present simulation while the lines refer to the dataset by Sahin and Owens [5]

no-penetration BCs at the lower and upper walls of the channel. Figure 62.1 provides the drag coefficient, $C_d = \frac{F_d}{\frac{1}{2}\rho_f U_{\max}^2 D}$, versus the Reynolds number, $Re_c = \rho_f U_{\max} D/\mu_f$, with F_d the drag force per unit depth, $U_{\max} = 1.5U_b$ the centerline velocity at inflow, U_b the bulk inflow velocity and D the cylinder diameter. Three different blockage ratios D/H are considered, letting the Reynolds number varying in the range 0.1–200. The present results are compared with the dataset by Sahin and Owens [5] showing the ability of the solver in dealing with immersed bodies.

The second benchmark concerns only the peridynamic solver: a 2D thin rectangular plate of length $L = 0.1$ m and height $H = 0.04$ m is subjected to a uniform tensile load, $\sigma_y = 12$ MPa, applied along the upper and lower horizontal edges. A cut of length 0.05 m is present along the half of the horizontal axis of symmetry of the plate. The material properties are set to that of *soda-lime glass*, $E = 72$ GPa, $\rho_s = 2440$ kg/m^3, $G_0 = 135$ J/m^2, and $\nu_s = 0.33$ for the Young's modulus, density, fracture energy release rate, and Poisson's ratio, respectively. The plate discretization is made of 16,000 Lagrangian particles. Figure 62.2 provides the local damage level at time $t = 25$ μs and $t = 40$ μs, damage level being a scalar index varying between 0 (no damage) and 1 (completely detached particle). A comparison between the present simulation (right side) and the results obtained by Dipasquale et al. [3] (left side) is also provided.

The third benchmark concerns a 2D elastic cylinder in a laminar channel flow. The domain extends for $H \times 6.5H$ in the wall-normal and flow direction, respectively, and is discretized by 128×832 grid nodes along the same directions. The cylinder center is fixed at $x_c/H = 0.7$ while the bulk Reynolds number is set to $Re_b = \rho_f U_b H/\mu_f = 19$. The same BCs used for the rigid cylinder test case are prescribed while the blockage ratio is set to $D/H = 0.7$. The solid discretization uses an equispaced distribution of 6,400 Lagrangian particles. Figure 62.3 provides a comparison between the longitudinal displacement field in the solid material predicted by the present solver and the results obtained using Ansys Fluent coupled with Ansys

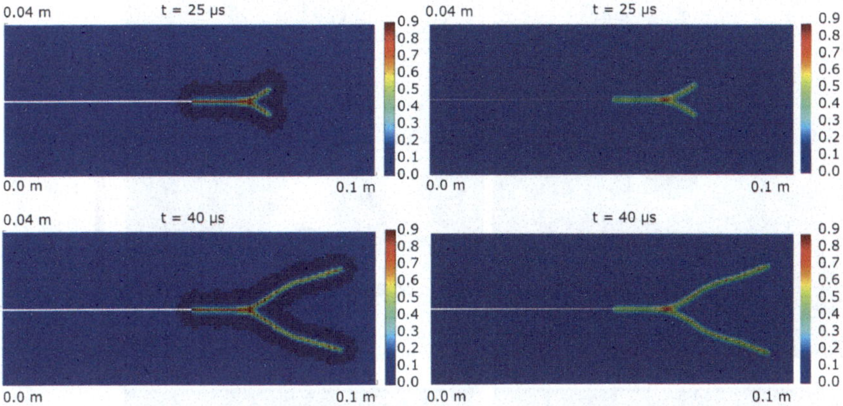

Fig. 62.2 Crack branching in a thin 2D rectangular plate. Comparison between present simulation (right) and results by Dipasquale et al. [3] (left) for the damage level computed at $t = 25$ μs and $t = 40$ μs

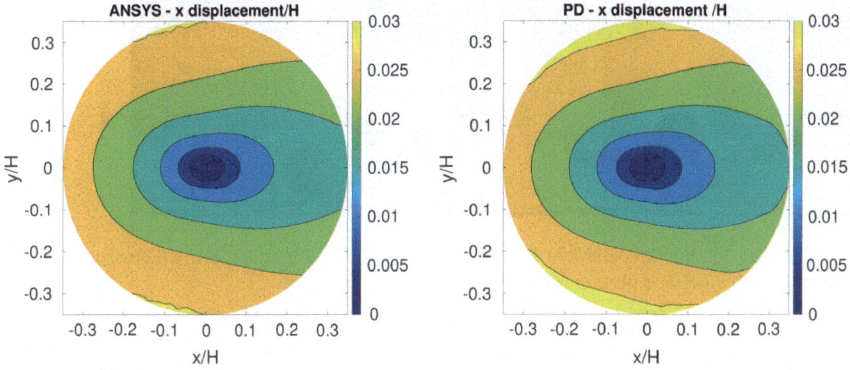

Fig. 62.3 Fully coupled FSI benchmark. Contour plot of the axial displacement in a 2D elastic cylinder in a laminar channel flow

Mechanical APDL solver. Finally, a qualitative test case of a fully coupled FSI problem with solid fragmentation is reported. The 2D simulation reproduces a cylinder of the solid elastic material of diameter $D = 0.015$ m falling in a gas-filled environment. The cylindrical body falls due to the gravitational acceleration, $g = -9.8$ m/s^2, and collides with an elastic plate of the same material of length $L = 0.06$ m and height $H = 0.004$ m, clamped at the two sides. The material parameters are set to $E = 10^8$ N/m^2, $\nu_s = 0.33$, $\rho_s = 1000$ kg/m^3, and $G_0 = 1$ J/m^2. The fluid is initially at rest. Kinematic viscosity and density of the fluid are $\nu_f = 10^{-5}$ m^2/s and $\rho_f = 1$ kg/m^3. The domain is $L_x \times L_y = 0.1 \times 0.1$ m large and is discretized with $N_x \times N_y = 240 \times 240$ nodes. Figure 62.4 provides the instantaneous velocity field of the fluid and the instantaneous damage level of the solid (Fig. 62.4).

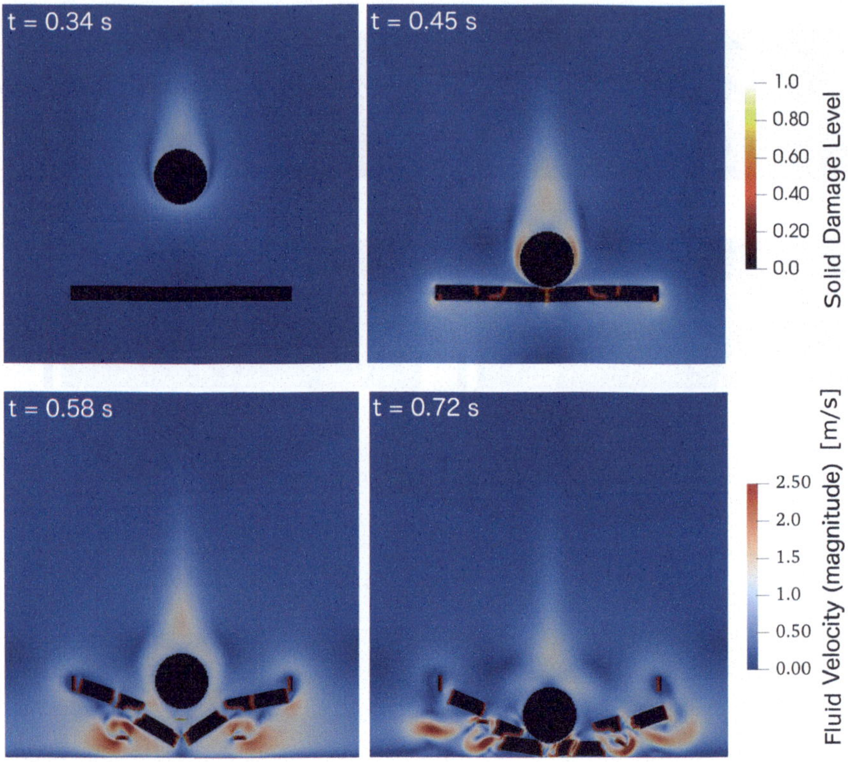

Fig. 62.4 Impact of a 2D cylinder of solid elastic material falling in a gas-filled environment and impacting with an elastic plate clamped at the opposite sides

Conclusions

A novel numerical methodology to address hydraulic fracture problems is proposed. The method is based on a NS, IBM, and peridynamics framework. An in-house MPI massive parallel solver for hydraulic fracture was developed. The novel approach takes advantage of the reliable capability of peridynamics to reproduce fracture mechanics. Moreover, the usage of IBM allows an efficient way to prescribe wall BCs across complex and time evolving fluid–solid interfaces. Some benchmarks are presented together with a qualitative test of a fully coupled FSI problem where solid fragmentation is accounted for. Even if we cannot provide quantitative validation for the results of this simulation, it appears that the method is capable to reproduce fracture of solid objects within a fluid, detecting crack formation and automatically reproducing crack branching.

Acknowledgements We would like to acknowledge the CINECA for providing high performance computing resources and support under the ISCRA-C grant (project HyPerNS HP10CG825K).

References

1. Breugem, W.P.: A second-order accurate immersed boundary method for fully resolved simulations of particle-laden flows. J. Comput. Phys. **231**(13), 4469–4498 (2012)
2. Costa, P.: A FFT-based finite-difference solver for massively-parallel direct numerical simulations of turbulent flows. Comput. Math. Appl. **76**(8), 1853–1862 (2018)
3. Dipasquale, D., Zaccariotto, M., Galvanetto, U.: Crack propagation with adaptive grid refinement in 2D peridynamics. Int. J. Fract. **190**(1–2), 1–22 (2014)
4. Le, Q.V., Bobaru, F.: Surface corrections for peridynamic models in elasticity and fracture. Comput. Mech. **61**(4), 1–20 (2018)
5. Sahin, M., Owens, R.G.: A numerical investigation of wall effects up to high blockage ratios on two-dimensional flow past a confined circular cylinder. Phys. Fluids **16**(5), 1305–1320 (2004)
6. Silling, S.A.: Reformulation of elasticity theory for discontinuities and long-range forces. J. Mech. Phys. Solids **48**(1), 175–209 (2000)

Chapter 63
Computational Modeling of Right Ventricle Flow Dynamics in Congenital Heart Disease

F. Capuano, Y. H. Loke, L. Olivieri and E. Balaras

Introduction

The evaluation of intracardiac blood flow dynamics is a powerful approach to improve the understanding and treatment of cardiovascular disease [1]. Most studies have focused on the left ventricle (LV), with important results of fundamental and clinical significance. In contrast, flow patterns within the right side of the heart received much less attention. Previous works of imaging [2] and computational [3] nature highlighted the complex, three-dimensional nature of the right ventricular flow, but additional studies are needed, especially in diseased conditions.

As shown in Fig. 63.1a, the pulmonary circuit starts as the deoxygenated blood carried from the inferior (IVC) and superior venae cavae (SVC) enters the right atrium (RA) and is pumped into the right ventricle (RV) through the tricuspid valve (TV). Upon RV contraction, the pulmonary valve (PV) opens and the blood volume is injected into the main pulmonary artery (MPA), which bifurcates into the left (LPA) and right (RPA) branches. After gas exchange takes place in the lungs, the oxygenated blood returns to the left atrium (LA) via the pulmonary veins.

F. Capuano (✉)
Department of Industrial Engineering, Università degli Studi di Napoli Federico II, Naples, Italy
e-mail: francesco.capuano@unina.it

Y. H. Loke · L. Olivieri
Division of Cardiology, Children's National Health System, Washington DC, USA
e-mail: yloke@childrensnational.org

L. Olivieri
e-mail: lolivieri@childrensnational.org

E. Balaras
Department of Mechanical and Aerospace Engineering, George Washington University, Washington DC, USA
e-mail: balaras@gwu.edu

M. García-Villalba et al. (eds.), *Direct and Large Eddy Simulation XII*, ERCOFTAC Series 27,
https://doi.org/10.1007/978-3-030-42822-8_63

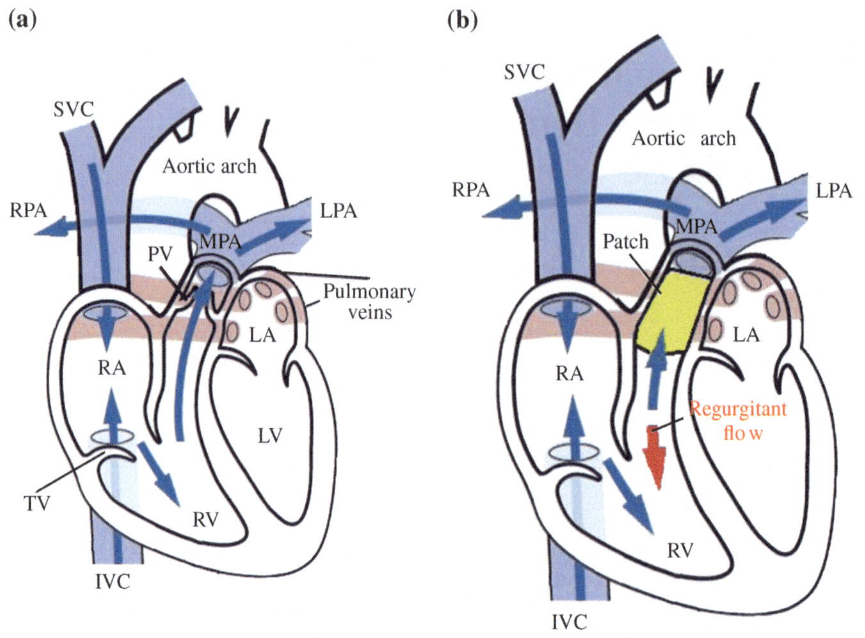

Fig. 63.1 **a** Schematic of the human heart with focus on the right side and paths of blood flow in normal conditions. **b** Pulmonary regurgitation after transannular patch repair of PV in Tetralogy of Fallot

This work is especially motivated by congenital heart disease (CHD), a form of the cardiac disorder that affects 8/1000 of the population [4]. After early life-saving interventions, many patients develop significant long-term complications with a currently unmet need to optimize health management. For instance, in Tetralogy of Fallot (ToF), the most common cyanotic CHD, part of the repair involves inserting a transannular patch across the PV to relieve pulmonary stenosis (one of the peculiar features of ToF). As a result, regurgitant flow occurs during the filling phase (see Fig. 63.1b), leading to progressive RV enlargement and dysfunction, and eventually requiring PV replacement (PVR). The timing of PVR is crucial since valve replacements have limited durability. However, a persistent knowledge gap remains on the best method to monitor ToF patients and determine the timing and benefits of PVR, as conventional clinical indices do not capture local biomechanical effects created from patient-specific cardiac anatomy [5]. Recent in-vivo studies based on time-resolved 3D imaging ("4D flow") has proposed vorticity and turbulent kinetic energy as potential biomarkers to guide the timing of PVR [6, 7].

A detailed quantification of local flow phenomena may help to understand the pathophysiology and to guide therapy planning. To this aim, we developed a complete pipeline from cardiac magnetic resonance (CMR) images to direct numerical simulation of the flow within the right ventricle and the pulmonary arteries. This work describes the components of this pipeline and provides preliminary results for the validation of the computational framework.

Modeling Approach

The proposed methodology is based on three steps, which are listed as follows:

1. Patient-specific geometries of the RV and the pulmonary arteries, as well as the RV wall dynamics are extracted from CMR images;
2. An image-registration algorithm is used to derive a space-and time-resolved representation of the three-dimensional model;
3. The moving geometry is immersed in a Cartesian domain wherein the flow equations are solved via an immersed boundary method.

The first step relies on standard multi-slice CMR datasets. In this work, clinical images were acquired at the cardiac MRI suite at Childrens National Medical Center (Washington DC, USA). They consist of several contours of the RV walls, extracted from slices perpendicular to the major axis of the ventricle (short-axis stack), as well as from additional longitudinal sections (3-and 4-chamber views), with a temporal resolution of 30 phases per cardiac cycle. Feature-tracking techniques available into the commercial software QStrain 2.0 (Medis) are used to automatically extract the time history of 48 evenly-spaced points along each endocardial contour. The data is further complemented by a fine three-dimensional triangulated surface mesh of the diastolic phase geometry obtained through a semi-automated segmentation process.

Starting from such input data, step 2 involves the reconstruction of a space-resolved, time-continuous representation of the moving walls to be used as a boundary condition for numerical simulations. This task is particularly challenging (compared to the corresponding effort required for the left ventricle) due to the complex shape and composite contraction pattern of the RV, the latter involving both in-plane and longitudinal motions. The approach undertaken in this work relies on an image-registration technique based on a control-point instance of the Large Deformation Diffeomorphic Geometric Mapping (LDDMM) implemented into the open-source software Deformetrica [8]. The basic idea is to *guide* the motion of the surface mesh by means of the endocardial contours trajectories. The latter is used to compute mappings (diffeomorphisms) that transform a given reference phase (e.g., diastole) into each of the 30 phases of the cardiac cycle. Of note, two different sets of transformations are computed: one for the in-plane motion, and one for the longitudinal motion. The final transformation is obtained by linear superposition of the two mappings. This novel procedure allowed proper incorporation of the longitudinal shortening of the RV, a parameter of great clinical and physiological relevance. The computed diffeomorphisms are eventually applied to the three-dimensional triangulated surface mesh, to yield 30 meshes with the same connectivity properties and with point correspondence. Full-time trajectories of each spatial point are obtained by trigonometric interpolation. This procedure provides the position and velocity of each point,

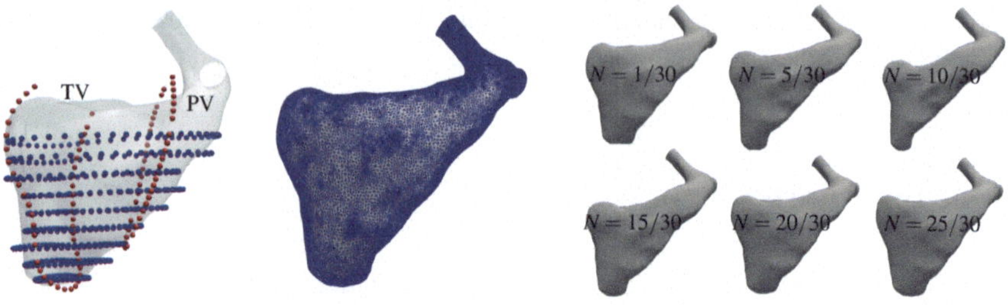

Fig. 63.2 Examplary data involved in the reconstruction of the moving geometry, here shown for a subject-specific healthy case: point cloud of the endocardial contours including short-axis (blue) and longitudinal (red) slices, triangulated surface of the diastolic phase geometry (middle), results of the deformation algorithm at different phases N of the cardiac cycle (right)

as well as the time history of the RV volume. Figure 63.2 shows exemplary data and results involved in steps 1 and 2.

Direct numerical simulations (step 3) are performed using an immersed boundary method (IBM) to solve the incompressible Navier-Stokes equations, assuming a Newtonian model for the blood. Spatial discretization relies on a staggered second-order scheme, while time-advancement employs a fractional step method coupled with an explicit third-order Runge-Kutta scheme. The specific IBM algorithm is the moving least-squares approach proposed in [9], which is especially suited for this kind of simulations. The TV and PV are represented by simple planes whose permeability is modulated depending on the phase of the cardiac cycle. The position of these planes is guided by clinical images. The inclusion of leaflet details will be considered for future work. Additional details for this step are provided in section "Results".

Results

In this section, we report preliminary results of a validation study conducted on a 13-year-old patient with ToF, who underwent surgical repair at 2 months of age. For this patient, in-vivo flow data obtained using 4D flow imaging were also available and are used here for the purpose of comparison with computational results.

Figure 63.3 shows a schematic of how the boundary conditions are imposed for this case. The position and velocity of the body walls (including the RV, the MPA, and portion of the pulmonary arteries) are assigned through the immersed boundary method as previously explained. The PV is practically absent in this case, and therefore, it is entirely excluded from the model; the TV is instead modeled as a plane that approximates the tricuspid annulus as detected from clinical images. The velocity on this plane is assigned on the basis of mass conservation: given the RV volume rate of change and the flow rate through the MPA (which was obtained in-vivo via phase-contrast magnetic resonance imaging), the TV flow rate is found as

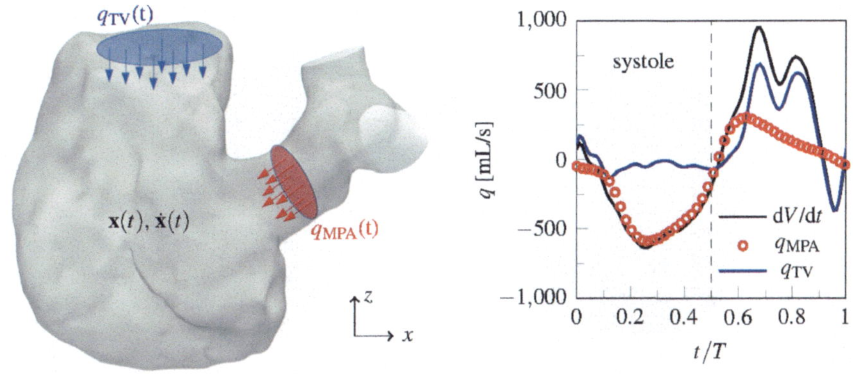

Fig. 63.3 Schematic showing the imposition of the boundary conditions on the body

$$q_{TV} = \frac{dV}{dt} - q_{MPA} , \tag{1}$$

where V is the RV volume. The flow rate q_{TV} is then uniformly distributed as a normal velocity on the plane. For this patient the pulmonary regurgitation due to the absence of the PV is quite strong, accounting for about 50% of the early diastolic filling. The end planes of the pulmonary arteries are left open to serve as outlets (or inlets). It is worth to note that the very good agreement between q_{MPA} and dV/dt during systole is an indirect validation of the kinematics reconstruction procedure.

The moving body is immersed into a parallelepiped whose sides are approximately 1.8 times larger than the RV characteristic sizes along each direction. The whole body volume occupies approximately 1/6 of the cube. Periodic boundary conditions are used along the directions x and y (refer to the sketch reported in the left portion of Fig. 63.3 for the reference frame), while a slip condition is employed along z. The grid is uniform with spacing $\Delta = 0.0625$ cm, leading to approximately 2 million cells inside the body during the diastolic phase. The time step is 1×10^{-4} s.

Results are reported in Fig. 63.4. Volume-rendered contours of the instantaneous velocity magnitude are displayed for two phases of the cardiac cycle (systole and diastole) and compared to the corresponding phase-averaged flow fields obtained in-vivo by 4D flow at the same phases. The agreement is overall satisfactory, taking into account the limitations of this comparison. In systole, the flow organization in the MPA is well captured, including a distinct recirculation area near the RPA. In diastole, the filling of the ventricle is characterized by a peculiar collision between two jets: a physiological one coming from the TV, and an abnormal one coming from the regurgitant flow, due to the absence of the PV. The flow field resulting from this collision is highly sensitive to various factors, including the relative magnitude and entrance timing of the two jets, which are affected by a certain degree of uncertainty coming from the clinical data. Work is currently underway to improve the comparison with in-vivo data by acting on these parameters. As compared to the normal filling pattern of the RV [2, 3], this abnormal diastolic re-organization is supposed to

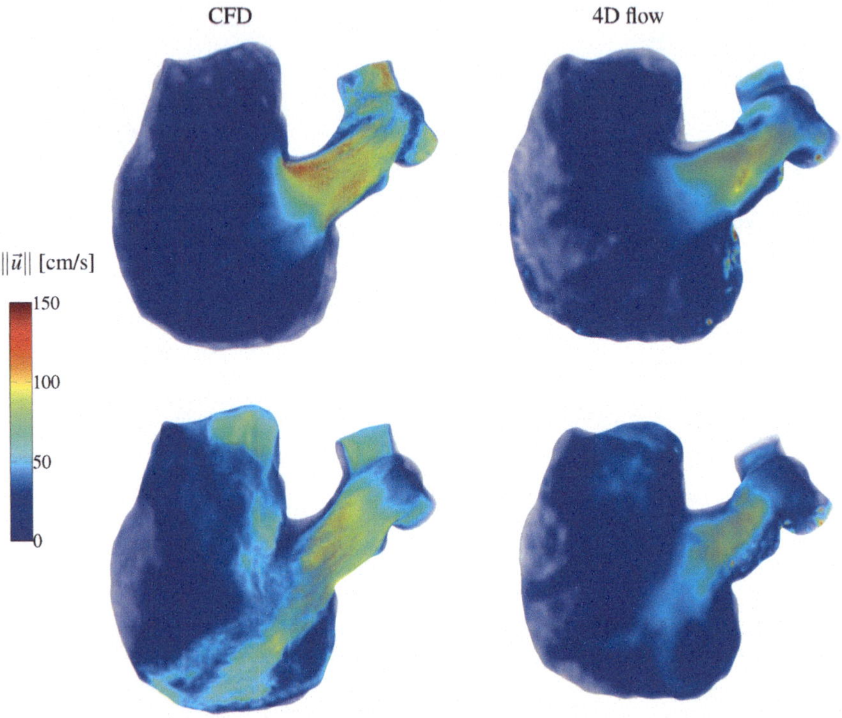

Fig. 63.4 Volume-rendered contours of the velocity magnitude at systole $t/T = 0.25$ (top row) and diastole $t/T = 0.75$ (bottom row), for CFD (left column) and 4D flow (right column)

significantly alter the RV pumping efficiency, and may act as a potential biomarker to guide PVR. Ultimately, CFD may also serve as a *virtual surgery* tool to simulate the effects of PVR.

Acknowledgements This work was funded by the Translation-Acceleration Pilot (TAP) Grant UL1TR001876 from the NIH National Center for Advancing Translational Sciences.

References

1. Bermejo, J., Martínez-Legazpi, P., del Álamo, J.C.: The clinical assessment of intraventricular flows. Annu. Rev. Fluid Mech. **47**, 315–342 (2015)
2. Fredriksson, A.G., Zajac, J., Eriksson, J., Dyverfeldt, P., Bolger, A.F., Ebbers, T., Carlhäll, C.J.: 4-D blood flow in the human right ventricle. Am. J. Physiol. Heart Circ. Physiol. **301**, H2344–H2350 (2011)
3. Mangual, J.O., Domenichini, F., Pedrizzetti, G.: Describing the highly three dimensional right ventricle flow. Ann. Biomed. Eng. **40**, 1790–1801 (2012)
4. Hoffman, J.I., Kaplan, S.: The incidence of congenital heart disease. J. Am. Coll. Cardiol. **39**, 1890–1900 (2002)

5. Geva, T.: Indications for pulmonary valve replacement in repaired tetralogy of fallot: the quest continues. Circulation **128**, 1855–1857 (2013)
6. Hirtler, D., Garcia, J., Barker, A.J., Geiger, J.: Assessment of intracardiac flow and vorticity in the right heart of patients after repair of tetralogy of Fallot by flow-sensitive 4D MRI. Eur. Radiol. **26**, 3598–3607 (2016)
7. Sjöberg, P., Bidhult, S., Bock, J., Heiberg, E., Arheden, H., Gustafsson, R., Nozohoor, S., Carlsson, M.: Disturbed left and right ventricular kinetic energy in patients with repaired tetralogy of Fallot: pathophysiological insights using 4D-flow MRI. Eur. Radiol. **28**, 4066–4076 (2018)
8. Bône, A., Louis, M., Martin, B., Durrleman, S.: Deformetrica 4: an open-source software for statistical shape analysis. In: Reuter, M., Wachinger, C., Lombaert, H., Paniagua, B., Lüthi, M., Egger, B. (eds.) International Workshop on Shape in Medical Imaging, pp. 3–13. Springer, Cham (2018)
9. Vanella, M., Balaras, E.: A moving-least-squares reconstruction for embedded-boundary formulations. J. Comput. Phys. **228**, 6617–6628 (2009)

Chapter 64
Evaluation of Blood Stasis in the Left Atrium Using Patient-Specific Direct Numerical Simulations

O. Flores, L. Rossini, A. Gonzalo, D. Vigneault, J. Bermejo, A. M. Kahn, E. McVeigh, M. García-Villalba and J. C. del Álamo

Introduction

Over the past two decades, cardiologists and engineers have increasingly worked together to describe and understand the blood flow in the left ventricle (LV) [1]. These studies have shown that there are remarkable differences in the blood flow of healthy and diseased LVs, a fact that helps in explaining their functional differences (ejection fraction, efficiency, etc.). These studies have also connected specific flow patterns and the stasis of the blood in the LV with the generation of clots (i.e., thrombogenesis), and the corresponding risk of stroke [2].

In recent years, a growing interest in performing similar analyses in the left atrium (LA) has arisen [3, 4]. One of the reasons for this interest is that patients with atrial fibrillation (AF, i.e., irregular beating of the LA) have a fivefold higher risk of stroke than healthy individuals. Increased blood stasis in the left atrial appendage (LAA) of AF patients promotes the formation of clots that may travel to the brain, causing stroke [5, 6]. Given the high prevalence of AF (35 million people worldwide) and

O. Flores (✉) · M. García-Villalba
Bioengineering and Aerospace Engineering Department,
Universidad Carlos III de Madrid, Leganés, Spain
e-mail: oflores@ing.uc3m.es

L. Rossini · A. Gonzalo · J. C. del Álamo
Mechanical and Aerospace Engineering Department,
University of California San Diego, San Diego, USA

D. Vigneault · E. McVeigh
Bioengineering Department, University of California San Diego, San Diego, USA

J. Bermejo
Hospital General Universitario Gregorio Marañón, Madrid, Spain

A. M. Kahn
Department of Cardiology, University of California San Diego, San Diego, USA

M. García-Villalba et al. (eds.), *Direct and Large Eddy Simulation XII*,
ERCOFTAC Series 27,
https://doi.org/10.1007/978-3-030-42822-8_64

its high incidence in the elderly population ($\sim$9% of people over 65 have AF), the evaluation of stroke risk in AF patients is of high clinical relevance.

Anticoagulation drugs reduce the risk of strokes in patients with AF, but these drugs also increase the risk of bleeding, including brain haemorrhage. Thus, anticoagulation therapy is only beneficial to patients for whom the risk of stroke outweighs the bleeding risk. The decision to apply anticoagulation therapies in AF patients is based on risk scores (like $CHAD_2S_2$-VASc), which only take into account demographical and clinical data of the patient, mainly ignoring the anatomy and function of each patient's heart (e.g., LA and LAA shapes, LA contractility, blood flow). Indeed, recent studies suggest that the performance of stroke risk scores improves by taking into account the morphological properties of the LA [7] or the LAA [8]. However, these approaches do not address the underlying pathophysiology—the increased blood stasis due to abnormal blood flow in the LA and LAA—directly.

The present work reports an efficient computational framework to evaluate blood flow and stasis in patient-specific anatomies, paying particular attention to the computation of blood residence time, and the imposition of appropriate boundary conditions at the pulmonary veins. Then, we illustrate and discuss some preliminary results obtained with this new computational framework.

Methodology

The computational framework developed in this work combines medical image techniques and Direct Numerical Simulations (DNS). Blood was assumed to be an incompressible, Newtonian fluid, with $\mu = 4 \times 10^{-3}$ Pa·s and $\rho = 1$ kg/m^3. The blood flow was computed with TUCAN [9, 10], an in-house solver for the Navier-Stokes equations of the incompressible flow. The flow solver uses a fractional step method, performing the time integration with a semi-implicit, 3-stage, low-storage Runge-Kutta. The spatial discretization implemented in TUCAN employs a staggered Cartesian mesh and 2nd-order centred finite differences. The motion of the LA walls was imposed using the Immersed Boundary method (IBM) of Uhlmann [11], specifying at each time step the position and velocity of the Lagrangian points that discretize the LA wall. The motion of these Lagrangian points was obtained from medical imaging, with the segmentation and registration processes described below.

The medical images were obtained from Computed Tomography (CT) scans. For the cases presented in section "Results", these were acquired using a GE Medical Systems Revolution CT scan, with a spatial resolution of 0.32–0.48 mm in the x–y plane and 0.5–1 mm in the z-direction. The temporal resolution of the CT scan was 20 frames per cardiac cycle. The segmentation of the CT images was performed using the semi-automatic active contour segmentation method in the itk-SNAP open source software. Additional manual input was needed to crop the pulmonary veins (PVs) and locate the mitral annulus. A surface description consisting of triangular faces and vertices was extracted from the volumetric data. Then, this surface was automatically decimated and smoothed in Matlab to match the spatial resolution requirements of the DNS. The open-source Medical Imaging Registration Toolbox for Matlab was used to discretize the surface in time, and a Fourier interpolation with 6 modes was used to increase the temporal resolution as required for time integration of the Navier-Stokes equations. The segmentation and registration process also provided the time histories of the volumes in the left atrium (V_{LA}) and the left ventricle (V_{LV}).

The flows through the PV inlets, which serve as boundary conditions to the simulation, were obtained from applying conservation of mass to $V_{LV}(t)$ and $V_{LA}(t)$. The flow rate through each pulmonary vein was estimated by evenly splitting the total flow rate through the pulmonary veins. Preliminary simulations showed that the results obtained in this way are not significantly different from those obtained by assuming that the averaged velocity in all pulmonary veins is the same.

To impose the inlet boundary conditions at the PVs, linear forcing was used in cylindrical buffer regions extruded from the PV inlet planes. Following a procedure analogous to the Immersed Boundary method, the Navier-Stokes equations in the buffer region were replaced by

$$\frac{\partial \mathbf{v}}{\partial t} = (\mathbf{v} - \mathbf{v}_T)/t_c, \tag{64.1}$$

where $\mathbf{v}$ is the blood velocity and $\mathbf{v}_T$ is the prescribed velocity at the PV inlets. The relaxation time t_c was set to $10\Delta t$, where Δt is the time step. It was found that this value of T_c was enough to prevent the generation of strong oscillations in the velocities upstream of the buffer region, at the cost of incurring in a small relative error between $\mathbf{v}_T$ and the actual velocity at the PV inlets (i.e, 5–10%, depending on the case and the time instant in the cardiac cycle).

Following Rossini et al. [12], the residence time of the blood in the LA was evaluated with the transport equation of a passive scalar,

$$\frac{\partial T_R}{\partial t} + \mathbf{v} \cdot \nabla T_R = 1, \tag{64.2}$$

where T_R is the residence time. T_R was set to zero at the PV inlets, so that the value of $T_R(\mathbf{x}, t)$ represents the time elapsed since the fluid on $(\mathbf{x}, t)$ entered the LA. The discretization of Eq. (64.2) was performed using a WENO scheme [13] and the same Runge-Kutta scheme used for the blood velocity. The boundary conditions at the PV inlets were imposed with a linear forcing analogous to that described earlier for the velocities.

Results

In this section, we present preliminary results for two subjects enrolled from a database of volunteers. The main physiological parameters of the LA of these two subjects are shown in Table 64.1. Case H corresponds to a subject with normal

Table 64.1 Cases and main physiological parameters, including the maximum LV and LA volumes, maximum and minimum LAA volumes, and the ejection fractions (EF) of LA and LV

	$V_{LV,\max}$ (ml)	$V_{LV,\min}$ (ml)	EF_{LV} (%)	$V_{LA,\max}$ (ml)	$V_{LA,\min}$ (ml)	EF_{LA} (%)	$V_{LAA,\max}$ (ml)	$V_{LAA,\min}$ (ml)
H	147	49	67	145	87	40	17.9	10.2
D	181	105	42	165	150	9	17.4	13.8

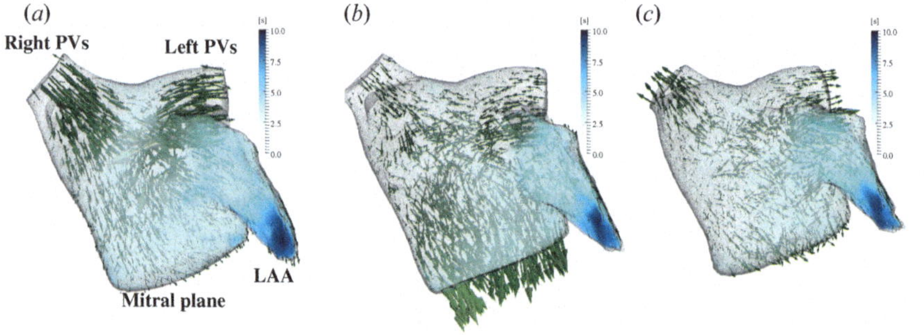

Fig. 64.1 Flow velocity (green) and residence time (blue) for case H. **a** During the atrial diastole ($t = 0.16$ s), when the inflow through the PV is maximum and the mitral annulus is closed. **b** During the early filling of the LV ($t = 0.44$ s), when the flow through the mitral annulus is maximum. **c** During the atrial systole ($t = 0.83$ s).

values of the LA and LV volumes and ejection fractions. On the other hand, case D corresponds to a subject with a dilated LA and impaired atrial function: the ejection fraction of the LA is only 9%, and the volume of the LAA changes little over the cardiac cycle. Indeed, the CT images of case D show a thrombus at the LAA, which was digitally removed prior to performing the simulations.

For both subjects, the simulations were run with a cardiac period equal to 1 s (i.e., 60 b.p.m). The simulations were run for 10 cardiac cycles with low resolution ($\Delta = 0.090$ cm, $\Delta t = 10^{-4}$ s), until the flow is developed. Then, another 10 cycles were run at nominal resolution ($\Delta = 0.051$ cm, $\Delta t = 5 \times 10^{-5}$ s), always keeping a $CFL \leq 0.1$.

In order to evaluate the differences in the blood flow of cases H and D, Figs. 64.1 and 64.2 show $\mathbf{v}$ and T_R at three representative time instants of the cardiac cycle. During the atrial diastole of case H, Fig. 64.1a shows strong jets at the PV while the LA expands. For case D (Fig. 64.2a), the LA does not expand and as a consequence, there is little flow through the PVs. During the early filling of the LV, the LA of case H contracts, producing a larger flow through the mitral annulus than the inflow through the PVs (Fig. 64.1b). For case D, the reduced contractility of the LA results in very strong jets at the PV to produce a similar flow through the mitral annulus (Fig. 64.2b). Finally, during the atrial systole, Fig. 64.1c shows blood flow exiting the LA trough the mitral valve and the PV (backflow) for case H. However, the limited contractility of case D results in virtually no flow in Fig. 64.2c.

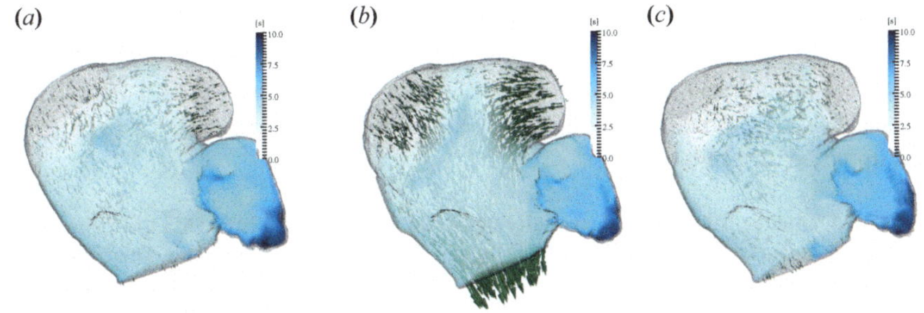

Fig. 64.2 Flow velocity (green) and residence time (blue) for case D. **a** During the atrial diastole ($t = 0.09$ s), when the inflow through the PV is maximum and the mitral annulus is closed. **b** During the early filling of the LV ($t = 0.52$ s), when the flow through the mitral annulus is maximum. **c** During the atrial systole ($t = 0.86$ s).

Focusing on the flow in the LAA, we can observe that the velocities in the LAA for case H are larger than in case D. In particular, the contractility of the LAA in case H results in blood flowing into the LAA during the atrial diastole (Fig. 64.1a), and in blood flowing back into the LA body during the early filling of the LV (Fig. 64.1b) and during the atrial systole (Fig. 64.1c). These blood flows are not observed in case D in Fig. 64.2. The differences between cases H and D are consistent with the maximum and minimum volumes of the LAA reported in Table 64.1. As a consequence of these differences in the flow patterns, the values of T_R in the LAA are larger for case D than for case H, as it can be observed in Figs. 64.1 and 64.2. Indeed, the average residence time (over space and time) in the LAA of case H is $T_R \approx 2$ s, while it increases to $T_R \approx 4$ s for case D.

The results presented here suggest that medical imaging and CFD can be used to perform a personalised analysis of the blood flow in patients, and that T_R shows potential as a surrogate for the risk of thrombogenesis.

Conclusions

This paper describes a computational framework that combines medical imaging and computational fluid dynamics to perform a patient-specific evaluation of the blood flow in the LA. This computational framework includes the segmentation of the CT scan images, and the registration and discretization of the resulting surfaces to provide a moving LA wall for the flow solver. We use an immersed boundary formulation to model the motion of the LA walls, on a standard Cartesian solver for the Navier-Stokes equations. The salient features of the flow solver are the boundary conditions at the PV inlets, and the advection equation for the residence time.

We have provided proof of principle of the computational framework by preliminary analysis of the blood flow and stasis for two subjects, one with normal LA function and another with impaired LA function. Our results suggest that the normal blood flow patterns of the left atrium are markedly altered in patients with impaired atrial function, both in the LA body and the LAA. Our preliminary data also indicates that these alterations result in significantly higher blood stasis in the LAA in patients with impaired atrial function.

Acknowledgements This work was partially supported by the Comunidad de Madrid (Sinergias), Cátedra Excelencia UC3M-Santander, MECD (Spain), NIH UC-CAI grant, and the American Heart Association. Computational time provided by XSEDE (Comet) and RES (Altamira) is gratefully acknowledged.

References

1. Bermejo, J., Martínez-Legazpi, P., del Álamo, J.C.: The clinical assessment of intraventricular flows. Annu. Rev. Fluid Mech. **47**, 315–342 (2015)
2. Martínez-Legazpi, P., et al.: Stasis mapping using ultrasound: a prospective study in acute myocardial infarction. JACC: Cardiovasc. Imag. **2360** (2017)
3. Koizumi, R., et al.: Numerical analysis of hemodynamic changes in the left atrium due to atrial fibrillation. J. Biomech. **48**, 472–478 (2015)
4. Masci, A., et al.: The impact of left atrium appendage morphology on stroke risk assessment in atrial fibrillation: a computational fluid dynamics study. Frontiers Physiol. **9** (2018)
5. Al-Saady, M.N., Obel, O.A., Camm, A.J.: Left atrial appendage: structure, function, and role in thromboembolism. Heart **82**(5), 547–554 (1999)
6. Fogelson, A.L., Neeves, K.B.: Fluid mechanics of blood clot formation. Annu. Rev. Fluid Mech. **47**, 377–403 (2015)
7. Bisbal, F., et al.: Left Atrial Geometry Improves Risk Prediction of Thromboembolic Events in Patients With Atrial Fibrillation. J. Cardiovasc. Electrophys. **27**(7), 804–810 (2016)
8. di Biase, L., et al.: Does the Left Atrial Appendage Morphology Correlate With the Risk of Stroke in Patients With Atrial Fibrillation? J. Am. Coll. Cardio. **60**(6), 531–538 (2012)
9. Moriche, M., Flores, O., García-Villalba, M.: On the aerodynamic forces on heaving and pitching airfoils at low Reynolds number. J. Fluid Mech. **828**, 395–423 (2017)
10. Gonzalo, A., et al.: From flapping to heaving: A numerical study of wings in forward flight. J. Fluids Struct. **83**, 293–309 (2018)
11. Uhlmann, M.: An immersed boundary method with direct forcing for the simulation of particulate flows. J. Comput. Phys. **209**(2), 448–476 (2005)
12. Rossini, L., et al.: A clinical method for mapping and quantifying blood stasis in the left ventricle. J. Biomech. **49**(11), 2152–2161 (2016)
13. Shu, C.W.: High-order finite difference and finite volume WENO schemes and discontinuous Galerkin methods for CFD. Int. J. Comp. Fluid Dyn. **17**(2), 107–118 (2003)

Chapter 65
On Shallow Mixing Interfaces and Their Relevance for Understanding Mixing at River Confluences

G. Constantinescu

Abstract In many environmental fluid mechanics applications, the spatial development of mixing layers is significantly affected by the friction at the channel bottom. This motivates the present study of mixing interfaces developing under shallow flow conditions between parallel and nonparallel streams with a velocity ratio larger/smaller than one (mixing layer mode) and with a velocity ratio close to one (wake mode). Such mixing interfaces are very important to understand flow and mixing at river confluences where shallow conditions are generally observed. Given that at most river confluences the incoming streams have different temperatures and suspended sediment loads, the paper also discusses stratification induced by density differences between the incoming streams. For sufficiently small densimetric Froude numbers, a spatially developing lock-exchange-like flow develops away from the confluence apex. In such cases, mixing at large distance from the confluence apex is controlled by the lock-exchange-like flow rather than the vertically oriented, mixing interface vortices.

Keywords Mixing layers · Bed friction effects · Shallow flows

Introduction

Confluent flows and the associated mixing interface (MI) region occur in many engineering applications including in geophysical systems (e.g., river networks), mechanical systems (e.g., networks of ducts or pipes), and biological systems (e.g., arterial networks). The main differences between confluent flows in river networks and the other types of confluent flows are the flow regime (open channel vs. pressurized flow) and the length scale (flow depths of up to hundreds of meters in large rivers vs. diameters of arteries of couple of millimeters).

River confluences are main components of natural drainage networks [10]. They regulate the movement of sediments through braided river systems. Moreover, the

G. Constantinescu (✉)
University of Iowa, Iowa City, IA 52242, USA
e-mail: sconstan@engineering.uiowa.edu

M. García-Villalba et al. (eds.), *Direct and Large Eddy Simulation XII*,
ERCOFTAC Series 27,
https://doi.org/10.1007/978-3-030-42822-8_65

dynamics of mixing controls how the tributary inputs of organic matter, nutrients, and contaminants are dispersed in the river. This is why there is lots of interest in understanding mixing mechanisms and quantitatively predicting the distance from the confluence apex over which the two streams will mix. The main flow and geometrical parameters affecting flow hydrodynamics and mixing are the velocity ratio of the two incoming streams, the angles between the two incoming streams and the main channel and the density ratio of the two incoming streams [2, 4, 5]. The presence of high channel curvature close to the confluence apex and/or of a large discordance between the bed levels in the two incoming channels can also greatly affect mixing [2, 4]. However, these two effects will not be discussed in the present paper.

Once the two streams come into contact, they start mixing along the MI. Large-scale coherent structures form inside and around the MI [4, 10]. These structures control to a large degree how fast the two streams are mixing. In the case of river confluences, flow conditions are fairly shallow. This means that the dynamics of the coherent structures inside the MI is significantly influenced by bed friction at large distances from the confluence apex. The MI is characterized by the formation of vertically oriented vortices. As discussed by Constantinescu et al. [6], if the velocity ratio between the mean streamwise velocities in the two streams is much larger, or much smaller than one, then the MI will contain co-rotating vortices forming because of the mean horizontal shear acting on the MI. The MI is a shallow mixing layer with vortex pairing occurring at least at relatively small distances from the confluence apex. These MIs are referred as MIs in the KH mode. However, if the velocity ratio is fairly close to one, the MI contains counter-rotating vortices forming because of the interactions between the separated shear layers on the two sides of the confluence apex. The mechanism for the formation of these vortices is similar to the one responsible for the generation of the von Karman street in the wake of bluff bodies. Such MIs are in the wake mode. The two types of MIs are illustrated in Fig. 65.1.

Additional large-scale coherent structures may form in the vicinity of the MI in the case the angle between the two streams is relatively large [4]. As the two streams come into contact, they have to lose their transverse momentum with respect to the orientation of the MI. The potential energy increases and the free surface elevation is higher inside this part of the MI. This generates a secondary flow from the free surface toward the channel bottom. This flow is then diverted away from the MI and from these back toward the free surface. This results in the formation of one or multiple streamwise-oriented vortical (SOV) cells beneath the free surface.

Finally, if the density difference between the fluids in the two streams is relatively large (e.g., large values of the Richardson number, Ri, defined with the mean velocity in the main channel, mean channel flow depth and the density difference between the two fluids or, equivalently, low values of the densimetric Froude number $Fr = 1/\mathrm{Ri}^2$), then stratification effects can have a significant effect on the structure of the MI and on mixing, especially at large distances from the confluence apex [2]. This is because a vertical density interface is present near the confluence apex where the two streams of unequal densities come into contact. The result is the development

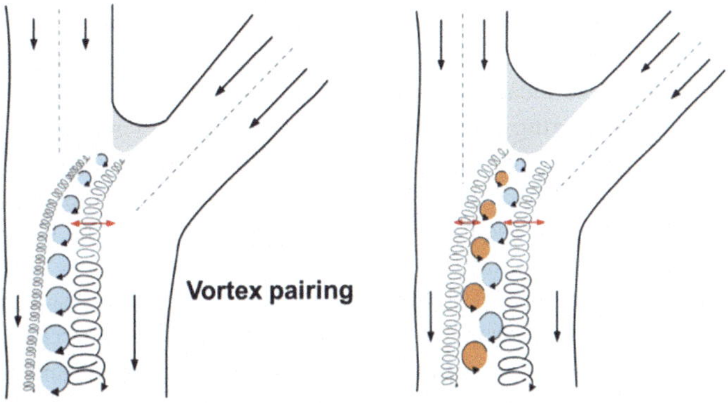

Fig. 65.1 Sketch showing coherent structures forming inside and around mixing interfaces at the confluence of two streams. In the left frame, the MI is in the Kelvin–Helmholtz (KH) mode and contains only co-rotating vertically oriented vortices. In the right frame, the MI is the wake mode and contains counter-rotating, vertically oriented vortices. If the angle between the two streams is sufficiently large, streamwise-oriented vortical (SOV) cells form beneath the free surface close to the MI

of a lock-exchange-like flow in spanwise planes that grows with increasing distance from the confluence apex [2, 4].

As such, a first goal of the present paper is to describe the structure and spatial development of MIs in shallow channels between parallel and nonparallel streams of constant depth and the dynamics of the other large-scale coherent structures playing an important role in mixing. A second goal, is to investigate mixing at natural river confluences with concordant beds and to determine to what extent the physics of MIs in simplified environments remains the same in natural environments with highly deformed bathymetry.

The present paper uses a numerical approach based on eddy-resolving numerical simulations performed on sufficiently fine meshes (up to 50 million grid cells) to resolve the dynamics of the relevant coherent structures and to avoid the use of wall functions. Details on the numerical code, Detached Eddy Simulation (DES) model, and boundary conditions are given in [1, 4–7]. The free surface is treated as a free-shear rigid lid while the instantaneous velocity flow fields fed at the two inlet boundaries are obtained from preliminary DES solutions of fully developed flow in a channel.

Mixing Interfaces in Simplified Geometries

In contrast to deep mixing layers that are characterized by a linear growth, the width of shallow mixing layers varies in a nonlinear way with the distance from the apex once bed friction effects become important. The main parameter describing the spatial development of a shallow mixing layer is the bed friction number, $S = \frac{\bar{c}_f \delta(x)}{2D} \frac{U(x)}{\Delta U(x)}$

where D is the flow depth, c_f is the mean bed friction coefficient of the two incoming streams calculated using the mean bed shear stress and the streamwise velocity in each stream, $U(x)$ is the mean velocity of the two streams, $\Delta U(x)$ is the velocity difference across the mixing layer, and δ is the mixing layer width [3, 11]. This variable can be seen as the ratio between the stabilizing effect of bottom friction and the destabilizing effect of mean shear across the mixing layer. Using experimental results and stability theory, one can show that for $S > 0.08$–0.1, the growth of the instabilities inside the mixing layer is affected by bottom friction. As this happens, vortex pairing ceases and the KH billows start losing their coherence. This is why the rate of increase of a shallow mixing layer is less than that of a deep mixing layer. Eventually, as S approaches zero, the shallow mixing layer reaches an equilibrium regime where the width of the mixing layer becomes constant (zero-growth). The same is true for the shift of the mixing layer centerline. The entrainment coefficient $\alpha = \frac{1}{\Delta U}\frac{\mathrm{d}\delta}{\mathrm{d}t} \sim \frac{U(x)}{\Delta U(x)}\frac{\mathrm{d}\delta(x)}{\mathrm{d}x}$ describes the rate of increase of the mixing layer width in nondimensional form. For a deep mixing layer, $\alpha \sim 0.1$, while for a shallow mixing layer α decreases with increasing S.

Numerical simulations were conducted for a shallow mixing layer developing in a wide channel with both a flat horizontal bed and with 2D dunes at the channel bottom to investigate the changes in the evolution of the mixing layer with increasing shallowness (Fig. 65.2). The mean Reynolds number was 15,000 and the velocity ratio of the incoming streams was $VR = 2.2$ ($D = 0.067$ m). Figure 65.3 uses a passive scalar ($C = 100$) introduced at the confluence apex to visualize the structure of the mixing layer. In the flatbed case, KH billows are generated inside the upstream part

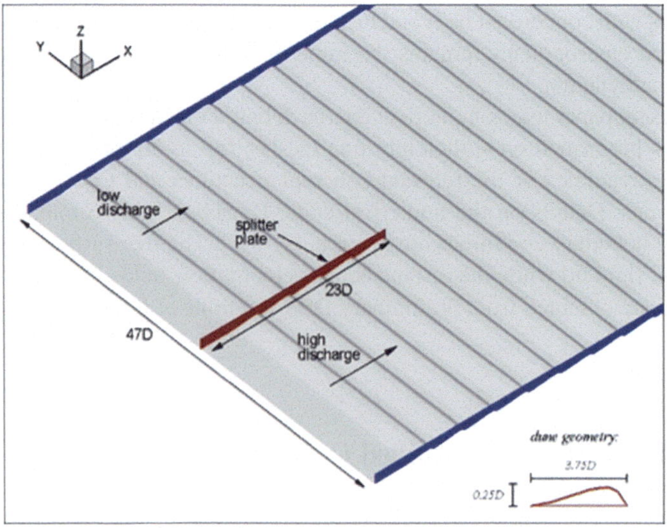

Fig. 65.2 Sketch showing the channel configuration used in the simulations of shallow mixing layers forming between parallel streams. From [9], reproduced with permission from the International Association of Hydraulic Research

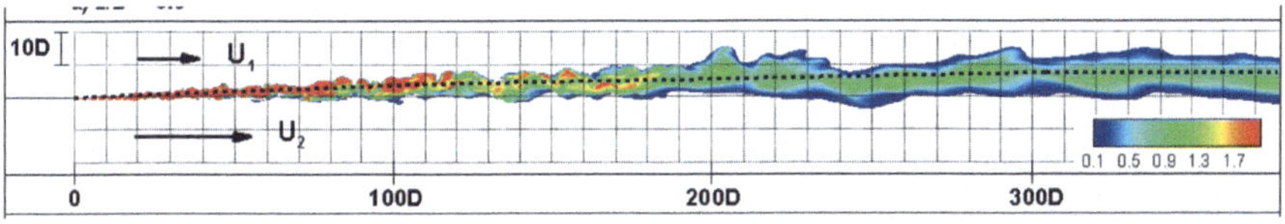

Fig. 65.3 Visualization of coherent structures forming inside the mixing layer for the case of a flat bed channel. The scalar concentration is shown in a plane ($z/D = 0.9$) situated close to the free surface plane

of the mixing layer and vortex pairing is observed until about $x = 100D$ where the mean diameter of these vortices is about $13D$ [8]. A clear decrease in the coherence of these vortices is observed for $x > 200D$. Moreover, the shape of the mixing layer becomes undulatory (Fig. 65.4).

In both simulations, the centerline of the mixing layer shifts toward the low-speed side. The velocity equalization proceeds on a length scale of the order of $D/2c_f$. This is why the mixing layer reaches faster the equilibrium regime in the simulations in which dunes are placed at the channel bed. The shift is larger for the shallower case but the mixing layer width at equilibrium is smaller compared to the deeper case. The plot of α versus S for the shallower case in which the equilibrium regime is reached around $x = 350D$ shows that the linear decay regime proposed in [3] is valid only until $S \sim 0.05$. Once the mixing layer assumes an undulatory shape, the decay of α with S is less than linear.

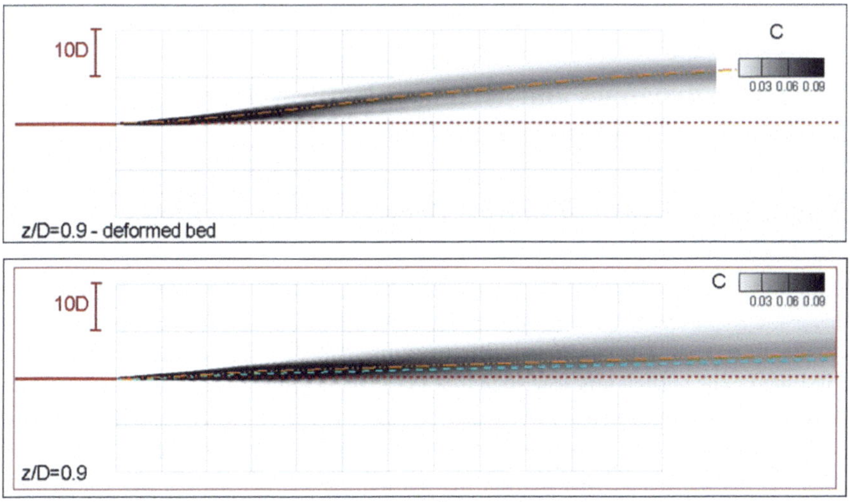

Fig. 65.4 Visualization of shallow mixing layer in the mean flow. The mean (time-averaged) scalar concentration is shown in a plane ($z/D = 0.9$) situated close to the free surface plane. The dash-dot orange line visualizes the mixing layer centerline based on the mean streamwise velocity profiles. From [8, 9], reproduced with permission from the International Association of Hydraulic Research

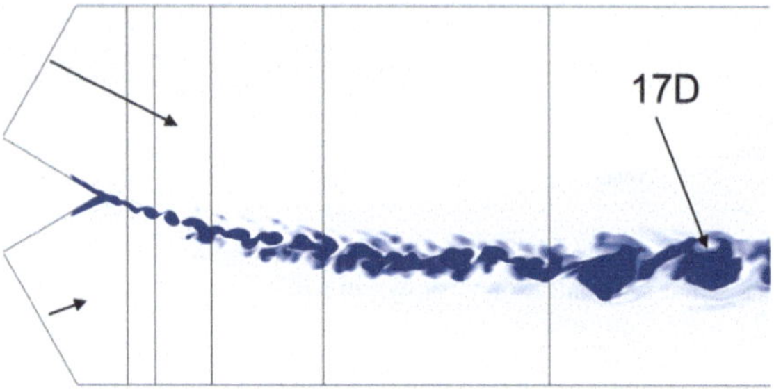

Fig. 65.5 Visualization of coherent structures forming inside the shallow mixing layer for the case the confluence angle is 60° and the channel bed is flat ($VR = 2.2$). A passive scalar with $C = 100$ is introduced at the confluence apex. The scalar concentration is shown in a plane ($z/D = 0.9$) situated close to the free surface plane. Reproduced from Constantinescu [4]

Additional simulations were performed for symmetrical and asymmetrical confluences with an angle of 60° between the two channels. The structure of the mixing layer was found to be similar to that observed in the corresponding zero angle cases. However, the KH billows are more coherent and attain a larger size (e.g., diameter of $17D$ for the $VR = 2.2$ case) before the start of the transition to the equilibrium regime. The mixing layer also assumed an undulatory shape starting in the later stages of the transition regime ($x > 250D$ for the $VR = 2.2$ case illustrated in Fig. 65.5). Moreover, strongly coherent SOV cells formed on both sides of the mixing layer (see sketch for the KH mixing later mode in Fig. 65.1).

Flow Hydrodynamics and Mixing at Natural River Confluences with a Concordant Bed

Cases with no-Density-Contrast Between Incoming Flows

Simulations were performed for the asymmetrical (KRCS) confluence between the Kaskaskia River and Copper Slough in Illinois that made the subject of several field investigations [10]. The confluence angle is close to 60°. The temperature difference is assumed to be sufficiently low such that $Fr > 10$, meaning that stratification effects on flow hydrodynamics and mixing can be neglected. The structure of the MI near the free surface in such cases is visualized in Fig. 65.6 for two cases with $VR \sim 1$ and $VR = 5.5$. The Reynolds number defined with the mean velocity of the two streams, U, and the mean depth of the channel downstream of the apex, D, was about 100,000–300,000. As expected, the MI is in the wake mode in the $VR \sim 1$ case as it contains eddies with opposite sense of rotation (Fig. 65.6). A wake region is clearly seen next to the confluence apex. By contrast, the MI contains primarily co-rotating eddies in the $VR = 5.5$ simulation, indicating the KH mode dominates (Fig. 65.6).

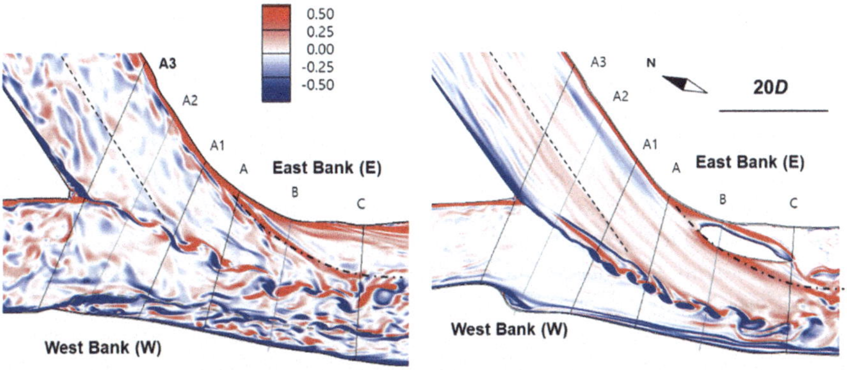

Fig. 65.6 Visualization of the MI at the KRCS confluence using the vertical vorticity, $\omega_z D/U$, in the instantaneous flow at the free surface. The left frame shows results for the $VR \sim 1$ case in which the MI is in the wake mode. The right frame shows results for the $VR = 5.5$ case in which the MI is in the Kelvin–Helmholtz mode. From Constantinescu et al. [6]. Copyright 2012 American Geophysical Union. Reproduced by permission of American Geophysical Union

The other main feature of the vortical flow fields is the presence of SOV cells on both sides of the MI. These cells are visualized in Fig. 65.7 for the $VR \sim 1$ case. In the $VR = 5.5$ case, the coherence of the SOV cells is larger on the high momentum side that also makes a larger angle with the MI. The strength (e.g., circulation) of the SOV cells is proportional to that of the adverse pressure gradients generated away from the MI [4–6].

Constantinescu et al. [5, 6] have shown that the largest bed shear stresses inside the main channel are induced beneath the SOV cells. Moreover, these cells strongly

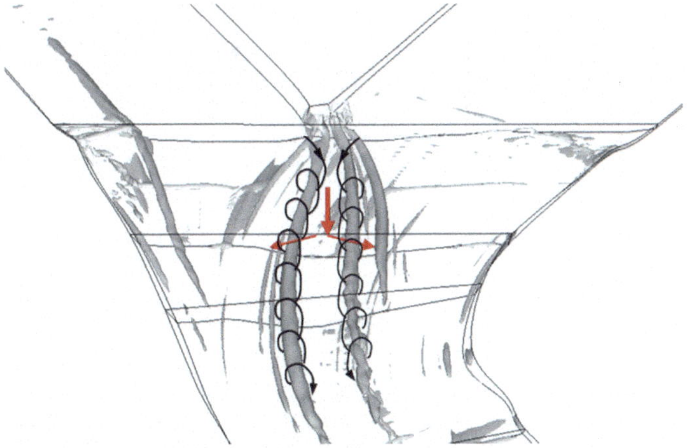

Fig. 65.7 Visualization of the SOV cells in the mean flow using the Q criterion for the KRCS confluence with $VR \sim 1$. The vertical red arrow points toward the MI

enhance mixing. The SOV cells can extract fluid from the MI when they interact with the MI vertically oriented eddies. The reason why the SOV cells are so efficient in inducing large bed shear stress values and also mixing is that their cores can be subject to large-scale bimodal oscillations toward and away from the MI. As a result a large amplification of the turbulent kinetic energy is observed inside the region where the core of the SOV oscillates, as observed in Fig. 65.8 for the *VR* ~ 1 case on the higher momentum side. The region where bimodal oscillations are present is generally limited to the one where the two streams collapse and the largest adverse pressure gradients are induced with respect to one or both sides of the MI [4–6]. The bimodal oscillations can be detected by plotting the velocity histograms at points situated inside the core of the SOV in the mean flow (e.g., see Fig. 65.9 for the SVI1 cell forming on the high momentum side). If bimodal oscillations are present, the velocity histograms display a two-peak shape corresponding to the state when the SOV cell is interacting with the MI eddies (MI mode) and to the state when the SOV cell moves toward the bank (bank mode). The low-frequency, sweeping motions between the two modes are the main reason for the large efficiency of some of the

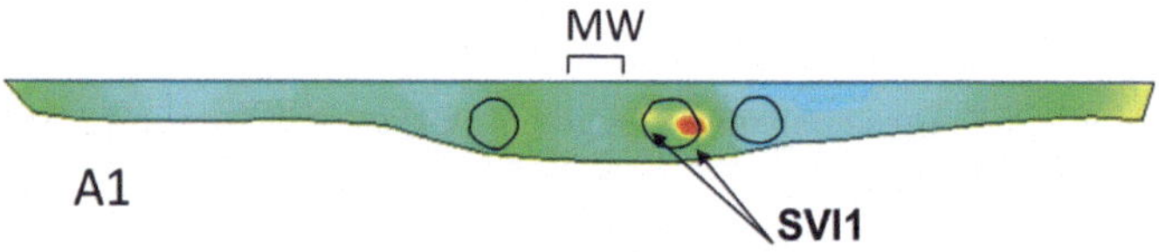

Fig. 65.8 Distribution of the mean pressure fluctuations at section A1 of the KRCS confluence for the *VR* ~ 1 case. Also shown are the positions of the main SOV cells in the mean flow. SVI1 is the most coherent SOV cell

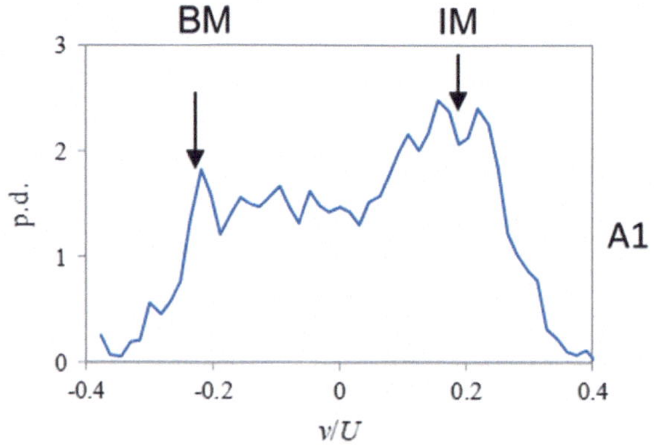

Fig. 65.9 Vertical velocity histogram at a point located inside the core of the main SOV cell at section A1 (SVI1m *VR* ~ 1 case, KRCS confluence) showing a two-peak distribution. The two peaks correspond to the bank mode (BM) and to the interface mode (IM). From Constantinescu et al. [6]. Copyright 2012 American Geophysical Union. Reproduced by permission of American Geophysical Union

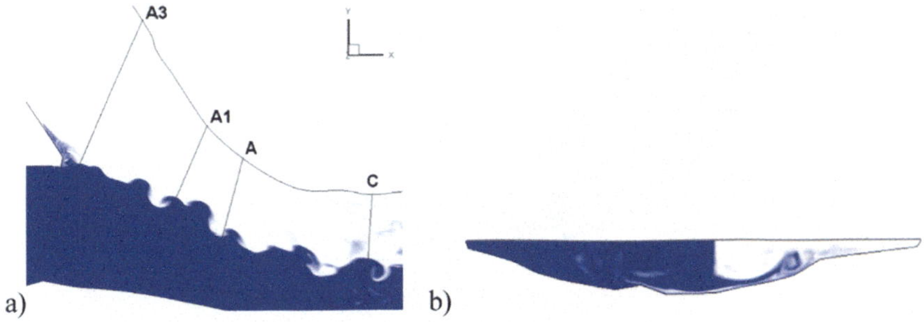

Fig. 65.10 Visualization of the mixing between the two streams at the KRCS confluence using temperature contours in the instantaneous flow. **a** free surface; **b** cross section A. The nondimensional temperature is equal to 1 (blue color) in the CS stream and is equal to 0 in the KR stream (white color)

SOV cells to entrain sediment and drive the formation of the confluence scour hole [4–6].

Figure 65.10 illustrates the main mixing modes at large-angle confluences with $Fr \gg 1$. In horizontal planes, mixing is driven primarily by stirring induced by the vertically oriented MI eddies. However, at locations where strongly coherent SOV cells form, streaks of mixed fluid are extracted near the bed by the SOV cell during the MI mode and then advected laterally toward the corresponding bank. All this mixing takes place well beneath the free surface.

Cases with Strong Density Contrast Between Incoming Flows

To investigate flow stratification effects on mixing at natural river confluences, additional simulations were conducted for the KRCS confluence with Boussinesq coupling between the momentum equations and the nondimensional temperature/density equation. Figures 65.11 and 12 illustrate instantaneous nondimensional temperature/density distributions for the $VR = 5.5$ case obtained in the no-density-effects case and then for stratified flow conditions with $Fr = 3.3$. The patterns of mixing are starkly different in the two cases. As a result of flow stratification, the MI moves faster toward the west bank in the $Fr = 3.3$ case and the distance at which the KH billows form increases from about $25D$ in the no-density-effects case to about $65D$ in the $Fr = 3.3$ case (Fig. 65.11). Moreover, in the $Fr = 3.3$ case, a region containing mixed fluid forms in the middle of the channel, downstream of section S2. Such a region cannot form in the absence of stratification effects. To understand its formation, one has to examine the temperature distribution in the spanwise-vertical planes. Figure 65.12 shows that in the $Fr = 3.3$ case, the heavier fluid on the west side of the confluence plunges down and forms a gravity current like flow. This bottom current first advances over the positive bottom slope then moves upwards over a region of negative slope and eventually reaches the free surface some distance downstream of

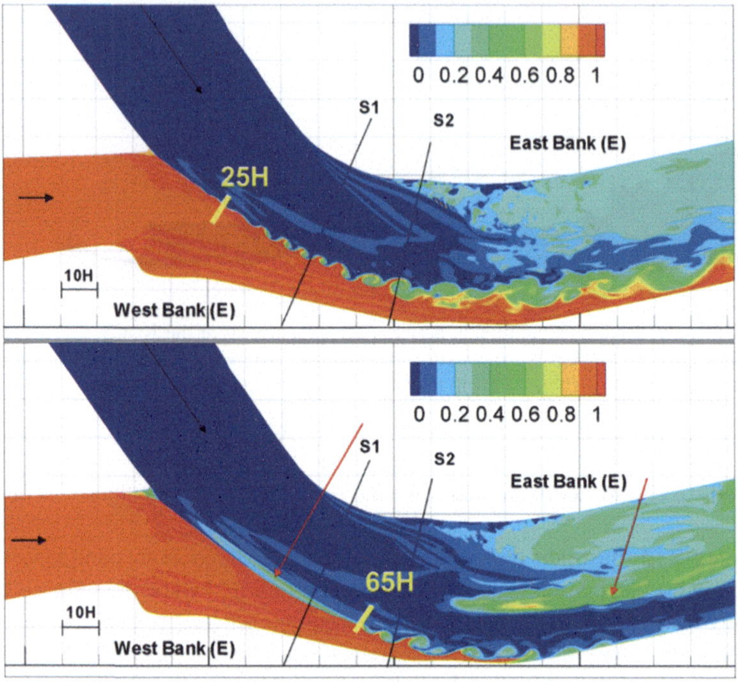

Fig. 65.11 Visualizaton of the free surface mixing between the KR and CS streams using instantaneous nondimensional temperature contours. The top frame shows results for the no-density-effects $VR = 5.5$ case. The bottom frame shows results for the $Fr = 3.3$, $VR = 5.5$. Reproduced from Constantinescu [4]

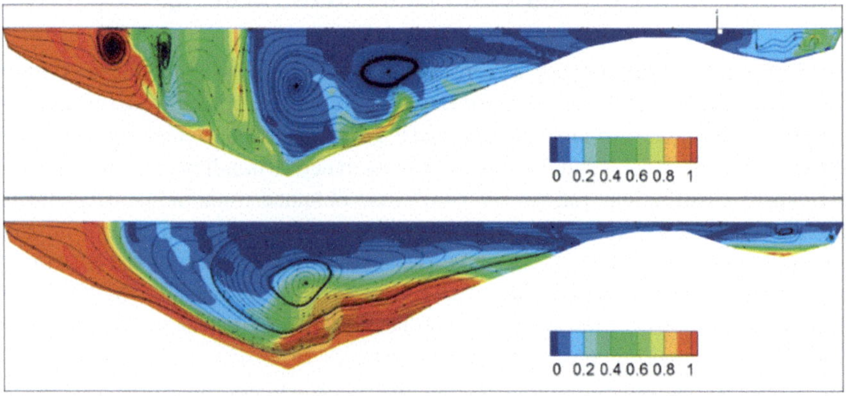

Fig. 65.12 Visualizaton of mixing at Section S2 between the KR and CS streams using instantaneous nondimensional temperature contours. The top frame shows results for the no-density-contrast $VR = 5.5$ case. The bottom frame shows results for the $Fr = 3.3$, $VR = 5.5$ case

section S2. This results in an inversion of the regions containing heavier and lighter fluid in the downstream part of the main channel compared to the flow in the two tributaries. The spatial development of a lock-exchange-like flow away from the confluence apex is a general characteristic of confluent flows for sufficiently low densimetric Froude numbers [2, 3].

Conclusions

Eddy-resolving simulations performed in wide and very long channels allowed describing in a quantitative way the transition of a shallow mixing layer developing between parallel and nonparallel streams toward the equilibrium regime. The physics of the shallow mixing layer was essentially the same for low and high confluence angles. Higher-coherence KH billows formed over the upstream part of the mixing layer in the latter cases along with strong SOV cells in the vicinity of the MI. In the case of natural river confluences with a large angle between the tributaries and no-density-contrast between the incoming streams, the SOV cells were found to play an important role in mixing regardless of whether or not the MI was in the KH mode or in the wake mode. For cases with a large density contrast between the incoming streams ($Fr < 6$), the development of a spatially developing bottom propagating current in the cross flow direction was the main mechanism driving mixing between the two streams at large distances from the confluence apex. Mixing driven by the vertically oriented MI eddies is just one mechanism responsible for mixing of confluent flows.

Acknowledgements The author would like to thank Dr. G. Kirkil, Dr. Z. Cheng, Dr. S. Miyawaki and Dr. D. Horna-Munoz who generated most of the results presented in the paper. The author would also like to acknowledge the contributions of his long-term collaborators on the topic of mixing at river confluences, Prof. B. Rhoads from University of Illinois Urbana Champaign and Dr. A. Sukhodolov from IGB Berlin, Germany.

References

1. Chang, K., Constantinescu, G., Park, S.O.: Assessment of predictive capabilities of Detached Eddy Simulation to simulate flow and mass transport past open cavities. ASME J. Fluids. Eng. **129**(11), 1372–1383 (2007)
2. Cheng, Z., Constantinescu, G.: Stratification effects on flow hydrodynamics and mixing at a confluence with a highly discordant bed and a relatively low velocity ratio. Water Resour. Res. **54**(7), 4537–4562 (2018)
3. Chu, V.H., Babarutsi, S.: Confinement and bed friction effects in shallow turbulent mixing layers. J. Hydraul. Eng. **114**, 1257–1274 (1988)
4. Constantinescu, G.: LE of shallow mixing interfaces: a review. Environ. Fluid Mech. **14**, 971–996 (2014)

5. Constantinescu, G.S., Miyawaki, S., Rhoads, B., Sukhodolov, A., Kirkil, G.: Structure of turbulent flow at a river confluence with momentum and velocity ratios close to 1: insights from an eddy-resolving numerical simulation. Water Resour. Res. **47**, W05507 (2011). https://doi.org/10.1029/2010WR010018
6. Constantinescu, G., Miyawaki, S., Rhoads, B., Sukhodolov, A.: Numerical analysis of the effect of momentum ratio on the dynamics and sediment entrainment capacity of coherent flow structures at a stream confluence. J. Geophys. Res. Earth Surf. **117**, F04028 (2012). https://doi.org/10.1029/2012JF002452
7. Keylock, C.J., Constantinescu, G., Hardy, R.J.: The application of computational fluid dynamics to natural river channels: eddy resolving versus mean flow approaches. Geomorphology **179**, 1–20 (2012)
8. Kirkil, G., Constantinescu, S.G.: A numerical study of shallow mixing layers between parallel streams. In: 2nd International Symposium on Shallow Flows, Hong Kong (2008)
9. Kirkil, G., Constantinescu, S.G.: A numerical study of a shallow mixing layer developing over dunes. In: XXXIIIrd International Association Hydraulic Research Congress, Vancouver, Canada (2009)
10. Rhoads, B.L., Sukhodolov, A.N.: Field investigation of three-dimensional flow structure at stream confluences: 1. Thermal mixing and time-averaged velocities. Water Res. Res. **37**(9), 2411–2424 (2001)
11. Uijttewaal, W.S.: Booij R Effects of shallowness on the development of free-surface mixing layers. Phys. Fluids **12**(2), 392–402 (2000)

Chapter 66
Large Eddy Simulation of Contact Tanks for Disinfection in Drinking Water Treatment

P. Bruno, G. Di Bella and M. De Marchis

Introduction

In the last two decades, Computational Fluid Dynamics (CFD) has proven to be a powerful and efficient tool to study turbulent phenomena occurring in the units of water treatment plants [1]. Specifically, some literature studies focused the attention on the hydraulic and sanitary efficiency of disinfection reactors and on flow and tracer transport in those tanks. Fluid motion in serpentine contact tanks, like most of fluid dynamic processes in real systems is a turbulent process (i.e. a non-stationary, irregular and apparently chaotic motion) characterized by the presence of eddies of different sizes and speeds, which continuously change. To date, the most commonly used disinfection methods are those that use chemical disinfectants and, in particular, chlorine and its compounds, ozone, halogens and hydrogen peroxide. In chemical disinfection treatments, the most widely used type of reactor is undoubtedly the plug-flow. In plug-flow systems, longitudinal dispersion is almost zero, and a uniform distribution of the tracer in each cross-section is verified. Moreover, in such configurations, Hydraulic Residence Time (HRT), defined as the ratio between the volume (V) and the flow rate (Q), is uniform and equal to theoretical mean residence time. Plug-flow reactors are, therefore, reactors that ensure an optimal hydraulic and sanitary performance in traditional chemical processes related to water disinfection operations [2, 3].

The traditional contact tanks, being structurally constituted by rectangular chambers and vertical baffles, present several zones with a whirling motion. Specifically, the flow field is characterized by large areas in which water speed decreases or even vanishes, or where stable recirculation phenomena occur. These turbulent phenomena inside disinfection basins must be avoided since they modify flow from the ideal plug-flow conditions. Moreover, in real contact tanks, real hydraulic residence time

P. Bruno · G. Di Bella · M. De Marchis (✉)
Faculty of Engineering and Architecture, Kore University of Enna, Enna, Italy
e-mail: mauro.demarchis@unikore.it

M. García-Villalba et al. (eds.), *Direct and Large Eddy Simulation XII*,
ERCOFTAC Series 27,
https://doi.org/10.1007/978-3-030-42822-8_66

can be significantly different from the theoretical one. Numerical simulations are fundamental to evaluate the optimization process in terms of efficiency obtained through the realization of new technical details and geometrical devices. The application of CFD simulations applied to water treatment plants can, therefore, represents a powerful tool for evaluation of the hydrodynamics within real tanks; design of new systems. Even though, Reynolds Averaged Navier-Stokes Simulation (RANSS) is able to predict the mean flow from chamber to chamber, it was demonstrated that [4] RANSS is not able to predict smaller scale eddies causing the entrapment of the disinfectant. In light of this, in the present analysis, Large Eddy Simulations (LESs) are applied to investigate the disinfection efficiency in multi-chamber contact tanks.

Numerical Simulations

The ability to study elements of treatment plants, through fluid dynamic simulations, is nowadays a fundamental point for the scientific community working in these areas.

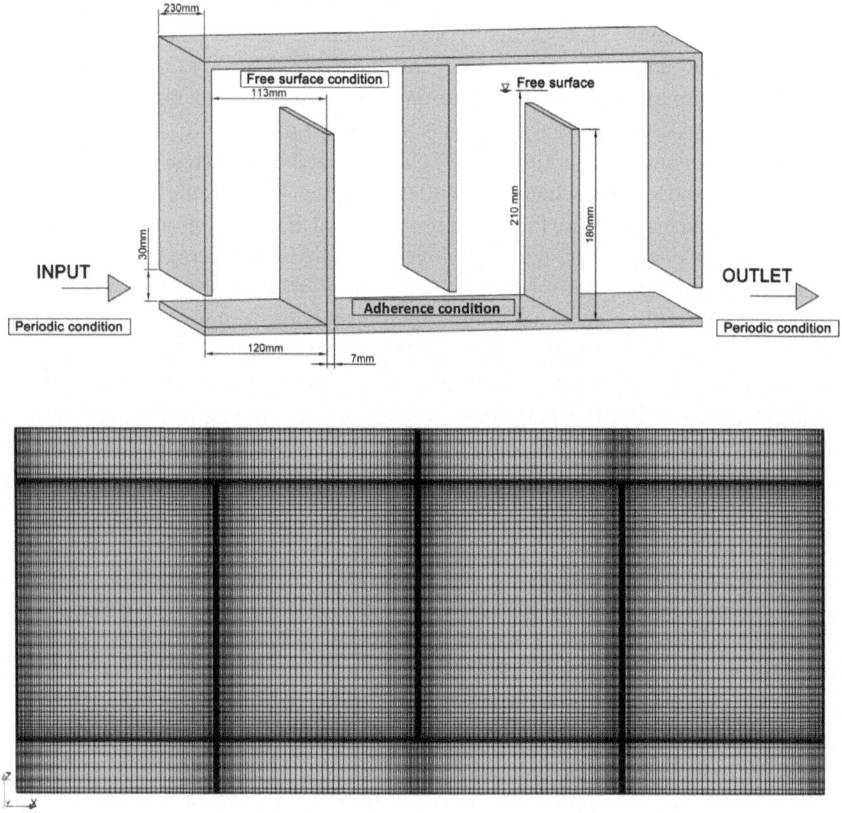

Fig. 66.1 Top: Layout of the computational domain of the ozone contactors. Bottom: Layout of the mesh in the streamwise direction

The application of numerical techniques allows, in fact, to reduce design time and to verify the optimal configurations from a hydrodynamic point of view. In the proposed research, turbulent flows are resolved using the LES approach, based on the numerical solution of the filtered mass and momentum conservation. The governing equations are resolved using an in-house finite-volume numerical code PANORMUS, which is second-order accurate both in time and space (available at the site www.panormus3d.org). The Adams-Bashfort method is used for the time advancement of the solution, with the classical fractional-step technique. The pressure Poisson equation is solved through a line-SOR technique in conjunction with a multigrid V-cycle accelerator [5]. Figure 66.1 shows a sketch of the multi-chamber contact tank analysed. The computational domain is decomposed using $256 \times 64 \times 64$ cells in the streamwise, spanwise and wall-normal directions, respectively. Due to the specific geometry of the chamber here a Cartesian grid is used. The mesh size was chosen by means of a grid sensitivity analysis. The grids were refined close to the bottom and baffle walls in order to resolve steep velocity gradients. Basically, the grid is refined thus to have the first grid at a distance from the solid boundary, in wall units (y^+), less than 1. In Fig. 66.1 we report a sketch of the computational domain of the contact tank [6, 7] located in the traditional water treatment plants.

Results

The classical disinfection tank reported in Fig. 66.1, was modelled through the PANORMUS solver looking at the flow field and at the distribution in time and space of a conservative passive tracer, thus simulating the presence of a disinfection process. Figure 66.2 shows the distribution of the time-averaged absolute velocity and streamlines obtained for the classical contactor.

The picture shows that the typical acceleration of flow, near the bottom or near the free surface, due to the presence of the baffles. This area is, therefore, characterized by the presence of a sort of high speed *jet* extending from one room to another. The high flow velocity determines the passage of a portion of water through the entire length of the contact tank in a time much shorter than the average hydraulic residence time, with consequent short-circuit, and therefore, a reduction in disinfection efficiency. Moreover, recirculation regions developing within the chambers contribute to this short-circuiting effect [2–4, 6]. A first large recirculation zone develops in the centre of the chamber; a small secondary recirculation zone occurs, instead, at the corners between the bottom and the walls of the baffles on both sides, while this secondary recirculation is not observed near the water surface; finally, a third recirculation cell is observed on the downstream side of the baffle.

The hydraulic efficiency of the classical reactor in terms of disinfection capacity was evaluated considering the flow of a passive scalar tracer injected inside the disinfection tank from the lower face of the inlet light for a time of 2.5 s. After this time, the tracer input was stopped and its transport was evaluated through the various compartments of the reactor.

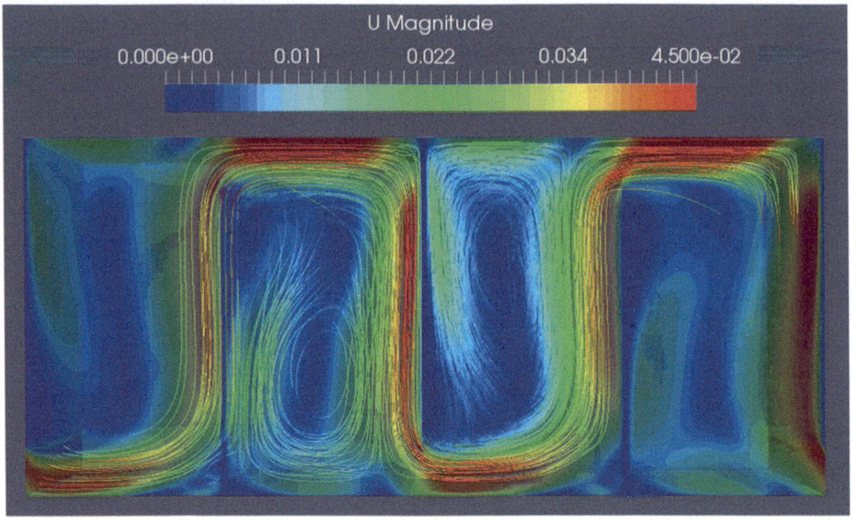

Fig. 66.2 Distribution of the time-averaged absolute velocity and streamlines in the centre-plane (z/L = 0.5) of the tank

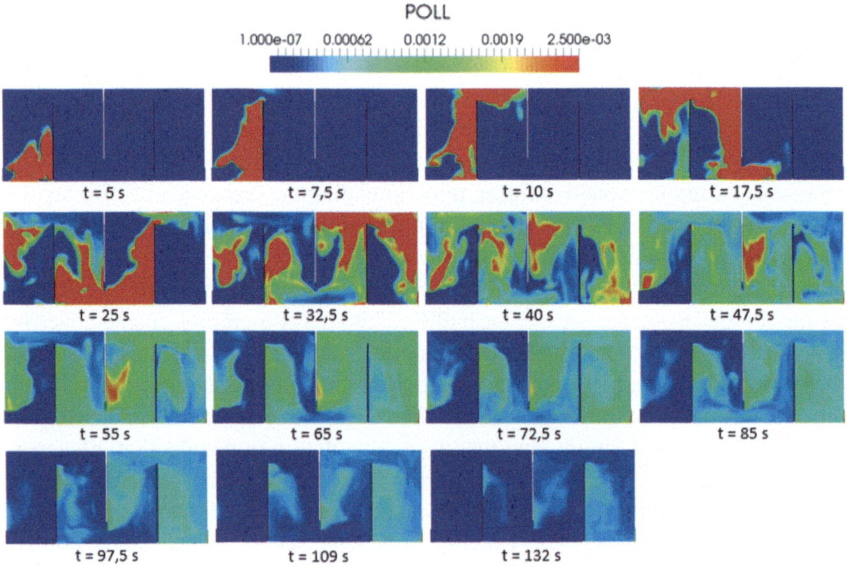

Fig. 66.3 Tracer concentration snapshots at different times, shown on the x–y plane at mid span (z = L/2)

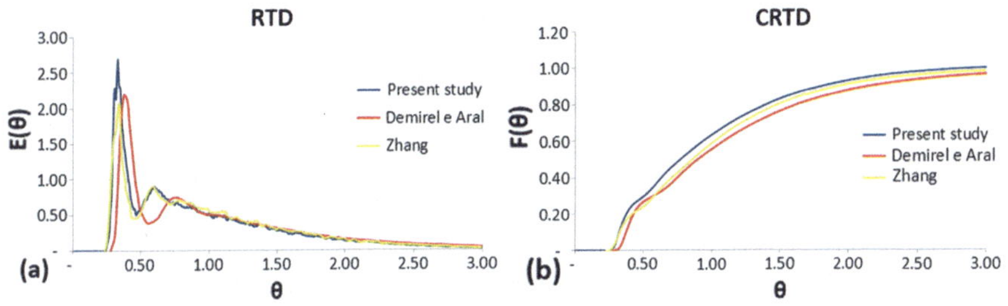

Fig. 66.4 Tracer analysis results: **a** normalized RTD plot; time is nondimensionalized using $\theta = t/\tau$ where τ is the theoretical mean residence time of the tank; tracer concentration is nondimensionalized using $E(\theta) = C/C_{init}T_{release}/\tau$, where the tracer pulse at the inlet is $C_{init} = 1$, and the injection period is $T_{release} = 2.5s$; **b** cumulative normalized RTD plot $F(\theta) = \int_o^{\infty} E(\theta)d\theta$

Figure 66.3 shows instantaneous snapshots of the distribution of the tracer concentration in time. LES, able to reproduce the different turbulence zones, allows a deeper knowledge of the tracer course within the reactor. In fact, the conservative tracer inserted in the tank at the initial instant is monitored, point by point, along the whole serpentine. Figure 66.3 shows that the highest concentrations of tracer are found at the walls of the baffles and near the free surface, while the lower concentrations are at the centre of the recirculation zones. The concentration distribution confirms again the nonideal flow (far from the plug-flow condition) inside the reactor, as they demonstrate the occurrence of short circuits and large recirculation regions in each chamber. After a certain time lap it is observed that even though a significant quantity of disinfectant has already reached the third and fourth chambers and is close to the reactor outlet, in the first chamber there are still non-negligible tracer concentration levels. This behaviour is mainly due to the entrapment of the tracer inside the recirculation zones (dead zones). Furthermore, it is possible to observe that what remains of the tracer persists inside the reactor for long periods after its initial release, resulting in efficiency losses of the disinfection treatment.

The tracer distribution analysis was based on the examination of the RTD (Residence Time Distribution) and CRTD (Cumulative Residence Time Distribution) curves. These distributions are plotted in Fig. 66.4, where the results achieved in the present analysis are compared with recent literature findings of [2–4, 6]. The results confirm a good agreement, thus validating the proposed study.

The first peak of the normalized RTD curve is a typical indicator of the degree of the short-circuit effect, since a large amount of the injected tracer remains inside the contact tank for a short period and leaves the reactor after a shorter time than the theoretical hydraulic residence time τ. The second peak, on the other hand, is associated with the delayed release of the tracer concentration trapped in the recirculation zones, so that the lower the value reached at this second peak, the smaller the amount of tracer trapped in the reactor [3]. The common Hydraulic Efficiency Indicators (HEI) indexes were used to evaluate the hydraulic and mixing efficiencies for the classical design and, as it is possible to observe from the values in the Table 66.1, the results obtained in the present experimentation are in agreement

Table 66.1 Comparison of hydraulic efficiency indicators

Model	t_{10}/τ	t_{90}/τ	Mo	σ^2
Present study	0.325	1.800	5.535	0.163
Demirel and Aral	0.375	2.117	5.646	0.398
Zhang	0.338	1.887	5.582	0.140

with the data present in the literature of the sector. This confirms that the traditional tanks with baffles, used for the treatment of disinfection, suffer from problems of short-circuits and recirculation, which limits mixing within the different chambers and require more energy to guide the flow through the multi-chamber contactor.

Conclusions

In conclusion, this study suggests that LES is a powerful model to accurately simulate and predict big and small recirculation zones in the tank. Simulations highlighted that traditional disinfection tanks suffer from inadequate flow conditions (recirculation, short circuits, dead zones) that cause nonideal disinfection. In addition, it was possible to observe that the flow pattern inside the reactor depends fundamentally on the geometry of the system and the disinfection efficiency can be optimized towards plug-flow. For future development, efforts are certainly required to apply CFD modelling to all the water treatment units, in order to obtain a model capable of simulating and improving hydrodynamics within all the plant's units.

References

1. Angeloudis, A., Stoesser, T., Kim, D., Falconer, R.A.: CFD study of flow and transport characteristics in baffled disinfection tanks. In: Proceedings of 2013 IAHR Congress 2013 Tsinghua University Press, Beijing, Sept. 2013
2. Demirel, E., Aral, M.M.: Unified Analysis of multi-chamber contact tanks and mixing efficiency based on vorticity field part I: hydrodynamic analysis. Water, **8**(11), 495 (2016)
3. Demirel, E., Aral, M.M.: Unified Analysis of multi-chamber contact tanks and mixing efficiency based on vorticity field part II: transport analysis. Water. **8**(11), 537 (2016)
4. Zhang, J., Tejada-Martínez, A.E., Zhang, Q.: Reynolds-averaged Navier-Stokes simulation of the flow and tracer transport in a multichambered ozone contactor. J. Environ. Eng. **139**(3), 450–454 (2013b)
5. De Marchis, M., Milici, B., Napoli, E.: Numerical observations of turbulence structure modification in channel flow over 2D and 3D rough walls. Int. J. Heat Fluid Flow **56**, 108–123 (2015)
6. Zhang, J., Tejada-Martínez, A.E., Zhang, Q.: Hydraulic efficiency in RANS of the flow in multi-chambered contactors. J. Hydraul. Eng. **139**(11), 1150–1157 (2013a)
7. Kim, D., Kim, D., Kim, J.: Stoesser, T: large eddy simulation of flow and tracer transport in multichamber ozone contactors. J. Environ. Eng. **136**(1), 22–31 (2010)

The manufacturer's authorised representative in the EU is Springer Nature Customer Service Centre GmbH, Europaplatz 3, 69115 Heidelberg, Germany. If you have any concerns regarding our products, please contact ProductSafety@springernature.com

Printed and bound by CPI Group (UK) Ltd, Croydon, CR0 4YY
30/09/2026
02225697-0014